Linear Models for the Prediction of the Genetic Merit of Animals

4th Edition

Linear Models for the Prediction of the Genetic Merit of Animals

4th Edition

Raphael A. Mrode
Scotland's Rural College, Edinburgh, UK
(Former Scottish Agricultural College); and
International Livestock Research Institute, Nairobi, Kenya

Ivan Pocrnic
The University of Edinburgh, Edinburgh, UK

Gregor Gorjanc (Chapter 1)
The University of Edinburgh, Edinburgh, UK

Robin Thompson (Chapter 17)
Rothamsted Research, Harpenden, UK

CABI

CABI is a trading name of CAB International

CABI
Nosworthy Way
Wallingford
Oxfordshire OX10 8DE
UK

Tel: +44 (0)1491 832111
E-mail: info@cabi.org
Website: www.cabi.org

CABI
200 Portland Street
Boston
MA 02114
USA

Tel: +1 (617)682-9015
E-mail: cabi-nao@cabi.org

A catalogue record for this book is available from the British Library, London, UK.

ISBN-13: 9781800620483 (paperback)
9781800620490 (ePDF)
9781800620506 (ePub)

DOI: 10.1079/9781800620506.0000

Commissioning Editor: Alexandra Lainsbury
Editorial Assistant: Lauren Davies
Production Editor: Shankari Wilford

Typeset by SPi, Pondicherry, India
Printed and bound in the UK by Severn, Gloucester

Contents

Supplementary R codes for key examples for this book can be accessed at: https://www.cabidigitallibrary.org/doi/book/10.1079/9781800620506.0000

Preface

Best Linear Unbiased Prediction (BLUP) has become the most widely accepted method for genetic evaluation of domestic livestock. Since the method was first published by Henderson (1949), it has evolved in terms of its application in models for genetic evaluation, from sire, sire and maternal grandsires models in the early years, followed by univariate and multivariate animal models, random regression models for the analysis of longitudinal data and more recently, the analysis of the genomic data (SNP-BLUP or GBLUP or ssGBLUP or Bayesian methods). Advances in computational methods and computing power have enhanced this development. Currently, most national conventional genetic evaluation systems for several domestic livestock species are based on animal or random regression models using BLUP. However, in the last ten years, genomic prediction has become the method of choice for prediction of breeding values and selection with rapid movement towards the application of ssGBLUP even in large animal populations.

Despite these developments and the proliferation of information in the literature, especially in methods of genomic prediction, there is no simple and straightforward text on the application of linear models to the prediction of breeding values. Moreover, in developing countries, where access to journals is limited, there is a basic lack of practical information on the subject area. This book has been written with a good balance of theory and application to fill this gap. It places in the hand of the reader the application of BLUP and genomic prediction models in modelling several genetic situations in a single text. The book has been compiled from various publications and with the benefit of experience gained from several colleagues in the subject area and from involvement in several national evaluation schemes over the last 31 years. Relevant references are included to indicate sources of some of the materials.

In Chapter 1, the main components of the phenotype and the basic structure of the genome are examined, and various modes of genetic inheritance are defined. Then, the basic model and assumptions governing genetic evaluation are presented together with simple situations involving prediction of breeding values from the records of an individual. This is followed by the introduction and use of selection indices to predict genetic merit combining information on several traits and individuals. Then, the general framework on the application of BLUP in genetic evaluation in a univariate and multivariate situation is presented in Chapters 4–6. The simplification of multivariate evaluations by means of several transformations is also examined, followed by maternal trait and social interaction models. Random regression models for the analysis of longitudinal data are discussed in Chapter 10, followed by a chapter on genomic prediction models and genomic selection. Non-additive genetic animal models are discussed with more emphasis on genomic prediction methods for dominance and epistasis effects. The challenge of genetic and genomic prediction in a multibreed and crossbreed situation is then presented. Next, threshold and survival models are discussed. In Chapters 16 and 17, the basic concepts for variance-component estimation are introduced followed by the application of the

Gibbs sampler in estimation of genetic parameter and evaluations for univariate and multivariate models. Finally, computing strategies for solving mixed-model equations are examined, with a presentation of the several formulae governing iterative procedures on the data. A knowledge of basic matrix algebra is needed to understand the principles of genetic evaluation discussed in the text. For the benefit of those not familiar with matrix algebra, a section on introductory matrix algebra has been incorporated as Appendix A. It is also assumed that the reader is familiar with the basic principles of quantitative genetics.

Several examples have been used to illustrate the various models for genetic and genomic evaluation covered in the text and attempts have been made to present formulae which explain how the solutions for random and fixed effects in the models were obtained from the mixed-model equations. This illustrates to the reader how the various pieces of information are weighted to obtain the genetic merit of an animal for various models.

This edition has, for the first time, provided R scripts for the examples in the textbook on an online depository to which readers will have access.

Every attempt has been made to ensure accuracy of materials in the text; however, in the event of errors being discovered, please inform the authors.

Professor R. Thompson contributed the chapter on estimation of variance components despite his busy schedule and helped review the manuscript. His contribution is immensely acknowledged. Dr Gregor Gorjanc contributed the first chapter on phenotypes and the genome. In addition, he helped shape the structure of the edition of the book and reviewed several sections of the manuscript. We are greatly indebted to him. The chapter on genomic selection in previous and current editions were reviewed by Professor Ben Hayes (The University of Queensland, Australia), Dr Ricardo Pong-Wong (The University of Edinburgh, UK), Professor John A. Woolliams (The University of Edinburgh, UK), and Dr Fernando de Oliveira Bussiman (University of Georgia, USA) and we are grateful for their valuable input. Drs Andres Legarra and Zulma Vitezica kindly reviewed Chapter 13 on genomic models for non-additive models at short notice and we are greatly indebted to them for their immense contribution. Drs Gabor Mészáros and Sue Brotherstone reviewed the chapter on survival analysis within a very tight schedule and we acknowledged their contribution. We are grateful to the late Professor Denny Cruz and Dr Victor Olori for reviewing the chapters on social interaction and on reducing the dimension of multivariate analysis. We are greatly indebted to the late Professor W.G Hill and Mr G. Swanson for reviewing the manuscript of earlier editions; their comments and suggestions resulted in substantial improvements in the text. Dr Martin Lidauer (Luke, Finland) and Professor Ismo Stranden (Luke, Finland) read specific chapters or sections. The contribution from Professor Ismo Stranden immensely shaped Chapter 14 on multibreed evaluations and we acknowledge his useful suggestions. Professor Sammy Aggrey (University of Georgia, USA) reviewed Chapter 1 and his contribution is acknowledged. The assistance of Dr Sebastian Mucha in preparing the graphs in the text is greatly acknowledged. In addition, experience gained from working with Professor Mike Coffey (SRUC, UK) and the late Professors C. Smith and B.W. Kennedy has been valuable in writing this book.

Feedback from various colleagues: Dr Per Madsen (Aarhus University, Denmark), the late Dr Deniz Koyuncu, Dr Maksim Struchalin (ComputeBio LLC, New Zealand) and Dr Jorge Hidalgo (University of Georgia, USA) on previous editions have resulted in improving the quality of this edition.

I also wish to express my thanks to Professor R.L. Quaas for permission to use information from his unpublished note on inbreeding algorithm, the Animal Genetics and Breeding

Unit, University of New England, Australia, for allowing me to adopt some materials from *BLUP Handbook* for Chapter 2 of the text and Professor Rohan L. Fernando to use some his material from the Iowa State University 2010 summer course.

Ivan would like to thank his wife Blanka for love and support, and his daughter Fiona for eternal motivation. Raphael would like to express his sincere gratitude to his wife, Doris, for her immense support and for typing part of the manuscript. Special thanks to Kevwe, Joshua and Esther for their co-operation, especially when I have taken off time to prepare the manuscript and to many dear friends who were of great encouragement. Finally, to God be all the glory.

R.A. Mrode
Scotland's Rural College
(Former Scottish Agricultural College)
Edinburgh, UK; and International Livestock Research Institute,
Nariobi, Kenya and The University of Edinburgh

Ivan Pocrnic
The University of Edinburgh
Edinburgh, UK

Abbreviations

ADG	average daily gain
ARHS	adjusted right-hand side
BFAT	backfat thickness
BLUE	best linear unbiased estimator
BLUP	best linear unbiased prediction
BV	breeding value
BW	birth weight
CF	covariance function
CR	correlated response
DBV	direct breeding value
DGV	direct genomic breeding values
DIM	days in milk
DRB	deregressed breeding values
DRP	deregressed proofs
DSP	durable performance sum
DYD	daughter yield deviation
EBV	estimated breeding value
EDC	effective daughter contributions
EM	expectation maximization
ETA	estimated transmitting ability
FA	factor analysis
GEBV	genomic breeding values
GLS	generalized least squares
GR	growth rate
HTD	herd–test–day
HYS	herd–year–season
IBD	identical by descent
IGE	indirect genetic effects
INET	index net
LD	linkage disequilibrium
LGR	lean growth rate
LP	lean per cent
LPL	length of productive life
LS	lifespan
LSE	least squares equations
MACE	multi-trait across-country evaluation
MAS	marker-assisted selection
MBLUP	multivariate best linear unbiased prediction
MCMC	Markov chain Monte Carlo

MGD	maternal granddams
MGS	maternal grandsire
ML	marker locus
MME	mixed-model equations
MQTL	quantitative trait locus (linked to a marker locus)
MS	muscle score
PC	progeny contribution
PCG	preconditioned conjugate gradient
PEC	prediction error covariance
PEV	prediction error variance
PIN	production index
PLI	profitable life index
PPA	probable producing ability
PTA	predicted transmitting ability
PWG	post-weaning gain
PYD	progeny yield deviation
QTL	quantitative trait locus
RAM	reduced animal model
REML	restricted (or residual) maximum likelihood
RHS	right-hand side
RP	residual polygenic
RR	random regression
RRM	random regression model
RRS	risk ratios
SBV	associative breeding value
SCC	somatic cell count
SEP	standard error of prediction
SNP	single nucleotide polymorphism
SSR	simple sequence repeats
TBV	total breed value
WWG	pre-weaning gain
XFA	extended factor analysis
YSP	year–season–parity

1 The Genome and Phenotypes

GREGOR GORJANC

University of Edinburgh, Edinburgh, UK

1.1 Introduction

This book is about methods for analysing variation between animals with the aim of estimating their genetic value. For this estimation, we use statistical models, as we will show in the following chapters. Before we delve into these statistical models, it is instructive to overview the biological processes that generate the data we are analysing. In this chapter, we will also describe models, but these are data-generation models upon which the theory of quantitative and statistical genetics is built (Falconer and Mackay, 1996; Lynch and Walsh, 1998). Although these data-generation models are often similar to the statistical models we use in our data analysis, it is important to note the following three interrelated points. First, all models are an abstraction of complex biology, the true model, that generates variation between animals. Second, simple models can often adequately describe complex phenomena with a small number of parameters. Third, although a very good attempt is usually made to match data-generation and statistical models, we cannot fully unravel complex biology. This is so because we typically have only a limited amount of data, or the data resolution is too coarse to decipher this complexity. Continued advancements in data-recording technologies will allow us to decipher more and more biology in future years.

The remainder of this chapter is organized into the following five sections. First, we continue this introduction and conceptualize variation between animals, including the definition of traits and underlying genetic, environmental and other sources. Second, we describe the molecule that encodes genetic information, the DNA, and its organization in the genome. We discuss the major DNA variation sources and how we encode this variation for quantitative genetic analysis. Third, we delve into a model that generates variation between animals from genetic and environmental effects. Fourth, we describe the inheritance of DNA from parents to offspring and how this process generates variation in a new generation. Understanding the processes that generate variation between animals and how this variation is inherited between generations is essential for understanding what we estimate with pedigree-based and genome-based statistical models described in the following chapters. Fifth, we point to the different types of traits, multiple traits, genotype-by-environment interactions, and additive and non-additive genetic effects.

It is well known that most, if not all, traits vary between animals and that this variation is due to many effects. But what is a trait? Any characteristic you can see or measure on animals may be called a trait. For example, weight, height, colour, and so on. All the traits we observe are called phenotypes of that animal. Derived from the Greek *pheno*, meaning to show, and *type*, meaning, well, "type".

There are many ways to organize traits into various groups. For example, milk yield, number of laid eggs, and body weight are often called production traits. The number of days between two calvings is an example of a reproduction trait, and so on. In this book,

DOI: 10.1079/9781800620506.0001

we will be most interested in grouping traits by their phenotypic expression or how we record this expression. Without being exhaustive, let us look at continuous, ordinal and binary traits. For example, milk yield is a continuous trait we usually record in kilograms, such as 7812.4 kg per cow's lactation. Looking at the distribution of such continuous traits (Fig. 1.1a), we will generally see a spread of recorded values around the central value with a decaying frequency towards the tails of this distribution. Another group is ordinal traits. Ordinal traits represent traits whose expression we count and hence have distinct categories. The frequency of animals recorded in each category can vary significantly between different systems. For example, the distribution of the number of progenies in a litter in pigs (Fig. 1.1b) or in sheep (Fig. 1.1c) are both ordinal. Binary traits are an extreme example, with only two categories, such as healthy or diseased (Fig. 1.1d). Ordinal and binary traits are often called discrete or categorical traits to distinguish them from continuous traits. Also, sometimes a trait has a continuous phenotype expression, but we record it as a categorical trait. This book focuses on continuous traits because most traits have such distributions. It is also generally recommended that recording is continuous to capture full trait variation. When this is not the case, we can use methods described in Chapter 15.

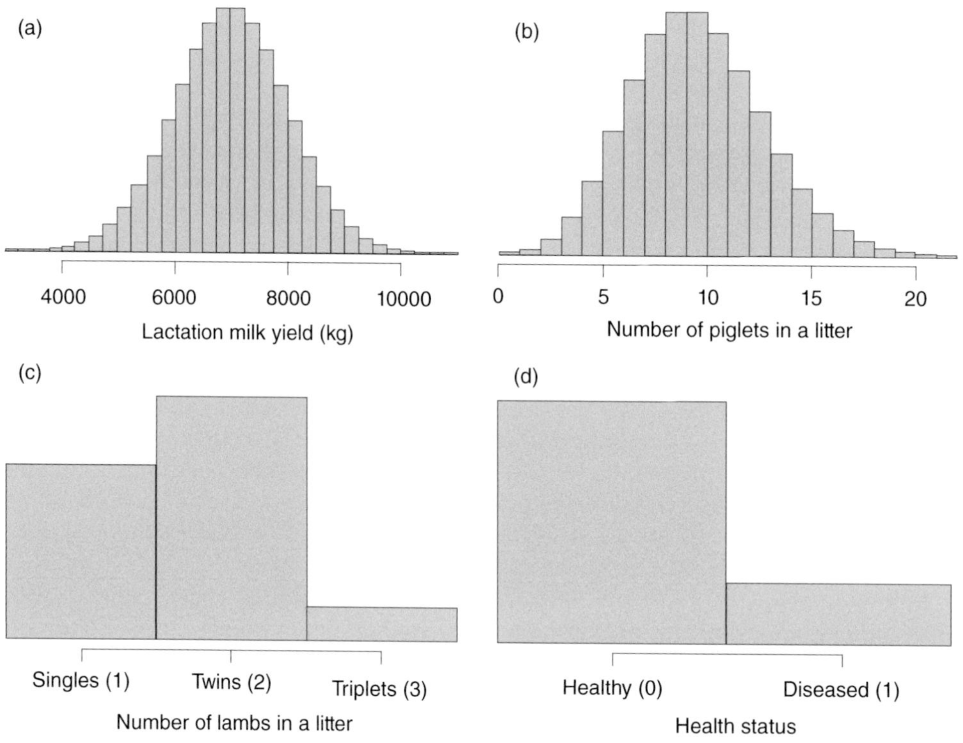

Fig. 1.1. Examples of distributions for: a) cow milk yield per lactation (continuous trait with mean of 7000 and standard deviation of 1000); b) litter size in pigs (ordinal trait with a mean of 10 and variance of 10); c) litter size in sheep (ordinal trait with mean of 1.7 and standard deviation of 0.6; given the small number of categories we can also report 38% singles, 54% twins and 8% triplets); and d) health status (binary trait encoded as 0 for healthy and 1 for diseased, with a mean of 0.2, standard deviation of 0.4 and 20% diseased animals) (CC-BY 4.0).

What drives this phenotype variation between animals? You might have heard of the concept 'nature versus nurture'. An animal's phenotype is a result of a combination of genes inherited from parents, known as the genotype, the environment they live in, and other factors:

$$Phenotype = f\left(Genotype, Environment, Other\ factors\right).$$

We will look at the genotype and its effect on phenotype in the next two sections. The environment includes the amount and quality of feed consumed, temperature, humidity, etc. Other factors include sex, the type of recording device, the data recording technician, the farmer's knowledge of animal husbandry, etc. We have loosely mentioned phenotype, genotype, environment and other factors. More concretely, an animal's recorded phenotype *value* is a function of the *effect* of the animal's genotype, the *effect* of the environment where the animal lives, and the *effect* of other factors. We emphasize the concepts of values and effects because they enable us to quantify the contribution of different sources to phenotype variation. If we knew the effect of genotype, environment and other factors, and their functional relationships, we would fully understand sources of variation in phenotype values. We never know these effects and their functional relationships. We use collected data and statistical models to estimate these effects and their functional relationships. While the collected data that we feed into the statistical models will vary substantially between animal systems, they will generally include phenotype values, associated descriptors (such as animal identification, sex, farm identification, etc.), pedigree and, increasingly, genomic data. The following chapters will show examples of such datasets.

1.2 Variation in DNA

Variation in the composition of an animal's genetic material, its genotype, is determined by the DNA inherited from its parents. This DNA instructs biological functions, such as the growth and reproduction of an animal, in all the trillions of cells ($\sim 10^{12+}$). Inside each cell is a nucleus with a complete copy of the inherited DNA. DNA is a long molecule that looks like a twisted ladder. The rungs of this ladder are smaller molecules called nucleotides or bases. There are four bases: Adenine (A), Cytosine (C), Guanine (G) and Thymine (T) (Fig. 1.2). These bases bind in pairs forming the twisted ladder, the double helix. Adenine (A) binds with Thymine (T), while Cytosine (C) binds with Guanine (G).

The complete collection of DNA molecules in a cell is called a genome. The structure and size of the genome vary between species. For example, in cattle, the genome is organized into 30 chromosomes. Cattle are diploid, meaning they have two copies of each chromosome, in total 60 DNA molecules. Each of the copies is inherited from one parent. We call each chromosome copy a haplotype, and the combination of two chromosome copies a genotype. The total length of the cattle genome is about 3 billion base pairs ($\sim 3x10^9$). This is the length of one copy of 30 chromosomes. Hence, each cell in cattle has about 6 billion base pairs. Some genomes are much smaller. For example, the honeybee genome has only about 250 million base pairs organized in 16 chromosomes.

Most of the genome is the same across all chromosome copies in a population. We are, however, most interested in the parts of the genome that differ between chromosome copies. This variation can be present both within one animal and between animals. These variable parts of the genome are called segregating/polymorphic sites or loci. If a locus is

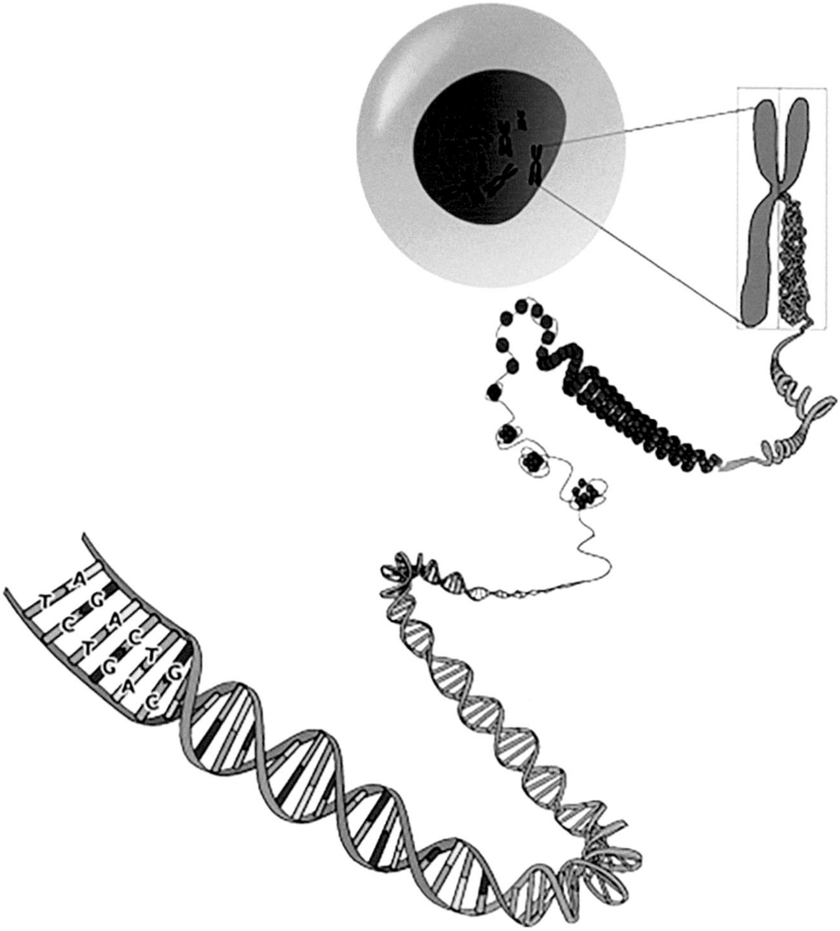

Fig. 1.2. Diagram zooming in from the cell's nucleus to an animal chromosome and to the unwinding of DNA double helical molecule with its bases (© OpenClipart-Vectors (2013) CC0).

segregating, it means that there is variation in DNA at that position within and between animals and their families. In other words, DNA variation (polymorphism) exists at that position. For example, some chromosome copies in a population have the A-T base pair at that locus while other chromosome copies have the G-C base pair. These loci show variation because, at some point in the past, one of the chromosomes has been copied with an error, with a mutation. If a mutation occurs in germline (reproductive) cells, it can be passed from parents to their progeny.

We refer to the different base pair sequences at a locus as alleles. A mutation is usually called a derived allele, while the original allele is called the ancestral allele. In the context of reference genomes, the genome that other genomes are compared to, we often use the term reference allele, which denotes the allele present in the reference genome. Alleles that differ from the reference allele are usually called alternative alleles. Variation at the single base pair mentioned above is called a Single Nucleotide Polymorphism (SNP). There are additional types of DNA polymorphism, such as deletions, insertions, repetitions and inversions, at a small scale involving few base pairs or at a large scale involving chromosome regions or whole chromosomes. Because the DNA molecule has a direction (that is, it is read from the 5' end towards the 3' end), we can observe four possible SNP alleles: A-T, T-A, G-C, and C-G (Fig. 1.3). Of the billions of DNA base pairs, most studies have found tens to hundreds of millions of SNPs (~10–100 x 10^6 = ~10^7–10^8) and other types of DNA polymorphisms (Hayes and Daetwyler, 2019; Halldorsson *et al.*, 2022; Ros-Freixedes *et al.*, 2022). This suggests that every 100th to 10th base pair in a genome could show polymorphism in a population. Many of these loci will have very low frequencies of mutated alleles.

The two most important technologies for generating genomic data are sequencing and SNP arrays. Sequencing simply means reading the DNA. There are two phases in sequencing. Initially, we must *de-novo* sequence the genome of one animal and build the so-called reference genome. Then, further animal genomes are re-sequenced against the reference genome. Most modern sequencing techniques involve high-molecular-weight DNA isolation, cutting the genome into smaller fragments, repeatedly sequencing these fragments, aligning the sequence reads to the reference genome and, finally, calling the alleles and genotypes of an animal. The accuracy of the resulting data depends on the quality of all the steps. For example, good quality DNA isolation is critical, as is repeated sequencing of DNA fragments to capture variation at both chromosome copies and to distinguish sequencing errors from real DNA variation. The advantage of sequencing is that it captures most of the genomic variation, all the millions of SNP loci and some structural variants. Note that some reference genomes do not contain all variation within a species, so re-sequencing against such a reference misses that variation. SNP arrays (also called SNP chips) are conceptually doing the same as sequencing but are using previously designed array probes to capture variation at a selected set of SNP markers. Most 'standard' SNP arrays have a density of ~60,000 (60K) markers. In cattle, this gives a marker every ~50,000 (50K) base pairs, of which about 500–1000 are expected to be polymorphic, yet represented by a single marker. The selection of SNP markers aims for a uniform spread along chromosomes and allele frequency spectrum, as well as reliable genotype calling across batches of animals. The cost of SNP array genotyping is generally lower than sequencing and has a lower DNA isolation quality requirement, but it captures less DNA variation.

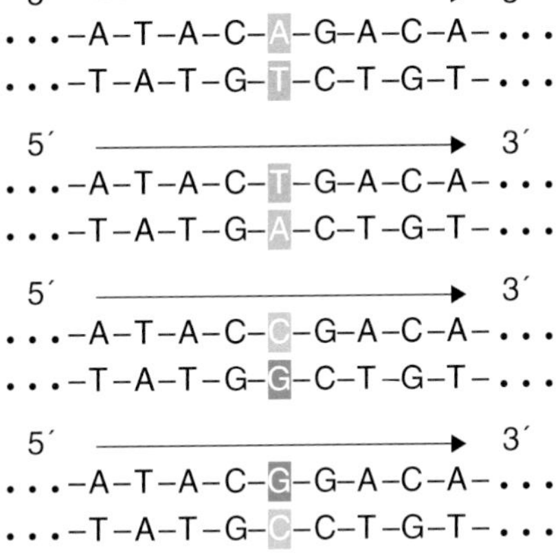

Fig. 1.3. Four possible SNP alleles at one base pair in four DNA fragments taking DNA read direction into account (CC-BY 4.0).

We often focus on biallelic SNPs, those with two alleles. The reason for this focus is that transition mutations between Adenine (A) and Guanine (G) and between Cytosine (C) and Thymine (T) are much more common than transversion mutations between Adenine (A) and Cytosine (C) or Thymine (T) and between Guanine (G) and Cytosine (C) or Thymine (T). This is driven by the molecular structure of the bases. Hence, if we have the A-T base pair as the ancestral allele, the frequency of the alternative C-G and T-A base pairs will be lower than that of the G-C base pair (considering the DNA orientation). Ultimately, the frequency of each mutation will depend on their spread between generations. We often focus on biallelic SNPs to avoid a mix-up between potential data generation errors and rare mutations. The accumulation of vast genomic data in recent years will likely broaden this focus. When calculating with biallelic SNPs, we numerically encode the two alleles in a computer with numbers: 0 represents the reference allele, and 1 represents the alternative allele. In diploid species, we can observe three possible genotypes at a biallelic SNP: homozygote for allele 0 (genotype 0/0), heterozygote for allele 0 and 1 (genotypes 0/1 or 1/0) and homozygote for allele 1 (genotype 1/1) (Fig. 1.4). Following the numerical encoding, the homozygote 0/0 is encoded as 0+0=0, heterozygotes 0/1 or 1/0 are encoded as 0+1=1+0=1, and homozygote 1/1 is encoded as 1+1=2 (Fig. 1.4). The numerical codes 0, 1 or 2 for the three genotypes mean that an animal has, respectively, 0, 1 or 2 alternative alleles. These codes are usually called allele dosage.

We can now write a sequence of biallelic SNPs along a chromosome as a series of zeros (0) and ones (1). We will write such sequences of SNP alleles from one chromosome or chromosome region (haplotype) in rows. Figure 1.5 shows two haplotypes of an animal across six SNPs and the corresponding genotype as a sum of the two haplotypes. The top haplotype has three alternative alleles in total. The bottom haplotype has four alternative alleles in total. Hence, the genotype has seven alternative alleles in total.

There are many ways to summarize DNA variation across animals and across loci. The simplest way is to calculate the frequency of alleles at a locus. Assume we have a

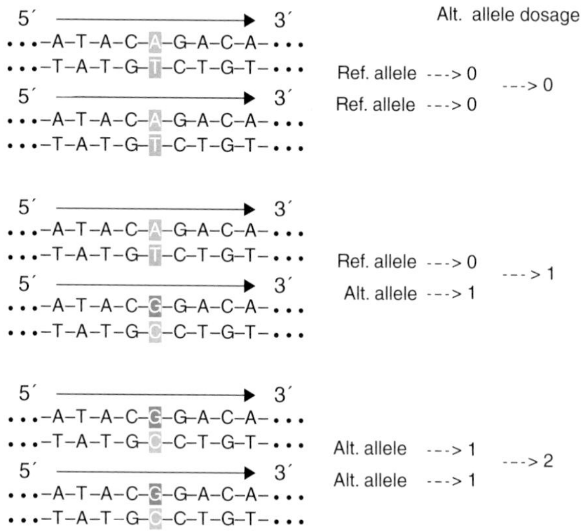

Fig. 1.4. Three possible genotype combinations at a biallelic SNP with the corresponding allele dosage encoding of the alleles and genotypes (CC-BY 4.0).

matrix of genotype allele dosages for biallelic SNPs where animals are represented in rows and loci in columns. Then, locus allele frequencies are calculated as column means divided by 2 – each column will give allele frequency p_l for the corresponding locus l. Next, we can calculate the frequency of genotypes at a locus by tabulating the frequency of three allele dosages: 0, 1 and 2. There are many other ways to summarize variation in DNA (Table 1.1). Statistics used in Chapters 11 and 12 are the expected allele dosage at a locus, the variance of allele dosages at a locus, and the correlation between allele dosages at two loci. We can estimate the expected allele dosage at a locus by multiplying allele frequency at a locus by 2: $2*p_l$. If allele frequency p_l is 0.2, we would expect an average genotype allele dosage of 2*0.2=0.4 (see SNP2 in Table 1.1). This means that the frequency of genotypes 0, 1 and 2 will be such that their average will be 0.4. We can estimate the variance of allele dosages by calculating the variance of observed allele dosages in our data set. It is common to compare the observed and expected genotype variation according to the Hardy-Weinberg equilibrium. This equilibrium is achieved primarily by random mating of parents and avoiding selection between their progeny. Under such conditions, we expect that the frequency of alleles in parents and progeny will be the same, and the frequency of genotypes 2, 1 and 0 in progeny will be, respectively, p_l^2, $2*p_l*q_l$ and q_l^2, where p_l is the frequency of the alternative allele 1 and q_l=1-p_l is the frequency of the reference allele 0. Following the binomial sampling of alleles under such conditions (this is the mathematical way of describing random mating), the variance of allele dosages is $2*p_l*q_l$. This quantity is often referred to as (expected) heterozygosity, the proportion of heterozygotes, as well as genic variance (expected variance of allele dosages at one locus, that is, one gene, hence the term genic) under Hardy-Weinberg equilibrium. Finally, we can calculate the correlation between allele dosages at two loci to study covariation between different genome regions.

Haplotype 1	0	1	1	0	0	1
Haplotype 2	1	1	1	1	0	0
Genotype	1	2	2	1	0	1

Fig. 1.5. Example of allele dosage encoding for two haplotypes across six SNPs of an animal and the corresponding genotype (CC-BY 4.0).

Table 1.1. Genotype allele dosages at two loci in five animals and corresponding summary.

Animal	SNP1	SNP2
1	0	0
2	2	1
3	2	0
4	1	1
5	0	0
Mean	1.00	0.40
Standard deviation	1.00	0.55
Variance	1.00	0.30
Allele frequency	0.50	0.20
Genic variance	0.50	0.32
Correlation	0.46	

This quantity is referred to as linkage-disequilibrium because a non-zero correlation suggests that alleles appear together more often than we would expect by chance. This can happen when loci are physically linked (placed on the same chromosome) or influenced by selection, population stratification, or admixture processes.

1.3 Variation in Phenotype Values

How is the DNA variation related to variation in phenotype values between animals? We generally do not know which DNA loci affect traits. We know that DNA gene regions are translated into RNA molecules, which are further transcribed into proteins that perform biological functions. Polymorphic loci in or around the genes drive genetic differences between animals' phenotype values. We call such loci causal loci or quantitative trait loci (QTL). Some traits are affected only by the genotype of an animal. We call such traits Mendelian traits. Mendelian traits are commonly affected by only a few DNA loci. When a trait is affected by one gene (or one locus), we call it a monogenic trait. When a trait is affected by several genes (or several loci), we call it an oligogenic trait. Most traits are affected by many DNA loci. We call such traits polygenic traits. Some traits, especially polygenic traits, are also affected by the environment in which animals live. We call such traits complex traits. This complex situation is the basis for the 'nature versus nurture' concept and recognition that an animal's observed phenotype value is a function of the effect of animal's genotype, the *effect* of the environment where the animal lives and the *effect* of other factors.

How many DNA loci affect polygenic traits? We do not know; but we can make an educated estimate. There are about 20,000 genes in the genomes of many species. Let us assume that each gene affects a polygenic trait and has at least one causal SNP. In this case, the number of causal SNP loci will be about 20,000. This might seem like a large number. However, note that traits related to biological processes such as growth or lactation are incredibly complex and involve most, if not all, body functions and hence all proteins and their upstream genes in one way or another.

While we do not know the form of this phenotype generation function, nor its effects, we will use Fisher's quantitative genetics framework to reason about the effects and later estimate them (Fisher, 1919). Fisher assumed that the observed phenotype value of an individual (y_i) can be partitioned into the effect of various factors that capture deviation of the phenotype value from the baseline of a population (μ); most importantly, the genetic value of the individual (g_i) and environmental effect (e_i) plus, possibly, interaction between the genotype and environment ($g_i \times e_i$):

$$y_i = \mu + g_i + e_i + g_i \times e_i.$$

An important simplification here is that this phenotype generation function is assumed to be linear, where we add up the effects of different factors. Here we ignore the highly non-linear and interconnecting biochemical pathways, metabolic processes, etc. All this biological complexity is swept under the 'model carpet'. We simply associate changes in phenotype values with changes in the genetic composition of individuals, which we quantify with genetic values, while the remainder is assumed to be due to environmental effects. Such linear models can be seen as a first-order (local) approximation of the highly complex biological system. This is the simplest possible, yet informative, approximation.

Following the same linear framework, we can further assume that the total genetic value of an individual (g_i) is a sum of the genetic values of that individual across causal loci ($g_{i,1}$, $g_{i,2}$, ..., $g_{i,l}$, ..., $g_{i,k}$, where k is the number of causal loci and $g_{i,l}$ takes as many values as there are genotypes observed at the locus l) and possibly their epistatic interactions ($g_{i,1} \times g_{i,2}$ + ...):

$$g_i = g_{i,1} + g_{i,2} + \ldots + g_{i,k} + g_{i,1} \times g_{i,2} + \ldots$$

Fisher also assumed possible interactions between alleles within a locus, which further decomposes genetic values into the additive genetic value ($a_{i,l}$) and the dominance genetic value ($d_{i,l}$) of an individual at each causal locus. We leave the topic of additive and non-additive (dominance and epistasis) genetic effects for Chapter 13. From this point onwards, we will assume additive allele effects only; hence, genetic values g_i will be additive genetic values, often called breeding values.

The following example demonstrates the decomposition of genetic value across loci. Assume that the baseline value of a population is ten units and that there is a single causal locus l with an additive effect a_l. The effect is such that substituting the reference allele 0 with the alternative allele 1 increases the phenotype value for one unit. Hence, substituting two reference alleles will increase the phenotype value for two units. With the three possible genotypes at a biallelic SNP (encoded as $x_{i,l}$ = 0, 1 and 2), we, respectively expect the following three phenotype values:

$$E\left(y_i|,x_{i,l} = 0|,a_l = 1\right) = \mu + x_{i,l}a_l = 10 + 0 \times 1 = 10,$$

$$E\left(y_i|,x_{i,l} = 1|,a_l = 1\right) = \mu + x_{i,l}a_l = 10 + 1 \times 1 = 11, \text{ and}$$

$$E\left(y_i|,x_{i,l} = 2|,a_l = 1\right) = \mu + x_{i,l}a_l = 10 + 2 \times 1 = 12.$$

Observed phenotype values will deviate from these expectations due to environmental effects. Assuming that environmental effects come from a normal distribution with mean zero and standard deviation (σ_e) of 0.5 unit, we can expect variation in phenotype values as shown in Fig. 1.6.

Figure 1.6 shows two sources of variation in phenotype values – genetic differences and environmental differences between animals. For every animal, we can write the following phenotype generation model:

$$y_i = \mu + g_i + e_i.$$

Assuming that environmental effects are normally distributed and that we know the genotype values (under the data-generation model), we can write the model in a probabilistic form for a specific genotype as:

$$y_i|g_i \sim N\left(\mu + g_i, \sigma_e^2\right),$$

where we see that the expectation of this data-generation process is $E(y_i|g_i) = \mu + g_i$ and the variance of this process is $Var\left(y_i|g_i\right) = \sigma_e^2$, the environmental variance. Note that if we do not know the genotypes and their effect, i.e. we are looking at phenotype variation across all the genotypes together, the probabilistic form changes to:

$$y \sim N\left(\mu, \sigma_g^2 + \sigma_e^2\right),$$

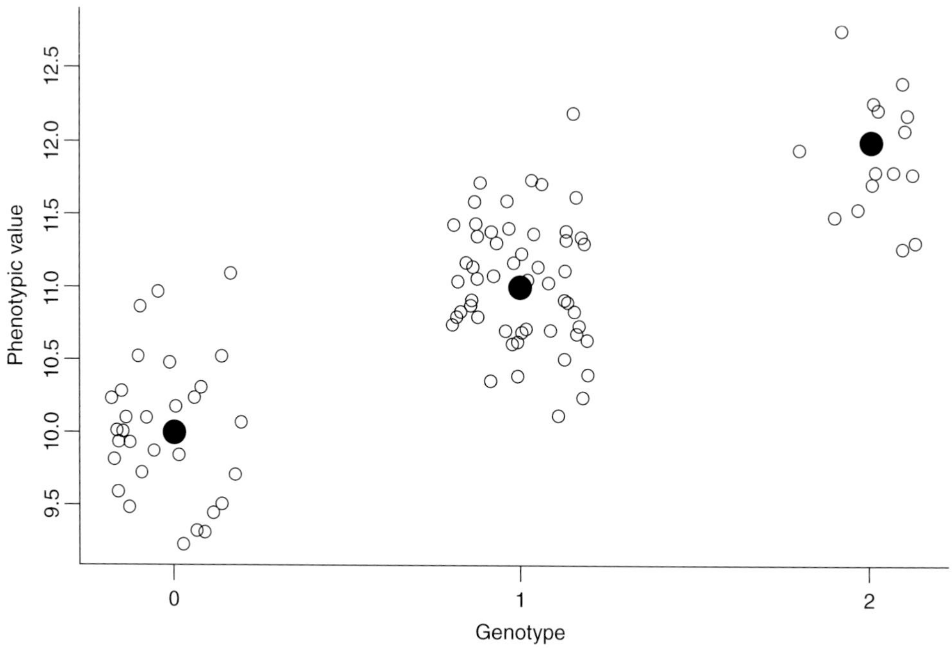

Fig. 1.6. Example of expected (large full circle) and observed (small empty circle) phenotype values as a function of three genotypes at a causal biallelic SNP locus (jittered to improve the display of points) with an allele substitution effect of one unit and normal environmental effects with a standard deviation of 0.5 units (CC BY 4.0).

where σ_g^2 is genetic variance, i.e. the variance between genetic values of individuals $Var(g_i) = \sigma_g^2$; and $\sigma_g^2 + \sigma_e^2 = \sigma_y^2$ is the phenotypic variance, i.e. the variance between phenotypic values of individuals $Var(y_i) = \sigma_y^2$, which is driven by genetic and environmental variation.

Until now, we have omitted a description of allele substitution effects: size, sign and distribution. While there is a growing body of literature on this topic, the field is still grappling with the challenge of identifying causal loci among all the loci. Namely, there are tens to hundreds of millions of SNPs and additional types of polymorphisms. We expect that only a fraction of these loci is causal (perhaps of the order of hundreds, thousands or tens of thousands). Whatever the distribution of allele substitution effects, once we add up these effects across loci, the resulting distribution of whole-genome genetic values will tend towards a normal distribution due to the central limit theorem. We demonstrate this by showing the distribution of a sample of the population baseline plus genetic values (expected phenotype value) in Fig. 1.7 for the trait affected by one, two, three or ten biallelic SNPs – all having the allele frequency of 0.5 and being on different chromosomes. We assumed that the baseline value of the population is ten units and that each alternative allele has an effect of *1/k* units, where *k* is the number of causal SNPs (this ensures that the scale of genetic values is comparable between the four examples, but note that this also scales genetic variance).

As seen in Fig. 1.7, the number of distinct genetic values is growing quickly with the number of causal loci, and the distribution is rapidly approaching a continuous normal-like

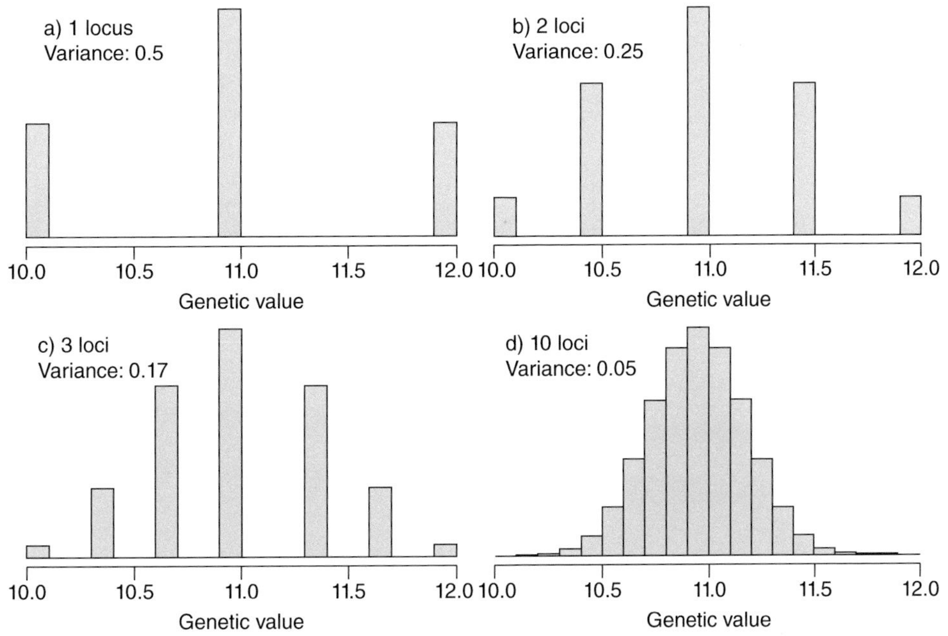

Fig. 1.7. Distribution of a sample of a population baseline plus genetic values for a trait that is affected by a) one, b) two, c) three or d) ten biallelic SNP – each sub-plot reports the corresponding variance of genetic values between individuals, the genetic variance $\left(\sigma_g^2\right)$ (CC BY 4.0).

distribution. This is not surprising since the total possible number of genotype combinations across k biallelic SNP is 3^k: $3^1 = 3$ for one SNP, $3^2 = 9$ for two SNPs, $3^3 = 27$ for three SNPs, and $3^{10} = 59{,}049$ for ten SNPs. With 500 SNPs, the number of genotype combinations grows to a whopping $\sim 10^{238}$, which is more than the number of atoms in the universe ($\sim 10^{80}$)! This is one of the reasons that early quantitative genetics work used the term *infinitesimal*, as in the infinitesimal model, to indicate that the contribution of one locus to total genetic variance is infinitely small.

To demonstrate how we generate genetic and phenotypic values for traits that are affected by multiple causal SNPs, take the example from Fig. 1.5, assuming a trait has a population baseline of ten units and is affected by the six biallelic SNPs with additive allele effects. At these SNPs, substituting the reference allele 0 with the alternative allele 1 changes the genetic value for +1 unit at the first SNP, +2 units at the second SNP, −1 unit at the third SNP, +1 unit at the fourth SNP, +1 unit at the fifth SNP, and −2 units at the sixth SNP. In Fig. 1.8 we show how we generate genetic value from an animal's haplotype and genotype allele dosages. At the top are allele dosages for the animal's two haplotypes and the corresponding genotype. At the bottom are values of alleles and corresponding genotypes alongside the six SNPs and their sums on the right. We obtain these allele and genotype values by multiplying the allele dosages with the effects and then summing the values along the SNPs. This animal has one haplotype with value −1 unit, another haplotype with value +3 units, which gives the genetic value of +2 units. If we now add the population baseline of 10 units and assume that the animal experienced a positive environment with an effect

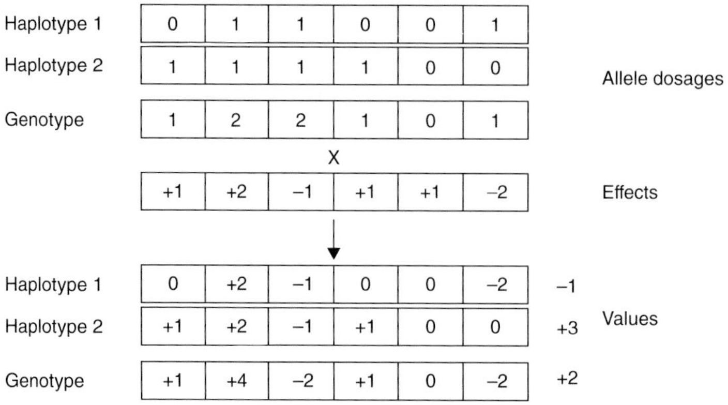

Fig. 1.8. Example of generating genetic value of one individual where the trait is affected by six biallelic SNPs (CC BY 4.0).

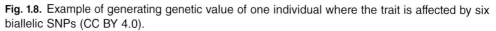

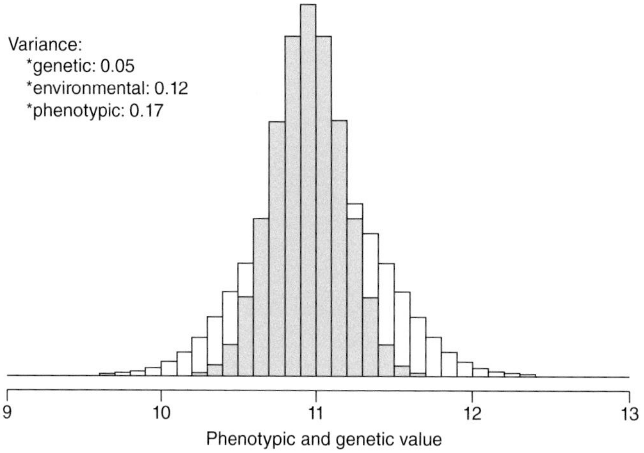

Fig. 1.9. Distribution of a sample of genetic values (dark bars) and phenotypic values (light bars) for the trait that is affected by ten biallelic SNP and environmental effects such that heritability is 0.3 – the plot reports corresponding genetic variance (σ_g^2), environmental variance (σ_e^2) and phenotypic variance (σ_y^2) (CC BY 4.0).

of +2 units and that there was no genotype-by-environment effect or non-additive genetic effects, then the phenotype value of this animal would be 10+2+2=14 units.

For the environmental effects, it is also reasonable to assume that many sources affect complex traits. We do not know all these sources and their effects, but if there are many, the distribution of their total effect will also tend towards a normal distribution due to the central limit theorem. The variance of these total environmental effects is the environmental variance (σ_e^2). Fig. 1.9 repeats the distribution of a sample of genetic values from Fig. 1.7 for the trait affected by ten SNPs with the addition of environmental effects so that the heritability of phenotype values is 0.3, that is, $h^2 = \sigma_g^2 / \sigma_y^2 = 0.3$, where $\sigma_y^2 = \sigma_g^2 + \sigma_e^2$ is the phenotypic variance.

1.4 DNA Lottery

We will now look at the randomness of DNA inheritance between parents and progeny (the DNA lottery) and how this process drives variation and resemblance between genomes of relatives and their genetic and phenotypic values. This variation comes from mitosis, which involves mutations, and from meiosis, which involves mutation, recombination and segregation.

Mutations may be the source of variation most people are familiar with. Mutations are occasional mistakes made by the DNA replication machinery each time a cell divides. If these mistakes are made in the germline (reproductive) cells, these mutations can be inherited. With billions of DNA bases, it is impressive that mutations only happen at about one mutation per chromosome in the germline (Goriely, 2016). A newborn animal will hence have about n de-novo mutations from each parent, where n is the number of chromosomes. In cattle, this could mean about 60 de-novo mutations. These de-novo mutations are in addition to mutations that parents have inherited from their parents (and so on from older ancestors) and have transmitted to their progeny. See the next paragraph on how recombination and segregation affect this transmission. With most chromosomes having about 10^8 base pairs, this number of mutations per chromosome means that the rate of mutations is about $1 \text{x} 10^{-8}$ per DNA base pair per generation in the germline of many animals. The somatic mutation rate seems at least one order of magnitude higher ($\sim 1 \text{x} 10^{-7}$) than the germline mutation rate (Lynch, 2016). With ~100 inherited germline mutations and many somatic cells ($\sim 10^{12+}$), every animal is expected to carry $\sim 10^{15+}$ mutations, with most of the genome mutated many times in many cells (Lynch, 2016). While somatic mutations are not inherited, they can affect phenotypes like germline mutations if they occur in key genome regions. Cancer is likely the most prominent example caused by somatic mutations. The effect of mutations can be either negative or positive, and this effect can depend on the environment.

While mutation is the source of new DNA variation, recombination and segregation create new combinations from existing DNA in parents and pass these combinations to offspring. Recombination and segregation happen during meiosis, the process through which germline cells produce gametes, such as sperm and ova. Each gamete contains half of the original set of DNA molecules. Which half a gamete receives is random and referred to as Mendelian sampling. Recombination and segregation can generate many combinations, which enable a continued response to selection from year to year. However, this large variation created in every new generation makes the estimation of the genetic values of newborn animals challenging. Fig. 1.10 shows a diploid cell going through meiosis. For simplicity, we show only one chromosome pair in a cell, colour each chromosome instance differently, assume the chromosome is only six base pairs long, and omit the actual DNA base pairs. While there are several steps in meiosis, we only show four. In the first step, each chromosome copy is doubled into two chromatids. In the second step, crossovers and recombinations occur, where chromatids can exchange DNA. When DNA is exchanged between paternally and maternally derived chromatids, we have recombination. Which chromatids and which parts are exchanged are largely random events. In the third step, the cell divides into two diploid cells. In the fourth step, each diploid cell splits, and we get four haploid gametes. The gametes are now ready for the final act of DNA lottery, segregation. Namely, which of the generated gametes will give rise to an offspring is again down to random events. Note that at DNA replication and recombination steps, there is a chance for germline mutations to occur.

To appreciate the power of combining existing DNA variation, let us first look at the number of chromosome combinations we can get in gametes from segregation only, without recombination. With one chromosome pair, we can get two chromosome combinations in gametes – [a light one = 1L] and [a dark one = 1D] (Fig. 1.10). The abbreviation '1L' refers to the first chromosome and its lightly coloured copy. With two chromosome pairs, we can get four chromosome combinations in gametes – [1L, 2L], [1L, 2D], [1D, 2L], and [1D, 2D]. With three chromosome pairs, we can get eight chromosome combinations in gametes. With n chromosomes, we can get 2^n chromosome combinations in gametes. For example, cattle have 30 chromosome pairs, giving 2^{30} = 1,073,741,824 (more than a billion, ~10^9) possible chromosome combinations in gametes. This is the number of possible chromosome combinations in gametes in one parent, assuming no recombination.

As we saw in Fig. 1.10, some chromosomes had no recombinations and some had one. We usually get about one recombination per generated chromosome of ~10^8 bases, but we can get no recombination or more than one. Hence, the recombination rate is about 1×10^{-8} per base pair per generation, like the germline mutation rate. The number and placement of recombinations are random events. While the number of recombinations is not large, random placement adds many possible chromosome combinations in gametes on top of segregation.

The seemingly simple process of meiosis can generate a staggering amount of DNA variation by recombining and segregating the parental chromosomes. This process drives genetic relationships between animals. To put this into the context of relatives, we show a three-generation pedigree in Fig. 1.11. This figure shows two siblings (G and H), their two parents (E and F), and their four grandparents (A, B, C and D). As before, we show only one chromosome pair with six loci. We have four diploid grandparents, hence eight chromosome instances. To simplify tracking of genetic inheritance within this pedigree, we have coloured the chromosomes and numbered their loci according to grandparental origin. Also, we use a convention that the top chromosome is of paternal origin, and the bottom chromosome is of maternal origin. There is no such ordering in an actual cell. Do not confuse this 'descent-based' encoding of alleles (and the related concept of identity-by-descent) with the 'state-based' encoding (and the related concept

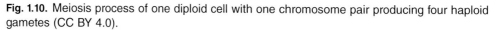

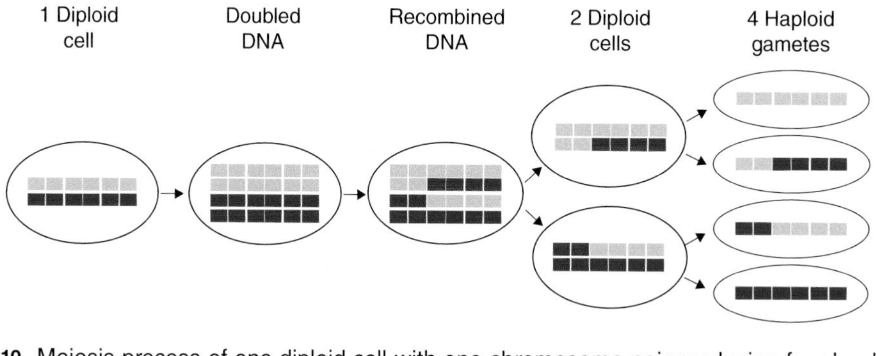

Fig. 1.10. Meiosis process of one diploid cell with one chromosome pair producing four haploid gametes (CC BY 4.0).

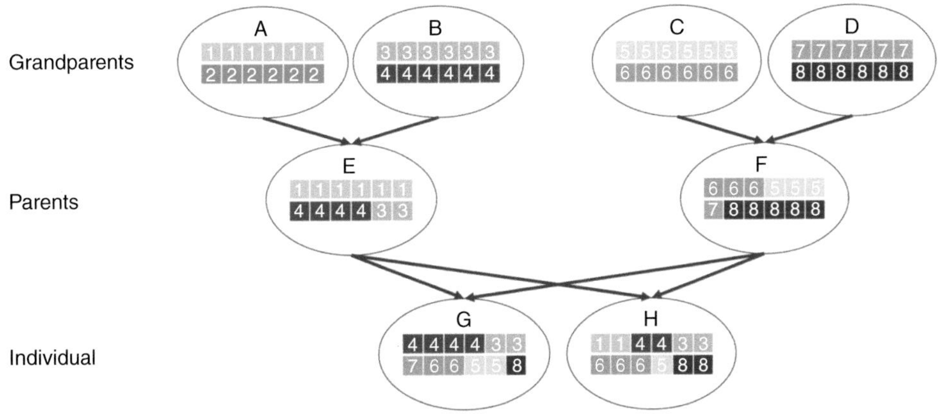

Fig. 1.11. Inheritance of DNA between generations of a pedigree – alleles are represented by descent-based encoding (CC BY 4.0).

of identity-by-state) that we have used up to now. Behind the numbers 1, 2, ..., 8 are DNA base pairs with the corresponding allele codes 0 and 1, respectively, for reference and alternative alleles.

Inspecting Fig. 1.11, we can see that an animal always receives 50% of DNA from each parent, but there is variation in which half is received due to recombination and segregation. Due to this sampling, an animal might not inherit DNA from one of his grandparents in a particular genome region. This means that recombination and segregation are sampling different ancestral lineages along the genome for every newborn animal. For example, animal G inherited no DNA from grandparent A. Across multiple chromosome pairs, we expect that a grandchild inherits 25% of DNA from each grandparent. Still, there is variation around this expectation due to recombination and segregation. Similarly, we expect that siblings share 50% of DNA, but there is variation around this expectation due to recombination and segregation.

We can formalize the above observations by relating the genetic value of an individual g_i with the genetic value of its father (sire) $g_{f(i)}$ and its mother (dam) $g_{m(i)}$:

$$g_i = \frac{1}{2} g_{f(i)} + \frac{1}{2} g_{m(i)} + r_i,$$

where $\frac{1}{2} g_{f(i)} + \frac{1}{2} g_{m(i)}$ is the *parent average*, the expected genetic value of an individual given the genetic values of its parents $E(g_i | g_{f(i)}, g_{m(i)}) = \frac{1}{2} g_{f(i)} + \frac{1}{2} g_{m(i)}$, and r_i is the *Mendelian sampling deviation*, the deviation of individuals' genetic value from the parent average $r_i = g_i - (\frac{1}{2} g_{f(i)} + \frac{1}{2} g_{m(i)})$. The above model is sometimes referred to as pedigree regression, where we regress the genetic value of an individual to the genetic values of its parents to get the expected value, while deviations from the regression lines are due to Mendelian sampling. To further connect this formalism with Fig. 1.11, note that the genetic value of an individual is a sum of the genetic values of its two chromosome (genome) copies ($g_{i,1}$ and $g_{i,2}$). Each chromosome (genome) copy originates from a parent, which also has two chromosome (genome) copies and passes a combination of these to its progeny. Hence, we can split parent average and Mendelian sampling terms per parent as:

$$g_i = g_{i,1} + g_{i,2},$$

$$g_{i,1} = \frac{1}{2} g_{f(i),1} + \frac{1}{2} g_{f(i),2} + r_{i,1},$$

$$g_{i,2} = \frac{1}{2} g_{m(i),1} + \frac{1}{2} g_{m(i),2} + r_{i,2}$$

Because the genetic values of individuals are a sum of their genetic values across the causal loci, we can show the connection between the genetic values of an individual and its parents along the causal loci as a sum of locus genetic values for each parental chromosome (genome) (giving the parent average) and a sum of locus deviations (giving the Mendelian sampling term):

$$g_i = \sum_{l=1}^{k} \left(g_{i,1,l} + g_{i,2,l} \right),$$

$$g_{i,1} = \sum_{l=1}^{k} \left(\frac{1}{2} g_{f(i),1,l} + \frac{1}{2} g_{f(i),2,l} + r_{i,1,l} \right),$$

$$g_{i,2} = \sum_{l=1}^{k} \left(\frac{1}{2} g_{m(i),1,l} + \frac{1}{2} g_{m(i),2,l} + r_{i,2,l} \right)$$

where the summation is across the causal loci *1, 2, ..., k*, $g_{i,s,l}$ is the genetic value of individual i in genome set s at the causal locus l, $r_{i,s,l}$ is the corresponding Mendelian

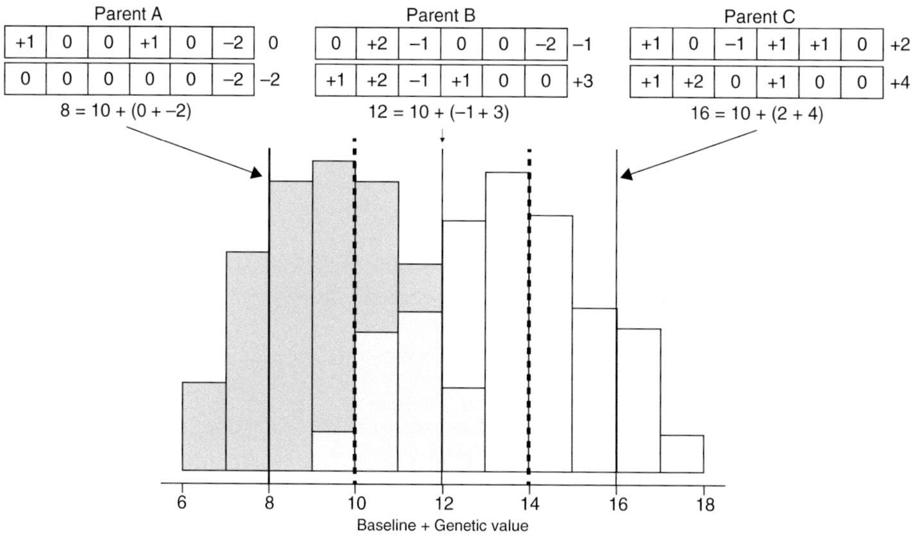

Fig. 1.12. Example of genetic variation between and within two half-sib families due to variation in parental genetic values (parent averages) and Mendelian sampling. At the top are parental haplotypes with allele values and associated haplotype and genotype values, while at the bottom are two distributions of progeny genetic values in the families (dark bars represent progeny from crossing parent A with B and light bars represent progeny from crossing parent C with B), with overlaid vertical lines denoting parental genetic values (full line) and parent averages (dashed line). All genetic values have the baseline value of 10 added (CC BY 4.0).

sampling deviation, and k is the number of causal loci. The above formulation shows how parent average and Mendelian sampling deviation of a genetic value result from DNA inheritance between generations.

Understanding the variation of genetic values between families (i.e. between family parent averages) and within families (i.e. between Mendelian sampling deviations within families) is important for various breeding operations. Fig. 1.12 demonstrates such variation for two half-sib families originating from crossing parent A with B and parent C with B. Here we assume that such crosses can produce many progenies. If this is not possible, the figure shows the extent of possible variation between potential progenies. There are three notable observations from Fig. 1.12. First, progeny genetic values are distributed around their parent average in line with the above-mentioned theory. Second, there is substantial within-family variation due to Mendelian sampling. Third, some progeny genetic values are below or above parental genetic values, again indicating the extent of Mendelian sampling. Importantly, this has been generated from only six causal loci on one chromosome. Many more causal loci across multiple chromosomes will influence many traits, generating even more variation.

To further appreciate the magnitude of between- and within-family variation, we can evaluate how much genetic variance is due to variation between and within families. We will address this topic more extensively in Chapter 3, but here we give the standard result by decomposing the genetic variance according to the pedigree regression. In the following, we assume that parents are randomly sampled from a population, not inbred, and unrelated. Under such conditions, genetic variation in a population is 50% due to between-family variation and 50% due to within-family variation. This result is important because it shows the extent of variation we can expect from combining parental genomes (=between-family or parent average variation) and from recombining and segregating parental genomes (=within-family or Mendelian sampling variation).

$$
\begin{aligned}
Var(g_i) &= Var\left(\tfrac{1}{2}g_{f(i)} + \tfrac{1}{2}g_{m(i)} + r_i\right) \\
\sigma_g^2 &= Var\left(\tfrac{1}{2}g_{f(i)}\right) + Var\left(\tfrac{1}{2}g_{m(i)}\right) + Var(r_i) \\
&= \tfrac{1}{4}Var\left(g_{f(i)}\right) + \tfrac{1}{4}Var\left(g_{m(i)}\right) + Var(r_i) \\
&= \tfrac{1}{2}\sigma_g^2 + \tfrac{1}{2}\sigma_g^2
\end{aligned}
$$

Finally, when we combine between- and within-family genetic variation with environmental variation (Fig. 1.13), we start to appreciate the challenge of estimating the unknown genetic values of individuals from data. Namely, Fig. 1.13 shows the variation of a sample of phenotypic values based on genetic values from Fig. 1.12. From the genetic values of parents equal to 8, 12 and 16 units, we generated progeny phenotype values with a range between 0 and 25 units. And this phenotype variation does not yet include other factors, such as sex, farm and other effects, which would increase the phenotypic variance even more. This book describes methods to make the best estimates of genetic values from a combination of collected phenotypic, pedigree and genomic data.

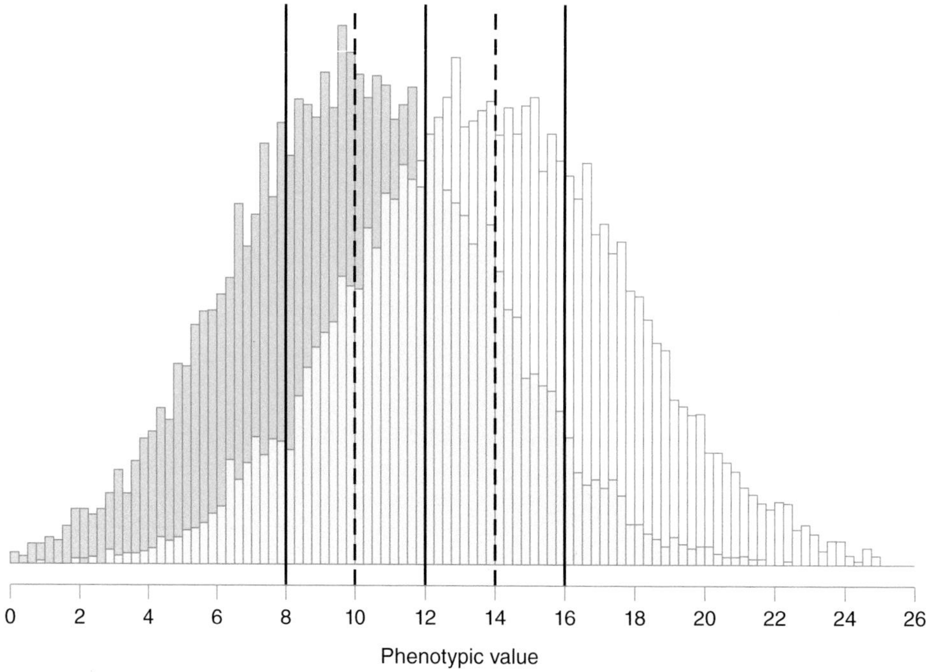

Fig. 1.13. Variation of a sample of phenotypic values with heritability of 0.5 between and within two half-sib families relative to the genetic values of parents (full vertical lines) and corresponding parent average (dashed vertical lines) (see also Fig. 1.12) (CC BY 4.0).

1.5 Additional Points

To close this chapter, we point to the different types of traits, multiple traits, genotype-by-environment interactions, and biological versus statistical interpretation of genetic effects. In the introduction to this chapter, we mentioned traits that do not have a continuous distribution. The above-presented genomic and phenotypic data-generation processes can also be used for traits with other distributions. When we used Fisher's linear phenotype decomposition, we implicitly used the normal (Gaussian) distribution, assuming a linear link function between the phenotype values and their components. We can relax this assumption with generalized linear models that work with additional distributions and corresponding link functions. An example of such a model is shown in Chapter 15. Further, we have described the data-generation model for a single trait only, but the same framework can also be used for multiple traits. The key extension for multiple traits is the addition of covariance between trait genetic effects, as well as for environmental effects. When we measure the same trait in different environments, we can consider the trait expression in different environments as multiple traits with environment-specific genetic effects representing genotype-by-environment interactions. Finally, in this chapter we focused solely on additive genetic effects. As described, we do not know the true biological model that generates phenotype values because of the complex underlying biology. Above, we have described a conceptual data-generation model following Fisher's linear decomposition of phenotypic values. In these models, the additive genetic effects capture most of the genetic variance (Hill *et al.*, 2008). However, these additive genetic effects are so-called statistical effects – estimated statistically from the data at hand – and are hence data-dependent. This means that statistical additive genetic effects likely capture additive genetic variance and a part of non-additive genetic variance. In Chapter 13, we describe models for estimating additive and non-additive genetic effects.

2 Genetic Evaluation with Different Sources of Records

2.1 Introduction

The prediction of breeding values constitutes an integral part of most breeding programmes for genetic improvement. Crucial to the accurate prediction of breeding value is the availability of records. In a population, data available at the initial stages are usually on individual animals, which may or may not be related, and, later, on offspring and other relatives. Thus, initially, the prediction of breeding values may be based on the records of individuals and few relatives. In this chapter, the use of individual records and information from other related sources in the prediction of breeding value is addressed. Also, the principles for the calculation of selection indices combining information from different sources and relatives are discussed.

2.2 The Basic Model

Every phenotypic observation on an animal is determined by environmental and genetic factors and may be defined by the following model:

Phenotypic observation = environmental effects + genetic effects + residual effects

or

$$y_{ij} = \mu_i + g_i + e_{ij} \tag{2.1}$$

where y_{ij} is the record j of the ith animal; μ_i refers to the identifiable non-random (fixed) environmental effects such as herd management, year of birth or sex of the ith animal; g_i is the sum of the additive, dominance and epistatic genetic values of the genotype of animal i; and e_{ij} is the sum of random environmental effects affecting animal i.

The additive genetic value (a_i) component of the g_i term above represents the average additive effects of genes an individual receives from both parents and is called the breeding value. Each parent contributes a sample half of its genes to its progeny. The average effect of the sample half of genes that a parent passes to its progeny is called the transmitting ability of the parent and corresponds to half of its additive genetic value. The breeding value of the progeny therefore is the sum of the transmitting abilities of both parents. Since the additive genetic value is a function of the genes transmitted from parents to progeny, it is the only component that can be selected for and therefore the main component of interest. In most cases, dominance and epistasis, which represent intra-locus and inter-loci interactions, respectively, are assumed to be of little significance and are included in the e_{ij} term of the model as:

$$y_{ij} = \mu_i + a_i + e^*_{ij} \tag{2.2}$$

© R.A. Mrode and I. Pocrnic 2023. *Linear Models for the Prediction of the Genetic Merit of Animals, 4th Edition* (R.A. Mrode and I. Pocrnic)
DOI: 10.1079/9781800620506.0002

with e^*_{ij} being the sum of the random environmental effects, dominance and epistatic genetic values. In matrix form, Eqn 2.2 can be written as:

$$\mathbf{y} = \mathbf{Xb} + \mathbf{Za} + \mathbf{e}^*$$

where: **y** is the vector of observations, **b** is a vector of fixed effects and, in this case, it is just μ; **a** and **e*** are vectors of random additive genetic and residual effects, respectively; X and **Z** are incidence matrices relating records to the fixed and random genetic effects, respectively (see Chapter 4).

It is assumed that the expectations (E) of the variables are:

$$\mathrm{E}(\mathbf{y}) = \mathbf{Xb};\ \mathrm{E}(\mathbf{a}) = \ \mathrm{E}(\mathbf{e}^*) = 0$$

Also, it is assumed that that **y** follows a multivariate normal distribution, implying that traits are determined by infinitely many additive genes of infinitesimal effect at unlinked loci, the so-called infinitesimal model (Fisher, 1918; Bulmer, 1980). Random residual effects are assumed to be independently distributed with variance matrix, var(**e***) = $\mathbf{I}\sigma^2_{e^*}$, and the variance matrix of **a** (var(**a**)) = $\mathbf{I}\sigma^2_a$, assuming that animals are unrelated. Finally, it assumed that there is no correlation between **a** and **e*** (cov(**a**, **e***)) is a matrix with zero elements.

Equation 2.2 constitutes the linear model usually employed in most problems of breeding value prediction in animal breeding. Also μ, which is used subsequently in this chapter to represent the mean performance of animals in the same management group, for instance animals reared under the same management system, of the same age and sex, is assumed known. From Eqn 2.2, the problem of predicting breeding value reduces to that of adjusting phenotypic observations for identifiable non-random environmental effects and appropriately weighting the records of animals and their available relatives.

From the earlier explanation, if a_i is the breeding value of animal i, then:

$$a_i = \left(\tfrac{1}{2}\right) a_s + \left(\tfrac{1}{2}\right) a_d + m_i$$

where a_s and a_d are the breeding values of the sire and dam, respectively, and m_i is the deviation of the breeding value of animal i from the average breeding value for both parents, that is, Mendelian sampling. The sampling nature of inheritance implies that each parent passes only a sample one-half of their genes to their progeny. There is, therefore, genetic variation between offspring of the same parents since all offspring do not receive exactly the same genes. Mendelian sampling could be regarded as the deviation of the average effects of additive genes an individual receives from both parents from the average effects of genes from the parents common to all offspring.

The accurate prediction of breeding value constitutes an important component of any breeding programme since genetic improvement through selection depends on correctly identifying individuals with the highest true breeding value. The method used to predict breeding value depends on the type and amount of information available on candidates for selection. The next section discusses the prediction of breeding value using different sources of information. It should be noted that many applications of genetic evaluation deal with the prediction of transmitting ability, usually referred to as predicted transmitting ability (PTA) or estimated transmitting ability (ETA), which is half of the predicted breeding value.

2.3 Breeding Value Prediction from the Animal's Own Performance

2.3.1 Single record

When one phenotypic record (y_i) is the only available information on animal i, the estimated breeding value (EBV) (a_i) for animal i can be calculated as:

$$\hat{a}_i = b(y_i - \mu_i) \quad (2.3)$$

where b is the regression of true breeding value on phenotypic performance and μ_i is the mean performance of the management group for ith animal and is assumed to be known. Thus:

$$b = \text{cov}(a_i, y_i)/\text{var}(y_i) = \text{cov}(a_i, a_i + e_i^*)/\text{var}(y_i)$$
$$= \sigma_a^2 / \sigma_y^2 = h^2$$

noting that, for all individuals, $\text{var}(y_i) = \sigma_y^2$.

The prediction is simply the adjusted record multiplied by the heritability (h^2). The correlation between the selection criterion, in this case the estimated breeding value, and the true breeding value is known as the accuracy of prediction. It provides a means of evaluating different selection criteria because the higher the correlation, the better the criterion as a predictor of breeding value. In some cases, the accuracy of evaluations is reported in terms of reliability ($r_{a,\hat{a}}^2$), which is the squared correlation between the selection criterion and the true breeding value. With a single record per animal, the accuracy is:

$$r_{a,\hat{a}} = \text{cov}(a_i, by_i) / [\sigma_a \sqrt{var(by_i)}]$$
$$= bcov(a_i, y_i) / [\sigma_a \sqrt{b^2 \sigma_y^2}]$$
$$= h^2 \sigma_a^2 / (\sigma_a \sqrt{h^2 \sigma_a^2}) = h$$

and reliability equals h^2.

Expected response (R) to selection based on a single record per individual (Falconer and Mackay, 1996) is:

$$R = ib\sigma_y = ih^2\sigma_y = ir_{a,\hat{a}}\sigma_a$$

where i, the intensity of selection, refers to the superiority of selected individuals above population average expressed in phenotypic standard deviation.

The variance among EBVs is of practical value, giving us an indication of the difference in EBV between the highest and lowest animals. It can be computed as follows.

The var of EBV_i ($\text{var}(\hat{a}_i)$) is:

$$\text{var}(\hat{a}_i) = \text{var}(by_i) = \text{var}(h^2 y_i)$$
$$= h^4 \sigma^2_y$$
$$= r_{a,\hat{a}}^2 h^2 \sigma^2_y = r_{a,\hat{a}}^2 \sigma^2_a \quad (2.4)$$

Also of practical importance is the prediction error variance (PEV), which provides an insight into the amount of error associated with the prediction of the breeding value. For the ith individual,

$$\text{PEV}_i = \text{Var}(a_i - \hat{a}_i) = (1 - r_{a,\hat{a}}^2)\sigma_a^2$$

Note that PEV_i decreases as accuracy increases, and from the above equation,

$$\text{Var}(\hat{a}_i) + \text{PEV}_i \text{ is equal to } \sigma_a^2 \; : r_{a,\hat{a}}^2 \sigma_a^2 + (1 - r_{a,\hat{a}}^2)\sigma_a^2 = \sigma_a^2$$

The standard error of prediction (SEP_i) can be computed from the PEV_i, and SEP_i is useful in computing the confidence interval of EBV_i around the true breeding value:

$$\text{SEP}_i = (\text{PEV}_i)^{1/2} = ((1 - r_{a,\hat{a}}^2)\sigma_a^2)^{1/2}$$

Therefore, a 95% confidence interval (CI) for the true breeding value (TBV_i) can be computed as:

$$\text{CI}_{(0.95)} \text{for TBV}_i = \text{EBV}_i \pm t_\alpha * \text{SEP}_i$$

where t_α is the t value from the Student's t-distribution table at 0.05 level.

Example 2.1

Assume that the genetic variance of yearling weight is 80.5kg^2 with a heritability of 0.45. If the yearling weight of a heifer raised in a management group is 320 kg with a mean of 250 kg, predict her breeding value, the accuracy of the EBV, standard error of prediction and a 95% confidence interval for the true breeding value.

From Eqn 2.3:

$$\hat{a}_i = 0.45(320 - 250) = 31.50\text{kg}$$

and

$$r_{a,\hat{a}} = \sqrt{0.45} = 0.67$$

$$\text{SEP}_i = ((1 - 0.67^2)80.5)^{1/2} = 6.66$$

and

$$\text{CI}_{(0.95)} \text{ for TBV}_i = 31.50 \pm 1.96 * 6.66 = 31.50 \pm 13.05$$

This means that the TBV_i is in the range of 18.45 and 44.55 with 95% confidence.

2.3.2 Repeated records

When multiple measurements on the same trait, such as milk yield in successive lactations, are recorded on an animal, its breeding value may be predicted from the mean of these records. With repeated measurements it is assumed that there is additional resemblance between records of an individual due to environmental factors or circumstances that affect the records of the individual permanently. In other words, there is an additional covariance between records of an individual due to non-genetic permanent environmental effects. Thus the between-individual variance is partly genetic and partly environmental (permanent environmental effect). The within-individual variance is attributed to differences between successive measurements of the individual arising from temporary environmental variations from one parity to the other. The variance of observations for the *i*th individual could therefore be partitioned as:

$$\text{var}(y_i) = \text{var}(g_i) + \text{var}(pe_i) + \text{var}(te_i)$$

where var(g_i) = genetic variance including additive and non-additive, var(pe_i) = variance due to permanent environmental effect, and var(te_i) = variance due to random temporary environmental effect.

Noting that, for all individuals, $\text{var}(g_i) = \sigma_g^2$, $\text{var}(pe_i) = \sigma_{pe}^2$ and $\text{var}(te_i) = \sigma_{pe}^2$, the intra-class correlation (t), which is the ratio of the between-individual variance to the phenotypic

$$t = (\sigma_g^2 + \sigma_{pe}^2) / \sigma_y^2 \tag{2.5}$$

is usually called the repeatability and measures the correlation between the records of an individual.

From Eqn (2.5):

$$\sigma_{te}^2 / \sigma_y^2 = 1 - t \tag{2.6}$$

With this model, it is always usually assumed that the repeated records on the individual measure the same trait, i.e. there is a genetic correlation of 1 between all pairs of records. Also, it is assumed that all records have equal variance and that the environmental correlations between all pairs of records are equal. Let $\bar{y}_i$ represent the mean of n records for animal i. The breeding value may be predicted as:

$$\hat{a}_i = b(\bar{y}_i - \mu_i) \tag{2.7}$$

where $b = \text{cov}(a_i, \bar{y}_i) / \sigma_y^2$ and μ_i is the mean performance of the management group for ith animal and is assumed to be known.

Now:

$$\text{cov}(a_i, \bar{y}) = \text{cov}(a_i, g_i + pe_i + \textstyle\sum_{j=1}^{n} te_{ij})/n)$$

and

$$\sigma_{\bar{y}}^2 = \sigma_g^2 + \sigma_{pe}^2 + \sigma_{te}^2 / n$$

From Eqns 2.5 and 2.6, $t\sigma_y^2 = \sigma_g^2 + \sigma_{pe}^2$ and $(1-t)\sigma_y^2 = \sigma_{te}^2$
Therefore, $\sigma_{\bar{y}}^2$ can be expressed as:

$$\sigma_{\bar{y}}^2 = [t + (1-t)/n]\sigma_y^2$$

and

$$\begin{aligned} b &= \sigma_a^2 / [t + (1-t)/n]\sigma_y^2 \\ &= nh^2 / [1 + (n-1)t] \end{aligned}$$

Note that b now depends on heritability, repeatability and the number of records.

As mentioned earlier, the difference between repeated records of an individual is assumed to be due to temporary environmental differences between successive performances. However, if successive records are known to be affected by factors that influence performance, these must be corrected for. For instance, differences in age at calving in first and second lactations may influence milk yield in first and second lactations. Such age differences should be adjusted for before using the means of both lactations for breeding-value prediction.

The accuracy of the EBV is:

$$r_{a,\hat{a}} = cov(a_i, b\bar{y}_i) / [\sigma_a \sqrt{var(b y_i)}]$$

Now:

$$cov(a_i, b\bar{y}_i) = b cov(a_i, \bar{y}_i)$$

and

$$var(b y_i) = b^2 var(y_i) = b cov(a_i, \bar{y}_i)$$

Therefore:

$$r_{a,\hat{a}} = b cov(a_i, \bar{y}_i) / \left[\sigma_a \sqrt{b cov(a_i, \bar{y}_i)}\right] = \sqrt{b cov(a_i, \bar{y}_i)} / \sigma_a$$
$$= \sqrt{(nh^2 / [1 + (n-1)t])\sigma_a^2} / \sigma_a = \sqrt{b}$$

Compared with single records, there is a gain in the accuracy of prediction with repeated records from the above equation, which is dependent on the value of repeatability and the number of records. This gain in accuracy results mainly from the reduction in temporary environmental variance (within-individual variance) as the number of records increases. When t is low, this gain is substantial as the number of records increases. When t is high, there is little gain in accuracy with repeated records compared with using only single records. The gain in accuracy from repeated records compared with selection on single records can be obtained as the ratio of accuracy from repeated records (r_n) to that from single records (r_k):

$$(r_n / r_k) = \sqrt{h^2 / (t + (1-t)/n)} / h = \sqrt{n / (1 + (n-1)t)}$$

Using the above equation, the gain in accuracy from repeated records compared with selection on single records is given in Table 2.1. The increase in accuracy with four measurements at a low t of 0.4 was 35%, but this dropped to only 8% when t equalled 0.8. In general, the rate of increase dropped rapidly as the number of records exceeded four, and it is seldom necessary to record more than four measurements.

Expected response to selection based on mean of repeated records is:

$$R = ib\sigma_{\bar{y}} = ir_{a,\hat{a}}\sigma_a$$

Table 2.1. Percentage increase in accuracy of prediction with repeated records compared with single records.

	Number of records				
Repeatability (*t*)	2	4	6	8	10
0.4	20	35	41	45	47
0.6	12	20	22	24	25
0.8	5	8	10	10	10

Example 2.2

Assume that a cow has a mean yield of 8000 kg of milk for first and second lactations. If the phenotypic standard deviation and heritability of milk yield in the first two lactations are 600 kg and 0.30, respectively, and the correlation between first and second lactation yields is 0.5, predict the breeding value of the cow for milk yield in the first two lactations and its accuracy. Assume that the herd mean for first and second lactations is 6000 kg.

From Eqn 2.7:

$$\hat{a}_{cow} = b(8000 - 6000)$$

with

$$b = 2(0.3)/(1 + (2 - 1)0.5)) = 0.4$$

Therefore:

$$\hat{a}_{cow} = 0.4(8000 - 6000) = 800 \text{ kg}$$

and

$$r_{a,\hat{a}} = \sqrt{0.4} = 0.632$$

2.4 Breeding Value Prediction from Progeny Records

For traits where records can be obtained only on females, the prediction of breeding values for sires is usually based on the mean of their progeny. This is typical of the dairy cattle situation, where bulls are evaluated on the basis of their daughters. Let $\bar{y}_i$ be the mean of single records of n progeny of sire i and assume that the progeny are only related through the sire (paternal half-sibs), and so the breeding value of sire i is:

$$\hat{a}_i = b(\bar{y}_i - \mu_i) \tag{2.8}$$

where $b = \text{cov}(a_i, \bar{y}_i)/\sigma^2_{\bar{y}}$ and μ_i is the mean of single records of n progeny of sire i and it is assumed to be known.

Now:

$$\text{cov}(a_i, \bar{y}_i) = \text{cov}\left(a_i, \left(\frac{1}{2}\right)a_s + \left(\frac{1}{2}\right)a_d + \sum\nolimits_{j=1}^{n} e_{ij})/n\right)$$
$$= \left(\frac{1}{2}\right)\text{cov}(a_s, a_s) = \left(\frac{1}{2}\right)\sigma^2_a$$

Noting that i = s and a_s is the sire breeding value and a_d represents the mean breeding value for the dams mated to the sire.

Using the same principles as in Section 2.3.2:

$$\sigma^2_{\bar{y}} = [t + (1 - t)/n)]\sigma^2_y$$

assuming there is no environmental covariance between the half-sib records and t, the intra-class correlation between half-sibs, is $\left(\frac{1}{4}\right)\sigma^2_a / \sigma^2_y = \left(\frac{1}{4}\right)h^2.$

Therefore:

$$b = \left(\frac{1}{2}\right)\sigma_a^2 / [\mathrm{t} + (1-t)/n]\sigma_y^2$$
$$= \left(\frac{1}{2}\right)h^2\sigma_y^2 / \left[\left(\frac{1}{4}\right)\mathrm{h}^2 + (1 - \left(\frac{1}{4}\right)h^2)/n\right]\sigma_y^2$$
$$= 2nh^2 / \left[nh^2 + (4-h^2)\right]$$
$$= 2n / \left[n + (4-h^2)/h^2)\right]$$
$$= 2n / [n+k]$$

with

$$k = (4-h^2)/h^2)$$

The term k is constant for any assumed heritability. The weight (b) depends on the heritability and number of progeny and approaches 2 as the number of daughters increases.

Using the same arguments as in Section 2.3.2, the accuracy of the EBV is:

$$r_{a,\hat{a}} = cov(a_i, b\overline{y}_i) / [\sigma_a \sqrt{var(b\overline{y}_i)}\,]$$
$$= bcov(a_i, \overline{y}_i) / [\sigma_a \sqrt{bcov(a_i, \overline{y}_i)}] = \sqrt{bcov(a_i, \overline{y}_i)}/\sigma_a$$
$$= \sqrt{b\left(\frac{1}{2}\right)\sigma_a^2} \,/\sigma_a = \sqrt{\left(\frac{1}{2}\right)b} = \sqrt{n/(n+k)}$$

which approaches unity (1) as the number of daughters becomes large. Reliability of the predicted breeding value therefore equals $n/(n + k)$.

The equation for expected response when selection is based on the mean of half-sibs is the same as that given in Section 2.3.2 for the mean of repeated records but with t now referring to the intra-class correlation between half-sibs.

The performance of any future daughters of the sire can be predicted from the mean performance of the present daughters. The breeding value of a future daughter ($\hat{a}_{dtr}$) of the sire can be predicted as:

$$\widehat{a_{dtr}} = b(\overline{y}_i - u_i)$$

with $\overline{y}_i$ and μ_i as defined in Eqn 2.8, respectively, and:

$$b = \mathrm{cov}(a_{dtr}, \overline{y}_i) / \sigma_{\overline{y}}^2$$

Now:

$$\mathrm{cov}(a_{dtr}, \overline{y}_i) = \mathrm{cov}\left(\left(\frac{1}{2}\right)a_s + \left(\frac{1}{2}\right)a_{d*}, \left(\frac{1}{2}\right)a_s + \left(\frac{1}{2}\right)a_d + (\textstyle\sum_{i=1}^{n} e_i)/n\right)$$

where the subscript d^* refers to the dam of the future daughter, which is assumed to be unrelated to dams (d) of present daughters. Therefore:

$$\mathrm{cov}(a_{dtr}, \overline{y}_i) = \mathrm{cov}\left(\left(\frac{1}{2}\right)a_s, \left(\frac{1}{2}\right)a_s\right) = \left(\frac{1}{4}\right)\mathrm{cov}(a_s, a_s) = \left(\frac{1}{4}\right)\sigma_a^2$$

Therefore:

$$b = \left(\frac{1}{4}\right)\sigma_a^2 / [t + (1-t)/n]\sigma_y^2$$

Using the same calculations for obtaining b in Eqn 2.8:

$$b = n / (n + k)$$

The b value is half of the value of b in Eqn 2.8, thus the predicted breeding value of a future daughter of the sire is equal to half the EBV of the sire. The performance of a future daughter of the sire can be predicted as:

$$y = u_i + \hat{a}_{dtr}$$

where u_i is the management mean.

The accuracy of the predicted breeding value of the future daughter is:

$$r_{a,\hat{a}} = cov(a_{dtr}, b\bar{y}_i) / [\sigma_a \sqrt{var(b\bar{y}_i)}]$$

Using arguments in previous sections:

$$r_{a,\hat{a}} = bcov(a_{dtr}, \bar{y}_l) / [\sigma_a \sqrt{bcov(a_{dtr}, \bar{y}_i)}] = \sqrt{bcov(a_{dtr}, \bar{y}_l)} / \sigma_a$$

$$= \sqrt{b(\frac{1}{4})\sigma_a^2} / \sigma_a = \sqrt{\frac{1}{4}b} = \left(\frac{1}{2}\right)\sqrt{n / (n + k)}$$

which is equal to half of the accuracy of the predicted breeding value of the sire. Reliability of the predicted breeding value equals n/(4(n+k)), which is one-quarter of the reliability of the bull proof.

Example 2.3

Suppose the fat yield of 25 half-sib progeny of a bull averaged 250 kg in the first lactation. Assuming a heritability of 0.30 and herd mean of 200 kg, predict the breeding value of the bull for fat yield and its accuracy. Also predict the performance of a future daughter of this bull for fat yield in this herd.

From Eqn 2.6:

$$\hat{a}_{bull} = b(250 - 200)$$

with

$$b = 2n / (n + (4 - h^2) / h^2) = 2(25) / (25 + (4 - 0.3) / 0.3) = 1.34$$

$$\hat{a}_{bull} = 1.34(250 - 200) = 67 \text{ kg}$$

$$r_{a,\hat{a}} = \sqrt{n / (n + k))} = \sqrt{[25 / (25 + (4 - 0.3) / 0.3))]} = 0.82$$

The future performance of the daughter of the bull is:

$$y = (0.5)a_{bull} + \text{herd mean}$$

$$= 33.5 + 200 = 233.5 \text{ kg}$$

2.5 Breeding Value Prediction from Pedigree

When an animal has no record, its breeding value can be predicted from the evaluations of its sire (s) and dam (d). Each parent contributes half of its genes to their progeny, and so the predicted breeding value of progeny (o) is:

$$\hat{a}_o = (\hat{a}_s + \hat{a}_d) / 2 \qquad (2.9)$$

Let $f = (\hat{a}_s + \hat{a}_d)/2$, then the accuracy of the predicted breeding value is:

$$r_{\hat{a}_o, f} = cov(a_o, \left(\frac{1}{2}\right)\hat{a}_s + \left(\frac{1}{2}\right)\hat{a}_d) / \sqrt{\sigma_a^2\, var\left[\left(\frac{1}{2}\right)\hat{a}_s + \left(\frac{1}{2}\right)\hat{a}_d\right]}$$

Now:

$$\text{cov}\left(a_o, \left(\frac{1}{2}\right)\hat{a}_s + \left(\frac{1}{2}\right)\hat{a}_d\right) = \text{cov}\left(a_o, \left(\frac{1}{2}\right)\hat{a}_s\right) + \text{cov}\left(a_o, \left(\frac{1}{2}\right)\hat{a}_d\right)$$

$$= \text{cov}\left(\left(\frac{1}{2}\right)a_s + \left(\frac{1}{2}\right)a_d, \left(\frac{1}{2}\right)\hat{a}_s\right) + \text{cov}\left(\left(\frac{1}{2}\right)a_s + \left(\frac{1}{2}\right)a_d, \left(\frac{1}{2}\right)\hat{a}_d\right)$$

Assuming sire and dam are unrelated:

$$\text{cov}\left(a_o, \left(\frac{1}{2}\right)\hat{a}_s + \left(\frac{1}{2}\right)\hat{a}_d\right) = \left(\frac{1}{4}\right)\text{cov}(a_s, \hat{a}_s) + \left(\frac{1}{4}\right)\text{cov}(a_d, \hat{a}_d)$$

$$= \frac{1}{4}\text{var}(\hat{a}_s) + \frac{1}{4}\text{var}(\hat{a}_d)$$

Substituting the solution for the variance of EBV in Eqn 2.4:

$$\text{cov}\left(a_o, \left(\frac{1}{2}\right)\hat{a}_s + \left(\frac{1}{2}\right)\hat{a}_d\right) = \left(\frac{1}{4}\right)(r_s^2 + r_d^2)\sigma_a^2$$

From the calculation above, the term $\text{var}((\frac{1}{2})\hat{a}_s + (\frac{1}{2})\hat{a}_d)$ in the denominator for the equation for $r_{\hat{a}_o,f}$ above is also equal to $\frac{1}{4}(r_s^2 + r_d^2)\sigma_a^2$, assuming random mating and the absence of joint information in the sire and dam proofs. Therefore:

$$r_{\hat{a}_o,f} = \left(\frac{1}{4}\right)(r_s^2 + r_d^2)\sigma_a^2 / \sqrt{\sigma_a^2\left(\frac{1}{4}\right)\left[r_s^2 + r_d^2\right]\sigma_a^2} = \sigma_f/\sigma_a = (\frac{1}{2})\sqrt{\left[r_s^2 + r_d^2\right]}$$

where

$$\sigma_f = \sqrt{var\left[\left(\frac{1}{2}\right)\hat{a}_s + \left(\frac{1}{2}\right)\hat{a}_d\right]}$$

From the above equation, the upper limit for r when prediction is from pedigree is $\frac{1}{2}\sqrt{2} = 0.7$; that is, the accuracy of the proof of each parent is unity. Note that when the prediction is only from the sire proof, for instance, then:

$$r_{\hat{a}_o, \frac{1}{2}\hat{a}_s} = \left(\frac{1}{2}\right)\sqrt{r_s^2} = \left(\frac{1}{2}\right)\sqrt{n/(n+k)}$$

The accuracy of the predicted breeding value of a future daughter of the sire as shown in Section 2.4.

Example 2.4

Suppose that the EBVs for the sire and dam of a heifer are 180 kg and 70 kg for yearling body weight, respectively. Given that the accuracy of the proofs are 0.97 for the sire and 0.77 for the dam, predict the breeding value of the heifer and its accuracy for body weight at 12 months of age.

From Eqn 2.9:

$$\hat{a}_{heifer} = 0.5(180 + 70) = 125 \text{ kg}$$

The accuracy is:

$$r_{\hat{a}_{heifer}, a} = 0.5\sqrt{\left[0.97^2 + 0.77^2\right]} = 0.62$$

2.6 Breeding Value Prediction for One Trait from Another

The breeding value of one trait may be predicted from the observation on another trait if the traits are genetically correlated. If y is the observation on animal i from one trait, its breeding value for another trait x is:

$$\hat{a}_{ix} = b(y - \mu) \qquad (2.10)$$

with

$$b = \text{cov}(a_x, \text{ measurement on } y) / \text{var(measurement on } y) \qquad (2.11)$$

The genetic correlation between traits x and y (r_{axy}) is:

$$r_{axy} = \text{cov}(a_x, a_y) / (\sigma_{ax}\sigma_{ay})$$

Therefore:

$$\text{cov}(a_x, a_y) = r_{axy}\sigma_{ax}\sigma_{ay} \qquad (2.12)$$

Substituting Eqn 2.12 into Eqn 2.11:

$$b = r_{axy}\sigma_{ay}\sigma_{ax} / \sigma_y^2 \qquad (2.13)$$

If the additive genetic standard deviations for x and y in Eqn 2.13 are expressed as the product of the square root of their individual heritabilities and phenotypic variances, then:

$$\begin{aligned} b &= r_{axy}\sigma_y\sigma_x h_x h_y / \sigma_y^2 \\ &= r_{axy} h_x h_y \sigma_x / \sigma_y \end{aligned} \qquad (2.14)$$

The weight depends on the genetic correlation between the two traits, their heritabilities and phenotypic standard deviations.

The accuracy of the predicted breeding value is:

$$\begin{aligned} r_{axy} &= \text{cov}(a_x, \text{ measurement on } y) / \sigma_{ax}\sigma_y \\ &= r_{axy}\sigma_{ay}\sigma_{ax} / (\sigma_{ax}\sigma_y) \\ &= r_{axy} h_y \end{aligned}$$

The accuracy depends on the genetic correlation between the two traits and heritability of the recorded trait.

Correlated response (CR) in trait x as a result of direct selection on y (Falconer and Mackay, 1996) is:

$$\text{CR}x = ih_x h_y \, r_{axy}\sigma_y$$

Example 2.5

Suppose the standard deviation for growth rate (GR) (g/day) to 400 days in a population of beef cattle was 80, with a heritability of 0.43. The standard deviation for lean growth rate (LGR) (g/day) for the same population was 32, with a heritability of 0.45. If the genetic correlation between both traits is 0.95 and the herd management mean for growth rate is 887 g/day, predict the breeding value for LGR for an animal with a GR of 945 g/day.

Using Eqn 2.10:

$$\hat{a}_{\text{LGR}} = b(945 - 887)$$

with

$$b = \text{cov(GR, LGR)} / \text{var(GR)}$$

From Eqn 2.13:

$$b = (0.95(0.656)(0.671)(32)) / 80 = 0.167$$
$$\hat{a}_{\text{LGR}} = 0.167(945 - 887) = 9.686$$

The accuracy of the prediction is:

$$r = 0.95\left(\sqrt{0.43}\right) = 0.623$$

2.7 Selection Index

The selection index is a method for estimating the breeding value of an animal combining all information available on the animal and its relatives. It is the best linear prediction of an individual breeding value. The numerical value obtained for each animal is referred to as the index (I) and it is the basis on which animals are ranked for selection. Suppose y_{i}, y_{j} and y_{k} are phenotypic values for animal i and its jth sire and kth dam and μ_{i} ,μ_{j} and μ_{k} are the corresponding mean performance of the management group, respectively, then the index for this animal using this information would be:

$$I_i = \hat{a}_i = b_1(y_{\text{i}} - \mu_{\text{i}}) + b_2(y_{\text{j}} - \mu_{\text{j}}) + b_3(y_{\text{k}} - \mu_{\text{k}}) \tag{2.15}$$

where b_1, b_2, b_3 are the factors by which each measurement is weighted. The determination of the appropriate weights for the several sources of information is the main concern of the selection index procedure. In Eqn 2.15, the index is an estimate of the true breeding value of animal i.

Properties of a selection index are:

1. It minimizes the average square prediction error, i.e. it minimizes the average of all $(a_i - \hat{a}_i)^2$.
2. It maximizes the correlation ($r_{a,\hat{a}}$) between the true breeding value and the index. The correlation is often called the accuracy of prediction.
3. The probability of correctly ranking pairs of animals on their breeding value is maximized.

The **b** values in Eqn 2.15 are obtained by minimizing $(a_i - I_i)^2$, which is equivalent to maximizing r_{aI}. This is the same procedure employed in obtaining the regression coefficients in multiple linear regression. Thus, the **b** values could be regarded as partial regression coefficients of the individual's breeding value on each measurement. The minimization results in a set of simultaneous equations similar to the normal equations of multiple linear regression, which are solved to obtain the **b** values. The set of equations to be solved for the **b** values is:

$$\begin{aligned} p_{11}b_1 + p_{12}b_2 + \cdots + p_{1m}b_m &= g_1 \\ p_{21}b_1 + p_{22}b_2 + \cdots + p_{2m}b_m &= g_2 \\ &\vdots \\ p_{m1}b_1 + p_{m2}b_2 + \cdots + p_{mm}b_m &= g_m \end{aligned} \tag{2.16}$$

where p_{ii} with $i = 1$ to m are the phenotypic variances for individuals or traits i; p_{ij}, with $i \neq j$ are the phenotypic covariances among individuals or traits i and j; and g_i is the genetic covariances between the ith measured traits and the trait of genetic interest.

In matrix form, Eqn 2.16 is:

$$\mathbf{Pb} = \mathbf{g}$$

$$\mathbf{b} = \mathbf{P}^{-1}\mathbf{g}$$

where **P** is the variance and covariance matrix for observations, and **g** is the covariance matrix between observations and breeding value to be predicted. If n traits are measured on animals, for instance, for the prediction of the index value and m is the number of traits in the index, then **P** is of size n by n and **g** is of size m by 1.

Therefore, the selection index equation is:

$$I = \hat{\mathbf{a}} = \mathbf{g}'\mathbf{P}^{-1}(\mathbf{y} - \boldsymbol{\mu}) \tag{2.17}$$

$$= \mathbf{b}'(\mathbf{y} - \boldsymbol{\mu}) \tag{2.18}$$

where $\boldsymbol{\mu}$ refers to estimates of environmental influences on observations, assumed to be known without error. The application of the selection index to some data therefore involves setting up Eqn 2.17. From Eqn 2.18 it is obvious that the previous methods for predicting breeding values discussed in Sections 2.3–2.6 are no different from a selection index and they could be expressed as in Eqn 2.17.

2.7.1 Accuracy of index

As before, the accuracy ($r_{a,I}$) of an index is the correlation between the true breeding value and the index. The higher the correlation, the better the index as a predictor of breeding value. It provides a means of evaluating different indices based on different observations to find out, for instance, whether a particular observation is worth including in an index or not.

From the definition above:

$$r_{a,I} = \mathrm{cov}(a,\ I)/(\sigma_a\sigma_I)$$

First, we need to calculate σ_I^2 and $\mathrm{cov}(a, I)$ in the above equation. Using the formula for the variance of predicted breeding value in Section 2.3.1:

$$\sigma_I^2 = \mathrm{var}(b_1y_1 + \mathrm{var}(b_2y_2 + \ldots + 2b_1b_2\,\mathrm{cov}(y_1,\ y_2) + \ldots$$

$$= b_1^2\,\mathrm{var}(y_1) + b_2^2\,\mathrm{var}(y_2) + \ldots + 2b_1b_2\,\mathrm{cov}(y_1,\ y_2) + \ldots$$

$$\sigma_I^2 = b_1^2p_{11} + b_2^2p_{22} + \ldots + 2b_1b_2p_{12} + \ldots$$

or, in general:

$$\sigma_I^2 = \sum_{i=1}^{m} b_i^2 p_{ii} + \left(\sum_{i=1}^{m}\sum_{j=1}^{m} b_ib_jp_{ij};\ i \neq j\right)$$

where m is the number of traits or individuals in the index.

In matrix notation:

$$\sigma_I^2 = \mathbf{b}'\mathbf{Pb}$$

Now, $\mathbf{b} = \mathbf{P}^{-1}\mathbf{g}$, substituting this value for $\mathbf{b}$:

$$\sigma_I^2 = \mathbf{g}'\mathbf{P}^{-1}\mathbf{g} \tag{2.19}$$

The covariance between the true breeding value for trait or individual i and index is:

$$\begin{aligned}\text{cov}(a_i, I) &= \text{cov}(a_i, b_1 y_1) + \text{cov}(a_i, b_2 y_2) + \ldots + \text{cov}(a_i, b_j y_j) \\ &= b_1 \text{cov}(a_i, y_1) + b_2 \text{cov}(a_i, y_2) + \ldots + b_j \text{cov}(a_i, y_j)\end{aligned}$$

or, in general:

$$\text{cov}(a_i, I) = \sum_{j=1}^{m} b_j g_{ij} \tag{2.20}$$

where g_{ij} is the genetic covariance between traits or individuals i and j and m is the number of traits or individuals in the index.

In matrix notation:

$$\text{cov}(a_i, I) = \mathbf{b}'\mathbf{g}$$

Substituting $\mathbf{P}^{-1}\mathbf{g}$ for $\mathbf{b}$

$$= \mathbf{g}'\mathbf{P}^{-1}\mathbf{g} = \sigma^2{}_I$$

Thus, as previously, the regression of breeding value on predicted breeding values is unity. Therefore:

$$r_{a,I} = \sigma_I^2 / (\sigma_a \sigma_I) = \sigma_I / \sigma_a$$

For calculation purposes, r is better expressed as:

$$r_{a,I} = \sqrt{\frac{\sum_{j=1}^{m} b_j g_{ij}}{\sigma_a^2}} \tag{2.21}$$

Response to selection on the basis of an index is:

$$\begin{aligned}R &= i r_{a,I} \sigma_a \\ &= i \sigma_I\end{aligned}$$

2.7.2 Examples of selection indices using different sources of information

Data available on correlated traits

Example 2.6

Assume the following parameters were obtained for average daily gain (ADG) from birth to 400 days and lean per cent (LP) at the same age in a group of beef calves:

	Heritability	Standard deviation
ADG (g/day)	0.43	80.0
LP (%)	0.30	7.2

The genetic and phenotypic correlations (r_g and r_p) between ADG and LP are 0.30 and –0.10, respectively. Construct an index to improve growth rate in the beef calves. Assuming ADG as trait 1 and LP as trait 2, then from the given parameters:

$$p_{11} = 80^2 = 6400$$
$$p_{22} = 7.2^2 = 51.84$$
$$p_{12} = r_p\sqrt{(p_{11}p_{22}} = -0.1\sqrt{(6400)(51.84)} = -57.6$$
$$g_{11} = h^2(p_{11}) = 0.43(64000) = 2752$$
$$g_{22} = h^2(p_{22}) = 0.30(51.84) = 15.552$$
$$g_{12} = r_g\sqrt{(g_{11}g_{22}} = 62.064$$

The index equations to be solved are:

$$\begin{bmatrix} b_1 \\ b_2 \end{bmatrix} = \begin{bmatrix} p_{11} & p_{12} \\ p_{21} & p_{22} \end{bmatrix}^{-1} \begin{bmatrix} g_{11} \\ g_{21} \end{bmatrix}$$

Inserting appropriate values gives:

$$\begin{bmatrix} b_1 \\ b_2 \end{bmatrix} = \begin{bmatrix} 6400.00 & -57.60 \\ -57.60 & 51.84 \end{bmatrix}^{-1} \begin{bmatrix} 2752.000 \\ 62.064 \end{bmatrix}$$

The solutions are b_1 = 0.445 and b_2 = 1.692.

The index therefore is:

$$I = 0.445(\text{ADG} - \mu_{\text{ADG}}) + 1.692(\text{LP} - \mu_{\text{LP}})$$

where μ_{ADG} and μ_{LP} are herd averages for ADG and LP. Using Eqn 2.21:

$$r = \sqrt{\left[\left(0.445(2752) + 1.692(62.064)\right)/2752\right]} = 0.695$$

Using single records on individual and relatives

Example 2.7

Suppose the ADG for a bull calf (y_1) is 900 g/day and the ADG for his sire (y_2) and dam (y_3) are 800 g/day and 450 g/day, respectively. Assuming all observations were obtained in the same herd and using the same parameters as in Example 2.6, predict the breeding value of the bull calf for ADG and its accuracy.

From the parameters given:

$$p_{11} = p_{22} = p_{33} = \sigma_y^2 = 6400$$
$$p_{12} = \text{cov}(y_1, y_2) = \left(\frac{1}{2}\right)\sigma_a^2 = \left(\frac{1}{2}\right)(2752) = 1376$$
$$p_{13} = p_{12} = 1376$$
$$p_{23} = 0$$
$$g_{11} = \sigma_a^2 = 2752$$
$$g_{12} = g_{13} = \left(\frac{1}{2}\right)\sigma_a^2 = 1376$$

The index equations are:

$$\begin{bmatrix} b_1 \\ b_2 \\ b_3 \end{bmatrix} = \begin{bmatrix} 6400 & 1376 & 1376 \\ 1376 & 6400 & 0 \\ 1376 & 0 & 6400 \end{bmatrix}^{-1} \begin{bmatrix} 2752 \\ 1376 \\ 1376 \end{bmatrix}$$

Solutions to the above equations are $b_1 = 0.372$, $b_2 = 0.135$ and $b_3 = 0.135$. The index is:

$$I = 0.372(900 - \mu) + 0.135(800 - \mu) + 0.135(450 - \mu)$$

where μ is the herd average. The accuracy is:

$$r = \sqrt{[(0.372(2752) + 0.135(1376) + 0.135(1376)) / 2752]} = 0.712$$

The high accuracy is due to the inclusion of information from both parents.

Using means of records from animal and relatives

Example 2.8

It is given that average protein yield for the first two lactations for a cow $(\bar{y}_1)$ called Zena is 230 kg and the mean protein yield of five other cows $(\bar{y}_2)$, each with two lactations, is 300 kg. If all cows are all daughters of the same bull and no other relationship exists among them, predict the breeding value of Zena, assuming a heritability of 0.25, a repeatability (t) of 0.5, standard deviation of 34 kg and herd average of 250 kg for protein yield in the first two lactations.

From the given parameters:

$$g_{11} = \sigma_a^2 = h^2\sigma_y^2 = 0.25(34^2) = 289$$

and

$$g_{12} = \text{covariance between half-sibs} = (\tfrac{1}{4})(\sigma_a^2) = (\tfrac{1}{4})(289) = 72.25.$$

From calculations in Section 2.3.2:

$$p_{11} = \sigma_{\overline{y1}}^2 = [\mathrm{t} + (1-t)/n]\sigma_y^2 = (0.5 + (1 - 0.5)/2)34^2 = 867$$

Using similar arguments:

$$p_{22} = \sigma_{\overline{y2}}^2 = \sigma_B^2 + 1/\mathrm{n}(\sigma_w^2)$$

where σ_B^2 is the between-cow variance and $1/n(\sigma_w^2)$ is the mean of the within-cow variance.

From Section 2.4:

$$\sigma_B^2 = (\tfrac{1}{4})\sigma_a^2$$

and for cow i in the group of five cows:

$$\sigma_w^2 = (\sigma_{\overline{y2i}}^2 - \sigma_B^2)$$

where $\bar{y}_{2i}$ is the mean of the first two lactations for cow i. Since all five cows each have two records, like Zena:

$$\sigma_W^2 = (p_{11} - (\tfrac{1}{4})\sigma_a^2)$$

and

$$(1/n)(\sigma^2_w) = (1/n)(p_{11} - (\tfrac{1}{4})\sigma^2_a)$$

Therefore:

$$p_{22} = (\tfrac{1}{4})\sigma^2_a + (1/n)(p_{11} - (\tfrac{1}{4})\,\sigma^2_a)$$

$$= (\tfrac{1}{4})\,(289) + (\tfrac{1}{5})(867 - (\tfrac{1}{4})(289)) = 231.2$$

The index equations are:

$$\begin{bmatrix} b_1 \\ b_2 \end{bmatrix} = \begin{bmatrix} 867 & 72.25 \\ 72.25 & 231.2 \end{bmatrix}^{-1} \begin{bmatrix} 289 \\ 72.25 \end{bmatrix}$$

The solutions are $b_1 = 0.316$ and $b_2 = 0.213$ and the index is:

$$I = 0.316(230 - 250) + 0.213(300 - 250)$$

The accuracy of the index is:

$$r = \sqrt{\left[(0.316(289) + 0.213(72.5)/289)\right]} = 0.608$$

2.7.3 Prediction of aggregate genotype

At times, the aim is not just to predict the breeding value of a single trait but that of a composite of several traits evaluated in economic terms. The aggregate breeding value (H) or merit for such several or n traits can be represented as:

$$H = w_1a_1 + w_2a_2 + \ldots + w_na_n$$

where a_i is the breeding value of the ith trait and w_i the weighting factor, which expresses the relative economic importance associated with the ith trait. The construction of an index to predict or improve H is based on the same principles as those discussed earlier except that it includes the relative economic weight for each trait.

Thus:

$$I = \mathbf{w'G'P^{-1}}(\mathbf{y} - \boldsymbol{\mu}) \qquad (2.22)$$

where $\mathbf{w}$ is the vector of economic weights size n by 1 with n being the number of traits of economic value and $\mathbf{G}$ is of size m by n representing genetic covariances between the measured traits and the traits of economic value and all other terms are as defined in Eqn 2.17 with $\mathbf{P}$ of size m by m, where m is the number of measured traits.

The equations to be solved to get the weights ($\mathbf{b}$ values) to be used in the index are:

$$p_{11}b_1 + p_{12}b_2 + \ldots + p_{1m}b_m = g_{11}w_1 + g_{12}w_2 + \ldots + g_{1n}w_n$$

$$p_{21}b_1 + p_{22}b_2 + \ldots + p_{2m}b_m = g_{21}w_1 + g_{22}w_2 + \ldots + g_{2n}w_n$$

$$\vdots \qquad\qquad \vdots$$

$$p_{m1}b_1 + p_{m2}b_2 + \ldots + p_{mm}b_m = g_{m1}w_1 + g_{m2}w_2 + \ldots + g_{mn}w_n$$

In matrix notation these equations are:

$$\mathbf{Pb} = \mathbf{Gw}$$

$$\mathbf{b} = \mathbf{P^{-1}Gw}$$

It should be noted that it is possible there are some traits in the index that are not in the aggregate breeding value but may be correlated with other traits in *H*. Conversely, some traits in the aggregate breeding value may be difficult to measure or occur late in life and may therefore not be in the index. Such traits may be replaced in the index with other highly correlated traits that are easily measurable or occur early in life. Consequently, the vector of economic weights may not necessarily be of the same dimension as traits in the index, as indicated in the equations for *b* above. Each trait in the index is weighted by the economic weight relevant to the breeding value of the trait it is predicting in the aggregate breeding value.

Accuracy of the index

Using the same principle in Section 2.7.1, the accuracy of the selection index (r_{HI}) is:

$$r_{HI} = cov(H,I) / \sqrt{var(H),var(I)}$$
$$= \mathbf{b'Gw} / \sqrt{(\mathbf{w'Dw})(\mathbf{b'Pb})}$$

where w'**D**w is the genetic variance of *H* with **D** being the genetic variance and covariance matrix for traits in the aggregate breeding goal. Therefore:

$$r_{HI} = \sqrt{(\mathbf{w'Dw})(\mathbf{b'Gw})}$$

which is equivalent to Eqn 2.21 for a simple index as $\mathbf{w'Dw} = \sigma_a^2$
Expected response in *H* from selecting on the index is:

$$R = i\mathbf{b'Gw} / \sqrt{(\mathbf{b'Pb}} = i\sqrt{(\mathbf{b'Gw}} = i\sigma_I$$

The expected correlated response (CR_j) for the *j*th trait in *H* is:

$$CR_j = ib_{jI}\sigma_I = i(\mathbf{b'G}_j w_j) / \sqrt{(\mathbf{b'Pb})}$$

where $\mathbf{G}_j$ is *the* *j*th column of **G** and w_j is the economic value for *j*th trait.
Therefore:

$$CR_j = i\left(\mathbf{b'G}_j w_j\right) / \sigma_I$$

The index calculated using Eqn 2.22 implies that the same economic weights are applied to the traits in the aggregate genotype across the whole population. A change in the economic weight for one of the traits would imply recalculating the index. An alternative formulation of Eqn 2.22 involves calculating a sub-index for each trait in *H* without the economic weights. The final index in Eqn 2.23 is obtained by summing the sub-indices for each trait weighted by their respective economic weights. Thus:

$$I = \sum_{i=1}^{m} I_i w_i \qquad (2.23)$$

where $I_i = \mathbf{g'}_i \mathbf{P}^{-1}(\mathbf{y} - \boldsymbol{\mu})$, the sub-index for trait *i* in *H* and w_i = economic weight for trait *i*.

With Eqn 2.23, a change in the economic weights of any of the traits in the index can easily be implemented without recalculating the index.

To demonstrate that Eqns 2.22 and 2.23 are equivalent, assume that there are two traits in *H*, then Eqn 2.23 becomes:

$$I = I_1 w_1 + I_2 w_2$$
$$= w'_1 \mathbf{g'}_1 \mathbf{P}^{-1}(\mathbf{y} - \boldsymbol{\mu}) + w'_2 \mathbf{g'}_2 \mathbf{P}^{-1}(\mathbf{y} - \boldsymbol{\mu})$$

where $\mathbf{g}_i$ is the covariance matrix between trait i and all traits in the index.

$$= (w'_1\mathbf{g}'_1 + w'_2\mathbf{g}'_2)\mathbf{P}^{-1}(\mathbf{y} - \boldsymbol{\mu})$$
$$= w'\mathbf{g}'\mathbf{P}^{-1}(\mathbf{y} - \boldsymbol{\mu})$$

which is the same as Eqn 2.22.

Example 2.9

Assume the economic weights for ADG and LP are £1.5 and £0.5 per an increase of 1 kg in ADG and 1% increase in LP, respectively. Using the genetic parameters in Example 2.6, construct an index to select fast-growing lean beef calves using Eqn 2.22. Repeat the analysis using Eqn 2.23.

Using Eqn 2.22, index equations are:

$$\begin{bmatrix} b_1 \\ b_2 \end{bmatrix} = \begin{bmatrix} p_{11} & p_{12} \\ p_{21} & p_{22} \end{bmatrix}^{-1} \begin{bmatrix} w_1g_{11} + w_2g_{12} \\ w_1g_{21} + w_2g_{22} \end{bmatrix}$$

Inserting the appropriate values:

$$\begin{bmatrix} b_1 \\ b_2 \end{bmatrix} = \begin{bmatrix} 6400.00 & -57.60 \\ -57.60 & 51.84 \end{bmatrix}^{-1} \begin{bmatrix} 1.5(2752) + 0.5(62.064) \\ 1.5(62.064 + 0.5(15.552) \end{bmatrix}$$

Solutions for b_1 and b_2 from the above equations are 0.674 and 2.695, respectively. The index therefore is:

$$I = 0.674(\text{ADG} - \mu_{\text{ADG}}) + 2.694(\text{LP} - \mu_{\text{LP}})$$

Applying Eqn 2.23, the sub-index for ADG is the same as that calculated in Example 2.6 with b_1 = 0.445 and b_2 = 1.692. The sub-index for LP is:

$$b_1p_{11} + b_2p_{12} = g_{12}$$
$$b_1p_{21} + b_2p_{22} = g_{22}$$

which gives:

$$\begin{bmatrix} b_1 \\ b_2 \end{bmatrix} = \begin{bmatrix} 6400.00 & -57.60 \\ -57.60 & 51.84 \end{bmatrix}^{-1} \begin{bmatrix} 62.064 \\ 15.552 \end{bmatrix}$$

The solutions are b_1 = 0.0125 and b_2 = 0.314. Multiplying the sub-indices by their respective weights gives:

$$I_{\text{ADG}} = 0.445(1.5)(\text{ADG} - \mu_{\text{ADG}}) + 1.692(1.5)(\text{LP} - \mu_{\text{LP}})$$
$$= 0.668(\text{ADG} - \mu_{\text{ADG}}) + 2.538(\text{LP} - \mu_{\text{LP}})$$

and

$$I_{\text{LP}} = 0.0125(0.5)(\text{ADG} - \mu_{\text{ADG}}) + 0.314(0.5)(\text{LP} - \mu_{\text{LP}})$$
$$= 0.006(\text{ADG} - \mu_{\text{ADG}}) + 0.157(\text{LP} - \mu_{\text{LP}})$$

Summing the b terms from the two sub-indices, the final b terms are:

$$b_1 = 0.668 + 0.006 = 0.674$$
$$b_2 = 2.538 + 0.157 = 2.695$$

Therefore, the final index is:

$$I = 0.675(\text{ADG} - \mu_{\text{ADG}}) + 2.695(\text{LP} - \mu_{\text{LP}})$$

which is the same as calculated using Eqn 2.22.

2.7.4 Overall economic indices using predicted genetic merit

Overall economic indices that combine PTAs or EBVs calculated by best linear unbiased prediction (BLUP, see Chapter 4) have become very popular in the last decade. In addition to the recognition that more than one trait contributes to profitability, the broadening of selection goals has also been due to the need to incorporate health and welfare traits to accommodate public concerns. Examples of indices constructed with PTAs or BVs of several traits and used in genetic improvement of dairy cattle include production index (PIN), combining PTAs for milk, fat and protein in the UK, production life index (PLI), which is PIN plus PTAs for longevity and somatic cell count in the UK; and in the Netherlands, index net (INET), which combines BVs for milk, fat and protein and durable performance sum (DPS), which is INET plus durability (Interbull, 2000). The principles for calculating these indices are similar to those outlined in previous sections. Given that the PTAs or BVs are from a complete multivariate analysis, the optimal index weights (**b**) are the sum of the partial regression coefficients of each goal trait on each index trait, weighted by the economic value of the goal trait (Veerkamp *et al.*, 1995). Thus, given n traits in the selection goal and m traits in the index, the partial regressions can be calculated as:

$$\mathbf{R} = \mathbf{G}^{-1}\mathbf{G}_{ig}$$

and

$$\mathbf{b} = \mathbf{R}\mathbf{w}$$

where **R** is a matrix of partial genetic regression, $\mathbf{G}_{ig}$ is the matrix of genetic covariance between n goal and m index traits, **G** is the genetic covariance matrix between the index traits, and **w** is the vector of economic weights. It is obvious that when goal and index traits are the same, $\mathbf{G}_{ig} = \mathbf{G}$ and $\mathbf{b} = \mathbf{w}$. In the case where the index and goal traits are not the same, **R** can be estimated directly from a regression of phenotype on the EBVs for the index traits (Brotherstone and Hill, 1991). However, if PTAs or BVs are from a univariate analysis, rather than from a multivariate analysis, the use of **b** above results only in minimal loss of efficiency in the index (Veerkamp *et al.*, 1995).

Selection based on breeding values from BLUP is usually associated with an increased rate of inbreeding as it favours the selection of closely related individuals. Quadratic indices can be used to optimize the rate of genetic gain and inbreeding. This does not fall within the main subject area of this text and interested readers should see the work by Meuwissen (1997) and Grundy *et al.* (1998).

2.7.5 Restricted selection index

Restricted selection index is used when the aim is to maximize selection for a given aggregate genotype, subject to the restriction that no genetic change is desired in one or more of the traits in the index for H. This is achieved by the usual index procedure and setting the covariance between the index and the breeding value ($\text{cov}(I, a_i)$ for the ith trait specified not to change to zero. It was Kempthorne and Nordskog (1959) who introduced the idea of imposing restrictions on the general index procedure.

For instance, consider the aggregate genotype composed of two traits:

$$H = w_1a_1 + w_2a_2$$

However, it is desired that there should be no genetic change in trait 2; thus, effectively:

$$H = w_1a_1$$

and the index to predict H is:

$$I = b_1y_1 + b_2y_2$$

To ensure that there is no genetic change in trait 2, $\text{cov}(I, a_2)$ must be equal to zero. From Eqn 2.20:

$$\text{cov}(I,\ a_2) = b_1g_{12} + b_2g_{22} = 0$$

This is included as an extra equation to the normal equations for the b values, and a dummy unknown, the so-called Lagrange multiplier, is added to the vector of solutions for the index weights (Ronningen and Van Vleck, 1985). The equations for the index therefore are:

$$\begin{bmatrix} b_1 \\ b_2 \\ \lambda \end{bmatrix} = \begin{bmatrix} p_{11} & p_{12} & g_{12} \\ p_{21} & p_{22} & g_{22} \\ g_{12} & g_{22} & 0 \end{bmatrix}^{-1} \begin{bmatrix} g_{11} \\ g_{12} \\ 0 \end{bmatrix} \qquad (2.24)$$

Example 2.10

Using the same data and parameters as in Example 2.6, construct an index to improve the aggregate genotype for fast-growing lean cattle using an index consisting of GR and LP but with no genetic change in LP.

From Eqn 2.23, the index equations are:

$$\begin{bmatrix} 6400 & -57.600 & 62.064 \\ -57.600 & 51.800 & 15.552 \\ 62.064 & 15.552 & 0 \end{bmatrix} \begin{bmatrix} b_1 \\ b_2 \\ \lambda \end{bmatrix} = \begin{bmatrix} 2752 \\ 62.064 \\ 0 \end{bmatrix}$$

The solutions for b_1 and b_2 from solving the above equations are 0.325 and –1.303. Therefore, the index is:

$$I = 0.325(\text{ADG} - \mu_{\text{ADG}}) + (-1.303(\text{LP} - \mu_{\text{LP}}))$$

The accuracy of this index (Eqn 2.21) is:

$$r = \sqrt{\left[(0.325)(2752) + (-1.303(62.064)) / 2752\right]} = 0.544$$

which is lower than the accuracy for the equivalent index in Example 2.6, but with no restriction on LP, and is also lower than the accuracy of prediction of breeding value for ADG on the basis of single records. The imposition of a restriction on any trait in the index will never increase the efficiency of the index but usually reduces it unless $I_i = 0$ for the constrained trait.

2.7.6 Index combining breeding values from phenotype and genetic marker information

Consider a situation in which one or more genes affecting a trait with a large impact on profit have been identified to be linked to a genetic marker (see Chapter 11). If genetic prediction based only on marker information is available in addition to the conventional BV estimated without marker information, then both sources of information can be combined into an index (Goddard, 1999). It is also possible that the conventional BV is based on a subset of traits in the breeding goal and marker information is available on other traits that are not routinely measured, such as meat quality traits.

A selection index could be used to combine both sources of information and the increase in accuracy from including marker information could be computed (Goddard, 1999). Given r as the accuracy of the conventional breeding BV and d as the proportion of genetic variance explained by the marker information, the covariance between the two sources of information is dr^2. If m is the BV based on marker information and a the BV from phenotypic information, then:

$$\text{var}\begin{pmatrix} m \\ a \end{pmatrix}\begin{pmatrix} d & dr^2 \\ dr^2 & r^2 \end{pmatrix}$$

Let g be the true breeding value to be predicted, then cov(g, m) = d and cov(g, a) = r^2. The normal index equations are:

$$\begin{pmatrix} b_1 \\ b_2 \end{pmatrix} = \begin{pmatrix} d & dr^2 \\ dr^2 & r^2 \end{pmatrix}^{-1}\begin{pmatrix} d \\ r^2 \end{pmatrix}$$

Solving the above equations gives the following index weights:

$$\mathrm{b}_1 = (1-\mathrm{r}^2)/(1-\mathrm{dr}^2) \text{ and } \mathrm{b}_2 = (1-\mathrm{d})/(1-\mathrm{dr}^2)$$

The variance of the index = reliability (r_I^2) is:

$$r_I^2 = [(1-r^2)d + (1-d)r^2]/(1-dr^2)$$

The increase in reliability (r_{inc}^2) from incorporating marker information therefore is:

$$r_{inc}^2 = (r_I^2 - r^2) = d/(1-dr^2)[(1-r^2)^2]$$

For example, given that r^2 of the conventional BV is 0.34 and marker information accounts for 25% of the genetic variance, then r_I^2 is 0.459, an increase in reliability of 0.12. However, if r^2 is 0.81, then r_I^2 is 0.83 and r_{inc}^2 is only 0.02. Thus, the usefulness of marker information is greater when reliability is low, such as in traits of low heritability and also traits that cannot be measured in young animals such as carcass traits (Goddard and Hayes, 2002).

3 Genetic Covariance Between Relatives

3.1 Introduction

Of fundamental importance in the prediction of breeding values is the genetic relationship among individuals. From Chapter 2 it was found that the use of the selection index to predict breeding values requires the genetic covariance between individuals to construct the genetic covariance matrix. Genetic evaluation using best linear unbiased prediction (BLUP), the subject of Chapter 4, is heavily dependent upon the genetic covariance among individuals, both for higher accuracy and unbiased results. The genetic covariance among individuals is comprised of three components: the additive genetic covariance; the dominance covariance; and the epistatic covariance. This chapter addresses the calculation of the additive genetic relationships among individuals and how to determine the level of inbreeding to model additive genetic covariance between individuals. Dominance and epistasis genetic relationships are considered in Chapter 13, which deals with non-additive models.

The calculation of the genetic relationship among individuals is usually based on the assumptions that base or founder animals are unrelated. However, due to the finite size of populations, base or founder animals might also be related. The chapter also deals with how related base animals may be handled in computing the genetic relationship among animals and introduces the concept of metafounders.

3.2 Identity-by-State Versus Identity-by-Descent

In this chapter, we use the concepts of identity-by-state (IBS) and identity-by-descent (IBD) and their application in the computation of the genetic relationship among individuals. These two concepts are key to understanding the relatedness between individuals utilized in many genetic analyses, including methods that are this book's topic. In this context, the IBS denotes the similarity between two DNA fragments regarding their sequence of base pairs. We get IBS similarity through the duplication of DNA fragments and their segregation through a population. Recombinations and mutations change the sequence of base pairs of DNA fragments and hence change IBS similarity. Before the advent of genomics, we could not observe DNA sequence and hence could not evaluate IBS. IBD denotes the similarity between two DNA fragments regarding their ancestry (descent). As such, IBD is always defined relative to some base population or time point – for example, the founders of a studied population. Two DNA fragments are IBD through the duplication of DNA fragments and their segregation, the same processes as for IBS. Hence, two DNA fragments that are IBD are also IBS, except for occasional mutation. The key applied difference between the IBS and IBD is that we require genomic data to evaluate IBS or IBD, while we can use pedigree data to evaluate the probability of IBD within the pedigree without genomic data.

DOI: 10.1079/9781800620506.0003

Specifically, genetic inheritance (recombination and segregation) distributes DNA between generations. This process gives realized IBS similarities between DNA fragments and realized IBD similarities between DNA fragments relative to a base population. With pedigree data, we can evaluate expected IBD similarity between individuals, with the probability of IBD, relative to a base population. When the two DNA fragments are from two individuals, the probability of IBD denotes the coefficient of coancestry (kinship) between the individuals – the probability that these two individuals have both inherited the same DNA fragment from a common ancestor. When the two DNA fragments are from one individual, we evaluate the coefficient of coancestry of an individual with itself – the probability that this individual has inherited the same DNA fragments from a common ancestor. The additive genetic relationship between individuals is equal to twice their coefficient of coancestry. In this chapter, we will present methods to build pedigree-based additive genetic relationship matrices (usually denoted as A), while we will present methods to build genome-based additive genetic relationship matrices (usually denoted as **G**) in Chapter 11.

3.3 The Numerator Relationship Matrix

The probability of identical genes by descent occurring in two individuals is termed the coancestry or the coefficient of kinship (Falconer and Mackay, 1996) and the additive genetic relationship between two individuals is twice their coancestry. The matrix that indicates the additive genetic relationship among individuals is called the numerator relationship matrix (**A**). It is symmetric and its diagonal element for animal i (a_{ii}) is equal to $1 + F_i$, where F_i is the inbreeding coefficient of animal i (Wright, 1922). The diagonal element represents twice the probability that two gametes taken at random from animal i will carry identical alleles by descent. The off-diagonal element, a_{ij}, equals the numerator of the coefficient of relationship (Wright, 1922) between animals i and j. When multiplied by the additive genetic variance $\left(\sigma_u^2\right)$, $\mathbf{A}\sigma_u^2$ is the covariance among breeding values. Thus, if u_i is the breeding value for animal i, $\text{var}(u_i) = a_{ii}\sigma_u^2 = (1+F_i)\sigma_u^2$. The matrix **A** can be computed using path coefficients, but a recursive method that is suitable for computerization was described by Henderson (1976). Initially, animals in the pedigree are coded 1 to n and ordered such that parents precede their progeny. The following rules are then employed to compute **A**. If both parents (s and d) of animal i are known:

$$a_{ji} = a_{ij} = 0.5\left(a_{js} + a_{jd}\right); \quad j = 1 \text{ to} \left(i-1\right)$$
$$a_{ii} = 1 + 0.5\left(a_{sd}\right)$$

If only one parent **s** is known and assumed unrelated to the mate:

$$a_{ji} = a_{ij} = 0.5\left(a_{js}\right); \quad j = 1 \text{ to} \left(i-1\right)$$
$$a_{ii} = 1$$

If both parents are unknown and are assumed unrelated:

$$a_{ji} = a_{ij} = 0; \quad j = 1 \text{ to} \left(i-1\right)$$
$$a_{ii} = 1$$

For example, assume that the data in Table 3.1 are the pedigree for six animals. The numerator relationship matrix for the example pedigree is:

	1	2	3	4	5	6
1	1.00	0.00	0.50	0.50	0.50	0.25
2	0.00	1.00	0.50	0.00	0.25	0.625
3	0.50	0.50	1.00	0.25	0.625	0.563
4	0.50	0.00	0.25	1.00	0.625	0.313
5	0.50	0.25	0.625	0.625	1.125	0.688
6	0.25	0.625	0.563	0.313	0.688	1.125

Table 3.1. Pedigree for six animals.

Calf	Sire	Dam
3	1	2
4	1	Unknown
5	4	3
6	5	2

For instance:

$$a_{11} = 1 + 0 = 1$$

$$a_{12} = 0.5(0 + 0) = 0 = a_{12}$$

$$a_{22} = 1 + 0 = 1$$

$$a_{13} = 0.5(a_{11} + a_{12}) = 0.5(1.0 + 0) = 0.5 = a_{31}$$

$$a_{23} = 0.5(a_{12} + a_{22}) = 0.5(0 + 1.0) = 0.5 = a_{32}$$

.

.

.

$$a_{34} = 0.5(a_{13}) = 0.5(0.5 + 0) = 0.25 = a_{43}$$

.

.

.

$$a_{66} = 1 + 0.5(a_{52}) = 1 + 0.5(0.25) = 1.125$$

From the above calculation, the inbreeding coefficient for calf 6 is 0.125.

3.4 Decomposing the Relationship Matrix

The relationship matrix can be expressed (Thompson, 1977a) as:

$$\mathbf{A} = \mathbf{TDT}' \tag{3.1}$$

where **T** is a lower triangular matrix and **D** is a diagonal matrix. This relationship has been used to develop rules for obtaining the inverse of **A**. A non-zero element of the matrix **T**, say t_{ij}, is the coefficient of relationship between animals *i* and *j* if *i* and *j* are direct relatives or *i* = *j* and it is assumed that there is no inbreeding. Thus, the matrix **T** traces the flow of genes from one generation to the other; in other words, it accounts only for direct (parent–offspring) relationships. It can easily be computed applying the following rules. For the *i*th animal:

$$t_{ii} = 1$$

If both parents (*s* and *d*) are known:

$$t_{ij} = 0.5\left(t_{sj} + t_{dj}\right)$$

If only one parent (s) is known:

$$t_{ij} = 0.5\left(t_{sj}\right)$$

If neither parent is known:

$$t_{ij} = 0$$

The diagonal matrix **D** is the variance and covariance matrix for Mendelian sampling. The Mendelian sampling (m) for an animal i with breeding value u_i and u_s and u_d as breeding values for its sire and dam, respectively, is:

$$m_i = u_i - 0.5\left(u_s + u_d\right) \tag{3.2}$$

D has a simple structure and can easily be calculated. From Eqn 3.2, if both parents of animal i are known, then:

$$\begin{aligned}\text{var}(m_i) &= \text{var}\left(u_i\right) - \text{var}\left(0.5u_s + 0.5u_d\right)\\ &= \text{var}\left(u_i\right) - \text{var}\left(0.5u_s\right) - \text{var}\left(0.5u_d\right) - 2\,\text{cov}\left(0.5u_s, 0.5u_d\right)\\ &= \left(1 + F_i\right)\sigma_u^2 - 0.25a_{ss}\sigma_u^2 - 0.25a_{dd}\sigma_u^2 - 0.5a_{sd}\sigma_u^2\end{aligned}$$

where a_{ss}, a_{dd} and a_{sd} are elements of the relationship matrix **A** and F_i is the inbreeding coefficient of animal I:

$$\text{var}\left(m_i\right) / \sigma_u^2 = d_{ii} = \left(1 + F_i\right) - 0.25a_{ss} - 0.25a_{dd} - 0.5a_{sd}$$

since $F_i = 0.5a_{sd}$

$$\begin{aligned}d_{ii} &= 1 - 0.25\left(1 + F_s\right) - 0.25\left(1 + F_d\right)\\ &= 0.5 - 0.25\left(F_s + F_d\right)\end{aligned}$$

where F_s and F_d are the inbreeding coefficients of its sire and dam, respectively. If only one parent (s) is known, the diagonal element is:

$$\begin{aligned}d_{ii} &= 1 - 0.25(1 + F_s)\\ &= 0.75 - 0.25(F_s)\end{aligned}$$

and if no parent is known:

$$d_{ii} = 1$$

For the pedigree in Table 3.1, the matrix **T** is:

	1	2	3	4	5	6
1	1.0	0.0	0.0	0.0	0.0	0.0
2	0.0	1.0	0.0	0.0	0.0	0.0
3	0.5	0.5	1.0	0.0	0.0	0.0
4	0.5	0.0	0.0	1.0	0.0	0.0
5	0.5	0.25	0.5	0.5	1.0	0.0
6	0.25	0.625	0.25	0.25	0.5	1.0

and **D** is:

$$\mathbf{D} = \text{diag}\left(1.0, 1.0, 0.5, 0.75, 0.5, 0.469\right)$$

For instance, animal 4 has only the sire known, which is not inbred, therefore:

$$d_{44} = 0.75 - 0 = 0.75$$

and

$$d_{66} = 0.5 - 0.25(0.125 + 0) = 0.469$$

because both parents are known and the sire has an inbreeding coefficient of 0.125.

3.5 Computing the Inverse of the Relationship Matrix

The prediction of breeding value requires the inverse of the relationship matrix, $\mathbf{A}^{-1}$. This could be obtained by setting up **A** by the recursive method and inverting it. This is, however, not computationally feasible when evaluating a large number of animals. In 1976, Henderson presented a simple procedure for calculating $\mathbf{A}^{-1}$ without setting up **A**. The procedure and its principles are described below.

From Eqn 3.1, the inverse of **A** can be written as:

$$\mathbf{A}^{-1} = (\mathbf{T} - 1)'\mathbf{D}^{-1}\mathbf{T}^{-1} \tag{3.3}$$

The matrix $\mathbf{D}^{-1}$ is easy to obtain because **D** is a diagonal matrix. The diagonal elements of $\mathbf{D}^{-1}$ are simply the reciprocals of the diagonal elements of **D** computed in Section 3.4. $\mathbf{T}^{-1}$ is a lower triangular matrix with ones in the diagonals and the only non-zero elements to the left of the diagonal in the row for the animal *i* are −0.5 for columns corresponding to the known parents. It can be derived as **I** – **M**, where **I** is an identity matrix of the order of animals on the pedigree and **M** is a matrix of the contribution of gametes from parents to progeny (Kennedy, 1989). Since progeny *i* receives half of its genes from each parent, the only non-zero elements in row *i* of **M** are 0.5, corresponding to columns of known parents. Thus, if both parents of progeny *i* are unknown, all elements of row *i* are zero. For the pedigree in Table 3.1, $\mathbf{T}^{-1}$ can be calculated as:

$$\underset{(\mathbf{I})}{\begin{bmatrix} 1 & 0 & 0 & 0 & 0 & 0 \\ 0 & 1 & 0 & 0 & 0 & 0 \\ 0 & 0 & 1 & 0 & 0 & 0 \\ 0 & 0 & 0 & 1 & 0 & 0 \\ 0 & 0 & 0 & 0 & 1 & 0 \\ 0 & 0 & 0 & 0 & 0 & 1 \end{bmatrix}} - \underset{(\mathbf{M})}{\begin{bmatrix} 0.0 & 0.0 & 0.0 & 0.0 & 0.0 & 0.0 \\ 0.0 & 0.0 & 0.0 & 0.0 & 0.0 & 0.0 \\ 0.5 & 0.5 & 0.0 & 0.0 & 0.0 & 0.0 \\ 0.5 & 0.0 & 0.0 & 0.0 & 0.0 & 0.0 \\ 0.0 & 0.0 & 0.5 & 0.5 & 0.0 & 0.0 \\ 0.0 & 0.5 & 0.0 & 0.0 & 0.5 & 0.0 \end{bmatrix}}$$

$$= \underset{(\mathbf{T}^{-1})}{\begin{bmatrix} 1.0 & 0.0 & 0.0 & 0.0 & 0.0 & 0.0 \\ 0.0 & 1.0 & 0.0 & 0.0 & 0.0 & 0.0 \\ -0.5 & -0.5 & 1.0 & 0.0 & 0.0 & 0.0 \\ -0.5 & 0.0 & 0.0 & 1.0 & 0.0 & 0.0 \\ 0.0 & 0.0 & -0.5 & -0.5 & 1.0 & 0.0 \\ 0.0 & -0.5 & 0.0 & 0.0 & -0.5 & 1.0 \end{bmatrix}}$$

and

$$\mathbf{D}^{-1} = \text{diag}(1, 1, 2, 1.333, 2, 2.133).$$

3.5.1 Inverse of the numerator relationship matrix ignoring inbreeding

The relationship shown in Eqn 3.3 was used by Henderson (1976) to derive simple rules for obtaining $\mathbf{A}^{-1}$ without accounting for inbreeding. With inbreeding ignored, the diagonal elements of $\mathbf{D}^{-1}$ are either 2 or $\frac{4}{3}$ or 1, if both or one or no parents are known, respectively. Let a_i represent the diagonal element of $\mathbf{D}^{-1}$ for animal i. Initially, set $\mathbf{A}^{-1}$ to zero and apply the following rules. If both parents of the ith animal are known, add:

α_i to the (i,i) element

$-\alpha_i/2$ to the (s,i), (i,s), (d,i) and (i,d) elements

$\alpha_i/4$ to the (s,s), (s,d), (d,s) and (d,d) elements

If only one parent (s) of the ith animal is known, add:

α_i to the (i,i) element

$-\alpha_i/2$ to the (s,i) and (i,s) elements

$\alpha_i/4$ to the (s,s) element

If neither parent of the ith animal is known, add:

α_i to the (i,i) element

As an illustration, the inverse of the relationship matrix in Section 3.3 can be calculated as below. Initially, list all animals in the pedigree:

Calf	Sire	Dam
1	Unknown	Unknown
2	Unknown	Unknown
3	1	2
4	1	Unknown
5	4	3
6	5	2

Then set up a 6x6 table for the animals. For animals 1 and 2, both parents are unknown, therefore $\alpha_1 = \alpha_2 = 1$. Add 1 to their diagonal elements (1,1 and 2,2). For animal 3, both parents are known, therefore $\alpha_3 = 2$. Add 2 to the 3,3 element, –1 to the (3,1), (1,3), (3,2) and (2,3) elements and $\frac{1}{2}$ to the (1,1), (1,2), (2,1) and (2,2) elements. For animal 4, only one parent is known, therefore $\alpha_4 = \frac{4}{3}$. Add $\frac{4}{3}$ to the (4,4) element, $-\frac{2}{3}$ to the (4,1) and (1,4) elements and $\frac{1}{3}$ to the (1,1) element. After the first four animals, the table is:

	1	2	3	4	5	6
1	$1+\frac{1}{2}+\frac{1}{3}$	$\frac{1}{2}$	-1	$-\frac{2}{3}$		
2	$\frac{1}{2}$	$1+\frac{1}{2}$	-1			
3	-1	-1	2			
4	$-\frac{2}{3}$			$\frac{4}{3}$		
5						
6						

After applying the relevant rules to animals 5 and 6, the inverse of **A** then is:

	1	2	3	4	5	6
1	1.83	0.5	−1.0	−0.67	0.0	0.0
2	0.5	2.0	−1.0	0.0	0.5	−1.0
3	−1.0	−1.0	2.5	0.5	−1.0	0.0
4	−0.67	0.0	0.5	1.83	−1.0	0.0
5	0.0	0.5	−1.0	−1.0	2.5	−1.0
6	0.0	−1.0	0.0	0.0	−1.0	2.0

Using Eqn 3.3, the inverse of **A** can be calculated directly. If inbreeding is ignored, **D** for the pedigree is:

$$\mathbf{D} = \text{diag}(1.0, 1.0, 0.5, 0.75, 0.5, 0.5)$$

and

$$\mathbf{D}^{-1} = \text{diag}(1, 1, 2, 1.33, 2, 2)$$

Therefore, the inverse of the relationship matrix using Eqn 3.3 is:

$$\underbrace{\begin{bmatrix} 1.0 & 0.0 & -0.5 & -0.5 & 0.0 & 0.0 \\ 0.0 & 1.0 & -0.5 & 0.0 & 0.0 & -0.5 \\ 0.0 & 0.0 & 1.0 & 0.0 & -0.5 & 0.0 \\ 0.0 & 0.0 & 0.0 & 1.0 & -0.5 & 0.0 \\ 0.0 & 0.0 & 0.0 & 0.0 & 1.0 & -0.5 \\ 0.0 & 0.0 & 0.0 & 0.0 & 0.0 & 1.0 \end{bmatrix}}_{\left(\mathbf{T}^{-1}\right)'} \underbrace{\begin{bmatrix} 1.00 & 0.00 & 0.00 & 0.00 & 0.00 & 0.00 \\ 0.00 & 1.00 & 0.00 & 0.00 & 0.00 & 0.00 \\ 0.00 & 0.00 & 2.00 & 0.00 & 0.00 & 0.00 \\ 0.00 & 0.00 & 0.00 & 1.33 & 0.00 & 0.00 \\ 0.00 & 0.00 & 0.00 & 0.00 & 2.00 & 0.00 \\ 0.00 & 0.00 & 0.00 & 0.00 & 0.00 & 2.00 \end{bmatrix}}_{\mathbf{D}^{-1}}$$

$$\underbrace{\begin{bmatrix} 1.0 & 0.0 & 0.0 & 0.0 & 0.0 & 0.0 \\ 0.0 & 1.0 & 0.0 & 0.0 & 0.0 & 0.0 \\ -0.5 & -0.5 & 1.0 & 0.0 & 0.0 & 0.0 \\ -0.5 & 0.0 & 0.0 & 1.0 & 0.0 & 0.0 \\ 0.0 & 0.0 & -0.5 & -0.5 & 1.0 & 0.0 \\ 0.0 & -0.5 & 0.0 & 0.0 & -0.5 & 1.0 \end{bmatrix}}_{\left(\mathbf{T}^{-1}\right)}$$

$$= \underbrace{\begin{bmatrix} 1.83 & 0.50 & -1.00 & -0.67 & 0.00 & 0.00 \\ 0.50 & 2.00 & -1.00 & 0.00 & 0.50 & -1.00 \\ -1.00 & -1.00 & 2.50 & 0.50 & -1.00 & 0.00 \\ -0.67 & 0.00 & 0.50 & 1.83 & -1.00 & 0.00 \\ 0.00 & 0.50 & -1.00 & -1.00 & 2.50 & -1.00 \\ 0.00 & -1.00 & 0.00 & 0.00 & -1.00 & 2.00 \end{bmatrix}}_{\mathbf{A}^{-1}}$$

which is the same inverse obtained previously by applying the rules.

3.5.2 Inverse of the numerator relationship matrix accounting for inbreeding

The calculation of $\mathbf{A}^{-1}$ with inbreeding accounted for involves the application of the same rules outlined in Section 3.5.1 but **D** and, therefore, $\mathbf{D}^{-1}$ in Eqn 3.3 are calculated using the inbreeding coefficients of sires and dams (see Section 3.4). This implies that the diagonal elements of the relationship matrix are needed for $\mathbf{A}^{-1}$ to be properly calculated. This could be achieved by initially calculating the **A** for the group of animals and writing the diagonal elements to a file. The diagonal elements could be read from the file while computing $\mathbf{A}^{-1}$. For a large pedigree, this approach would require a large amount of memory for storage and be computationally demanding. However, Quaas (1976) presented a strategy for obtaining the diagonal elements of **A** while computing $\mathbf{A}^{-1}$ without setting up the relationship matrix.

Recall from Section 3.4 that **A** can be expressed as:

$$\mathbf{A} = \mathbf{TDT}'$$

$$\text{If } \mathbf{L} = \mathbf{T}\sqrt{\mathbf{D}}$$

$$\mathbf{A} = \mathbf{LL}' \tag{3.4}$$

where **L** is a lower triangular matrix and, since **D** is diagonal, $\sqrt{\mathbf{D}}$ refers to a matrix obtained by calculating the square root of the diagonal elements of **D**. Equation 3.4 implies that the diagonal element of **A** for animal *i* is:

$$a_{ii} = \sum_{m=1}^{i} l_{im}^2 \tag{3.5}$$

Thus, for a pedigree consisting of *m* animals:

$$a_{11} = l_{11}^2$$
$$a_{22} = l_{21}^2 + l_{22}^2$$
$$a_{33} = l_{31}^2 + l_{32}^2 + l_{33}^2$$
$$\vdots$$
$$a_{mm} = l_{m1}^2 + l_{m2}^2 + l_{m3}^2 + \mathrm{K} + l_{mm}^2$$

From the above, all the diagonal elements of **A** can be computed by calculating **L** one column at a time (Quaas, 1984). Only two vectors of dimension equal to the number of animals for storage will be required: one to store the column of **L** being computed and the second to accumulate the sum of squares of the elements of **L** for each animal. The matrices **L** and $\mathbf{A}^{-1}$ can be computed using the following procedure. From Eqn 3.4, the diagonal element of **L** for animal *i* is:

$$l_{ii} = \sqrt{d_i}$$
$$l_{ii} = \sqrt{\left[0.5 - 0.25\left(F_s + F_d\right)\right]}$$
$$l_{ii} = \sqrt{\left[1.0 - 0.25\left(a_{ss} + a_{dd}\right)\right]}; \quad \text{with } a_{ss} = 1 + F_{ss} \text{ and } a_{dd} = 1 + F_{dd}$$

Using Eqn 3.5:

$$l_{ii} = \sqrt{\left[1.0 - 0.25\left(\sum_{m=1}^{s} l_{sm}^2 + \sum_{m=1}^{s} l_{dm}^2\right)\right]}$$

To set up $\mathbf{A}^{-1}$ at the same time, calculate the diagonal element of $\mathbf{D}^{-1}(\alpha_i)$ for animal i as $\alpha_i = 1/l_{ii}^2$. Then compute the contribution of animal i to $\mathbf{A}^{-1}$, applying the usual rules for computing $\mathbf{A}^{-1}$ (see Section 3.5.1).

The off-diagonal elements of $\mathbf{L}$ to the left of the diagonal for animal i are calculated as:

$$l_{ij} = 0.5\left(l_{sj} + l_{dj}\right); s \text{ and } d \text{ equal to or greater than } j$$

For the example pedigree used in Section 3.5.1, the $\mathbf{L}$ matrix is:

	1	2	3	4	5	6
1	1.0	0.0	0.0	0.0	0.0	0.0
2	0.0	1.0	0.0	0.0	0.0	0.0
3	0.5	0.5	0.707	0.0	0.0	0.0
4	0.5	0.0	0.0	0.866	0.0	0.0
5	0.5	0.25	0.354	0.433	0.707	0.0
6	0.25	0.625	0.177	0.217	0.354	0.685

and $\mathbf{A}^{-1}$ with inbreeding accounted for is:

	1	2	3	4	5	6
1	1.833	0.5	–1.0	–0.667	0.0	0.0
2	0.5	2.033	–1.0	0.0	0.533	–1.067
3	–1.0	–1.0	2.5	0.5	–1.0	0.0
4	–0.667	0.0	0.5	1.833	–1.0	0.0
5	0.0	0.533	–1.0	–1.0	2.533	–1.067
6	0.0	1.067	0.0	0.0	–1.067	2.133

The calculation columns of $\mathbf{L}$ and α_i for the first three animals are illustrated below:

$$l_{11} = \sqrt{\left[1 - 0.25(0+0)\right]} = 1$$

$\alpha_1 = 1$ and its contribution to $\mathbf{A}^{-1}$ is computed using the rules in Section 3.5.1

$$l_{21} = 0$$
$$l_{31} = 0.5(l_{11} + l_{21}) = 0.5(1+0) = 0.5$$
$$l_{41} = 0.5(l_{11}) = 0.5$$
$$l_{51} = 0.5(l_{41} + l_{31}) = 0.5(0.5+0.5) = 0.5$$
$$l_{61} = 0.5(l_{51} + l_{21}) = 0.5(0.5+0) = 0.25$$
$$l_{22} = \sqrt{\left[1 - 0.25(0+0)\right]} = 1$$

$\alpha_2 = 1$ and its contribution to $\mathbf{A}^{-1}$ is computed using the rules in Section 3.5.1

$$l_{32} = 0.5(l_{12} + l_{22}) = 0.5(0+1) = 0.5$$
$$l_{42} = 0.5(l_{12}) = 0.5(0) = 0$$
$$l_{52} = 0.5(l_{42} + l_{32}) = 0.5(0+0.5) = 0.25$$
$$l_{62} = 0.5(l_{52} + l_{22}) = 0.5(0.25+1.0) = 0.625$$
$$l_{33} = \sqrt{\left[1 - 0.25(l_{11}^2) - 0.25(l_{21}^2 + l_{22}^2)\right]}$$
$$= \sqrt{\left[1 - 0.25(1) - 0.25(0+1)\right]} = 0.707$$

$\alpha_3 = 1/(0.707)^2 = 2.0$ and its contribution to $\mathbf{A}^{-1}$ is computed using the usual rules

$l_{43} = 0.5(l_{13}) = 0.5(0) = 0$

$l_{53} = 0.5(l_{43} + l_{33}) = 0.5(0 + 0.707) = 0.354$

$l_{63} = 0.5(l_{53} + l_{23}) = 0.5(0.354 + 0) = 0.177$

Faster algorithms for computing the inverse of **A** accounting for inbreeding based on the **L** matrix have been published by Meuwissen and Luo (1992) and Quaas (unpublished note, 1995), and these are presented in Appendix B.

3.6 Inverse of the Relationship Matrix for Sires and Maternal Grandsires

In some cases, the prediction of breeding value is only for sires and maternal grandsires, the so-called sire and maternal grandsire (MGS) model. In such cases, the $\mathbf{A}^{-1}$ to be incorporated in the mixed-model equations (MME) involves only sire and maternal grandsires and the rules for calculating $\mathbf{A}^{-1}$ are different from those discussed in the previous sections relating to pedigrees with individuals, sires and dams. With the MGS model, the relationship matrix **A** required pertains to males and can be approximated (Quaas, 1984) as:

$$a_{ii} = 1 + 0.25a_{sk} \tag{3.6}$$

$$a_{ij} = 0.5a_{sj} + 0.25a_{kj} \tag{3.7}$$

where s and k are the sires and maternal grandsires, respectively, for sire i. When all maternal granddams are unrelated (base animals) and there are no maternal half-sibs, the above will yield the exact **A**.

The inverse of approximate **A** can be calculated from a list of sires and maternal grandsires, applying Eqn 3.3. In this case, $\mathbf{T}^{-1}$ is a lower triangular matrix with ones in the diagonal and the only non-zero elements to the left of the diagonal in the row for the ith animal are −0.5 and −0.25 for the columns corresponding to the sire and maternal grandsire, respectively. The elements of **D**, and therefore $\mathbf{D}^{-1}$, can be calculated in a manner similar to that described in Sections 3.4 and 3.5. The diagonal elements of **D** (d_{ii}) for animal i are calculated by the following rules. If both sire (s) and maternal grandsire (k) are known:

$$d_{ii} = \left[\text{var}(u_i) - \text{var}\left(\frac{1}{2}u_s + \frac{1}{4}u_k\right)\right] / \sigma_u^2$$

where the u terms are breeding values. Following the same arguments as in Section 3.4:

$$d_{ii} = \frac{11}{16} - \frac{1}{4}F_s - \frac{1}{16}F_k$$

where F_s and F_k are inbreeding coefficients for sire and maternal grandsire, respectively.

When only the maternal grandsire is known:

$$d_{ii} = \left[\text{var}(u_i) - \text{var}\left(\frac{1}{4}u_k\right)\right] / \sigma_u^2$$

$$d_{ii} = \frac{15}{16} - \frac{1}{16}F_k$$

When only the sire is known or no parents are known, d_{ii} is as calculated in Section 3.4.

The elements of $\mathbf{D}^{-1}$ are reciprocals of $\mathbf{D}$, calculated above. Using Eqn 3.3, $\mathbf{A}^{-1}$ can be calculated on the basis of $\mathbf{T}^{-1}$ and $\mathbf{D}^{-1}$, defined above, as follows. Initially, set $\mathbf{A}^{-1}$ to zero.

If both sire (s) and maternal grandsire (k) of animal i are known, add:

d_{ii}^{-1} to the (i,i) element
$-d_{ii}^{-1}/2$ to the (s,i) and (i,s) elements
$-d_{ii}^{-1}/4$ to the (k,i) and (i,k) elements
$d_{ii}^{-1}/4$ to the (s,s) element
$d_{ii}^{-1}/8$ to the (s,k) and (k,s) elements
$d_{ii}^{-1}/16$ to the (k,k) element

Without inbreeding, $d_{ii}^{-1} = \frac{16}{11}$.

If only the maternal grandsire (k) of animal i is known, add:

d_{ii}^{-1} to the (i,i) element
$-d_{ii}^{-1}/4$ to the (k,i) and (i,k) elements
$d_{ii}^{-1}/16$ to the (k,k) element

Without inbreeding, $d_{ii}^{-1} = \frac{16}{15}$.

If only the sire (s) of animal i is known, add:

d_{ii}^{-1} to the (i,i) element
$-d_{ii}^{-1}/2$ to the (s,i) and (i,s) elements
$d_{ii}^{-1}/4$ to the (s,s) element

Without inbreeding, $d_{ii}^{-1} = \frac{4}{3}$ in this situation, as in Section 3.5.1.

When s and k are unknown, add:

d_{ii}^{-1} to the (i,i) element

and $d_{ii}^{-1} = 1$.

3.7 An Example of the Inverse of a Sire and Maternal Grandsire Relationship Matrix

A pedigree consisting of sires and maternal grandsires set up from the pedigree in Table 3.1 is:

Sire	Sire of sire	Maternal grandsire of sire
1	Unknown	Unknown
4	1	Unknown
5	4	1

Recoding sires 1 to n, the pedigree becomes:

1	Unknown	Unknown
2	1	Unknown
3	2	1

Using Eqns 3.6 and 3.7, $\mathbf{A} = \begin{bmatrix} 1.0 & 0.5 & 0.5 \\ 0.5 & 1.0 & 0.625 \\ 0.5 & 0.625 & 1.125 \end{bmatrix}$

Note that the relationship among sires is the same as in **A** calculated from the full pedigree in Section 3.3.

The $\mathbf{T}^{-1}$ matrix for the pedigree is: $\mathbf{T}^{-1} = \begin{bmatrix} 1.0 & 0.0 & 0.0 \\ -0.5 & 1.0 & 0.0 \\ -0.25 & -0.5 & 1.0 \end{bmatrix}$

and

$$\mathbf{D}^{-1} = \text{diag}\left(1, \frac{4}{3}, \frac{16}{11}\right)$$

Applying Eqn 3.3, $\mathbf{A}^{-1}$ is:

$$\mathbf{A}^{-1} = \begin{bmatrix} 1.0 & -0.5 & -0.25 \\ 0.0 & 1.0 & -0.5 \\ 0.0 & 0.0 & 1.0 \end{bmatrix} \begin{bmatrix} 1 & 0 & 0 \\ 0 & \frac{4}{3} & 0 \\ 0 & 0 & \frac{16}{11} \end{bmatrix} \begin{bmatrix} 1.0 & 0.0 & 0.0 \\ -0.5 & 1.0 & 0.0 \\ -0.25 & -0.5 & 1.0 \end{bmatrix}$$

$$= \begin{bmatrix} 1.424 & -0.485 & -0.364 \\ -0.485 & 1.697 & -0.727 \\ -0.364 & -0.727 & 1.455 \end{bmatrix}$$

To calculate the inverse of the sire and maternal grandsire relationship matrix, applying the rules given earlier, initially set $\mathbf{A}^{-1}$ to zero. The elements of $\mathbf{D}^{-1}$ have already been given above. Processing the first animal, add 1 (d_{11}^{-1}) to the diagonal element (1,1) of $\mathbf{A}^{-1}$. For the second animal, add $\frac{4}{3}$ (d_{22}^{-1}) to the diagonal element (2,2) of $\mathbf{A}^{-1}$, $\frac{1}{3}$ to the (1,1) element and $-\frac{2}{3}$ to the (1,2) and (2,1) elements. Finally, processing the third animal, add $\frac{16}{11}(d_{33}^{-1})$ to the (3,3) element of $\mathbf{A}^{-1}$, $-\frac{16}{11}$ to the (3,4) and (4,3) elements, $-\frac{16}{22}$ to the (1,3) and (3,1) elements, $\frac{16}{44}$ to the (4,4) element, $\frac{16}{88}$ to the (1,4) and (4,1) elements and $\frac{16}{176}$ to the (1,1) element. This gives the same $\mathbf{A}^{-1}$ as previously calculated using Eqn 3.3.

In Chapter 4, the incorporation of $\mathbf{A}^{-1}$ in the MME for the prediction of breeding value using BLUP is addressed.

3.8 Inferring Ancestral Relationships and Metafounders

The construction of **A** matrix in Section 3.3 is based on the assumptions that base or founder animals are unrelated and are sampled from one population with average breeding value of zero and common genetic variance σ_u^2. In practice, these assumptions may not hold as base animals may be sampled from populations of different genetic merit. In Chapter 4, Section 4.6, a procedure to account for differences in the genetic merit of base animals by assigning them to unknown parent groups is discussed and illustrated.

In addition, due to the finite size of populations, base or founder animals might also be related. The section deals with how related base animals may be handled in building the **A** matrix and its inverse. Christensen (2012) presented a method in which across-founder relationships can be estimated from genetic marker information in a subset of individuals and a pedigree. However, the extension of the method was not easy for estimating relationship among founders within and across populations. However, Legarra *et al.* (2015) presented the concept of metafounders to compute such relationships. This section, initially, discusses inferring relationships among ancestors or founders in a single population and across populations and the concept of metafounders is introduced and illustrated.

3.8.1 Ancestral relationship within and across individuals

Assume that the founders of a pedigree are drawn from a population referred to as the ancestral population (A), while the base population (B) refers to the pedigree founders (Fig. 3.1). First, consider a single population in which the base founders were sampled with replacement from a finite ancestral monoecious population with effective size Ne, and therefore 2Ne gametes. If two gametes were sampled from an ancestral population with replacement, then the second gamete will be identical to the first gamete by 1/(2Ne) of the times. Therefore, the probability of identity by descent (coancestry coefficient) between all pairs of gametes is $1/(2Ne) = \gamma/2$; with $\gamma = 1/Ne$ and for identity by descent with self, the probability will be 1.

Consider the relationship among two diploid individuals in the base population: X with gametes a and b and Y with gametes c and d (Fig. 3.2). These gametes were sampled from a pool of gametes where the probability of identical by descent, as stated above, is $\gamma/2$ across gametes and 1 with self; therefore the coefficient of coancestry between X and Y is the four-way average of probabilities of being identical for each possible pair of gametes, which sums

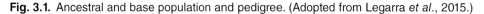

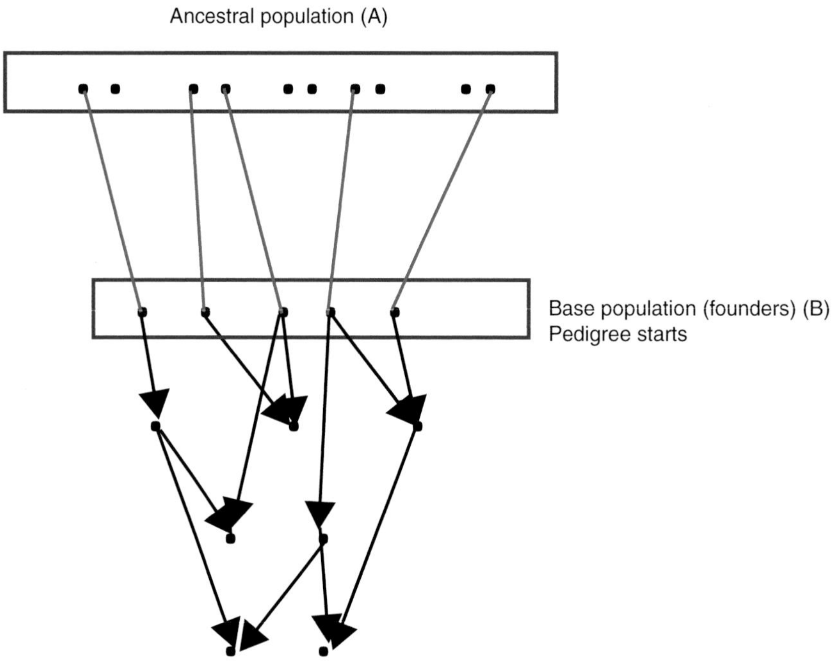

Fig. 3.1. Ancestral and base population and pedigree. (Adopted from Legarra *et al.*, 2015.)

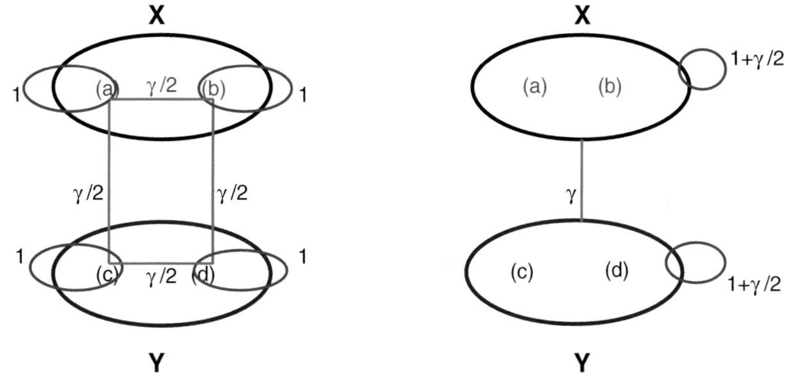

Fig. 3.2. Relationship (blue lines) in a related base population if two individuals X and Y seen as gametes (left) or individuals right). Individual X (Y) has gametes a and b (c and d). Relationships across and within gametes are γ/2 and 1 respectively; relationships across and within individuals are γ and 1 + γ/2. (Adopted from Legarra *et al*., 2015.)

to γ/2. Therefore, the additive relationship between X and Y is γ, i.e. twice the coancestry. If individual X is considered, the self-coancestry considers the four ways of sampling alleles a and b with replacement, and given that the probability of identical by descent between a and b equals γ/2, self-coancestry equals [2(1 + γ/2)]/4 = ½ + γ/4; therefore the additive self-relationship, which is twice the self-coancestry, equals 1+ γ/2 (Legarra *et al.*, 2015).

Pedigree relationship with related base population

Assuming the oth individual in the base population has breeding value u_o, then the variance of u_o, $\text{var}(u_o) = (1+\gamma/2)\sigma_u^2$, which is equal to $((1-\gamma/2)+\gamma)\sigma_u^2$. Therefore, pedigree relationships can be computed from related base populations following the tabular rules (Emik and Terrill, 1949) as:

$$\mathbf{A}^{\gamma} = \mathbf{A}(1-\gamma/2)+\gamma\mathbf{J}$$

where **A** is the usual numerator relationship matrix and **J** is a matrix of 1s.

While the computation of $\mathbf{A}^{\gamma}$ is easy, if inbreeding is to be accounted for in large populations, for instance, using the method of Quass (1976) or obtaining its inverse using the rules of Henderson (1976), such algorithms will need to modified to account for the non-zero relatedness among founders; which is rather complex, especially across several related base populations (Legarra *et al.*, 2015). This leads to the introduction of the concept of metafounders as a better framework to handle related base population animals.

3.8.2 Metafounders

Legarra *et al.* (2015) introduced the concept of the metafounders as an equivalent approach to account for related base animals in computing additive relationship among animals. A metafounder can be considered as a pseudo-individual who is the father and mother of all base animals (Fig. 3.3). The metafounders represent the ancestral population in Fig. 3.1 and can therefore be considered as a pseudo-individual that represents pools of gametes. Given that the gametes in the base animals (2 and 3) are sampled from the metafounder, using similar arguments as in section 3.8.1, then any two gametes drawn at random with replacement will have an across-gamete relationship of $\gamma/2$. Therefore the metafounder can be considered as having an additive self-relationship of $a_{11} = \gamma$ and an individual inbreeding coefficient of $F_1 = a_{11} - 1 = \gamma - 1$. As Legarra *et al.* (2015) explained, the negative inbreeding represents the heterozygosity of the pool of gametes. In a case where $\gamma = 0$ and $F = -1$, this implies any two random gametes are always different by descent and base population individuals are unrelated (complete heterozygosity by descent). In contrast, if $\gamma = 2$ with $F = 1$, this implies any two random gametes are always identical by descent (complete homozygosity) and all individuals in the base population are identical and completely inbred.

Computing additive relationship matrix and inbreeding with one metafounder

Considering one metafounder in the first instance, computing $\mathbf{A}^{\gamma}$ involves initially assigning a self-relationship γ for the metafounder and the following two rules used (Emilk and Terill, 1949):

$$a_{ij} = a_{ij} = 0.5(a_{js} + a_{jd}); \quad j = 1 \text{ to } i-1$$
$$a_{ij} = 1 + 0.5(a_{sd})$$

For example, given the pedigree in Table 3.2, the computation of A^{γ} is illustrated.

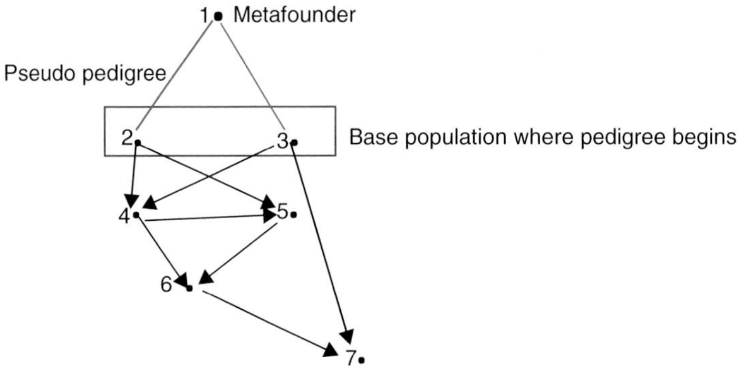

Fig. 3.3. Base population with one metafounder, two base animals and corresponding pedigree.

Table 3.2. Example pedigree with one metafounder (individual 1).

Calf	Sire	Dam
1	0	0
2	1	1
3	1	1
4	2	3
5	2	4
6	5	4
7	6	3

The steps for the first few animals are as follows:

$a_{11}(\text{metafounder}) = \gamma$

$a_{12} = 0.5(a_{11} + a_{11}) = \gamma;\ a_{22} = 1 + 0.5(a_{11}) = 1 + \gamma / 2$

$a_{13} = 0.5(a_{11} + a_{11}) = \gamma;\ a_{23} = 0.5(a_{12} + a_{12}) = \gamma; a_{33} = 1 + 0.5(a_{11}) = 1 + \gamma / 2$

$a_{14} = 0.5(a_{12} + a_{13}) = \gamma;\ a_{24} = 0.5(a_{22} + a_{23}) = 1 + \tfrac{3}{2}\gamma; a_{34} = 0.5(a_{32} + a_{33}) = 1 +$
$\tfrac{3}{2}\gamma;\ a_{34} = 1 + 0.5(a_{23}) = 1 + \gamma / 2$

If it is assumed that $\gamma = 0.2$, then:

$\mathbf{A}^{\gamma} =$

0.200						
0.200	1.100					
0.200	0.200	1.100				
0.200	0.650	0.650	1.100			
0.200	0.875	0.425	0.875	1.325		
0.200	0.763	0.538	0.988	1.100	1.438	
0.200	0.481	0.819	0.819	0.763	0.988	1.269

As an illustration for the first two animals; a_{11} (metafounder) = 0.2
$a_{12} = 0.5(0.2 + 0.2) = 0.2$; $a_{22} = 1 + 0.5(0.2) = 1.10$

Note that if you set $\gamma = 0$, you get the usual **A** matrix:

$\mathbf{A} =$

0.000						
0.000	1.000					
0.000	0.000	1.000				
0.000	0.500	0.500	1.000			
0.000	0.750	0.250	0.750	1.250		
0.000	0.625	0.375	0.875	1.000	1.375	
0.000	0.312	0.688	0.688	0.625	0.875	1.188

Computing inverse of $(A^{\gamma})^{-1}$ with one metafounder

The algorithm for computing the inverse of the numerator relationship matrix by Henderson (1976) with inbreeding accounted for, will also work with the inclusion of one metafounder provided that inbreeding for all individuals is computed previously. Inbreeding can be computed using either recursive algorithms or those based on the decomposition of **A**, such as those of Quaas (1976) and Meuwissen and Luo (1992). In the algorithm of Meuwissen and Luo (1992), see Appendix B, in the case of unknown parents, F_0 is set to -1. With one metafounder, the rules for computing of the Mendelian sampling terms do not change, so F_1 for metafounder should be set to $\gamma - 1$. In fact,

Legarra *et al.* (2015) indicated that the Meuwissen and Luo (1992) algorithm could be regarded as a situation with one metafounder and $\gamma = 0$.

The inverse of $\mathbf{A}^{\gamma}$ given for the example pedigree in Table 3.2, using usual rules of Henderson (1976) and $\gamma - 1$ as the inbreeding coefficient for the metafounder and the one minus the diagonal elements $\mathbf{A}^{\gamma}$ computed above for the other animals:

$\mathbf{A}^{(\gamma)-1} =$

7.222						
-1.111	2.222					
-1.111	0.556	2.351				
0.000	-0.556	-1.111	3.413			
0.000	-1.111	0.000	-0.476	2.857		
0.000	0.000	0.684	-1.270	-1.270	3.224	
0.000	0.000	-1.368	0.000	0.000	-1.368	2.736

As an illustration, the contribution for the first few animals is shown. First, the inbreeding coefficient for the metafounder equals 0.2 – 1 and for the rest of the animals the inbreeding coefficients are:

$F(2) = 0.1$, $F(3) = 0.1$, $F(4) = 0.1$, $F(5) = 0.325$, $F(6) = 0.438$, $F(7) = 0.269$

Commence by inverting $\gamma = 1/\gamma = 1/0.2 = 5$ and add 5 to the element $a^{(\gamma)-1}{}_{11}$. For animal 2, compute $d^{-1}{}_{22} = 1/[\ 0.5 - 0.25(F(1) + F(1))] = 1/[0.5 - 0.25(-0.8 + -0.8)] = 1.111$. Then add 1.111 to $a^{(\gamma)-1}{}_{22}$; $-1.112/2$ to $a^{(\gamma)-1}{}_{12}$, $a^{(\gamma)-1}{}_{21}$, and to $a^{(\gamma)-1}{}_{12}$ $a^{(\gamma)-1}{}_{21}$ again, since metafounder is both sire and dam; add 1.111/4 to $a^{(\gamma)-1}{}_{11}$ for four times, since metafounder is both sire and dam. For animal 3, the computation is exactly the same as for animal 2. Thus, after processing the animal the $a^{(\gamma)-1}{}_{11} = 7.111$.

The corresponding $\mathbf{A}^{-1}$ is given below when γ is set to zero and metafounder is treated as an unknown parent group:

$\mathbf{A}^{-1} =$

2.000						
-1.000	2.000					
-1.000	0.500	2.115				
0.000	-0.500	-1.000	3.071			
0.000	-1.000	0.000	-0.429	2.571		
0.000	0.000	0.615	-1.143	-1.143	2.901	
0.000	0.000	-1.231	0.000	0.000	-1.231	2.462

3.8.3 Computing relationships and inbreeding with several metafounders

Extension to several metafounders is straightforward, with metafounders considered to be related. Assuming two metafounders, A and B, their relationship ($\mathbf{\Gamma}$) can be written as:

$$\Gamma = \begin{pmatrix} \gamma^{A} & \gamma^{A,B} \\ \gamma^{B,A} & \gamma^{B} \end{pmatrix}$$

The relationship among the metafounders can be based on historical information about the populations or estimated from the genetic markers. The relationship matrix A^{Γ} for a pedigree with several metafounders can be computed with the tabular algorithm discussed in Section 3.8.2. Using the tabular approach, the initial step is to form $\mathbf{\Gamma}$ as the relationships of the metafounders and then apply the usual rules. Thus, the diagonal element for metafounders A and B will be γ^A and γ^B, respectively, and $\gamma^{B,A}$ as the lower off-diagonal element between both of them.

Computing the inverse $A^{\Gamma-1}$ involves, initially, inverting $\mathbf{\Gamma}$ to form a submatrix of $A^{\Gamma-1}$ for the metafounders and then using the rules of Henderson (1976) with the diagonal elements of D_{ii} (the variance and covariance matrix for Mendelian sampling, see Section 3.4) for all animals modified according to self-relationships of metafounders in the manner discussed in Section 3.8.2. As Legarra *et al.* (2015) indicated, if $\mathbf{\Gamma} = 0$, generalized inverse of $\mathbf{\Gamma}$ is 0 and this gives the $A^{\Gamma-1}$ as when base animals are assigned to unknown parent groups as in Thompson (1979) or Quaas (1988) (Chapter 4, Section 4.6), indicating that metafounders are a generalization of unknown parent groups.

For example, given the pedigree in Table 3.3, let us assume that:

$$\Gamma = \begin{pmatrix} 0.2 & 0.05 \\ 0.05 & 0.4 \end{pmatrix}$$

The computation of A^{γ} the first few animals i:

$$a_{11} = 0.2$$

$$a_{12} = 0.05\,;\ a_{22} = 0.4$$

$$a_{13} = 0.5(a_{11} + a_{11}) = 0.2;\ a_{23} = 0.5(a_{12} + a_{12}) = 0.05;\ a_{33} = 1 + 0.5(a_{11}) = 1.10$$

After processing all animals:

$\mathbf{A}^{\gamma} =$

0.200
0.050 0.400
0.200 0.050 1.100
0.000 0.400 0.050 1.200
0.100 0.225 0.125 0.200 1.000
0.050 0.312 0.087 0.700 0.600 1.100
0.125 0.181 0.594 0.375 0.363 0.594 1.044

Table 3.3. Example pedigree with two metafounders (individuals 1 and 2).

Calf	Sire	Dam
1	0	0
2	0	0
3	1	1
4	2	2
5	1	2
6	5	4
7	6	3

The inverse of $\mathbf{A}^{\gamma}$ given for the example pedigree in Table 3.3 involves initially computing Γ^{-1} which equals:

$$\boldsymbol{\Gamma}^{-1} = \begin{pmatrix} 6.791 & -0.849 \\ -0.849 & 3.396 \end{pmatrix}$$

Using the rules of Henderson (1976) and inbreeding coefficients for the 2 metafounders are: $F(1) = 0.2 - 1$ and $F(2)=0.4 - 1$. For other animals inbreeding coefficients are: $F(3)=0.1$, $F(4)=0.2$, $F(6)=0.1$ and $F(7)=0.044$, the inverse of $\mathbf{A}^{\gamma}$ is:

$\mathbf{A}^{(\gamma)\text{-}1} =$

8.196						
0.294	**4.940**					
-1.111	**0.000**	**1.667**				
0.000	**-1.250**	**0.000**	**1.806**			
-0.588	**-0.588**	**0.000**	**0.556**	**1.732**		
0.000	**0.000**	**0.566**	**-1.111**	**-1.111**	**2.778**	
0.000	**0.000**	**-1.111**	**0.000**	**0.000**	**-1.111**	**2.222**

As an illustration, the computation of $\mathbf{A}^{(\gamma)\text{-}1}$ for the first three animals is as follows: add 6.791 to the element $a^{(\gamma)\text{-}1}{}_{11}$, -0.849 to element $a^{(\gamma)\text{-}1}{}_{12}$ and 3.396 to element $a^{(\gamma)\text{-}1}{}_{22}$. Processing animal 3, compute $d^{-1}{}_{33} = 1/[0.5 - 0.25(F(1)+F(1))] = 1/[0.5 - 0.25(-0.8 + -0.8)] = 1.111$. Then add 1.111 to $a^{(\gamma)\text{-}1}{}_{33}$; $-1.112/2$ to $a^{(\gamma)\text{-}1}{}_{13}$, $a^{(\gamma)\text{-}1}{}_{31}$, and to $a^{(\gamma)\text{-}1}{}_{13}$ $a^{(\gamma)\text{-}1}{}_{31}$ again, since metafounder is both sire and dam; add 1.111/4 to $a^{(\gamma)\text{-}1}{}_{11}$ for four times, since metafounder is both sire and dam.

With large pedigree, the computation of inbreeding coefficients needed to calculate the diagonal elements of D_{ii} (the variance and covariance matrix for Mendelian sampling) can be done by the usual recursion algorithm or modifying the Meuwissen and Luo (1992 (Appendix B). Legarra *et al.* (2015) indicated that the algorithm of Meuwissen and Luo (1992) may need to be modified in some particular cases, such when a crossbred individual is the direct progeny of two related metafounders; that is, an F_1 crossbred individual with unknown parents. For instance, algorithm processes the ancestors of a given animal *i* and adds contributions $l_{ij}D_{jj}$ (which are elements of the lower triangular matrix of A, see Section 3.5.2 and Appendix B) to the inbreeding coefficient of *i*; then animal ***j*** is deleted from the list of ancestors and l_{ij} is set to zero. In the special case of a progeny being directly from two related metafounders, this will not work as contributions of all metafounders need to be processed simultaneously. In this situation, the contributions from the metafounders to A_{ii} are a sum over all metafounders: $\sum_k^{nmf}(L_{i,k}K_{:,k})^2$, where $\mathbf{K}_{:,k}$ is the *k*th column of K, the lower Cholesky decomposition of $\Gamma = \mathbf{KK}'$, and nmf is the number of metafounders (Legarra *et al.*, 2015). For the algorithm modifying the Meuwissen and Luo code to accommodate metafounders, see Legarra *et al.* (2015).

3.8.4 Estimation of metafounders' ancestral relationships from genomic data

Given that the within and across-founder relationships cannot be estimated from pedigree, Legarra *et al.* (2015) suggested using molecular markers. They presented a maximum likelihood approach and a method of moments which makes the first- and second-order statistics of genomic and pedigree relationships comparable. However, the use of the method of moments to compute the within- and across-founder relationships has not been used widely. However, the derivation given by Garcia-Baccino *et al.* (2017) based on the use of markers with $\gamma = 8\mathrm{Var}(p_i) = 8\sigma^2_p$; that is, γ for a single population is eight times the variance of allele frequencies (σ^2_p) in the base population has been used widely. Assuming two metafounders b and b′, the relationship between them, $\gamma_{b,b}{'} = 8\sigma_{pb,pb}{'}$, which is the covariance across loci between allele frequencies of two populations b and b′. Therefore, $\boldsymbol{\Gamma}$ is:

$$\Gamma = \begin{pmatrix} \sigma^2_{Pb} & \sigma^2_{Pbb'} \\ \sigma^2_{Pb'b} & \sigma^2_{pb'b'} \end{pmatrix}$$

The derivation for $\gamma = 8\sigma^2_p$ for a single population was given by Garcia-Baccino *et al.* (2017) and is briefly outlined. The relationship coefficient between pedigree founders or the self-relationship of the metafounder, $\gamma = 4\mathrm{Var}(p_i)/[2\mathrm{Var}(p_i)+\mathrm{E}(2p_iq_i)]$, where $p_i = 1 - q_i$ is the allele frequency at the *i*th random locus, assuming that reference alleles are chosen at random across loci, such that $\mathrm{E}(p_i) = \mathrm{E}(q_i) = 0.5$. Also, the parameter *s*, a measure of marker heterozygosity, equals $n(2\mathrm{Var}(p_i) + \mathrm{E}(2p_iq_i))$, where *n* is the number of markers.

The term $\mathrm{E}(2p_iq_i)$ in *s* can be expanded as:

$$\mathrm{E}(2p_iq_i) = 2\mathrm{E}(p_i)\mathrm{E}(q_i) \text{ and} \tag{3.8}$$

$$\mathrm{E}(p_i)\mathrm{E}(q_i) = \mathrm{E}(p_i(1-p_i)) = \mathrm{E}(p_i) - \mathrm{E}(p^2{}_i) \tag{3.9}$$

Given that $\mathrm{Var}(p_i) = \mathrm{E}(p_i^2) - \mathrm{E}(p_i)^2$, then

$$\mathrm{E}(p^2{}_{\mathrm{i}}) = \mathrm{Var}(p_{\mathrm{i}}) + \mathrm{E}(p_{\mathrm{i}})^2 \tag{3.10}$$

Substituting Eqn 3.10 into Eqn 3.9 gives:

$$\begin{aligned} \mathrm{E}(p_i)\mathrm{E}(q_i) &= \mathrm{E}(p_i) - \mathrm{Var}(p_i) - \mathrm{E}(p_i)^2 \\ &= \mathrm{E}(p_i)(1-\mathrm{E}(p_i)) - \mathrm{Var}(p_i) = \mathrm{E}(p_i)\mathrm{E}(q_i) - \mathrm{Var}(p_i) \end{aligned}$$

Therefore, (3.8) can be written as:

$$\mathrm{E}(2p_iq_i) = 2\mathrm{E}(p_{\mathrm{i}})\mathrm{E}(q_{\mathrm{i}}) - 2\mathrm{Var}(p_{\mathrm{i}}) = 0.5 - 2\mathrm{Var}(p_{\mathrm{i}})$$

assuming that reference alleles are chosen at random across loci, such that $\mathrm{E}(p_{\mathrm{i}}) = \mathrm{E}(q_{\mathrm{i}}) = 0.5$.

The parameter *s* can be expressed as:

$s = n(2\mathrm{Var}(p_i) + 0.5 - 2\mathrm{Var}(p_i)) = n(0.5) = n/2$, that is, number of markers divided by 2.

Similarly, substituting $\mathrm{E}(2p_iq_i) = 0.5 - 2\mathrm{Var}(p_i)$ into the expression for γ gives:

$$4\mathrm{Var}(p_{\mathrm{i}})/[2\mathrm{Var}(p_{\mathrm{i}}) + 0.5 - 2\mathrm{Var}(p_{\mathrm{i}})] = 8\mathrm{Var}(p_{\mathrm{i}}) = 8\sigma^2\mathrm{p}$$

In practice, the estimation of *s* is trivial as it is simply half the number of markers. Parameter γ is proportional to the variance of allele frequencies in the base population and should be trivial if base population individuals were genotyped. However, base population individuals are not usually available to be genotyped and various methods can be used to estimate their allele frequencies. The method of Gengler *et al.* (2007), based on pedigree information and linear mixed-model equations to account for selection and drift in allele frequencies across time, is commonly used. This is discussed and illustrated in Chapter 11.

4 Best Linear Unbiased Prediction of Breeding Value: Univariate Models with One Random Effect

4.1 Introduction

In Chapter 2, the use of the selection index (best linear prediction) for genetic evaluation was examined; however, it is associated with some major disadvantages. First, records may have to be pre-adjusted for fixed or environmental factors and these are assumed to be known. These are not usually known, especially when no prior data exist for new subclasses of fixed effect or new environmental factors. Second, solutions to the index equations require the inverse of the covariance matrix for observations and this may not be computationally feasible for large data sets.

Henderson (1949) developed a methodology called best linear unbiased prediction (BLUP), by which fixed effects and breeding values can be simultaneously estimated. The properties of the methodology are similar to those of a selection index and the methodology reduces to selection indices when no adjustments for environmental factors are needed. The properties of BLUP are more or less incorporated in the name:

- Best – it maximizes the correlation between true (a) and predicted breeding value $\hat{a}$ or minimizes prediction error variance (PEV) $(\text{var}(a - \hat{a}))$.
- Linear – predictors are linear functions of observations.
- Unbiased – estimation of realized values for a random variable such as animal breeding values, and of estimable functions of fixed effects are unbiased $(\text{E}(\hat{a}) = a)$.
- Prediction – involves prediction of true breeding value.

BLUP has found widespread usage in genetic evaluation of domestic animals because of its desirable statistical properties. This has been enhanced by the steady increase in computing power and has evolved in terms of its application to simple models, such as the sire model, in its early years, to more complex models such as the animal, maternal, multivariate and random regression models, in recent years. Several general purpose computer packages for BLUP evaluations such as PEST (Groeneveld *et al.*, 1990), BREEDPLAN, MiX99 (Lidauer *et al.*, 2011), BLUPf90 and a host of others have been written and made available. In this chapter, BLUP's theoretical background is briefly presented, considering a univariate animal model, and its application to several univariate models in genetic evaluation is illustrated.

4.2 Brief Theoretical Background

Consider the following equation for a mixed linear model:

$$\mathbf{y} = \mathbf{Xb} + \mathbf{Za} + \mathbf{e} \tag{4.1}$$

DOI: 10.1079/9781800620506.0004

where:

$\mathbf{y} = n \times 1$ vector of observations; n = number of records
$\mathbf{b} = p \times 1$ vector of fixed effects; p = number of levels for fixed effects
$\mathbf{a} = q \times 1$ vector of random animal effects; q = number of levels for random effects
$\mathbf{e} = n \times 1$ vector of random residual effects
$\mathbf{X}$ = design matrix of order $n \times p$, which relates records to fixed effects
$\mathbf{Z}$ = design matrix of order $n \times q$, which relates records to random animal effects

Both **X** and **Z** are termed incidence matrices.
It is assumed that the expectations (E) of the variables are:

$$E(y) = Xb; \; E(a) = E(e) = 0$$

and it is assumed that residual effects, which include random environmental and non-additive genetic effects, are independently distributed with variance σ_e^2; therefore, $\text{var}(\mathbf{e}) - \mathbf{I}\sigma_e^2 = \mathbf{R}$; $\text{var}(\mathbf{a}) - \mathbf{A}\sigma_a^2 = \mathbf{G}$ and $\text{cov}(\mathbf{a}, \mathbf{e}) = \text{cov}(\mathbf{e}, \mathbf{a}) = 0$, where **A** is the numerator relationship matrix. Then:

$$\begin{aligned} \text{var}(y) &= \mathbf{V} = \text{var}(\mathbf{Za} + \mathbf{e}) \\ &= \mathbf{Z}\text{var}(\mathbf{a})\mathbf{Z}' + \text{var}(\mathbf{e}) + \text{cov}(\mathbf{Za}, \mathbf{e}) + \text{cov}(\mathbf{e}, \mathbf{Za}) \\ &= \mathbf{ZGZ}' + \mathbf{R} + \mathbf{Z}\text{cov}(\mathbf{a}, \mathbf{e}) + \text{cov}(\mathbf{e}, \mathbf{a})\mathbf{Z}' \end{aligned}$$

Since $\text{cov}(\mathbf{a}, \mathbf{e}) = \text{cov}(\mathbf{e}, \mathbf{a}) = 0$, then:

$$\mathbf{V} = \mathbf{ZGZ}' + \mathbf{R} \tag{4.2}$$

$$\begin{aligned} \text{cov}(\mathbf{y}, \mathbf{a}) &= \text{cov}(\mathbf{Za} + \mathbf{e}, \mathbf{a}) \\ &= \text{cov}(\mathbf{Za}, \mathbf{a}) + \text{cov}(\mathbf{e}, \mathbf{a}) \\ &= \mathbf{Z}\text{cov}(\mathbf{a}, \mathbf{a}) \\ &= \mathbf{ZG} \end{aligned}$$

and

$$\begin{aligned} \text{cov}(\mathbf{y}, \mathbf{e}) &= \text{cov}(\mathbf{Za} + \mathbf{e}, \mathbf{e}) \\ &= \text{cov}(\mathbf{Za}, \mathbf{e}) + \text{cov}(\mathbf{e}, \mathbf{e}) \\ &= \mathbf{Z}\text{cov}(\mathbf{a}, \mathbf{e}) + \text{cov}(\mathbf{e}, \mathbf{e}) \\ &= \mathbf{R} \end{aligned}$$

The general problem with respect to Eqn 4.1 is to predict a linear function of **b** and **a**, that is, $\mathbf{k'b} + \mathbf{a}$ (predictand), using a linear function of **y**, say **L′y** (predictor), given that **k′b** is estimable. The predictor **L′y** is chosen such that it is unbiased (i.e. its expected value is equal to the expected value of the predictand) and PEV is minimized. This minimization leads to the BLUP of **a** (Henderson, 1973) as:

$$\hat{\mathbf{a}} = \text{BLUP}(\mathbf{a}) = \mathbf{GZ'V}^{-1}\left(\mathbf{y} - \mathbf{X}\hat{\mathbf{b}}\right) \tag{4.3}$$

and

$$\mathbf{L'y} = \mathbf{k'}\hat{\mathbf{b}} + \mathbf{GZ'V}^{-1}\left(\mathbf{y} - \mathbf{X}\hat{\mathbf{b}}\right)$$

where $\hat{\mathbf{b}} = (\mathbf{X}'\mathbf{V}^{-1}\mathbf{X})\mathbf{X}'\mathbf{V}^{-1}\mathbf{y}$, the generalized least square solution (GLS) for **b**, and $\mathbf{k}'\hat{\mathbf{b}}$ is the best linear unbiased estimator (BLUE) of **k′b**, given that **k′b** is estimable. BLUE is similar in meaning and properties to BLUP but relates to estimates of linear functions of fixed effects. It is an estimator of the estimable functions of fixed effects that has minimum sampling variance, is unbiased and is based on the linear function of the data (Henderson, 1984). An outline for the derivation of Eqn 4.3 and the equation for **L′y** above are given in Appendix C, Section C.1.

As mentioned in Section C.1, BLUP is equivalent to selection index with the GLS of $\hat{\mathbf{b}}$ substituted for **b** in Eqn 4.3. Alternatively, this could simply be illustrated (W.G. Hill, Edinburgh, 1995, personal communication) by considering the index to compute breeding values for a group of individuals with relationship matrix **A**, which have records with known mean. From Eqn 2.17, the relevant matrices are then:

$$\mathbf{P} = \mathbf{I}\sigma_e^2 + \mathbf{A}\sigma_a^2 \quad \text{and} \quad \mathbf{G} = \mathbf{A}\sigma_a^2$$

with:

$$\alpha = \sigma_e^2 / \sigma_a^2 \quad \text{or} \quad (1 - h^2)/h^2$$

Hence:

$$I = \mathbf{P}^{-1}\mathbf{G}\mathbf{y} = (\mathbf{I} + \alpha\mathbf{A}^{-1})^{-1}\mathbf{y}$$

which is similar to the BLUP (Eqn 4.3) assuming fixed effects are absent and with **Z** = **I**.

The solutions for **a** and **b** in Eqn 4.3 require $\mathbf{V}^{-1}$, which is not always computationally feasible. However, Henderson (1950) presented the mixed-model equations (MME) to estimate solutions **b** (fixed-effects solutions) and predict solutions for random effects (**a**) simultaneously without the need for computing $\mathbf{V}^{-1}$. The proof that solutions for **b** and **a** from MME are the GLS of **b** and the BLUP of **a** is given in Appendix C, Section C.2. The MME for Eqn 4.1 are:

$$\begin{pmatrix} \mathbf{X}'\mathbf{R}^{-1}\mathbf{X} & \mathbf{X}'\mathbf{R}^{-1}\mathbf{Z} \\ \mathbf{Z}'\mathbf{R}^{-1}\mathbf{X} & \mathbf{Z}'\mathbf{R}^{-1}\mathbf{Z} + \mathbf{G}^{-1} \end{pmatrix} \begin{pmatrix} \hat{\mathbf{b}} \\ \hat{\mathbf{a}} \end{pmatrix} = \begin{pmatrix} \mathbf{X}'\mathbf{R}^{-1}y \\ \mathbf{Z}'\mathbf{R}^{-1}y \end{pmatrix}$$

assuming that **R** and **G** are non-singular. Since $\mathbf{R}^{-1}$ is an identity matrix from the earlier definition of **R** in this section, it can be factored out from both sides of the equation to give:

$$\begin{bmatrix} \mathbf{X}'\mathbf{X} & \mathbf{X}'\mathbf{Z} \\ \mathbf{Z}'\mathbf{X} & \mathbf{Z}'\mathbf{Z} + \mathbf{A}^{-1}\alpha \end{bmatrix} \begin{bmatrix} \hat{\mathbf{b}} \\ \hat{\mathbf{a}} \end{bmatrix} = \begin{bmatrix} \mathbf{X}'\mathbf{y} \\ \mathbf{Z}'\mathbf{y} \end{bmatrix} \tag{4.4}$$

Note that the MME may not be of full rank, usually due to dependency in the coefficient matrix for fixed environmental effects. It may be necessary to set certain levels of fixed effects to zero when there is dependency to obtain solutions to the MME (see Section 4.6). However, the equations for **a** (Eqn 4.3) are usually of full rank since **V** is usually positive definite and **Xb** is invariant to the choice of constraint.

Some of the basic assumptions of the linear model for the prediction of breeding value were given in Section 2.2. The solutions to the MME give the BLUE of **k′b** and BLUP of **a** under certain assumptions, especially when data span several generations and may be subject to selection. These assumptions are:

1. Distributions of **y**, **a** and **e** are assumed to be multivariate normal, implying that traits are determined by many additive genes of infinitesimal effects at many infinitely unlinked

loci (infinitesimal model, see Section 2.2). With the infinitesimal model, changes in genetic variance resulting from selection, such as gametic disequilibrium (negative covariance between frequencies of genes at different loci), or from inbreeding and genetic drift, are accounted for in the MME through the inclusion of the relationship matrix (Sorensen and Kennedy, 1983), as well as assortative mating (Kemp, 1985).
2. The variances and covariances (**R** and **G**) for the base population are assumed to be known or at least known to proportionality. In practice, variances and covariances of the base population are never known exactly but, assuming the infinitesimal model, these can be estimated by restricted (or residual) maximum likelihood (REML) if data include information on which selection is based.
3. The MME can account for selection if based on a linear function of **y** (Henderson, 1975) and there is no selection on information not included in the data.

The use of these MME for the prediction of breeding values and estimation of fixed effects under an animal model is presented in the next section.

4.3 A Model for an Animal Evaluation (Animal Model)

Example 4.1
Consider the data set in Table 4.1 for the pre-weaning gain (WWG) of beef calves (calves assumed to be reared under the same management conditions).

The objective is to estimate the effects of sex and predict breeding values for all animals. Assume that $\sigma_a^2 = 20$ and $\sigma_e^2 = 40$, therefore $\alpha = \frac{40}{20} = 2$. The model to describe the observations is:

$$y_{ij} = p_i + a_j + e_{ij}$$

where y_{ij} = the WWG of the jth calf of the ith sex; p_i = the fixed effect of the ith sex; a_j = random effect of the jth calf; and e_{ij} = random error effect. In matrix notation, the model is the same as that described in Eqn 4.1.

4.3.1 Constructing the mixed-model equations

The matrix **X** in the MME relates records to fixed (sex) effects. For the example data set, its transpose is:

$$\mathbf{X}' = \begin{bmatrix} 1 & 0 & 0 & 1 & 1 \\ 0 & 1 & 1 & 0 & 0 \end{bmatrix}$$

The first row indicates that the first, fourth and fifth observations are from male calves and the second row shows the second and third records are from female calves.

Table 4.1. Pre-weaning gain (kg) for five beef calves.

Calves	Sex	Sire	Dam	WWG (kg)
4	Male	1	Unknown	4.5
5	Female	3	2	2.9
6	Female	1	2	3.9
7	Male	4	5	3.5
8	Male	3	6	5.0

The **Z** matrix relates records to all animals – those with or without yield records. In this case, animals 1 to 3 are parents with no records and animals 4 to 8 are recorded. Thus, for the example data, **Z** is:

$$\mathbf{Z}=\begin{bmatrix}0&0&0&1&0&0&0&0\\0&0&0&0&1&0&0&0\\0&0&0&0&0&1&0&0\\0&0&0&0&0&0&1&0\\0&0&0&0&0&0&0&1\end{bmatrix}$$

Note that the first three columns of **Z** are zeros and these correspond to the animals 1–3, which are parents without records.

The vector y is simply the vector of the observations. For the data set under consideration, it is:

$$\mathbf{y}'=\begin{bmatrix}4.5 & 2.9 & 3.9 & 3.5 & 5.0\end{bmatrix}$$

Having set up the matrices **X**, **Z** and **y**, the other matrices in the MME, such as **X′Z**, **Z′X**, **X′y** and **Z′y** are easily obtained by matrix multiplication. In practice, these matrices are not calculated through multiplication from the design matrices and vector of observations but are usually set up or computed directly. However, for the example data set, these matrices are:

$$\mathbf{X'Z}=\begin{bmatrix}0&0&0&1&0&0&1&1\\0&0&0&0&1&1&0&0\end{bmatrix}$$

and **Z′X** is the transpose of **X′Z**.

$\mathbf{X'y}=\begin{pmatrix}13.0\\6.8\end{pmatrix}$ and the transpose of **Z′y** is $(0\quad 0\quad 0\quad 4.5\quad 2.9\quad 3.9\quad 3.5\quad 5.0)$.

The matrix **Z′Z** is a diagonal matrix, with the first three diagonal elements zeros and the next five elements all ones.

The various matrices in the MME have been calculated, apart from $\mathbf{A}^{-1}\alpha$. With these matrices, we can set up what are known as the least squares equations (LSE) as:

$$\begin{bmatrix}\mathbf{X'X} & \mathbf{X'Z}\\ \mathbf{Z'X} & \mathbf{Z'Z}\end{bmatrix}\begin{bmatrix}\hat{\mathbf{b}}\\ \hat{\mathbf{a}}\end{bmatrix}=\begin{bmatrix}\mathbf{X'y}\\ \mathbf{Z'y}\end{bmatrix}$$

For the example data set, the LSE are:

$$\begin{bmatrix}3&0&0&0&0&1&0&0&1&1\\0&2&0&0&0&0&1&1&0&0\\0&0&0&0&0&0&0&0&0&0\\0&0&0&0&0&0&0&0&0&0\\0&0&0&0&0&0&0&0&0&0\\1&0&0&0&0&1&0&0&0&0\\0&1&0&0&0&0&1&0&0&0\\0&1&0&0&0&0&0&1&0&0\\1&0&0&0&0&0&0&0&1&0\\1&0&0&0&0&0&0&0&0&1\end{bmatrix}\begin{bmatrix}\hat{b}_1\\\hat{b}_2\\\hat{a}_1\\\hat{a}_2\\\hat{a}_3\\\hat{a}_4\\\hat{a}_5\\\hat{a}_6\\\hat{a}_7\\\hat{a}_8\end{bmatrix}=\begin{bmatrix}13.0\\6.8\\0\\0\\0\\4.5\\2.9\\3.9\\3.5\\5.0\end{bmatrix}$$

The addition of $\mathbf{A}^{-1}\alpha$ to $\mathbf{Z'Z}$ in the LSE yields the MME. Using the rules outlined in Section 3.4.1, $\mathbf{A}^{-1}$ for the example data is:

$$\mathbf{A}^{-1} = \begin{bmatrix} 1.833 & 0.500 & 0.000 & -0.667 & 0.000 & -1.000 & 0.000 & 0.000 \\ 0.500 & 2.000 & 0.500 & 0.000 & -1.000 & -1.000 & 0.000 & 0.000 \\ 0.000 & 0.500 & 2.000 & 0.000 & -1.000 & 0.500 & 0.000 & -1.000 \\ -1.667 & 0.000 & 0.000 & 1.833 & 0.500 & 0.000 & -1.000 & 0.000 \\ 0.000 & -1.000 & -1.00 & 0.500 & 2.500 & 0.000 & -1.000 & 0.000 \\ -1.000 & -1.000 & 0.500 & 0.000 & 0.000 & 2.500 & 0.000 & -1.000 \\ 0.000 & 0.000 & 0.000 & -1.000 & -1.000 & 0.000 & 2.000 & 0.000 \\ 0.000 & 0.000 & -1.000 & 0.000 & 0.000 & -1.000 & 0.000 & 2.000 \end{bmatrix}$$

and $\mathbf{A}^{-1}\alpha$ is easily obtained by multiplying every element of $\mathbf{A}^{-1}$ by 2, the value of α. Adding $\mathbf{A}^{-1}\alpha$ to $\mathbf{Z'Z}$, the MME for the example data are:

$$\begin{bmatrix} \hat{b}_1 \\ \hat{b}_2 \\ \hat{a}_1 \\ \hat{a}_2 \\ \hat{a}_3 \\ \hat{a}_4 \\ \hat{a}_5 \\ \hat{a}_6 \\ \hat{a}_7 \\ \hat{a}_8 \end{bmatrix} = \begin{bmatrix} 3.000 & 0.000 & 0.000 & 0.000 & 0.000 & 1.000 & 0.000 & 0.000 & 1.000 & 1.000 \\ 0.000 & 2.000 & 0.000 & 0.000 & 0.000 & 0.000 & 1.000 & 1.000 & 0.000 & 0.000 \\ 0.000 & 0.000 & 3.667 & 1.000 & 0.000 & -1.333 & 0.000 & -2.000 & 0.000 & 0.000 \\ 0.000 & 0.000 & 1.000 & 4.000 & 1.000 & 0.000 & -2.000 & -2.000 & 0.000 & 0.000 \\ 0.000 & 0.000 & 0.000 & 1.000 & 4.000 & 0.000 & -2.000 & 1.000 & 0.000 & -2.000 \\ 1.000 & 0.000 & -1.333 & 0.000 & 0.000 & 4.667 & 1.000 & 0.000 & -2.000 & 0.000 \\ 0.000 & 1.000 & 0.000 & -2.000 & -2.000 & 1.000 & 6.000 & 0.000 & -2.000 & 0.000 \\ 0.000 & 1.000 & -2.000 & -2.000 & 1.000 & 0.000 & 0.000 & 6.000 & 0.000 & -2.000 \\ 1.000 & 0.000 & 0.000 & 0.000 & 0.000 & -2.000 & -2.000 & 0.000 & 5.000 & 0.000 \\ 1.000 & 0.000 & 0.000 & 0.000 & -2.000 & 0.000 & 0.000 & -2.000 & 0.000 & 5.000 \end{bmatrix}^{-1} \begin{bmatrix} 13.0 \\ 6.8 \\ 0.0 \\ 0.0 \\ 0.0 \\ 4.5 \\ 2.9 \\ 3.9 \\ 3.5 \\ 5.0 \end{bmatrix}$$

Solving the MME by direct inversion of the coefficient matrix gives the following solutions:

Sex effects		Animals							
Males	Females	1	2	3	4	5	6	7	8
4.358	3.404	0.098	−0.019	−0.041	−0.009	−0.186	0.177	−0.249	0.183

The solutions indicate that male calves have a higher rate of gain up to weaning than female calves, which is consistent with the raw averages for males and females. From the first row in the MME (Eqn 4.4), the equations for sex effect are:

$$\begin{aligned} (\mathbf{X'X})\hat{\mathbf{b}} &= \mathbf{X'y} - (\mathbf{X'Z})\hat{\mathbf{a}} \\ \hat{\mathbf{b}} &= (\mathbf{X'X})^{-1}\mathbf{X}'(\mathbf{y} - \mathbf{Z}\hat{\mathbf{a}}) \end{aligned}$$

Thus, the solution for the ith level of sex effect may be written as:

$$\hat{b}_i = \left(\sum_j y_{ij} - \sum_j \hat{a}_{ij} \right) / \operatorname{diag}_i \qquad (4.5)$$

where y_{ij} is the record and $\hat{a}_{ij}$ is the solution of the jth animals within the sex subclass i and diag_i is the sum of observations for the sex subclass i. For instance, the solution for male calves is:

$$b_1 = [(4.5 + 3.5 + 5.0) - (-0.009 + -0.249 + 0.183)] / 3 = 4.358$$

The equations for animal effects from the second row of Eqn 4.4 are:

$$\begin{aligned}(\mathbf{Z'Z} + \mathbf{A}^{-1}\alpha)\hat{\mathbf{a}} &= \mathbf{Z'y} - (\mathbf{Z'X})\hat{\mathbf{b}} \\ (\mathbf{Z'Z} + \mathbf{A}^{-1}\alpha)\hat{\mathbf{a}} &= \mathbf{Z'}(\mathbf{y} - \mathbf{X}\hat{\mathbf{b}}) \\ (\mathbf{Z'Z} + \mathbf{A}^{-1}\alpha)\hat{\mathbf{a}} &= (\mathbf{Z'Z})\ \mathbf{YD} \end{aligned} \quad (4.6)$$

with $\mathbf{YD} = (\mathbf{Z'Z})^{-1}\mathbf{Z'}(\mathbf{y} - \mathbf{X}\hat{\mathbf{b}})$, where **YD** is the vector of yield deviations (YDs) and represents the yields of the animal adjusted for all effects other than genetic merit and error. The matrix $\mathbf{A}^{-1}$ has non-zero off-diagonals only for the animal's parents, progeny and mates (see Section 3.4), transferring off-diagonal terms to the right-hand side of Eqn 4.6 gives the equation for animal i with k progeny as:

$$(\mathbf{Z'Z} + u_{ii}\alpha)\hat{a}_i = \alpha u_{ip}(\hat{a}_s + \hat{a}_d) + (\mathbf{Z'Z})YD + \alpha \sum_k u_{im}(\hat{a}_{anim} - 0.5\hat{a}_m)$$

where u_{ip} is the element of the $\mathbf{A}^{-1}$ between animal i and its parents with the sign reversed, and u_{im} is the element of $\mathbf{A}^{-1}$ between the animal and the dam of the kth progeny.

Therefore:

$$(\mathbf{Z'Z} + u_{ii}\alpha)\hat{a}_i = \alpha u_{par}(PA) + (\mathbf{Z'Z})YD + 0.5\alpha \sum_k u_{prog}(2\hat{a}_{anim} - \hat{a}_m) \quad (4.7)$$

where PA is the parent average, $u_{par} = 2(u_{ip})$, with u_{ip} equal to 1, $\frac{2}{3}$; or $\frac{1}{2}$ if both, one or neither parents are known and $u_{prog} = u_{im}$, with u_{im} equal to 1 when the mate of animal i is known or $\frac{2}{3}$ when the mate is not known.

Multiplying both sides of the equation by $(\mathbf{Z'Z} + u_{ii}\alpha)^{-1}$ (VanRaden and Wiggans, 1991) gives:

$$a_i = n_i(PA) + n_2(YD) + n_3(PC) \quad (4.8)$$

where

$$PC = \sum_k u_{prog}(2\hat{a}_{anim} - \hat{a}_m) / \sum_k u_{prog}$$

is regarded as the progeny contribution and n_1, n_2 and n_3 are weights that sum to one. The derivation of the equation for PC is given in Appendix C, Section C.3. The numerators of n_1, n_2 and n_3 are αu_{par}, $\mathbf{Z'Z}$ (number of records the animal has) and $0.5\alpha\sum u_{prog}$, respectively. The denominator of all three n terms is the sum of the three numerators.

From Eqn 4.8, the breeding value for an animal is dependent on the amount of information available on that animal. For base animals, YD in the equation does not exist and $\hat{a}_s$ and $\hat{a}_d$ are zeros with no genetic groups in the model; therefore, the solutions for these animals are a function of the contributions from their progeny breeding values adjusted for the mate solutions *(PC)*. For instance, the proof for sire 1 in Example 4.1 can be calculated from the contributions from its progeny (calves 4 and 6) using Eqn 4.8 as:

$$\hat{a}_1 = n_1(0) + n_3\left[\left(\frac{2}{3}\right)(2\hat{a}_4) + (1)(2\hat{a}_6 - \hat{a}_2)\right] / \left(\frac{2}{3}+1\right)$$

$$\hat{a}_1 = n_1(0) + n_3\left[\left(\frac{2}{3}\right)(-0.018) + (1)(0.354 - (-0.019))\right] / \left(\frac{2}{3}+1\right)$$

$$\hat{a}_1 = n_3(0.2166) = 0.098$$

with $n_1 = \frac{\alpha}{3.667}$ and $n_3 = 0.5\alpha\left(\frac{2}{3}+1\right)/3.667$ and 3.667 is the sum of the numerators of n_1 and n_3. The higher breeding value for sire 1 compared with sire 3 is due to the fact that the progeny of sire 1 have higher proofs after correcting for the solutions of the mates.

The solutions for an animal with a record but with no progeny depend on the average contributions from its parents and its yield deviation. Equation 4.8 reduces to:

$$a_i = n_1(PA) + n_2(YD)$$

Thus, for progeny 8, its EBV can be calculated as:

$$\begin{aligned} a &= n_1(\hat{a}_3 + \hat{a}_6)/2 + n_2(y_8 - b_1) \\ &= n_1(0.068) + n_2(5.0 - 4.358) = 0.183 \end{aligned}$$

with $n_1 = \frac{2\alpha}{5}$, $n_2 = \frac{1}{5}$ and 5 is the sum of the numerators of n_1 and n_2.

It can also be demonstrated that for an animal with a record but with no progeny its solution is a function of an estimate of Mendelian sampling (m) and parent average. From Eqn c.8 in Appendix C, Section C.3, the solution for calf i can be written as:

$$(1 + u_{ii}\alpha)\hat{a}_i + \alpha u_{cs}\hat{a}_s + \alpha u_{cd}\hat{a}_d = y_i$$

Therefore:

$$\hat{a}_i = (1 + u_{ii}\alpha)^{-1}[y_i - \alpha u_{is}\hat{a}_s - \alpha u_{id}\hat{a}_d]$$

If there is no inbreeding, $u_{is} = u_{id} = -0.5\ u_{ii}$. Therefore:

$$\begin{aligned} \hat{a}_i &= (1 + u_{ii}\alpha)^{-1}[y_i + 0.5\ u_{ii}\alpha(\hat{a}_s + \hat{a}_d)] \\ &= (1 + u_{ii}\alpha)^{-1}[(y_i - 0.5(\hat{a}_s + \hat{a}_d)) + 0.5(1 + u_{ii}\alpha)(\hat{a}_s + \hat{a}_d)] \\ &= (1 + u_{ii}\alpha)^{-1}(y_i - 0.5(\hat{a}_s + \hat{a}_d)) + 0.5(\hat{a}_s + \hat{a}_d) \end{aligned}$$

$$\hat{\alpha}_i = 0.5(a_s + a_d) + m_i \tag{4.9}$$

where $m_i = k(y_i - 0.5\hat{a}_s - 0.5\hat{a}_d)$ is an estimate of Mendelian sampling, and $k = 1/(1 + d^{-1}\alpha)$, with $d = \frac{1}{2}$ if both parents of animal i are known or $\frac{3}{4}$ if only one parent is known. Alternatively, the weight (k) can also be derived as:

$$k = \text{cov}(m,\ y_c)/\text{var}(y_c) = \text{cov}(m,\ m + e)/(\text{var}(m) + \text{var}(e))$$

where y_c is the yield record corrected for fixed effects and parent average.

$$\begin{aligned} k &= \text{var}(m)/(\text{var}(m) + \text{var}(e)) \\ &= d\sigma_a^2/(d\sigma_a^2 + \sigma_e^2) \\ &= dh^2/(dh^2 + (1 - h^2)) \end{aligned}$$

where d, as defined earlier, equals $\frac{1}{2}$, $\frac{3}{4}$ or 1 if both, one or no parents are known, respectively. Using the parameters for Example 4.1 and assuming both parents are known, k = 10/(10 + 40) = 0.2.

Thus, for progeny 8, its EBV can be calculated as:

$$\begin{aligned}\hat{a}_8 &= 0.5(\hat{a}_3+\hat{a}_6)+k(y_3-b_1-0.5(\hat{a}_3+\hat{a}_6))\\ &= 0.5(-0.041+0.177)+0.2(5.0-4.358-0.5(-0.041+0.177))\\ &= 0.183\end{aligned}$$

Compared with calf 7, the proof of calf 8 is higher because it has a higher parent average solution and higher estimate of Mendelian sampling.

In the case of an animal with records and having progeny, there is an additional contribution from its offspring to its breeding value. Thus, the breeding values of progeny 4 and 6 using Eqn 4.8 are:

$$\begin{aligned}\hat{a}_4 &= n_1(\hat{a}_1/2)+n_2(y_4-b_1)+n_3(2(\hat{a}_7)-\hat{a}_5)\\ &= n_1(0.098/2)+n_2(4.5-4.38)+n_3(2(-0.249)-(-0.186)) = -0.009\end{aligned}$$

with $n_1 = 2\alpha(\frac{2}{3})/4.667$, $n_2 = 1/4.667$ and $n_3 = 0.\alpha/4.667$; 4.667 = the sum of the numerators of n_1, n_2 and n_3; and

$$\begin{aligned}\hat{a}_6 &= n_1((\hat{a}_1+\hat{a}_2)/2)+n_2(y_6-b_2)+n_3(2(\hat{a}_8)-\hat{a}_3)\\ &= n_1((0.098+-0.019)/2)+n_2(3.9+3.404)+n_3(2(0.183)-(-0.041))\\ &= 0.177\end{aligned}$$

with $n_1 = \frac{2\alpha}{6}$, $n_2 = \frac{1}{6}$ and $n_3 = \frac{0.5\alpha}{6}$; 6 = the sum of the numerators of n_1, n_2 and n_3.

Although contributions from parent average to both calves are similar, differences in progeny contributions resulted in a higher breeding value for calf 6, accounting for about 75% of the difference in the predicted breeding values between both calves.

4.3.2 Progeny (daughter) yield deviation

The yield deviation of a progeny contributes indirectly to the breeding value of its sire after it has been combined with information from parents and the offspring of the progeny (see Eqn 4.8). Thus, progeny contribution is a regressed measure and it is not an independent measure of progeny performance as information from parents and the progeny's offspring is included. VanRaden and Wiggans (1991) indicated that a more independent and unregressed measure of progeny performance is progeny yield deviation (PYD). However, they called it daughter yield deviation (DYD) as they were dealing with the dairy cattle situation and records were only available for daughters of bulls. PYD or DYD can simply be defined as a weighted average of corrected yield deviation of all progeny of a sire; the correction is for all fixed effects and the breeding values of the mates of the sire.

DYD has been used for various purposes in dairy cattle evaluation and research. It was used in the early 1990s for the calculation of conversion equations to convert bull evaluations across several countries (Goddard, 1985). It was initially the variable of choice for international evaluations of dairy bulls by Interbull, but, due to the inability of several countries to calculate DYD, de-regressed proofs were used (Sigurdsson and

Banos, 1995). In addition, Interbull methods for the validation of genetic trends in national evaluations prior to acceptance for international evaluations utilize DYDs (Boichard *et al.*, 1995). DYDs are also commonly employed in dairy cattle studies aimed at detecting quantitative trait loci using the granddaughter design (Weller, 2001). The equation for calculating DYD from univariate animal model evaluations was presented by VanRaden and Wiggans (1991) and its derivation is briefly outlined here.

For the progeny *(prog)* of a bull *i* that has no offspring of her own, Eqn 4.8 becomes:

$$\hat{a}_{prog} = n_{1prog}PA + n_{2prog}YD \tag{4.10}$$

Substituting Eqn 4.10 into the equation for PC in Eqn 4.8 gives:

$$\begin{aligned} PC &= \sum_k u_{prog}[2(n_{1prog}PA + n_{2prog}YD) - \hat{a}_{mi}] / \sum_k u_{prog} \\ &= \sum_k u_{prog}[n_{1prog}(\hat{a}_i + \hat{a}_{mi}) + n_{2prog}2YD - \hat{a}_{mi}] / \sum_k u_{prog} \end{aligned}$$

where n_{1prog} and n_{2prog} are the n_1 and n_2 of progeny. Since these progeny have no offspring of their own, n_{3prog} equals zero; therefore n_{1prog} equals $1 - n_{2prog}$. Then:

$$\begin{aligned} PC &= \sum_k u_{prog}[(1 - n_{2prog})(\hat{a}_i + \hat{a}_{mi}) + n_{2prog}2YD - \hat{a}_{mi}] / \sum_k u_{prog} \\ &= \sum_k u_{prog}[(1 + -n_{2prog})\,\hat{a}_i + n_{2prog}(2YD - \hat{a}_{mi})] / \sum_k u_{prog} \\ &= \hat{a}_i + \sum_k u_{prog}[n_{2prog}(-\hat{a}_i + 2YD - \hat{a}_{mi})] / \sum_k u_{prog} \end{aligned} \tag{4.11}$$

Substituting Eqn 4.11 into Eqn 4.8 and accumulating all terms involving $\hat{a}_i$ to the left side gives:

$$\begin{aligned} &\hat{a}_i - n_3\hat{a}_i + n_3\sum_k u_{prog}n_{2prog}\hat{a}_i / \sum_k u_{prog} \\ &= n_1PA + n_2YD + n_3\sum_k u_{prog}n_{2prog}(2YD - \hat{a}_{mi}) / \sum_k u_{prog} \end{aligned}$$

Therefore:

$$\begin{aligned} &\left(1 - n_3 + n_3\sum_k u_{prog}n_{2prog} / \sum_k u_{prog}\right)\hat{a}_i \\ &= n_1PA + n_2YD + n_3\sum_k u_{prog}n_{2prog}(2YD - \hat{a}_m) / \sum_k u_{prog} \end{aligned}$$

Substituting (n_1 + n_2) for 1 – n_3 and removing the common denominator of the *n* terms from both sides of the equation, with DYD as:

$$\text{DYD or PYD} = \sum_k u_{prog}n_{2prog}(2YD - \hat{a}_m) / \sum_k u_{prog}n_{2prog} \tag{4.12}$$

the breeding value of animal *i* can be expressed as:

$$\hat{a}_i = w_1PA + w_2YD + w_3\ \text{DYD} \tag{4.13}$$

where the weights w_1, w_2 and w_3 sum to unity. The numerators of w_1 and w_2 are equal to those of n_1 and n_2 in Eqn 4.8. The numerator of:

$$w_3 = 0.5\alpha\sum_k u_{prog}n_{2prog}$$

which is derived as n_3 times

$$\sum_k u_{prog} n_{2prog} / \sum_k u_{prog}$$

As VanRaden and Wiggans (1991) indicated, w_3 is always less than unity and therefore less than n_3, which reflects that PYD or DYD is an unregressed measure of progeny performance. Note that, for bulls with granddaughters, PYD or DYD does not include information from these granddaughters. Also, in the dairy cattle situation, the information from sons is not included in the calculation of DYD.

Illustrating the calculation of PYD or DYD

The computation of DYD is usually carried out in dairy cattle evaluations and it is illustrated later for a dairy data set in Example 4.1. Using the beef data in Example 4.1, the calculation of PYD is briefly illustrated for animal 3, using information on both female and male progeny, since observations are available on both sexes.

First, the YDs for both progeny of sire 3 are calculated:

$$\begin{aligned} YD_5 &= (y_5 - b_2) = (2.9 - 3.404) = -0.504 \\ YD_8 &= (y_8 - b_1) = (5.0 - 4.358) = 0.642 \end{aligned}$$

Therefore, using Eqn 4.12:

$$\begin{aligned} \text{PYD}_3 &= n_{2(5)}u_{(5)}(2YD_5 - \hat{a}_2) + n_{2(8)}u_{(8)}(2YD_8 - \hat{a}_6) / (n_{2(5)}u_{(5)} + n_{2(8)}u_{(8)}) \\ &= 0.2(1)(-1.008 - (-0.019) + 0.2(1)(1.284 - 0.177) / 0.2(1) + 0.2(1)) \\ &= 0.059 \end{aligned}$$

where $n_{2(j)}$ and $u_{(j)}$ are the n_2 and u for the jth progeny. Note that in calculating $n_{2(j)}$, it has been assumed that progeny j has no offspring. Thus $n_{2(5)} - 1/(1 + 2\alpha(1)) = 0.2$.

Using Eqn 4.12 to calculate the breeding value of sire 3 gives the value of 0.0098, with w_1 = 0.833 and w_2 = 0.167. This is different from the breeding value reported from solving the MME as the granddaughter information (calf 7) has not been included.

4.3.3 Accuracy of evaluations

The accuracy *(r)* of predictions is the correlation between true and predicted breeding values. However, in dairy cattle evaluations, the accuracy of evaluations is usually expressed in terms of reliability, which is the squared correlation between true and predicted breeding values (r^2). The calculation for r or r^2 requires the diagonal elements of the inverse of the MME, as shown by Henderson (1975).

If the coefficient matrix of the MME in Eqn 4.4 is represented as:

$$\begin{bmatrix} C_{11} & C_{12} \\ C_{21} & C_{22} \end{bmatrix}$$

and a generalized inverse of the coefficient matrix as: $\begin{bmatrix} C^{11} & C^{12} \\ C^{21} & C^{22} \end{bmatrix}$ Henderson (1975) showed that:

$$\text{PEV} = \text{var}(a - \hat{a}) = C^{22}\sigma_e^2 \tag{4.14}$$

Thus the diagonal elements of the coefficient matrix for animal equations are needed to calculate PEV for animals. The PEV could be regarded as the fraction of additive genetic variance not accounted for by the prediction. Therefore, for animal i, it could be expressed as:

$$\text{PEV}_i = C_i^{22}\sigma_e^2 = (1 - r^2)\sigma_a^2$$

where r^2 is the squared correlation between the true and EBVs. Thus:

$$d_i\sigma_e^2 = (1 - r^2)\sigma_a^2$$

where d_i is the ith diagonal element of C^{22}.

$$d_i\sigma_e^2 / \sigma_a^2 = 1 - r^2$$
$$r^2 = 1 - d_i\alpha$$

and the accuracy (r) is just the square root of reliability.

From Eqn 4.14 the standard error of prediction (SEP) is:

$$\begin{aligned}\text{SEP} &= \sqrt{\text{var}\,(a - \hat{a})}\\ &= \sqrt{d_i\sigma_e^2} \text{ for animal } i\end{aligned}$$

Note also that:

$$r^2 = 1 - (\text{SEP}^2 / \sigma_a^2)$$

The inverse of the coefficient matrix for Example 4.1 is:

$$\begin{bmatrix} 0.596 & 0.157 & -0.164 & -0.084 & -0.131 & -0.265 & -0.148 & -0.166 & -0.284 & -0.238\\ 0.157 & 0.802 & -0.133 & 0.241 & -0.112 & -0.087 & -0.299 & 0.306 & 0.186 & -0.199\\ -0.164 & -0.133 & 0.471 & 0.007 & 0.033 & 0.220 & 0.045 & 0.221 & 0.139 & 0.134\\ -0.084 & -0.241 & 0.007 & 0.0492 & -0.010 & 0.020 & 0.237 & 0.245 & 0.120 & 0.111\\ -0.131 & -0.112 & 0.033 & -0.010 & 0.456 & 0.048 & 0.201 & 0.023 & 0.126 & 0.218\\ -0.265 & -0.087 & 0.220 & 0.020 & 0.048 & 0.428 & 0.047 & 0.128 & 0.243 & 0.123\\ -0.148 & -0.299 & 0.045 & 0.237 & 0.201 & 0.047 & 0.428 & 0.170 & 0.220 & 0.178\\ -0.166 & -0.306 & 0.221 & 0.245 & 0.023 & 0.128 & 0.170 & 0.442 & 0.152 & 0.219\\ -0.284 & -0.186 & 0.139 & 0.120 & 0.126 & 0.243 & 0.220 & 0.152 & 0.442 & 0.168\\ -0.238 & -0.199 & 0.134 & 0.111 & 0.218 & 0.123 & 0.178 & 0.219 & 0.168 & 0.422 \end{bmatrix}$$

The r^2, r and SEP for animals in Example 4.1 are:

Animal	Diagonals of inverse	r^2	r	SEP
1	0.471	0.058	0.241	4.341
2	0.492	0.016	0.126	4.436
3	0.456	0.088	0.297	4.271
4	0.428	0.144	0.379	4.138
5	0.428	0.144	0.379	4.138
6	0.442	0.116	0.341	4.205
7	0.442	0.116	0.341	4.205
8	0.422	0.156	0.395	4.109

In the example, the reliabilities of animals with records are generally higher than those of ancestors since each has only two progeny. The two calves in the female sex subclass are progeny of dam 2 and this may explain the very low reliability for this ancestor as the effective number of daughters is reduced. The amount of information on calves 4 and 5 is very similar; each has a record, a common sire and parents of the same progeny, hence they have the same reliability. Calf 8 has the highest reliability and this is due to the information from the parents (its sire has another progeny and the dam has both parents known) and its record. The standard errors are large due to the small size of the data set but follow the same pattern as the reliabilities.

In practice, obtaining the inverse of the MME for large populations is not feasible and various methods have been used to approximate the diagonal element of the inverse. A methodology published by Meyer (1989) is presented in Appendix D and was used in the national dairy evaluation programme in Canada (Wiggans *et al.*, 1992) in the 1990s.

4.4 A Sire Model

The application of a sire model implies that only sires are being evaluated using progeny records. Most early applications of BLUP for the prediction of breeding values, especially in dairy cattle, were based on a sire model. The main advantage with a sire model is that the number of equations is reduced compared with an animal model since only sires are evaluated. However, with a sire model, the genetic merit of the mate (dam of progeny) is not accounted for. It is assumed that all mates are of similar genetic merit and this can result in bias in the predicted breeding values if there is preferential mating.

The sire model in matrix notation is:

$$\mathbf{y} = \mathbf{Xb} + \mathbf{Zs} + \mathbf{e} \tag{4.15}$$

All terms in Eqn 4.15 are as defined for Eqn 4.1 and s is the vector of random sire effects, Z now relates records to sires and:

$$\begin{aligned} \text{var(s)} &= \mathbf{A}\sigma_s^2 \\ \text{var(y)} &= \mathbf{ZAZ}'\sigma_s^2 + \mathbf{R} \end{aligned}$$

where **A** is the numerator relationship matrix for sires, $\sigma_s^2 = 0.25\sigma_a^2$ and $\mathbf{R} = \mathbf{I}\sigma_e^2$. The MME are exactly the same as in Eqn 4.4 except that $\alpha = \sigma_2^2 / \sigma_s^2 = (4 - h^2) / h^2$.

4.4.1 An illustration

Example 4.2

An application of a sire model is illustrated below using the same data as for the animal model evaluation in Table 4.1. Assigning records to sires, and including the pedigree for sires, the data can be presented as:

Sex of progeny	Sire	Sire of sire	Dam of sire	WWG (kg)
Male	1	–	–	4.5
Female	3	–	–	2.9
Female	1	–	–	3.9
Male	4	1	–	3.5
Male	3	–	–	5.0

The objective is to estimate sex effects and predict breeding values for sires 1, 3 and 4. Using the same parameters as in Section 4.3, $\sigma_s^2 = 0.25(20) = 5$ and $\sigma_e^2 = 60 - 5 = 55$, therefore $\alpha = 55/5 = 11$.

Setting up the design of matrices and MME

The design matrix **X** relating records to sex is as defined in Section 4.3.1. However, **Z** is different and its transpose is:

$$\mathbf{Z}' = \begin{bmatrix} 1 & 0 & 1 & 0 & 0 \\ 0 & 1 & 0 & 0 & 1 \\ 0 & 0 & 0 & 1 & 0 \end{bmatrix}$$

indicating that sires 1 and 3 have two records each while sire 4 has only one record. The vector of observations y is as defined in Section 4.3.1. The matrices **X′X**, **X′Z**, **Z′X**, **Z′Z**, **X′y** and **Z′y** in the MME can easily be calculated through matrix multiplication. Thus:

$$\mathbf{X'X} = \begin{bmatrix} 3 & 0 \\ 0 & 2 \end{bmatrix}, \mathbf{X'Z} = \begin{bmatrix} 1 & 1 & 1 \\ 1 & 1 & 0 \end{bmatrix}, \mathbf{Z'Z} = \text{diag}(2,2,1), \mathbf{X'y} \text{ is as in Section 4.3.1}$$

and the transpose of $\mathbf{Z'y} = (\mathbf{Z'y})' = [8.4 \quad 7.9 \quad 3.5]$

The LSE are:

$$\begin{bmatrix} 3 & 0 & 1 & 1 & 1 \\ 0 & 2 & 1 & 1 & 0 \\ 1 & 1 & 2 & 0 & 0 \\ 1 & 1 & 0 & 2 & 0 \\ 1 & 0 & 0 & 0 & 1 \end{bmatrix} \begin{bmatrix} \hat{b}_1 \\ \hat{b}_2 \\ \hat{s}_1 \\ \hat{s}_3 \\ \hat{s}_4 \end{bmatrix} = \begin{bmatrix} 13.00 \\ 6.80 \\ 8.40 \\ 7.90 \\ 3.50 \end{bmatrix}$$

Apart from the fact that sire 4 is the son of sire 1, no other relationships exist among the three sires. Therefore $\mathbf{A}^{-1}$ for the three sires is:

$$\mathbf{A}^{-1} = \begin{bmatrix} 1.333 & 0.0 & -0.667 \\ 0.000 & 1.0 & 0.000 \\ -0.667 & 0.0 & 1.333 \end{bmatrix}$$

The MME obtained after adding $\mathbf{A}^{-1}\alpha$ to $\mathbf{Z'Z}$ in the LSE are:

$$\begin{bmatrix} \hat{b}_1 \\ \hat{b}_2 \\ \hat{s}_1 \\ \hat{s}_3 \\ \hat{s}_4 \end{bmatrix} \begin{bmatrix} 3.000 & 0.000 & 1.000 & 1.000 & 1.000 \\ 0.000 & 2.000 & 1.000 & 1.000 & 0.000 \\ 1.000 & 1.000 & 16.666 & 0.000 & -7.334 \\ 1.000 & 1.000 & 0.000 & 13.000 & 0.000 \\ 1.000 & 1.000 & -7.334 & 0.000 & 15.666 \end{bmatrix}^{-1} = \begin{bmatrix} 13.00 \\ 6.80 \\ 8.40 \\ 7.90 \\ 3.50 \end{bmatrix}$$

The solutions to the MME by direct inversion of the coefficient matrix are:

Sex effects		Sires		
Males	Females	1	3	4
4.336	3.382	0.022	0.014	–0.043

The difference between solutions for sex subclasses, $\mathbf{L'b}$, where $\mathbf{L}$ is [1 – 1], is the same as in the animal model. However, sire proofs and differences between sire proofs $(s_i - s_j)$ are different from those from the animal model, although the ranking for the three sires is the same in both models. The differences in the proofs are due to the lack of adjustment for breeding values of mates in the sire model and differences in progeny contributions under both models. In this example, most of the differences in sire solutions under both models are due to differences in progeny contributions. The proofs for these sires under the animal model are based on their progeny contributions, since their parents are unknown. This contribution from progeny includes information from progeny yields and those of grand-offspring of the sires. However, in the sire model, progeny contributions include information from only male grand-offspring of the sires in addition to progeny yields. The effect of this difference on sire proofs under the two models is illustrated for two bulls below.

From the calculations in Section 4.3.1, the proportionate contribution of calves 4 and 6 to the proof of sire 1 in the animal model are –0.003 and 0.102, respectively. Using Eqn 4.8, the contribution of information from the different yield records to sire 1 under the sire model are as follows.

Contributions (CONT) from yields for calves 4 and 6 are:

$$\mathrm{CONT}_4 = n_2(0.082) = 0.010$$
$$\mathrm{CONT}_6 = n_2(0.259) = 0.031$$

where $n_2 = 2 / 16.667$.

Contributions from yield record for male grand-progeny (calf 7) through animal 4 (progeny) are:

$$\mathrm{CONT}_7 = n_3(-0.086) = -0.019$$

where $n_3 = 3.667/16.667$.
Therefore:

$$s_1 = \mathrm{CONT}_4 + \mathrm{CONT}_6 + \mathrm{CONT}_7 = 0.022$$

In the sire model the sum of CONT_4 and CONT_7 is equivalent to the contribution from calf 4 to the sire proof in the animal model. Thus, the main difference in the proof for sire 1 in the two models is due largely to the lower contribution of calf 6 in the sire model. This lower contribution arises from the fact that the contribution is only from the yield record in the sire model while it is from the yield and the progeny of calf 6 in the animal model.

Similar calculations for sire 3 indicate that the proportionate contributions from its progeny are –0.088 for calf 5 and 0.047 for calf 8 in the animal model. However, in the sire model the contributions are –0.037 and 0.051, respectively, from the yield of these calves. Again, the major difference here is due to the contribution from calf 5, which contains information from her offspring (calf 7) in the animal model. The similarity of the contributions of calf 8 to the proof of sire 3 in both models is because it is a non-parent and the contribution is slightly higher under the sire model due to the lack of adjustment for the breeding value of the mate.

4.5 Reduced Animal Model

In Section 4.2, the BLUP of breeding value involved setting up equations for every animal, that is all parents and progeny. Thus, the order of the animal equations was equal to the number of animals being evaluated. If equations were set up only for parents, this would greatly reduce the number of equations to be solved, especially since the number of parents is usually less than the number of progeny in most data sets. Breeding values of progeny can be obtained by back-solving from the predicted parental breeding values. Quaas and Pollak (1980) developed the reduced animal model (RAM), which allowed equations to be set up only for parents in the MME, and breeding values of progeny are obtained by back-solving from the predicted parental breeding values. This section presents the theoretical background for the RAM and illustrates its use for the prediction of breeding values.

4.5.1 Defining the model

The application of an RAM involves setting up animal equations for parents only and representing the breeding values of non-parents in terms of parental breeding value. Thus, for the non-parent i, its breeding value can be expressed as:

$$a_i = \frac{1}{2}(a_s - a_d) + m_i \tag{4.16}$$

where a_s and a_d are the breeding values of sire and dam and m_i is the Mendelian sampling. It was shown in Section 3.3 that:

$$\text{var}(m_i) = (0.5 - 0.25(F_s + F_d))\sigma_a^2$$

Let $F = (F_s + F_d)/2$, then:

$$\begin{aligned} \text{var}(m_i) &= (0.5 - 0.5(F))\sigma_a^2 \\ &= 0.5(1 - F)\sigma_a^2 \end{aligned} \tag{4.17}$$

The animal model applied in Section 4.3 was:

$$y_{ijk} = p_i + a_j + e_{ij} \tag{4.18}$$

In matrix notation:

$$\mathbf{y} = \mathbf{Xb} + \mathbf{Za} + \mathbf{e} \tag{4.19}$$

The terms in the above equations have been defined in Section 4.3.

Using Eqn 4.18, Eqn 4.19 can be expressed as:

$$y_{ijk} = p_i + \frac{1}{2}a_s + \frac{1}{2}a_d + m_j + e_{ijk} \tag{4.20}$$

For non-parents, the terms m_j and e_{ijk} can be combined to form a single residual term e^*_{ijk} as:

$$e^*_{ijk} = m_j + e_{ijk} \tag{4.21}$$

and

$$\text{var}(e^*_{ijk}) = \text{var}(m_j) + \text{var}(e_{ijk})$$

Using Eqn 4.19:

$$\text{var}(e^*_{ijk}) = \frac{1}{2}(1-F)\sigma_a^2 + \sigma_e^2$$

In general:

$$\text{var}(m_j) = d_j(1-F_j)\sigma_a^2 \qquad (4.22)$$

where d_j equals $\frac{1}{2}$ or $\frac{3}{4}$ or 1 if both, one or no parents are known, respectively, and F_j is the average inbreeding for both parents or, if only one parent is known, it is the inbreeding coefficient of the known parent. F_j equals zero when no parent is known. Ignoring inbreeding:

$$\text{var}(e^*_{ijk}) = \sigma_e^2 + d_j\sigma_a^2 = (1 + d_j\alpha^{-1}\sigma_e^2)$$

Equation 4.20 can be expressed in matrix notation as:

$$\mathbf{y} = \mathbf{X}_n\mathbf{b} + \mathbf{Z}_1\mathbf{a}_p + \mathbf{e}^* \qquad (4.23)$$

where $\mathbf{X}_n$ is the incidence matrix that relates non-parents' records to fixed effects, $\mathbf{Z}_1$ is an incidence matrix of zeros and halves identifying the parents of animals, and a_p is a vector of breeding values of parents.

The application of RAM involves applying the model:

$$\mathbf{y}_p = \mathbf{X}_p\mathbf{b} + \mathbf{Z}\mathbf{a} + \mathbf{e}$$

for parents and the model:

$$\mathbf{y}_n = \mathbf{X}_n\mathbf{b} + \mathbf{Z}_1\mathbf{a}_p + \mathbf{e}^*$$

for non-parents.

From the above two equations, the model for RAM analysis can be written as:

$$\begin{bmatrix}\mathbf{y}_p\\ \mathbf{y}_n\end{bmatrix} = \begin{bmatrix}\mathbf{X}_p\\ \mathbf{X}_n\end{bmatrix}\mathbf{b} + \begin{bmatrix}\mathbf{Z}\\ \mathbf{Z}_1\end{bmatrix}\mathbf{a}_p + \begin{bmatrix}\mathbf{e}\\ \mathbf{e}^*\end{bmatrix}$$

If:

$$\mathbf{X} = \begin{bmatrix}\mathbf{X}_p\\ \mathbf{X}_n\end{bmatrix}, \quad \mathbf{W} = \begin{bmatrix}\mathbf{Z}\\ \mathbf{Z}_1\end{bmatrix} \quad \text{and} \quad \mathbf{R} = \begin{bmatrix}\mathbf{R}_p\\ \mathbf{R}_n\end{bmatrix} = \begin{bmatrix}\mathbf{I}\sigma_e^2 & 0\\ 0 & \mathbf{I}\sigma_e^{2^*}\end{bmatrix} = \begin{bmatrix}\mathbf{I} & 0\\ 0 & \mathbf{I}+\mathbf{D}\alpha^{-1}\end{bmatrix}\sigma_e^2$$

Then:

$$\begin{aligned}\text{var}(\mathbf{y}) &= \mathbf{W}\mathbf{A}_p\mathbf{W}'\sigma_e^2 + \mathbf{R}\\ \text{var}(\mathbf{a}_p) &= \mathbf{A}_p\sigma_a^2\end{aligned}$$

where $\mathbf{A}_p$ is the relationship matrix among parents and $\mathbf{D}$ above is a diagonal matrix with elements as defined for d_j in Eqn 4.22.

The MME to be solved are:

$$\begin{bmatrix}\hat{\mathbf{b}}\\ \hat{\mathbf{a}}\end{bmatrix}\begin{bmatrix}\mathbf{X}'\mathbf{R}^{-1}\mathbf{X} & \mathbf{X}'\mathbf{R}^{-1}\mathbf{W}\\ \mathbf{W}'\mathbf{R}^{-1}\mathbf{X} & \mathbf{W}'\mathbf{R}^{-1}\mathbf{W} + \mathbf{A}^{-1}1/\sigma_a^2\end{bmatrix}\begin{bmatrix}\mathbf{X}'\mathbf{R}^{-1}\mathbf{y}\\ \mathbf{W}'\mathbf{R}^{-1}\mathbf{y}\end{bmatrix} \qquad (4.24)$$

Equation 4.24 can also be written as:

$$\begin{bmatrix}\mathbf{X}'_p\mathbf{R}^1_p\mathbf{X}_p + \mathbf{X}'_n\mathbf{R}_n^{-1}\mathbf{X}_n & \mathbf{X}'_p\mathbf{R}_p^{-1}\mathbf{Z} + \mathbf{X}'_n\mathbf{R}_n^{-1}\mathbf{Z}_1\\ \mathbf{Z}'\mathbf{R}_p^{-1}\mathbf{X}_p + \mathbf{Z}'_1\mathbf{R}_n^{-1}\mathbf{X}_n & \mathbf{Z}'\mathbf{R}_p^{-1}\mathbf{Z} + \mathbf{Z}'_1\ \mathbf{R}_n^{-1}\mathbf{Z}_1 + \mathbf{A}^{-1}1/\sigma_a^2\end{bmatrix} = \begin{bmatrix}\mathbf{X}'_p\mathbf{R}_p^{-1}\mathbf{y}_p + \mathbf{X}'_n\mathbf{R}_n^{-1}\mathbf{y}_n\\ \mathbf{Z}'\mathbf{R}_p^{-1}\mathbf{y}_p + \mathbf{Z}'_1\mathbf{R}_n^{-1}\mathbf{y}_n\end{bmatrix}$$

Multiplying the equations above by $\mathbf{R}_p$ gives:

$$\begin{bmatrix} \mathbf{X}'_p\mathbf{X}_p + \mathbf{X}'_n\mathbf{R}_v^{-1}\mathbf{X}_n & \mathbf{X}'_p\mathbf{Z} + \mathbf{X}'_n\mathbf{R}_v^{-1}\mathbf{Z}_1 \\ \mathbf{Z}'\mathbf{X}_p + \mathbf{Z}'_1\mathbf{R}_v^{-1}\mathbf{X}_n & \mathbf{Z}'\mathbf{Z} + \mathbf{Z}'_1\ \mathbf{R}_v^{-1}\mathbf{Z}_1 + \mathbf{A}^{-1}\alpha \end{bmatrix} = \begin{bmatrix} \mathbf{X}'_p\mathbf{y}_p + \mathbf{X}'_n\mathbf{R}_v^{-1}\mathbf{y}_n \\ \mathbf{Z}'\mathbf{y}_p + \mathbf{Z}'_1\mathbf{R}_v^{-1}\mathbf{y}_n \end{bmatrix} \quad (4.25)$$

where $\mathbf{R}_v^{-1}$ equals $1/(1+\mathbf{D}\alpha^{-1})$.

4.5.2 An illustration

Example 4.3

The application of RAM using Eqn 4.24 for the prediction of breeding values is illustrated below with the same data set (Table 4.1) as in Example 4.1 for the animal model evaluation. The genetic parameters are $\sigma_a^2 = 20.0$ and $\sigma_e^2 = 40.0$.

Constructing the MME

First, we need to set up **R**, the matrix of residual variances, and its inverse. In the example data set, animals 4, 5 and 6 are parents; therefore the diagonal elements in **R** corresponding to these animals are equal to σ_e^2, that is 40.0.

Calves 7 and 8 are non-parents, therefore the diagonal elements for these animals in **R** are equal to $\sigma_e^2 + d_i\sigma_a^2$, assuming that the average inbreeding coefficients of the parents of these animals equal zero. For each calf, d_i equals $\frac{1}{2}$ because both their parents are known, therefore $r_{77} = r_{88} = 40 + \frac{1}{2}(20) = 50$.

The matrix **R** for animals with records is:

$$\mathbf{R} = \text{diag}(40, 40, 50, 50)$$

and

$$\mathbf{R}^{-1} = \text{diag}(0.025, 0.025, 0.025, 0.020, 0.020)$$

The matrix X is the same as in Section 4.3.1 and relates records to sex effects.

Therefore:

$$\mathbf{X}'\mathbf{R}^{-1}\mathbf{X} = \begin{bmatrix} 0.065 & 0.000 \\ 0.000 & 0.050 \end{bmatrix}$$

For the matrix **W**, the rows for parents with records (animals 4, 5 and 6) consist of zeros except for the columns corresponding to these animals, which contain ones, indicating that they have records. However, the rows for non-parents with records (animals 7 and 8) contain halves in the columns that correspond to their parents, and otherwise zeros. Thus:

$$\mathbf{W} = \begin{bmatrix} 0.0 & 0.0 & 0.0 & 1.0 & 0.0 & 0.0 \\ 0.0 & 0.0 & 0.0 & 0.0 & 1.0 & 0.0 \\ 0.0 & 0.0 & 0.0 & 0.0 & 0.0 & 1.0 \\ 0.0 & 0.0 & 0.0 & 0.5 & 0.5 & 0.0 \\ 0.0 & 0.0 & 0.5 & 0.0 & 0.0 & 0.5 \end{bmatrix}$$

and

$$\mathbf{W'R^{-1}W} = \begin{bmatrix} 0.0 & 0.0 & 0.0 & 0.0 & 0.0 & 0.0 \\ 0.0 & 0.0 & 0.0 & 0.0 & 0.0 & 0.0 \\ 0.0 & 0.0 & 0.005 & 0.0 & 0.0 & 0.005 \\ 0.0 & 0.0 & 0.0 & 0.03 & 0.005 & 0.0 \\ 0.0 & 0.0 & 0.0 & 0.005 & 0.03 & 0.0 \\ 0.0 & 0.0 & 0.005 & 0.0 & 0.0 & 0.03 \end{bmatrix}$$

The transpose of the vector of observations, $\mathbf{y}$, is as defined in Section 4.3.1. The remaining matrices, $\mathbf{X'R^{-1}W}$, $\mathbf{W'R^{-1}X}$, $\mathbf{X'R^{-1}y}$ and $\mathbf{Z'R^{-1}y}$ can easily be calculated through matrix multiplication since $\mathbf{X}$, $\mathbf{R}^{-1}$, $\mathbf{W}$ and $\mathbf{y}$ have been set up. Therefore:

$$\mathbf{X'R^{-1}W} = \begin{bmatrix} 0.000 & 0.000 & 0.010 & 0.035 & 0.010 & 0.010 \\ 0.000 & 0.000 & 0.000 & 0.000 & 0.025 & 0.025 \end{bmatrix}$$

The matrix $\mathbf{W'R^{-1}X}$ is the transpose of $\mathbf{X'R^{-1}W}$.

$$\mathbf{X'R^{-1}y} = \begin{bmatrix} 0.282 \\ 0.170 \end{bmatrix} \quad \text{and} \quad \mathbf{W'R^{-1}y} = \begin{bmatrix} 0.000 \\ 0.000 \\ 0.050 \\ 0.148 \\ 0.107 \\ 0.148 \end{bmatrix}$$

The LSE are:

$$\begin{bmatrix} \hat{b}_1 \\ \hat{b}_2 \\ \hat{a}_1 \\ \hat{a}_2 \\ \hat{a}_3 \\ \hat{a}_4 \\ \hat{a}_5 \\ \hat{a}_6 \end{bmatrix} = \begin{bmatrix} 0.065 & 0.000 & 0.000 & 0.000 & 0.010 & 0.035 & 0.010 & 0.010 \\ 0.000 & 0.050 & 0.000 & 0.000 & 0.000 & 0.000 & 0.025 & 0.025 \\ 0.000 & 0.000 & 0.000 & 0.000 & 0.000 & 0.000 & 0.000 & 0.000 \\ 0.000 & 0.000 & 0.000 & 0.000 & 0.000 & 0.000 & 0.000 & 0.000 \\ 0.010 & 0.000 & 0.000 & 0.000 & 0.005 & 0.000 & 0.000 & 0.005 \\ 0.035 & 0.000 & 0.000 & 0.000 & 0.000 & 0.030 & 0.005 & 0.000 \\ 0.010 & 0.025 & 0.000 & 0.000 & 0.000 & 0.005 & 0.030 & 0.000 \\ 0.010 & 0.025 & 0.000 & 0.000 & 0.005 & 0.000 & 0.000 & 0.03 \end{bmatrix}^{-1} \begin{bmatrix} 0.282 \\ 0.170 \\ 0.000 \\ 0.000 \\ 0.050 \\ 0.148 \\ 0.107 \\ 0.148 \end{bmatrix}$$

The relationship matrix is only for parents, that is, animals 1 to 6. Thus:

$$\mathbf{A}^{-1} = \begin{bmatrix} 1.833 & 0.500 & 0.000 & -0.667 & 0.000 & -1.000 \\ 0.500 & 2.000 & 0.500 & 0.000 & -1.000 & -1.000 \\ 0.000 & 0.500 & 1.500 & 0.000 & -1.000 & 0.000 \\ -0.667 & 0.000 & 0.000 & 1.333 & 0.000 & 0.000 \\ 0.000 & -1.000 & -1.000 & 0.000 & 2.000 & 2.000 \\ -1.000 & -1.000 & 0.000 & 0.000 & 0.000 & \end{bmatrix}$$

Adding $\mathbf{A}^{-1}1/\sigma_a^2$ to the $\mathbf{W'R^{-1}W}$ of the LSE gives the MME, which are:

$$\begin{bmatrix} \hat{b}_1 \\ \hat{b}_2 \\ \hat{a}_1 \\ \hat{a}_2 \\ \hat{a}_3 \\ \hat{a}_4 \\ \hat{a}_5 \\ \hat{a}_6 \end{bmatrix} = \begin{bmatrix} 0.065 & 0.000 & 0.000 & 0.000 & 0.010 & 0.035 & 0.010 & 0.010 \\ 0.000 & 0.050 & 0.000 & 0.000 & 0.000 & 0.000 & 0.025 & 0.025 \\ 0.000 & 0.000 & 0.092 & 0.025 & 0.000 & -0.033 & 0.000 & -0.050 \\ 0.000 & 0.000 & 0.025 & 0.100 & 0.025 & 0.000 & -0.050 & -0.050 \\ 0.010 & 0.000 & 0.000 & 0.025 & 0.080 & 0.000 & -0.050 & 0.005 \\ 0.035 & 0.000 & -0.033 & 0.000 & 0.000 & 0.097 & 0.005 & 0.000 \\ 0.010 & 0.025 & 0.000 & -0.050 & -0.050 & 0.005 & 0.130 & 0.000 \\ 0.010 & 0.025 & -0.050 & -0.050 & 0.005 & 0.000 & 0.000 & 0.130 \end{bmatrix}^{-1} \begin{bmatrix} 0.282 \\ 0.170 \\ 0.000 \\ 0.000 \\ 0.050 \\ 0.148 \\ 0.107 \\ 0.148 \end{bmatrix}$$

The solutions are:

Sex effects		Animals					
Males	Females	1	2	3	4	5	6
4.358	3.404	0.098	−0.019	−0.041	−0.009	−0.186	0.177

The solutions for sex effects and proofs for parents are exactly as obtained using the animal model in Example 4.1. However, the number of non-zero elements in the coefficient matrix is 38 compared with 46 for an animal model in Section 4.3 on the same data set. This difference will be more marked in large data sets or in data sets where the number of progeny far exceeds the number of parents. This is one of the main advantages of the reduced animal model, as the number of equations, and therefore non-zero elements to be stored, are reduced. The solutions for non-parents can be obtained by back-solving, as discussed in the next section.

Solutions for non-parents

With the reduced animal model, solutions for non-parents are obtained by back-solving, using the solutions for the fixed effects and parents. Equation 4.9, derived earlier from the MME for an animal with its parents, can be used to back-solve for non-parent solutions. However, the $\mathbf{R}^{-1}$ has not been factored out of the MME in Eqn 4.25, and so the k term in Eqn 4.9 now equals:

$$k = r^{11} / r^{11} + d_i^{-1} g^{-1} \tag{4.26}$$

Solutions for non-parents in Example 4.3 can be solved using Eqn 4.9 but with k expressed as in Eqn 4.26. However, because there is a fixed effect in the model:

$$m_i = k(y_c - b_j - 0.5a_s - 0.5a_d).$$

In Example 4.3, both parents of non-parents (animals 7 and 8) are known. Therefore:

$$k = 0.025 / (0.025 + (2)0.05) = 0.20$$

Solution for calves 7 and 8 are:

$$\begin{aligned}\hat{a}_7 &= 0.5(-0.009 + -0.186) + 0.20(3.5 - 4.358 - 0.5(0.009 + -0.186) \\ &= -0.249) \\ \hat{a}_8 &= 0.5(-0.041 + 0.177) + 0.20(5.0 - 4.358 - 0.5(-0.041 + 0.177)) \\ &= 0.183\end{aligned}$$

Again, these solutions are the same for these animals as under the animal model.

4.5.3 An alternative approach

Note that, if the example data had been analysed using Eqn 4.25, the design matrices would be of the following form:

$$\mathbf{X}'_p = \begin{bmatrix} 1 & 0 & 0 \\ 0 & 1 & 1 \end{bmatrix}, \quad \mathbf{X}'_n = \begin{bmatrix} 1 & 1 \\ 0 & 0 \end{bmatrix}$$

Z including ancestors is:

$$\mathbf{Z} = \begin{bmatrix} 0 & 0 & 0 & 1 & 0 & 0 \\ 0 & 0 & 0 & 0 & 1 & 0 \\ 0 & 0 & 0 & 0 & 0 & 1 \end{bmatrix} \quad \text{and} \quad \mathbf{Z}_1 = \begin{bmatrix} 0 & 0 & 0 & 0.5 & 0.5 & 0 \\ 0 & 0 & 0.5 & 0 & 0 & 0.5 \end{bmatrix}$$

The remaining matrices can be calculated through matrix multiplication. The MME then are:

$$\begin{bmatrix} \hat{b}_1 \\ \hat{b}_2 \\ \hat{a}_1 \\ \hat{a}_2 \\ \hat{a}_3 \\ \hat{a}_4 \\ \hat{a}_5 \\ \hat{a}_6 \end{bmatrix} = \begin{bmatrix} 2.600 & 0.000 & 0.000 & 0.000 & 0.400 & 1.400 & 0.400 & 0.400 \\ 0.000 & 2.000 & 0.000 & 0.000 & 0.000 & 0.000 & 1.000 & 1.000 \\ 0.000 & 0.000 & 3.667 & 1.000 & 0.000 & -1.333 & 0.000 & -2.000 \\ 0.000 & 0.000 & 1.000 & 4.000 & 1.000 & 0.000 & -2.000 & -2.000 \\ 0.400 & 0.000 & 0.000 & 1.000 & 3.200 & 0.000 & -2.000 & 0.200 \\ 1.400 & 0.000 & -1.333 & 0.000 & 0.000 & 3.867 & 0.200 & 0.000 \\ 0.400 & 1.000 & 0.000 & -2.000 & -2.000 & 0.200 & 5.200 & 0.000 \\ 0.400 & 1.000 & -2.000 & -2.000 & 0.200 & 0.000 & 0.000 & 5.200 \end{bmatrix}^{-1} \begin{bmatrix} 11.300 \\ 6.800 \\ 0.000 \\ 0.000 \\ 2.000 \\ 5.900 \\ 4.300 \\ 5.900 \end{bmatrix}$$

and these give the same solutions as obtained from Eqn 4.24.

4.6 Animal Model with Groups

In Example 4.1, there were animals in the pedigree with unknown parents, usually called base population animals. The use of the relationship matrix in animal model evaluation assumes that these animals were sampled from a single population with average breeding value of zero and common variance σ_a^2. The breeding values of animals in subsequent generations are usually expressed relative to those of the base animals. However, if it is known that base animals were actually from populations that differ in genetic means – for instance sires from different countries – this must then be accounted for in the model. In the dairy cattle situation, due to differences in selection intensity, the genetic means for sires of bulls, sires of cows, dams of bulls and dams of cows may all be different. These various sub-population structures should be accounted for in the model to avoid bias in the prediction of breeding values. This can be achieved through a proper grouping of base animals using available information.

Westell and Van Vleck (1987) presented a procedure for grouping, which has generally been adopted. For instance, if sires have been imported from several countries over a period of time and their ancestors are unknown, these sires could be assigned to groups on the basis of the expected year of birth of the ancestors and the country of origin. The sires born within a similar time period in a particular foreign country are assumed to come from ancestors of similar genetic merit. Thus, each sire with one or both parents unknown is initially assigned phantom parents. Phantom parents are assumed to have had only one progeny each. Within each of the foreign countries, the phantom parents are grouped by the year of birth of their progeny and any other factor, such as sex of progeny. In addition, for the dairy cattle situation, the four selection paths – sire of sires, sire of dams, dam of sires and dam of dams – are usually assumed to be of different genetic merit and this is accounted for in the grouping strategy.

With groups, the model (Thompson, 1979) is:

$$y_{ij} = h_i + a_i + \sum_{k=1}^{n} t_{ik} g_k + e_{ij} \tag{4.27}$$

where h_j = effect of the jth herd, a_i = random effect of animal i, g_k = fixed group effect containing the kth ancestor, t_{ik} = the additive genetic relationship between the kth and ith animals and the summation is over all n ancestors of animal i, and e_{ij} = random environmental effect. From the model, it can be seen that the contribution of the group to the observation is weighted by the proportion of genes the ancestors in the group passed on to the animal with a record.

In matrix notation, the model can be written as:

$$\mathbf{y} = \mathbf{Xb} + \mathbf{ZQg} + \mathbf{Za} + \mathbf{e} \tag{4.28}$$

where:

$$\mathrm{Q} = \mathrm{TQ}^{*}$$

Q* assigns unidentified ancestors to groups and **T**, a lower triangular matrix, is obtained from **A** = **TDT′** (see Section 3.3). With this model the breeding value of an animal k is:

$$a_{k^*} = Q\hat{g} + \hat{a}_k$$

The MME are:

$$\begin{bmatrix} \mathrm{X'X} & \mathrm{X'Z} & \mathrm{X'ZQ} \\ \mathrm{Z'X} & \mathrm{Z'Z} + \mathrm{A}^{-1}\alpha & \mathrm{Z'ZQ} \\ \mathrm{Q'Z'X} & \mathrm{Q'Z'Z} & \mathrm{Q'Z'ZQ} \end{bmatrix} \begin{bmatrix} \hat{\mathrm{b}} \\ \hat{\mathrm{a}} \\ \hat{\mathrm{g}} \end{bmatrix} = \begin{bmatrix} \mathrm{X'y} \\ \mathrm{Z'y} \\ \mathrm{Q'Z'y} \end{bmatrix}$$

Solving the MME above will yield vectors of solutions for a and g but the ranking criterion (breeding value) is $\hat{\mathbf{a}}_{k^*} = \mathbf{Q}\hat{\mathbf{g}} + \hat{\mathbf{a}}_k$ for animal k. Modification of the MME

(Quaas and Pollak, 1981) and absorption of the group equations gave the following set of equations, which are usually solved to obtain $\hat{a}^*$ directly (Westell *et al.*, 1988):

$$\begin{bmatrix} X'X & X'Z & 0 \\ ZX & Z'Z + A^{-1}_{nm}\alpha & A^{-1}_{np}\alpha \\ 0 & A^{-1}_{pn}\alpha & A^{-1}_{pp}\alpha \end{bmatrix} \begin{bmatrix} \hat{b} \\ \hat{a} + \hat{Q}g \\ \hat{g} \end{bmatrix} = \begin{bmatrix} X'y \\ Z'y \\ 0 \end{bmatrix} \tag{4.29}$$

where n is the number of animals and p the number of groups.

Let:

$$A^{-1} = \begin{bmatrix} A^{1-}_{nn} & A^{-1}_{np} \\ A^{-1}_{pn} & A^{-1}_{pp} \end{bmatrix}$$

The matrix A^{-1} is obtained by the usual rules for obtaining the inverse of the relationship matrix outlined in Section 3.4.1. A list of pedigrees, consisting of only actual animals but with unknown ancestors assigned to groups, is set up. For the *i*th animal, calculate the inverse (b_i) of the variance of Mendelian sampling as:

b_i = 4/(2 + number of parents of animal *i* assigned to groups)

Then add:

b_i to the *(i,i)* element of A^{-1}

$-\frac{b_i}{2}$ to the *(i,s)*, *(i,d)*, *(s,i)* and *(d,i)* elements of A^{-1}

$\frac{b_i}{4}$ to the *(s,s)*, *(s,d)*, *(d,s)* and *(d,d)* elements of A^{-1}

Thus, for an animal *i* with both parents assigned to groups:

$b_i = 4/(2+2) = 1$

Then add:

1 to the *(i,i)* element of A^{-1}

$-\frac{1}{2}$ to the *(i,s)*, (i,d*)*, (s,i*)* and *(d,i)* elements of A^{-1}

$\frac{1}{4}$ to the *(s,s)*, *(s,d)*, *(d,s)* and *(d,d)* elements of A^{-1}

4.6.1 An illustration

Example 4.4

An animal model evaluation with groups is illustrated below using the same data set and genetic parameters as in Example 4.1. The aim is to estimate sex effects and predict breeding values for animals and phantom parents (groups). The model in Eqn 4.28 and the MME in Eqn 4.29 are used for the analysis. The pedigree file for the data set is:

Calf	Sire	Dam
1	Unknown	Unknown
2	Unknown	Unknown
3	Unknown	Unknown
4	1	Unknown
5	3	2
6	1	2
7	4	5
8	3	6

Assuming that males are of different genetic merit compared to females, the unknown sires can be assigned to one group (G1) and unknown dams to another group (G2). The pedigree file now becomes:

Calf	Sire	Dam
1	G1	G2
2	G1	G2
3	G1	G2
4	1	G2
5	3	2
6	1	2
7	4	5
8	3	6

Recoding G1 as 9 and G2 as 10:

Calf	Sire	Dam
1	9	10
2	9	10
3	9	10
4	1	10
5	3	2
6	1	2
7	4	5
8	3	6

Setting up the design matrices and MME

The design matrices X and Z, and the matrices **X′X, X′Z, Z′X, X′y** and **Z′y** in the MME are exactly as in Example 4.1. The MME without addition of the inverse of the relationship matrix for animals and groups are:

$$\begin{bmatrix} 3 & 0 & 0 & 0 & 0 & 1 & 0 & 0 & 1 & 1 & 0 & 0 \\ 0 & 2 & 0 & 0 & 0 & 0 & 1 & 1 & 0 & 0 & 0 & 0 \\ 0 & 0 & 0 & 0 & 0 & 0 & 0 & 0 & 0 & 0 & 0 & 0 \\ 0 & 0 & 0 & 0 & 0 & 0 & 0 & 0 & 0 & 0 & 0 & 0 \\ 0 & 0 & 0 & 0 & 0 & 0 & 0 & 0 & 0 & 0 & 0 & 0 \\ 1 & 0 & 0 & 0 & 0 & 1 & 0 & 0 & 0 & 0 & 0 & 0 \\ 0 & 1 & 0 & 0 & 0 & 0 & 1 & 0 & 0 & 0 & 0 & 0 \\ 0 & 1 & 0 & 0 & 0 & 0 & 0 & 1 & 0 & 0 & 0 & 0 \\ 1 & 0 & 0 & 0 & 0 & 0 & 0 & 0 & 1 & 0 & 0 & 0 \\ 1 & 0 & 0 & 0 & 0 & 0 & 0 & 0 & 0 & 1 & 0 & 0 \\ 0 & 0 & 0 & 0 & 0 & 0 & 0 & 0 & 0 & 0 & 0 & 0 \\ 0 & 0 & 0 & 0 & 0 & 0 & 0 & 0 & 0 & 0 & 0 & 0 \end{bmatrix} \begin{bmatrix} \hat{b}_1 \\ \hat{b}_2 \\ \hat{a}_1 \\ \hat{a}_2 \\ \hat{a}_3 \\ \hat{a}_4 \\ \hat{a}_5 \\ \hat{a}_6 \\ \hat{a}_7 \\ \hat{a}_8 \\ \hat{g}_1 \\ \hat{g}_2 \end{bmatrix} = \begin{bmatrix} 13.0 \\ 6.8 \\ 0 \\ 0 \\ 0 \\ 4.5 \\ 2.9 \\ 3.9 \\ 3.5 \\ 5.0 \\ 0 \\ 0 \end{bmatrix}$$

Using the procedure outline above, $\mathbf{A}^{-1}$ for the example data is:

	1	2	3	4	5	6	7	8	9	10
1	1.83	0.50	0.00	–0.67	0.00	–1.00	0.00	0.00	–0.50	–0.17
2	0.50	2.00	0.50	0.00	–1.00	–1.00	0.00	0.00	–0.50	–0.50
3	0.00	0.50	2.00	0.00	–1.00	0.50	0.00	–1.00	–0.50	–0.50
4	–0.67	0.00	0.00	1.83	0.50	0.00	–1.00	0.00	0.00	–0.67
5	0.00	–1.00	–1.00	0.50	2.50	0.00	–1.00	0.00	0.00	0.00
6	–1.00	–1.00	0.50	0.00	0.00	2.50	0.00	–1.00	0.00	0.00
7	0.00	0.00	0.00	–1.00	–1.00	0.00	2.00	0.00	0.00	0.00
8	0.00	0.00	–1.00	0.00	0.00	–1.00	0.00	2.00	0.00	0.00
9	–0.50	–0.50	–0.50	0.00	0.00	0.00	0.00	0.00	0.75	0.75
10	–0.17	–0.50	–0.50	–0.67	0.00	0.00	0.00	0.00	0.75	1.08

and $\mathbf{A}^{-1}\alpha$ is easily obtained by multiplying every element of $\mathbf{A}^{-1}$ by 2, the value of α. The matrix $\mathbf{A}^{-1}\alpha$ is added to equations for animal and group to obtain the MME, which are:

$$\begin{bmatrix} \hat{b}_1 \\ \hat{b}_2 \\ \hat{a}_1 \\ \hat{a}_2 \\ \hat{a}_3 \\ \hat{a}_4 \\ \hat{a}_5 \\ \hat{a}_6 \\ \hat{a}_7 \\ \hat{a}_8 \\ \hat{g}_1 \\ \hat{g}_2 \end{bmatrix} = \begin{bmatrix} 3.000 & 0.000 & 0.000 & 0.000 & 0.000 & 1.000 & 0.000 & 0.000 & 1.000 & 1.000 & 0.000 & 0.000 \\ 0.000 & 2.000 & 0.000 & 0.000 & 0.000 & 0.000 & 1.000 & 1.000 & 0.000 & 0.000 & 0.000 & 0.000 \\ 0.000 & 0.000 & 3.667 & 1.000 & 0.000 & -1.333 & 0.000 & -2.000 & 0.000 & 0.000 & -1.000 & -0.333 \\ 0.000 & 0.000 & 1.000 & 4.000 & 1.000 & 0.000 & -2.000 & -2.000 & 0.000 & 0.000 & -1.000 & -1.000 \\ 0.000 & 0.000 & 0.000 & 1.000 & 4.000 & 0.000 & -2.000 & 1.000 & 0.000 & -2.000 & -1.000 & -1.000 \\ 1.000 & 0.000 & -1.333 & 0.000 & 0.000 & 4.667 & 1.000 & 0.000 & -2.000 & 0.000 & 0.000 & 1.333 \\ 0.000 & 1.000 & 0.000 & -2.000 & -2.000 & 1.000 & 6.000 & 0.000 & -2.000 & 0.000 & 0.000 & 0.000 \\ 0.000 & 1.000 & -2.000 & -2.000 & 1.000 & 0.000 & 0.000 & 6.000 & 0.000 & -2.000 & 0.000 & 0.000 \\ 1.000 & 0.000 & 0.000 & 0.000 & 0.000 & -2.000 & -2.000 & 0.000 & 5.000 & 0.000 & 0.000 & 0.000 \\ 1.000 & 0.000 & 0.000 & 0.000 & -2.000 & 0.000 & 0.000 & -2.000 & 0.000 & 5.000 & 0.000 & 0.000 \\ 0.000 & 0.000 & -1.000 & -1.000 & -1.000 & 0.000 & 0.000 & 0.000 & 0.000 & 0.000 & 1.500 & 1.500 \\ 0.000 & 0.000 & -0.333 & -1.000 & -1.000 & -1.333 & 0.000 & 0.000 & 0.000 & 0.000 & 1.500 & 2.167 \end{bmatrix}^{-1} \begin{bmatrix} 13.0 \\ 6.8 \\ 0.00 \\ 0.00 \\ 0.00 \\ 4.50 \\ 2.90 \\ 3.90 \\ 3.50 \\ 5.00 \\ 0.00 \\ 0.00 \end{bmatrix}$$

There is dependency in the equations, i.e. all effects cannot be estimated; therefore the equation for the first group has been set to zero to obtain the following solutions:

Sex effects	
Males	5.458
Females	4.313
Animals	
1	–0.767
2	–0.923
3	–0.963
4	–1.268
5	–1.099
6	–0.728
7	–1.338
8	–0.768
Groups	
9	0.000
10	–1.769

The animal proofs above are generally lower than those from Example 4.1, the model without groups. In addition, the ranking for animals is also different. However, the relationship between the two sets of solutions can be shown by recalculating the vector of solutions for animals using the group solutions ($\hat{\mathbf{g}}$) above and the estimated breeding values ($\hat{\mathbf{a}}$) from Example 4.1 as:

$$\hat{\mathbf{a}}_* = \hat{\mathbf{a}} + \mathbf{Qg}$$

where $\mathbf{Q} = \mathbf{TQ}^*$, as defined earlier.

Assigning phantom parents (M1–M7) to animals with unknown ancestors, the pedigree for the example data can be written as:

Calf	Sire	Dam
1	M1	M2
2	M3	M4
3	M5	M6
4	1	M7
5	3	2
6	1	2
7	4	5
8	3	6

and the matrix **T** for the pedigree is:

	M1	M2	M3	M4	M5	M6	M7	1	2	3	4	5	6	7	8
M1	1.000	0.000	0.000	0.000	0.000	0.000	0.00	0.00	0.00	0.00	0.0	0.0	0.0	0.0	0.0
M2	0.000	1.000	0.000	0.000	0.000	0.000	0.00	0.00	0.00	0.00	0.0	0.0	0.0	0.0	0.0
M3	0.000	0.000	1.000	0.000	0.000	0.000	0.00	0.00	0.00	0.00	0.0	0.0	0.0	0.0	0.0
M4	0.000	0.000	0.000	1.000	0.000	0.000	0.00	0.00	0.00	0.00	0.0	0.0	0.0	0.0	0.0
M5	0.000	0.000	0.000	0.000	1.000	0.000	0.00	0.00	0.00	0.00	0.0	0.0	0.0	0.0	0.0
M6	0.000	0.000	0.000	0.000	0.000	1.000	0.00	0.00	0.00	0.00	0.0	0.0	0.0	0.0	0.0
M7	0.000	0.000	0.000	0.000	0.000	0.000	1.00	0.00	0.00	0.00	0.0	0.0	0.0	0.0	0.0
1	0.500	0.500	0.000	0.000	0.000	0.000	0.00	1.00	0.00	0.00	0.0	0.0	0.0	0.0	0.0
2	0.000	0.000	0.500	0.500	0.000	0.000	0.00	0.00	1.00	0.00	0.0	0.0	0.0	0.0	0.0
3	0.000	0.000	0.000	0.000	0.500	0.500	0.00	0.00	0.00	1.00	0.0	0.0	0.0	0.0	0.0
4	0.250	0.250	0.000	0.000	0.000	0.000	0.50	0.50	0.00	0.00	1.0	0.0	0.0	0.0	0.0
5	0.000	0.000	0.250	0.250	0.250	0.250	0.00	0.00	0.50	0.50	0.0	1.0	0.0	0.0	0.0
6	0.250	0.250	0.250	0.250	0.000	0.000	0.00	0.50	0.50	0.00	0.0	0.0	1.0	0.0	0.0
7	0.125	0.125	0.125	0.125	0.125	0.125	0.25	0.25	0.25	0.25	0.5	0.5	0.0	1.0	0.0
8	0.125	0.125	0.125	0.125	0.250	0.250	0.00	0.25	0.25	0.50	0.0	0.0	0.5	0.0	1.0

The transpose of the matrix $\mathbf{Q}^*$, which assigns phantom parents to groups, is:

$$\mathbf{Q}^{*\prime} = \begin{pmatrix} 1 & 0 & 1 & 0 & 1 & 0 & 0 & 0 & 0 & 0 & 0 & 0 & 0 & 0 & 0 \\ 0 & 1 & 0 & 1 & 0 & 1 & 0 & 0 & 0 & 0 & 0 & 0 & 0 & 0 & 0 \end{pmatrix}$$

and the transpose of $\mathbf{Q}$ ($\mathbf{Q} = \mathbf{TQ}^*$) is:

$$\mathbf{Q}' = \begin{pmatrix} 1 & 0 & 1 & 0 & 1 & 0 & 0 & 0.5 & 0.5 & 0.5 & 0.25 & 0.5 & 0.5 & 0.375 & 0.5 \\ 0 & 1 & 0 & 1 & 0 & 1 & 0 & 0.5 & 0.5 & 0.5 & 0.75 & 0.5 & 0.5 & 0.625 & 0.5 \end{pmatrix}$$

Therefore the vector of solutions using the EBVs from Example 4.1 is:

$$\hat{\mathbf{a}}_* = \hat{\mathbf{a}} + \mathbf{Q}\hat{\mathbf{g}} = \begin{bmatrix} 0.098 \\ -0.019 \\ -0.041 \\ -0.009 \\ -0.186 \\ 0.177 \\ -0.249 \\ 0.183 \end{bmatrix} + \begin{bmatrix} -0.898 \\ -0.898 \\ 0.898 \\ -1.346 \\ -0.898 \\ -0.898 \\ -1.122 \\ -0.898 \end{bmatrix} = \begin{bmatrix} -0.800 \\ -0.917 \\ -0.939 \\ -0.1355 \\ -1.084 \\ -0.721 \\ -1.371 \\ -0.715 \end{bmatrix}$$

These solutions are similar to those obtained in the model with groups. The slight differences are due to differences in sex solutions in the two examples and this is explained later. This indicates that when the solutions from the model without groups are expressed relative to the group solutions, similar solutions are obtained to those in the model with groups. Thus, the differences between the solutions in Examples 4.1 and 4.4 are due to the fact that the solutions in the former are expressed relative to base animals assumed to have an average breeding value of zero, while in the latter solutions are relative to the group solutions, one of which is lower than zero.

The inclusion of groups also resulted in a larger sex difference compared with Example 4.1. The solution for sex effect i can be calculated using Eqn 4.5. For instance, the solution for male calves in Example 4.4 is:

$$\hat{b}_1 = \left[(4.5 + 3.5 + 5.0) - (-1.268 - 1.338 - 0.768)\right] / 3 = 5.458$$

Since $\sum_j y_{ij}$ in Eqn 4.5 is the same in both examples, differences in $\sum_j \hat{a}_{ij}$ between the sexes in both models would result in differences in the linear function of **b**. The difference between average breeding values of male and female calves is –0.02 and –0.214, respectively, in Examples 4.1 and 4.4. The larger difference in the latter accounted for the higher sex difference in Example 4.4. Males had a lower breeding value in Example 4.4 due to the higher proportionate contribution of group two to their solutions (see the matrix **Q** above).

The basic principles involved in the application of BLUP for genetic evaluations and the main assumptions have been covered in this chapter, and its application to more complex models involves an extension of these principles. Equation 4.1 is a very general model and could include random animal effects for several traits (multivariate model), random environmental effects – such as common environmental effects affecting animals that are reared together – maternal effects (maternal model), non-additive genetic effects – such as dominance and epistasis (non-additive models) – and repeated data on individuals (random regression model). The extension of the principles discussed in this chapter under these various models constitutes the main subject area of the subsequent chapters.

5 Best Linear Unbiased Prediction of Breeding Value: Models with Random Environmental Effects

5.1 Introduction

In some circumstances, environmental factors constitute an important component of the covariance between individuals such as members of a family reared together (common environmental effects) or between the records of an individual (permanent environmental effects). Such environmental effects are usually accounted for in the model to ensure accurate prediction of breeding values. This chapter deals with models that account for these two main types of environmental effects in genetic evaluations.

5.2 Repeatability Model

The repeatability model has been employed for the analysis of data when multiple measurements on the same trait are recorded on an individual, such as litter size in successive pregnancies or milk yield in successive lactations (Interbull, 2000). The details of the assumptions and the components of the phenotypic variance have been given in Section 2.3.2. Briefly, the phenotypic variance comprises the genetic (additive and non-additive) variance, permanent environmental variance and temporary environmental variance. For an animal, the model usually assumes a genetic correlation of unity between all pairs of records, equal variance for all records and equal environmental correlation between all pairs of records. In practice, some of these assumptions do not hold in the analysis of real data. A more appropriate way of handling repeated measurements over time is by fitting a random regression model or a covariance function, and this is discussed in Chapter 10. This section has therefore been included to help illustrate the evolution of the model for the analysis of repeated records over time. The phenotypic structure for three observations of an individual under this model could be written (Quaas, 1984) as:

$$\text{var}\begin{bmatrix} y_1 \\ y_2 \\ y_{.3} \end{bmatrix} = \begin{bmatrix} \sigma_{t1}^2 + \sigma_{pe}^2 + \sigma_g^2 & \sigma_{pe}^2 + \sigma_g^2 & \sigma_{pe}^2 + \sigma_g^2 \\ \sigma_{pe}^2 + \sigma_g^2 & \sigma_{t2}^2 + \sigma_{pe}^2 + \sigma_g^2 & \sigma_{pe}^2 + \sigma_g^2 \\ \sigma_{pe}^2 + \sigma_g^2 & \sigma_{pe}^2 + \sigma_g^2 & \sigma_{t3}^2 + \sigma_{pe}^2 + \sigma_g^2 \end{bmatrix}$$

with: σ_{ti}^2 = temporary environmental variance specific to record *i*; σ_{pe}^2 = covariance due to permanent environmental effects (variances and covariances are equal); and σ_g^2 = genetic covariance (variances and covariances are equal). The correlation between records of an individual is referred to as repeatability and is $\left(\sigma_g^2 + \sigma_{pe}^2\right)/\sigma_y^2$. Genetic evaluation under this model is concerned not only with predicting breeding values but also with permanent environmental effects.

DOI: 10.1079/9781800620506.0005

5.2.1 Defining the model

The repeatability model is usually of the form:

$$\mathbf{y} = \mathbf{Xb} + \mathbf{Za} + \mathbf{Wpe} + \mathbf{e} \tag{5.1}$$

where **y** = vector of observations, **b** = vector of fixed effects, **a** = vector of random animal effects, **pe** = vector of random permanent environmental effects and non-additive genetic effects, and **e** = vector of random residual effect. **X**, **Z** and **W** are incidence matrices relating records to fixed animal and permanent environmental effects, respectively.

Note that the vector **a** only includes additive random animal effects; consequently, non-additive genetic effects are included in the **pe** term. It is assumed that the permanent environmental effects and residual effects are independently distributed with means of zero and variance σ^2_{pe} and σ^2_e, respectively. Therefore:

$$\text{var}(\mathbf{pe}) = \mathbf{I}\sigma^2_{pe}$$
$$\text{var}(\mathbf{e}) = \mathbf{I}\sigma^2_e = \mathbf{R}$$
$$\text{var}(\mathbf{a}) = \mathbf{A}\sigma^2_a$$

and

$$\text{var}(\mathbf{y}) = \mathbf{ZAZ}'\sigma^2_a + \mathbf{WI}\sigma^2_{pe}\mathbf{W}' + \mathbf{R}$$

The MME for the BLUE of estimable functions of **b** and for the BLUP of **a** and **pe** are:

$$\begin{bmatrix} \hat{\mathbf{b}} \\ \hat{\mathbf{a}} \\ \hat{\mathbf{pe}} \end{bmatrix} = \begin{bmatrix} \mathbf{X'R^{-1}X} & \mathbf{X'R^{-1}Z} & \mathbf{X'R^{-1}W} \\ \mathbf{Z'R^{-1}X} & \mathbf{Z'R^{-1}Z} + \mathbf{A}^{-1}1/\sigma^2_a & \mathbf{Z'R^{-1}W} \\ \mathbf{W'R^{-1}X} & \mathbf{W'R^{-1}Z} & \mathbf{W'R^{-1}W} + \mathbf{I}(1/\sigma^2_{pe}) \end{bmatrix}^{-1} \begin{bmatrix} \mathbf{X'R^{-1}y} \\ \mathbf{Z'R^{-1}y} \\ \mathbf{W'R^{-1}y} \end{bmatrix}$$

However, the MME with $\mathbf{R}^{-1}$ factored out from the above equations give the following equations, which are easier to set up:

$$\begin{bmatrix} \hat{\mathbf{b}} \\ \hat{\mathbf{a}} \\ \hat{\mathbf{pe}} \end{bmatrix} = \begin{bmatrix} \mathbf{X'X} & \mathbf{X'Z} & \mathbf{X'W} \\ \mathbf{Z'X} & \mathbf{Z'Z} + \mathbf{A}^{-1}\alpha_1 & \mathbf{Z'W} \\ \mathbf{W'X} & \mathbf{W'Z} & \mathbf{W'W} + \mathbf{I}\alpha_2 \end{bmatrix}^{-1} \begin{bmatrix} \mathbf{X'y} \\ \mathbf{Z'y} \\ \mathbf{W'y} \end{bmatrix} \tag{5.2}$$

where $\alpha_1 = \sigma^2_e / \sigma^2_a$ and $\alpha_2 = \sigma^2_e / \sigma^2_{pe}$

5.2.2 An illustration

Example 5.1

For illustrative purposes, assume a single dairy herd with the following data structure for five cows:

Cow	Sire	Dam	Parity	HYS	Fat yield (kg)
4	1	2	1	1	201
4	1	2	2	3	280

Continued

Continued.

Cow	Sire	Dam	Parity	HYS	Fat yield (kg)
5	3	2	1	1	150
5	3	2	2	4	200
6	1	5	1	2	160
6	1	5	2	3	190
7	3	4	1	1	180
7	3	4	2	3	250
8	1	7	1	2	285
8	1	7	2	4	300

HYS, herd–year–season.

It is assumed that $\sigma_a^2 = 20.0$, $\sigma_e^2 = 28.0$ and $\sigma_{pe}^2 = 12.0$, giving a phenotypic variance $\left(\sigma_y^2\right)$ of 60. From the given parameters, α_1 = 1.40, α_2 = 2.333 and repeatability is $\left(\sigma_a^2 + \sigma_{pe}^2\right)/\sigma_y^2 = (20+12)/60 = 0.53$. The aim is to estimate the effects of lactation number and predict breeding values for all animals and permanent environmental effects for cows with records. The above genetic parameters are proportional to estimates reported by Visscher (1991) for fat yield for Holstein Friesians in the UK for the first two lactations using a repeatability model. Later, in Section 5.4, this data set is re-analysed using a multivariate model assuming an unequal design with different herd–year–season (HYS) effects defined for each lactation using corresponding multivariate genetic parameter estimates of Visscher (1991).

Setting up the design matrices

The transpose of the matrix **X**, which relates records to HYS and parity is:

$$\mathbf{X}' = \begin{bmatrix} 1 & 0 & 1 & 0 & 0 & 0 & 1 & 0 & 0 & 0 \\ 0 & 0 & 0 & 0 & 1 & 0 & 0 & 0 & 1 & 0 \\ 0 & 1 & 0 & 0 & 0 & 1 & 0 & 1 & 0 & 0 \\ 0 & 0 & 0 & 1 & 0 & 0 & 0 & 0 & 0 & 1 \\ 1 & 0 & 1 & 0 & 1 & 0 & 1 & 0 & 1 & 0 \\ 0 & 1 & 0 & 1 & 0 & 1 & 0 & 1 & 0 & 1 \end{bmatrix}$$

The first four rows of **X′** relate records to HYS effects and the last two rows to parity effects. Considering only animals with records **Z′** and **W′** are equal and for the example data set:

$$\mathbf{Z}' = \begin{bmatrix} 1 & 1 & 0 & 0 & 0 & 0 & 0 & 0 & 0 & 0 \\ 0 & 0 & 1 & 1 & 0 & 0 & 0 & 0 & 0 & 0 \\ 0 & 0 & 0 & 0 & 1 & 1 & 0 & 0 & 0 & 0 \\ 0 & 0 & 0 & 0 & 0 & 0 & 1 & 1 & 0 & 0 \\ 0 & 0 & 0 & 0 & 0 & 0 & 0 & 0 & 1 & 1 \end{bmatrix}$$

Each row of **Z′** corresponds to each cow with records. The matrices **Z′Z** and **W′W** are both diagonal and equal and **Z′Z** is:

$$\mathbf{Z'Z} = \text{diag}(2, 2, 2, 2, 2)$$

Note, however, that it is necessary to augment $\mathbf{Z'Z}$ by three columns and rows of zeros to account for animals 1 to 3, which are ancestors. The remaining matrices in the MME apart from $\mathbf{A}^{-1}$ can easily be calculated through matrix multiplication. The inverse of the relationship matrix ($\mathbf{A}^{-1}$) is:

$$\mathbf{A}^{-1} = \begin{bmatrix} 2.50 & 0.50 & 0.00 & -1.00 & 0.50 & -1.00 & 0.50 & -1.00 \\ 0.50 & 2.00 & 0.50 & -1.00 & -1.00 & 0.00 & 0.00 & 0.00 \\ 0.00 & 0.50 & 2.50 & 0.50 & -1.00 & 0.00 & -1.00 & 0.00 \\ -1.00 & -1.00 & 0.50 & 2.50 & 0.00 & 0.00 & -1.00 & 0.00 \\ 0.50 & -1.00 & -1.00 & 0.00 & 2.50 & -1.00 & 0.00 & 0.00 \\ -1.00 & 0.00 & 0.00 & 0.00 & -1.00 & 2.00 & 0.00 & 0.00 \\ 0.50 & 0.00 & -1.00 & -1.00 & 0.00 & 0.00 & 2.50 & -1.00 \\ -1.00 & 0.00 & 0.00 & 0.00 & 0.00 & 0.00 & -1.00 & 2.00 \end{bmatrix}$$

and $\mathbf{A}^{-1}\alpha_1$ is added to the $\mathbf{Z'Z}$ to obtain the MME.

The MME are too large to be shown. There is dependency in the MME because the sum of equations for HYS 1 and 2 equals that of parity 1 and the sum of HYS 3 and 4 equals that for parity 2. The equations for HYS 1 and 3 were set to zero to obtain the following solutions by direct inversion of the coefficient matrix:

Effects	Solutions
HYS	
1	0.000
2	44.065
3	0.000
4	0.013
Parity	
1	175.472
2	241.893
Animal	
1	10.148
2	–3.084
3	–7.063
4	13.581
5	–18.207
6	–18.387
7	9.328
8	24.194
Permanent environment	
4	8.417
5	–7.146
6	–17.229
7	–1.390
8	17.347

The fixed-effect solutions for parity indicate that yield at second lactation is higher than that at first, which is consistent with the raw averages. From the MME, the solution for level *i* of the *n*th fixed effect can be calculated as:

$$\hat{b}_{in} = \sum_{f=1}^{\text{diag}_{in}} y_{inf} - \sum_j \hat{b}_{inj} - \sum_k \hat{a}_{ink} - \sum_l \hat{pe}_{inl} / \text{diag}_{in} \tag{5.3}$$

where y_{inf} is the record for animal f in level i of the nth fixed effect, diag_{in} is the number of observations for level i of the nth fixed effect, b_{inj}, $\hat{a}_{ink}$ and pe_{inl} are solutions for levels j, k and l of any other fixed effect, random animal and permanent environmental effects, respectively, within level i of the nth fixed effect. Thus, the solution for level two of HYS effect is:

$$\begin{aligned}\hat{b}_{21} &= \left[445 - \left(2\hat{b}_{12}\right) - (\hat{a}_6 + \hat{a}_8) - \left(\hat{pe}_6 + \hat{pe}_8\right)\right]/2 \\ &= \left[445 - 2(175.472) - 5.807 - (0.118)\right]/2 \\ &= 44.065\end{aligned}$$

Breeding values for animals with a repeatability model can also be calculated using Eqn 4.8, except that YD is now yield corrected for the appropriate fixed effects, permanent environmental effect and averaged. Thus, for animal 4:

$$\begin{aligned}\hat{a}_4 = {} & n_1\left[(\hat{a}_1 + \hat{a}_2)/2\right] + n_2\left[\left(\left(y_{41} - \hat{b}_1 - \hat{b}_5 - \hat{pe}_4\right) + \left(y_{42} - \hat{b}_3 - \hat{b}_6 - \hat{pe}_4\right)\right)/2\right] \\ & + n_3(2\hat{a}_7 - \hat{a}_3)\end{aligned}$$

where y_{ji} is yield for cow j in lactation i, n_1 = 2.8/5.5, n_2 = 2/5.5 and n_3 = 0.7/5.5 and 5.5 = the sum of the numerator of n_1, n_2 and n_3.

$$\begin{aligned}\hat{a}_4 &= n_1(3.532) + n_2[((201 - 0.0 - 175.472 - 8.417) \\ &\quad + (280 - 0.0 - 241.893 - 8.147))/2] + n_3(18.656 - (-7.063)) \\ &= 13.581\end{aligned}$$

The higher breeding value for sire 1 compared with sire 3 is due to the fact that, on average, the daughters of sire 1 were of higher genetic merit after adjusting for the breeding values of mates. The very high breeding value for cow 8 results from the high parent average breeding value and she has the highest yield in the herd, resulting in a large YD.

The estimate of pe for animal i could be calculated as:

$$\hat{pe}_i = \left[\left(\sum_f^{mi} Y_{ij} - \sum_j \hat{b}_{ij} - \sum_k \hat{a}_{ik}\right)\right] / (m_i + \alpha_2) \tag{5.4}$$

where m_i is the number of records for animal i $\alpha_2 = \sigma_e^2/\sigma_{pe}^2$ and other terms are as defined in Eqn 5.3. Thus, for animal 4:

$$\begin{aligned}\hat{pe}_4 &= [(201 - 0.0 - 175.472 - 13.581) \\ &\quad + (280 - 0.0 - 241.893 - 13.581)]/(2 + 2.333) \\ &= 8.417\end{aligned}$$

The estimate of permanent environment effect for an animal represents environmental influences and non-additive genetic effect, which are peculiar to the animal and affect its performance for life. These environmental influences could either be favourable – for instance, animal 8 has the highest estimates of **pe** and this is reflected by her high average yield – or could reduce performance (for example, cow 6 has a very negative estimate of

pe and low average yield). A practical example of such permanent environment effect could be the loss of a teat by a cow early in life due to infection. Thus, differences in estimates of **pe** represent permanent environmental differences between animals and could help the farmer, in addition to the breeding value, in selecting animals for future performance in the same herd. The sum of breeding value and permanent environment effect $(\hat{a}_i + pe_i)$ for animal i is termed the probable producing ability (PPA) and represents an estimate of the future performance of the animal in the same herd. If the estimate of the management level (M) for animal i is known, its future record (y_i) can be predicted as:

$$y_i = M + \text{PPA}$$

This could be used as a culling guide.

5.2.3 Calculating daughter yield deviations

As indicated in Section 4.3.3, daughter yield deviation (DYD) is commonly calculated for sires in dairy cattle evaluations. The calculation of DYD for sire 1 in Example 5.1 is hereby illustrated.

First, the yield deviations for the daughters (cows 4, 6 and 8) of sire 1 are calculated. Thus, for cow i, $YD_i = (\mathbf{Z'Z})^{-1}\mathbf{Z'}(\mathbf{y}_i - \mathbf{Xb} - \mathbf{Wpe})$. Therefore:

$$YD_4 = \tfrac{1}{2}\left[(201 - 175.472 - 0 - 8.417) + (280 - 241.893 - 0 - 8417)\right] = 23.4005$$
$$YD_6 = \tfrac{1}{2}\left[(160 - 175.472 - 44.065 - (-17.229)) + 190 - 241.893 - 0 - (-17.229)\right] = -38.486$$
$$YD_8 = \tfrac{1}{2}\left[(285 - 175.472 - 44.065 - 17.347) + (300 - 241.893 - 0.013 - 17.347)\right] = 44.432$$

Both parents of these daughters are known, therefore $n_{2prog} = 2/(2 + 2\alpha_1) = 0.4167$ and $u_{prog} = 1$ for each daughter. Using Eqn 4.12, DYD for sire 1 is:

$$\begin{aligned}\text{DYD}_1 &= \left[u_{(4)}n_{2(4)}(2YD_4 - \hat{a}_2) + u_{(5)}n_{2(6)}(2YD_6 - \hat{a}_5)\right.\\ &\quad \left. + u_{(8)}n_{2(8)}(2YD_8 - \hat{a}_7)\right] \Big/ \left(\sum_3 (u_{prog} + n_{2prog})\right)\\ \text{DYD}_1 &= \left[(1)(0.4167)(2(23.4005) - (-3.084)) + (1)0.4167(2(-38.486) - (-18.207))\right.\\ &\quad \left. + (1)0.4167(2(44.432) - 9.328)\right] / (3(1)(0.4167))\\ &= 23.552\end{aligned}$$

Calculating the proof of sire 1 using Eqn 4.13 and a DYD of 23.552 gives a breeding value of 9.058. It is slightly lower than the breeding value of 10.148 from solving the MME, as the contribution of the granddaughter through cow 4 is not included.

5.3 Model with Common Environmental Effects

Apart from the resemblance between records of an individual due to permanent environmental conditions, discussed in Section 5.2, environmental circumstances can also contribute to the resemblance between relatives. When members of a family are reared together, such as litters of pigs, they share a common environment and this contributes to the similarity between members of the family. Thus, there is an additional covariance between members of a family due to the common environment they share and this increases the variance between different families. The environmental variance may be partitioned, therefore, into the between-family or group component (σ_c^2), usually termed the common environment, which causes resemblance between members of a family, and the within-family or within-group variance (σ_e^2). Sources of common environmental variance between families may be due to factors such as nutrition and/or climatic conditions. All sorts of relatives are subject to an environmental source of resemblance, but most analyses concerned with this type of variation in animal breeding tend to account for the common environment effects associated with full-sibs or maternal half-sibs, especially in pig and chicken studies.

5.3.1 Defining the model

Genetic evaluation under this model is concerned with prediction of breeding values and common environmental effects and the phenotypic variance may be partitioned into:

1. Additive genetic effects resulting from additive genes from parents.
2. Common environmental effects affecting full-sibs or all offspring of the same dam. In the case of full-sibs, it may be confounded with dominance effects peculiar to offspring of the same parents. Further explanation is given later on the components of the common environmental effect.
3. Random environmental effects.

In matrix notation, the model, which is exactly the same as in Eqn 5.1, is:

$$\mathbf{y} = \mathbf{Xb} + \mathbf{Za} + \mathbf{Wc} + \mathbf{e}$$

where all terms are as given in Eqn 5.1 except **c**, which is the vector of common environmental effects and **W** now relates records to common environmental effects.

It is assumed that common environmental and residual effects are independently distributed with means of zero and variance σ_c^2 and σ_e^2, respectively. Thus $\text{var}(\mathbf{c}) = \mathbf{I}\sigma_c^2$, $\text{var}(\mathbf{e}) = \mathbf{I}\sigma_e^2$ and $\text{var}(\mathbf{a}) = \mathbf{A}\sigma_a^2$.

The MME for the BLUP of **a** and **c** and BLUE of estimable functions of **b** are exactly the same as Eqn 5.2 but with $\alpha_1 = \sigma_e^2/\sigma_a^2$ and $\alpha_2 = \sigma_e^2/\sigma_c^2$.

5.3.2 An illustration

Example 5.2

Consider the following data set on the weaning weight of piglets, which are progeny of three sows mated to two boars:

Piglet	Sire	Dam	Sex	Weaning weight (kg)
6	1	2	Male	90
7	1	2	Female	70
8	1	2	Female	65
9	3	4	Female	98
10	3	4	Male	106

Continued

Continued.

Piglet	Sire	Dam	Sex	Weaning weight (kg)
11	3	4	Female	60
12	3	4	Female	80
13	1	5	Male	100
14	1	5	Female	85
15	1	5	Male	68

The objective is to predict breeding values for all animals and common environmental effects for full-sibs. Given that $\sigma_a^2 = 20$, $\sigma_c^2 = 15$ and $\sigma_e^2 = 65$, then $\sigma_y^2 = 100$, $\alpha_1 = 3.25$ and $\alpha_2 = 4.333$.

The model for the analysis is that presented in Eqn 5.5 and, as mentioned earlier, the MME for the BLUP of **a** and **c** and BLUE of estimable functions of **b** are as given in Eqn 5.2, using α_1 and α_2 defined above.

Setting up the design matrices

The transpose of the matrix **X**, which relates records to sex effects in this example, is:

$$X' = \begin{bmatrix} 1 & 0 & 0 & 0 & 1 & 0 & 0 & 1 & 0 & 1 \\ 0 & 1 & 1 & 1 & 0 & 1 & 1 & 0 & 1 & 0 \end{bmatrix}$$

and **Z** = **I**, excluding parents. The transpose of matrix **W** that relates records to full-sibs is:

$$\mathbf{W}' = \begin{bmatrix} 1\,1\,1\,0\,0\,0\,0\,0\,0\,0 \\ 0\,0\,0\,1\,1\,1\,1\,0\,0\,0 \\ 0\,0\,0\,0\,0\,0\,0\,1\,1\,1 \end{bmatrix}$$

The MME can be set up as discussed in Example 5.1. The solutions to the MME by direct inversion of the coefficient matrix are:

Effects	Solutions
Sex	
Male	91.493
Female	75.764
Animals	
1	−1.441
2	−1.175
3	1.441
4	1.441
5	−0.266
6	−1.098
7	−1.667
8	−2.334
9	3.925
10	2.895
11	−1.141

Continued

Continued.

Effects	Solutions
12	1.525
13	0.448
14	0.545
15	–3.819
Common environment	
2	–1.762
4	2.161
5	–0.399

The equation for the solution of the *i* level of fixed, animal and common environmental effects under this model are the same as those given for fixed (Eqn 5.3), animal and permanent environmental effects (Eqn 5.4), respectively, in Example 5.1. The inclusion of common environmental effects in the model allows for accurate prediction of breeding values of animals. Assuming each dam reared her progeny and full-sib families were kept under similar environmental conditions, the estimates of common environmental effects indicate that dam 4 provided the best environment for her progeny compared with dams 2 and 5. Also, dam 4 has the highest breeding value among the dams and would therefore be the first dam of choice, whether selection is for dams of the next generation on the basis of breeding value only or selection is for future performance of the dams in the same herd, which will be based on some combination of breeding value and estimate of common environmental effect.

The environmental covariance among full-sibs or maternal half-sibs might be due to influences from the dam (mothering ability or maternal effect); therefore, differences in mothering ability among dams would cause environmental variance between families. For instance, resemblance among progeny of the same dam in body weight could be due to the fact that they share the same milk supply, and variation in milk yield among dams would result in differences between families in body weight. This variation in mothering ability of dams has a genetic basis and, to some degree, is due to genetic variation in some character of the dams. In Chapter 7, the genetic component of maternal effect is examined and the appropriate model that accounts for the genetic component in genetic evaluation is presented.

6 Best Linear Unbiased Prediction of Breeding Value: Multivariate Models

6.1 Introduction

Selection of livestock is usually based on a combination of several traits of economic importance that may be phenotypically and genetically related. Such traits may be combined into an index on which animals are ranked. A multiple-trait evaluation is the optimum methodology to evaluate animals on these traits because it accounts for the relationship between them. A multiple-trait analysis involves the simultaneous evaluation of animals for two or more traits and makes use of the phenotypic and genetic correlations between the traits. The first application of best linear unbiased prediction (BLUP) for multiple-trait evaluation was by Henderson and Quaas (1976).

One of the main advantages of multivariate best linear unbiased prediction (MBLUP) is that it increases the accuracy of evaluations. The gain in accuracy is dependent upon the absolute difference between the genetic and residual correlations between the traits. The larger the differences in these correlations, the greater the gain in accuracy of evaluations (Schaeffer, 1984; Thompson and Meyer, 1986). When, for instance, the heritability, genetic and environmental correlations for two traits are equal, multivariate predictions are equivalent, essentially, to those from univariate analysis for each trait. Moreover, traits with lower heritabilities benefit more when analysed with traits with higher heritabilities in a multivariate analysis. Also, there is an additional increase in accuracy with multivariate analysis resulting from better connections in the data due to residual covariance between traits (Thompson and Meyer, 1986).

In some cases, one trait is used to decide whether animals should remain in the herd and be recorded for other traits. For instance, only calves with good weaning weight may be allowed the chance to be measured for yearling weight. A single-trait analysis of yearling weight will be biased since it does not include information on weaning weight on which the selection was based. This is often called culling bias. However, a multi-trait analysis on weaning and yearling weight can eliminate this bias. Thus, MBLUP accounts for culling selection bias.

One of the disadvantages of a multiple-trait analysis is the high computing cost. The cost of multiple analysis of n traits is more than the cost of n single analyses. Second, a multiple-trait analysis requires reliable estimates of genetic and phenotypic correlations among traits and these may not be readily available.

In this chapter, MBLUP involving traits affected by the same effects (equal design matrices) and situations in which different traits are affected by different factors (non-identical design matrices) are discussed. In the next chapter, approximations of MBLUP when design matrices are equal, with or without missing records, are also examined.

DOI: 10.1079/9781800620506.0006

6.2 Equal Design Matrices and No Missing Records

Equal design matrices for all traits imply that all effects in the model affect all traits in the multivariate analysis and there are no missing records for any trait.

6.2.1 Defining the model

The model for a multivariate analysis resembles a stack of the univariate models for each of the traits. For instance, consider a multivariate analysis for two traits, with the model for each trait of the form given in Eqn 4.1, i.e. for trait 1:

$$\mathbf{y}_1 = \mathbf{X}_1\mathbf{b}_1 + \mathbf{Z}_1\mathbf{a}_1 + \mathbf{e}_1$$

and for trait 2:

$$\mathbf{y}_2 = \mathbf{X}_2\mathbf{b}_2 + \mathbf{Z}_2\mathbf{a}_2 + \mathbf{e}_2$$

If animals are ordered within traits, the model for the multivariate analysis for the two traits could be written as:

$$\begin{bmatrix} \mathbf{y}_1 \\ \mathbf{y}_2 \end{bmatrix} = \begin{bmatrix} \mathbf{X}_1 & 0 \\ 0 & \mathbf{X}_2 \end{bmatrix}\begin{bmatrix} \mathbf{b}_1 \\ \mathbf{b}_2 \end{bmatrix} + \begin{bmatrix} \mathbf{Z}_1 & 0 \\ 0 & \mathbf{Z}_2 \end{bmatrix}\begin{bmatrix} \mathbf{a}_1 \\ \mathbf{a}_2 \end{bmatrix} + \begin{bmatrix} \mathbf{e}_1 \\ \mathbf{e}_2 \end{bmatrix} \quad (6.1)$$

where $\mathbf{y}_i$ = vector of observations for the ith trait, $\mathbf{b}_i$ = vector of fixed effects for the ith trait, $\mathbf{a}_i$ = vector of random animal effects for the ith trait, ei = vector of random residual effects for the ith trait, and $\mathbf{X}_i$ and $\mathbf{Z}_i$ are incidence matrices relating records of the ith trait to fixed and random animal effects, respectively.

It is assumed that:

$$\text{var}\begin{bmatrix} a_1 \\ a_2 \\ e_1 \\ e_2 \end{bmatrix} = \begin{bmatrix} g_{11}\,\mathbf{A} & g_{12}\,\mathbf{A} & 0 & 0 \\ g_{21}\,\mathbf{A} & g_{22}\,\mathbf{A} & 0 & 0 \\ 0 & 0 & r_{11} & r_{12} \\ 0 & 0 & r_{21} & r_{22} \end{bmatrix}$$

where g_{ii} the elements of **G**, the additive genetic variance and covariance matrix for animal effect can be defined as: g_{11} = additive genetic variance for direct effects for trait 1; $g_{12} = g_{21}$ = additive genetic covariance between both traits; g_{22} = additive genetic variance for direct effects for trait 2; **A** is the relationship matrix among animals; and **R** = variance and covariance matrix for residual effects.

The MME are of the same form as in Section 4.2 and these are:

$$\begin{bmatrix} \mathbf{X}'\mathbf{R}^{-1}\mathbf{X} & \mathbf{X}'\mathbf{R}^{-1}\mathbf{Z}' \\ \mathbf{Z}'\mathbf{R}^{-1}\mathbf{X} & \mathbf{Z}'\mathbf{R}^{-1}\mathbf{Z} + \mathbf{A}^{-1}\mathbf{G}^{-1} \end{bmatrix}\begin{bmatrix} \hat{\mathbf{b}} \\ \hat{\mathbf{a}} \end{bmatrix} = \begin{bmatrix} \mathbf{X}'\mathbf{R}^{-1}\mathbf{y} \\ \mathbf{Z}'\mathbf{R}^{-1}\mathbf{y} \end{bmatrix} \quad (6.2)$$

where:

$$\mathbf{X} = \begin{bmatrix} \mathbf{X}_1 & 0 \\ 0 & \mathbf{X}_2 \end{bmatrix}, \quad \mathbf{Z} = \begin{bmatrix} \mathbf{Z}_1 & 0 \\ 0 & \mathbf{Z}_2 \end{bmatrix}, \quad \hat{\mathbf{b}} = \begin{bmatrix} \hat{\mathbf{b}}_1 \\ \hat{\mathbf{b}}_2 \end{bmatrix}, \quad \hat{\mathbf{a}} = \begin{bmatrix} \hat{\mathbf{a}}_1 \\ \hat{\mathbf{a}}_2 \end{bmatrix}, \quad \text{and} \quad \mathbf{y} = \begin{bmatrix} \mathbf{y}_1 \\ \mathbf{y}_2 \end{bmatrix}$$

Writing out the equations for each trait in the model separately, the MME become:

$$\begin{bmatrix}\hat{b}_1\\ \hat{b}_2\\ \hat{a}_1\\ \hat{a}_2\end{bmatrix} = \begin{bmatrix} X_1'R^{11}X_1 & X_1'R^{12}X_2 & X_1'R^{11}Z_1 & X_1'R^{12}Z_2 \\ X_2'R^{21}X_1 & X_2'R^{22}X_2 & X_2'R^{21}Z_1 & X_2'R^{22}Z_2 \\ Z_1'R^{11}X_1 & Z_1'R^{12}X_2 & Z_1'R^{11}Z_1 + A^{-1}g^{11} & Z_1'R^{12}Z_2 + A^{-1}g^{12} \\ Z_2'R^{21}X_1 & Z_2'R^{22}X_2 & Z_2'R^{21}Z_1 + A^{-1}g^{21} & Z_2'R^{22}Z_2 + A^{-1}g^{22}\end{bmatrix}^{-1} \begin{bmatrix} X_1'R^{11}y_1 + X_1'R^{12}y_2 \\ X_2'R^{21}y_1 + X_2'R^{22}y_2 \\ Z_1'R^{11}y_1 + Z_1'R_2^{12} \\ Z_2'R^{21}y_1 + Z_2'R_2^{22}\end{bmatrix} \quad (6.3)$$

where g^{ij} are r^{ij} elements of $\mathbf{G}^{-1}$ and $\mathbf{R}^{-1}$, respectively. It should be noted that if $\mathbf{R}^{12}$, $\mathbf{R}^{21}$, g^{12} and g^{21} were set to zero, the matrices in the equations above reduce to the usual ones computed when carrying out two single-trait analyses since the two traits become uncorrelated and there is no flow of information from one trait to the other.

6.2.2 An illustration

Example 6.1

Assume the data in Table 6.1 to be the pre-weaning gain (WWG) and post-weaning gain (PWG) for five beef calves. The objective is to estimate sex effects for both traits and to predict breeding values for all animals using a MBLUP analysis. Assume that the additive genetic covariance (G) matrix is:

$$\mathbf{G} = \begin{matrix}\text{WWG}\\ \text{PWG}\end{matrix}\begin{bmatrix}20 & 18\\ 18 & 40\end{bmatrix} \text{ and the residual covariance matrix } \mathbf{R} = \begin{matrix}\text{WWG}\\ \text{PWG}\end{matrix}\begin{bmatrix}40 & 11\\ 11 & 30\end{bmatrix}$$

The inverses of **G** and **R** are:

$$\mathbf{G}^{-1} = \begin{bmatrix}0.084 & -0.038\\ -0.038 & 0.042\end{bmatrix} \text{ and } \mathbf{R}^{-1} = \begin{bmatrix}0.028 & -0.010\\ -0.010 & 0.037\end{bmatrix}$$

Setting up the design matrices

The matrices $\mathbf{X}_1$ and $\mathbf{X}_2$ relate records for WWG and PWG, respectively, to sex effects. Both matrices are exactly the same as **X** in Section 4.3.1. Considering only animals with

Table 6.1. Pre-weaning gain (kg) and post-weaning gain (kg) for five beef calves.

Calves	Sex	Sire	Dam	WWG	PWG
4	Male	1	–	4.5	6.8
5	Female	3	2	2.9	5.0
6	Female	1	2	3.9	6.8
7	Male	4	5	3.5	6.0
8	Male	3	6	5.0	7.5

records, $\mathbf{Z}_1$ and $\mathbf{Z}_2$ relate records for WWG and PWG to animals, respectively. Both matrices are identity matrices since animals have only one record each for WWG and PWG. The matrix $\mathbf{y}$ is a vector of observations for WWG ($\mathbf{y}_1$) and PWG ($\mathbf{y}_2$). Thus, its transpose is:

$$\mathbf{y}' = \begin{bmatrix}\mathbf{y}_1' & \mathbf{y}_2'\end{bmatrix} = \begin{bmatrix}4.5 & 2.9 & 3.9 & 3.5 & 5.0 & 6.8 & 5.0 & 6.8 & 6.0 & 7.5\end{bmatrix}$$

The other matrices in the MME can then easily be calculated from the design matrices and vector of observations through matrix multiplication. Examples of some blocks of equations are given below.

From Eqns 6.2 and 6.3, the fixed effects by fixed effects block of equations for both traits in the coefficient matrix of the MME is:

$$\mathbf{X'R}^{-1}\mathbf{y} = \begin{bmatrix}\mathbf{X}_1'\mathbf{R}^{11}\mathbf{y}_1 + \mathbf{X}_2'\mathbf{R}^{12}\mathbf{y}_2 \\ \mathbf{X}_2'\mathbf{R}^{21}\mathbf{y}_1 + \mathbf{X}_2'\mathbf{R}^{22}\mathbf{y}_2\end{bmatrix} = \begin{bmatrix}0.084 & 0.000 & -0.03 & 0.00 \\ 0.00 & 0.056 & 0.00 & -0.02 \\ -0.03 & 0.00 & 0.101 & 0.00 \\ 0.00 & -0.02 & 0.00 & 0.074\end{bmatrix}$$

The right-hand side for the levels of sex effects for both traits is:

$$\mathbf{X'R}^{-1}\mathbf{y} = \begin{bmatrix}\mathbf{X}_1'\mathbf{R}^{11}\mathbf{y}_1 + \mathbf{X}_2'\mathbf{R}^{12}\mathbf{y}_2 \\ \mathbf{X}_2'\mathbf{R}^{21}\mathbf{y}_1 + \mathbf{X}_2'\mathbf{R}^{22}\mathbf{y}_2\end{bmatrix} = \begin{bmatrix}0.364 + (-0.203) \\ 0.190 + (-0.118) \\ -0.130 + 0.751 \\ -0.068 + 0.437\end{bmatrix}$$

The inverse of the relationship matrix for the example data is the same as that given in Example 4.1. The matrices $\mathbf{A}^{-1}g^{11}$, $\mathbf{A}^{-1}g^{12}$ and $\mathbf{A}^{-1}g^{22}$ are added to $\mathbf{Z'}_1\mathbf{R}^{11}\mathbf{Z}_1$, $\mathbf{Z'}_1\mathbf{R}^{12}\mathbf{Z}_2$ and $\mathbf{Z'}_2\mathbf{R}^{22}\mathbf{Z}_2$, respectively, to obtain the MME. For example, the matrix $\mathbf{Z'}_1\mathbf{R}^{12}\mathbf{Z}_2 + \mathbf{A}^{-1}g^{12}$ is:

$$\mathbf{Z}_1'\mathbf{R}^{12}\mathbf{Z}_2 + \mathbf{A}^{-1}g^{12} = \begin{bmatrix}-0.069 & -0.019 & 0.000 & 0.025 & 0.000 & 0.038 & 0.000 & 0.000 \\ -0.019 & -0.076 & -0.019 & 0.000 & 0.038 & 0.038 & 0.000 & 0.000 \\ 0.000 & -0.019 & -0.076 & 0.000 & 0.038 & -0.019 & 0.000 & 0.038 \\ 0.025 & 0.000 & 0.000 & -0.080 & -0.019 & 0.000 & 0.038 & 0.000 \\ 0.000 & 0.038 & 0.038 & -0.019 & -0.105 & 0.000 & 0.038 & 0.000 \\ 0.038 & 0.038 & -0.019 & 0.000 & 0.000 & -0.105 & 0.000 & 0.038 \\ 0.000 & 0.000 & 0.000 & 0.038 & 0.038 & 0.000 & -0.086 & 0.000 \\ 0.000 & 0.000 & 0.038 & 0.000 & 0.000 & 0.038 & 0.000 & -0.086\end{bmatrix}$$

The MME have not been presented because they are too large, but solving the MME by direct inversion of the coefficient matrix gives the solutions shown below. See also the solutions from a univariate analysis of each trait.

	Multivariate analysis traits		Univariate analysis traits	
Effects	WWG	PWG	WWG	PWG
Sex				
Male	4.361	6.800	4.358	6.798
Female	3.397	5.880	3.404	5.879

Continued

Continued.

Effects	Multivariate analysis traits		Univariate analysis traits	
	WWG	PWG	WWG	PWG
Animals				
1	0.151	0.280	0.098	0.277
2	−0.015	−0.008	−0.019	−0.005
3	−0.078	−0.170	−0.041	−0.171
4	−0.010	−0.013	−0.009	−0.013
5	−0.270	−0.478	−0.186	−0.471
6	0.276	0.517	0.177	0.514
7	−0.316	−0.479	−0.249	−0.464
8	0.244	0.392	0.183	0.384

The differences between the solutions for males and females for WWG and PWG in the multivariate analysis are more or less the same as those obtained in the univariate analyses of both traits. The solutions for fixed effects in the multivariate analysis from the MME can be calculated as:

$$\begin{bmatrix}\hat{b}_{1j}\\ \hat{b}_{2j}\end{bmatrix}=\begin{bmatrix}n_j r^{11} & n_j r^{12}\\ n_j r^{21} & n_j r^{22}\end{bmatrix}^{-1}\left[\mathbf{R}^{-1}\begin{bmatrix}y_{1j}-\hat{a}_{1j}-g^{12}\hat{a}_{2j}\\ y_{2j}-g^{21}\hat{a}_{1j}-\hat{a}_{2j}\end{bmatrix}\right] \tag{6.4}$$

where y_{ij} and $\hat{a}_{ij}$ are the sums of observations and EBVs, respectively, for calves for trait i in sex subclass j, $\hat{b}_{ij}$ is the solution for trait i in sex subclass j and n_j is the number of observations for sex subclass j. Using the above equation, the solutions for sex effects for males for WWG and PWG are:

$$\begin{bmatrix}\hat{b}_{11}\\ \hat{b}_{21}\end{bmatrix}=\begin{bmatrix}3r^{11} & 3r^{12}\\ 3r^{21} & 3r^{22}\end{bmatrix}^{-1}\left[\begin{bmatrix}r^{11} & r^{12}\\ r^{21} & r^{22}\end{bmatrix}\begin{bmatrix}13.0-(0.082)-g^{12}(-0.10)\\ 20.3-g^{21}(-0.082)-(-0.10)\end{bmatrix}\right]=\begin{bmatrix}4.361\\ 6.800\end{bmatrix}$$

6.2.3 Partitioning animal evaluations from multivariate analysis

An equation similar to Eqn 4.8 for the partitioning of evaluations from multivariate model was presented by Mrode and Swanson (2004) in the context of a random regression model (see Chapter 10). Since the yield records of animals contribute to the breeding values through the vector of yield deviations (**YD**), equations for calculating **YD** are initially presented. From Eqn 6.1, the equations for the breeding values of animals are:

$$\left(\mathbf{Z'R^{-1}Z}+\mathbf{A^{-1}G^{-1}}\right)\hat{\mathbf{a}}=\mathbf{Z'R^{-1}}\left(\mathbf{y}-\mathbf{X}\hat{\mathbf{b}}\right)$$

Therefore:

$$\left(\mathbf{Z'R^{-1}Z}+\mathbf{A^{-1}G^{-1}}\right)\hat{\mathbf{a}}=\left(\mathbf{Z'R^{-1}Z}\right)\mathbf{YD} \tag{6.5}$$

with:

$$\mathbf{YD}=\left(\mathbf{Z'R^{-1}Z}\right)^{-1}\left(\mathbf{Z'R^{-1}}\left(\mathbf{y}-\mathbf{X}\hat{\mathbf{b}}\right)\right) \tag{6.6}$$

Just as in the univariate model, **YD** is a vector of the weighted average of a cow's yield records corrected for all fixed effects in the model.

Transferring the left non-diagonal terms of $\mathbf{A}^{-1}$ in Eqn 6.5 to the right side of the equation (VanRaden and Wiggans, 1991) gives:

$$\left(\mathbf{Z'R}^{-1}\mathbf{Z}+\mathbf{G}^{-1}\alpha_{anim}\right)\hat{\mathbf{a}}_{anim} = \mathbf{G}^{-1}\alpha_{par}\left(\hat{\mathbf{a}}_{sire}+\hat{\mathbf{a}}_{dam}\right)+\left(\mathbf{Z'R}^{-1}\mathbf{Z}\right)\mathbf{YD}$$
$$+\mathbf{G}^{-1}\sum\alpha_{prog}\left(\mathbf{a}_{prog}-0.5\hat{\mathbf{a}}_{mate}\right)$$

where $\alpha_{par} = 1$, $\frac{2}{3}$ or $\frac{1}{2}$ if both, one or neither parents are known, respectively, and $\alpha_{prog} = 1$ if the animal's mate is known and $\frac{2}{3}$ if unknown. Note that $\alpha_{anim} = 2\,\alpha_{par} + 0.5\,\alpha_{prog}$.

The above equation can be expressed as:

$$\left(\mathbf{Z'R}^{-1}\mathbf{Z}+\mathbf{G}^{-1}\alpha_{anim}\right)\hat{\mathbf{a}}_{anim} = 2\mathbf{G}^{-1}\alpha_{par}\left(\mathbf{PA}\right)+\left(\mathbf{Z'R}^{-1}\mathbf{Z}\right)\mathbf{YD}$$
$$+\,0.5\mathbf{G}^{-1}\sum\alpha_{prog}\left(2\hat{\mathbf{a}}_{prog}-\hat{\mathbf{a}}_{mate}\right) \qquad (6.7)$$

where **PA** = parent average.

Pre-multiplying both sides of the equation by $(\mathbf{Z'R}^{-1}\mathbf{Z} + \mathbf{G}^{-1}\,\alpha_{anim})^{-1}$ gives:

$$\hat{\mathbf{a}}_{anim} = \mathbf{W}_1\mathbf{PA}+\mathbf{W}_2\mathbf{YD}+\mathbf{W}_3\mathbf{PC} \qquad (6.8)$$

with:

$$\mathbf{PC} = \sum\alpha_{prog}\left(2\hat{\mathbf{a}}_{prog}-\hat{\mathbf{a}}_{mate}\right)/\sum\alpha_{prog}$$

The weights $\mathbf{W}_1$, $\mathbf{W}_2$ and $\mathbf{W}_3 = \mathbf{I}$, with $\mathbf{W}_1 = (\mathbf{DIAG})^{-1}2\mathbf{G}^{-1}\alpha_{par}$, $\mathbf{W}_2 = (\mathbf{DIAG})^{-1}(\mathbf{Z'R}^{-1}\mathbf{Z})$ and $\mathbf{W}_3 = (\mathbf{DIAG})^{-1}0.5\mathbf{G}^{-1}\Sigma\alpha_{prog}$, where $(\mathbf{DIAG}) = (\mathbf{Z'R}^{-1}\mathbf{Z} + \mathbf{G}^{-1}\boldsymbol{\alpha}_{anim})$. Equation 6.8 is similar to Eqn 4.8 but the weights are matrices of the order of traits in the multivariate analysis. Equation 6.8 is illustrated below using calf 8 in Example 6.1.

Since $\mathbf{Z} = \mathbf{I}$ for calf 8, then Eqn 6.6 becomes $\mathbf{YD} = \mathbf{RR}^{-1}(\mathbf{y} - \mathbf{Xb}) = \mathbf{y} - \mathbf{Xb}$. Thus:

$$\begin{pmatrix} YD_{81} \\ YD_{82} \end{pmatrix} = \begin{pmatrix} y_{81}-\hat{b}_1 \\ y_{82}-\hat{b}_2 \end{pmatrix} = \begin{pmatrix} 5.0-4.361 \\ 7.5-6.800 \end{pmatrix} = \begin{pmatrix} 0.639 \\ 0.700 \end{pmatrix}$$

Both parents of calf 8 are known, therefore:

$$\mathbf{DIAG}_8 = \mathbf{R}^{-1}+2\mathbf{G}^{-1} = \begin{pmatrix} 0.1958 & -0.0858 \\ -0.0858 & 0.1211 \end{pmatrix}$$

and

$$\mathbf{W}_1 = \left(\mathbf{DIAG}\right)^{-1}2\mathbf{G}^{-1} = \begin{pmatrix} 0.8476 & -0.1191 \\ -0.0237 & 0.6092 \end{pmatrix} \quad \text{and}$$

$$\mathbf{W}_2 = \mathbf{I}-\mathbf{W}_1 = \begin{pmatrix} 0.1524 & 0.1191 \\ 0.0237 & 0.3908 \end{pmatrix}$$

Then, from Eqn 6.8:

$$\begin{pmatrix} \hat{a}_{81} \\ \hat{a}_{82} \end{pmatrix} = \mathbf{W}_1\begin{pmatrix} PA_{81} \\ PA_{82} \end{pmatrix}+\mathbf{W}_2\begin{pmatrix} YD_{81} \\ YD_{82} \end{pmatrix} = \mathbf{W}_1\begin{pmatrix} 0.099 \\ 0.1735 \end{pmatrix}+\mathbf{W}_2\begin{pmatrix} 0.639 \\ 0.700 \end{pmatrix}$$
$$= \begin{pmatrix} 0.06325 \\ 0.10335 \end{pmatrix}+\begin{pmatrix} 0.18075 \\ 0.28870 \end{pmatrix} = \begin{pmatrix} 0.244 \\ 0.392 \end{pmatrix}$$

In both traits, the contributions from **PA** accounted for about 26% of the breeding value of the calf.

In general, the estimates of breeding value for PWG from the multivariate analysis above are similar to those from the univariate analysis. The maximum difference between the multivariate and univariate breeding values is 0.008 kg (calf 8). The similarity of the evaluations for PWG from both models is due to the fact that genetic regression of WWG on PWG (0.45) is almost equal to the phenotypic regression (0.41) (Thompson and Meyer, 1986). However, the breeding values for WWG from the multivariate analysis are higher than those from the univariate analysis, with a maximum difference of 0.10 kg (calf 8) in favour of the multivariate analysis. Thus, much of the gain from the multivariate analysis is in WWG and this is due to its lower heritability, as mentioned earlier. However, there was only a slight re-ranking of animals for both traits in the multivariate analysis.

6.2.4 Accuracy of multivariate evaluations

One of the main advantages of MBLUP is the increase in the accuracy of evaluations. Presented below are estimates of reliabilities for the proofs for WWG and PWG from the multivariate analysis and the univariate analysis of each trait.

	Multivariate analysis					
	Diagonals[a]		Reliability		Univariate analysis reliability	
Animal	WWG	PWG	WWG	PWG	WWG	PWG
1	18.606	35.904	0.070	0.102	0.058	0.102
2	19.596	38.768	0.020	0.031	0.016	0.031
3	17.893	33.799	0.105	0.155	0.088	0.155
4	16.506	29.727	0.175	0.257	0.144	0.256
5	16.541	29.865	0.173	0.253	0.144	0.253
6	17.152	31.504	0.142	0.212	0.116	0.212
7	17.115	31.364	0.144	0.216	0.116	0.216
8	16.285	29.160	0.186	0.271	0.156	0.270

[a]Diagonal elements of the inverse of the coefficient matrix from multivariate analysis.

The reliability for the proof of animal i and trait $j\left(r_{ij}^2\right)$ in the multivariate analysis was calculated as $r_{ij}^2 = \left(g_{jj} - PEV_{ij}\right) / g_{jj}$, where PEV_{ij} is the diagonal element of the coefficient matrix pertaining to animal i and trait j. This formula is obtained by rearranging the equation given for reliability in Section 4.3.3. For instance, the reliabilities for the proofs for WWG and PWG for animal 1, respectively, are:

$$r_{11}^2 = (20 - 18.606) / 20 = 0.070$$

and

$$r_{21}^2 = (40 - 35.904) / 40 = 0.102$$

Similar to the estimates of breeding values, the reliabilities for animals for PWG from the multivariate analysis were essentially the same from the univariate analysis as $G_{ij} = r_p G_{jj}$ (Thompson and Meyer, 1986), where the jth trait is PWG and r_p is the phenotypic correlation. However, there was an increase of about 20% in reliability for WWG for each animal under the multivariate analysis compared with the univariate analysis. Again, much of the gain in accuracy from the multivariate analysis is observed in WWG.

6.2.5 Calculating daughter yield deviations in multivariate models

The equations for calculating daughter yield deviations (DYDs) with a multivariate model are similar to Eqn 4.12 for the univariate model except that the weights are matrices of order equal to the order of traits. The equations can briefly be derived (Mrode and Swanson, 2004) as follows.

Given the daughter (prog) of a bull, with no progeny of her own, Eqn 6.8 becomes:

$$\hat{\mathbf{a}}_{prog} = \mathbf{W}_{1\text{prog}}\mathbf{PA} + \mathbf{W}_{2\text{prog}}(\mathbf{YD}) \tag{6.9}$$

Let **PC** be expressed as in Eqn 6.7:

$$\mathbf{PC} = 0.5\mathbf{G}^{-1}\Sigma\alpha_{prog}\left(2\hat{\mathbf{a}}_{prog} - \hat{\mathbf{a}}_{mate}\right) \tag{6.10}$$

Substituting Eqn 6.9 into Eqn 6.10 gives:

$$\mathbf{PC} = 0.5\mathbf{G}^{-1}\Sigma\alpha_{prog}\left(\mathbf{W}_{1_{prog}}\hat{\mathbf{a}}_{anim} + \mathbf{W}_{1_{prog}}\hat{\mathbf{a}}_{mate} + \mathbf{W}_{2_{prog}}2\mathbf{YD} - \hat{\mathbf{a}}_{mate}\right)$$

Since the daughter has no offspring of her own, $\mathbf{W}_3 = 0$, therefore $\mathbf{W}_{1_{\text{prog}}} = \mathbf{I} - \mathbf{W}_{2_{\text{prog}}}$. Then:

$$\mathbf{PC} = 0.5\mathbf{G}^{-1}\Sigma\alpha_{prog}\left(\left(I - \mathbf{W}_{2_{prog}}\right)\hat{\mathbf{a}}_{anim} + \mathbf{W}_{2_{prog}}\left(2\mathbf{YD} - \hat{\mathbf{a}}_{mate}\right)\right) \tag{6.11}$$

Substituting Eqn 6.11 into Eq 6.7 and moving all terms involving $\hat{\mathbf{a}}_{anim}$ to the left-hand side gives:

$$\left(\mathbf{Z'R^{-1}Z} + 2\mathbf{G}^{-1}\alpha_{par} + 0.5\mathbf{G}^{-1}\Sigma\mathbf{W}_{2_{prog}}\alpha_{prog}\right)\hat{\mathbf{a}}_{anim}$$
$$= 2\mathbf{G}^{-1}\alpha_{par}\mathbf{PA} + \left(\mathbf{Z'R^{-1}Z}\right)\mathbf{YD} + 0.5\mathbf{G}^{-1}\Sigma\mathbf{W}_{2_{prog}}\alpha_{prog}\left(2\mathbf{YD} - \hat{\mathbf{a}}_{mate}\right)$$

Pre-multiplying both sides of the equation by the inverse coefficient matrix gives:

$$\hat{\mathbf{a}}_{anim} = \mathbf{M}_1(\mathbf{PA}) + \mathbf{M}_2(\mathbf{YD}) + \mathbf{M}_3(\mathbf{DYD}) \tag{6.12}$$

where:

$$\mathbf{DYD} = \Sigma\mathbf{W}_{2_{prog}}\alpha_{prog}\left(2\mathbf{YD} - \hat{\mathbf{u}}_{mate}\right) / \Sigma\mathbf{W}_{2_{\text{prog}}}\alpha_{prog} \tag{6.13}$$

and $\mathbf{M}_1 + \mathbf{M}_2 + \mathbf{M}_3 = \mathbf{I}$, with $\mathbf{M}_1 = (\mathbf{DIAG})^{-1}2\mathbf{G}^{-1}\alpha_{par}$, $\mathbf{M}_2 = (\mathbf{DIAG})^{-1}(\mathbf{Z'R^{-1}Z})$ and $\mathbf{M}_3 = (\mathbf{DIAG})^{-1}0.5\mathbf{G}^{-1}\Sigma\mathbf{W}_{2\text{prog}}\alpha_{prog}$ where $(\mathbf{DIAG}) = (\mathbf{Z'R^{-1}Z} + 2\mathbf{G}^{-1}\alpha_{par} + 0.5\mathbf{G}^{-1}\Sigma\mathbf{W}_{2\text{prog}}\alpha_{prog})$. The matrix $\mathbf{W}_{2\text{prog}}$ in the equation for DYD is not symmetrical and is of the order of traits and the full matrix has to be stored. This could make the computation of DYD cumbersome, especially with a large multivariate analysis or when a random regression model is implemented (see Chapter 10). For instance, in the Canadian test day model, which involves analysing milk, fat and protein yields and somatic cell count (SCC) in the first three lactations, it is a matrix of order 36 (Jamrozik *et al.*, 1997). Thus, for computational ease, pre-multiply $\mathbf{W}_{2prog}$ with $\mathbf{G}^{-1}$ and the equation for **DYD** becomes:

$$\text{DYD} = \sum \mathbf{G}^{-1}\mathbf{W}_{2prog}\boldsymbol{\alpha}_{prog}\left(2\text{YD} - \hat{\mathbf{u}}_{mate}\right) / \mathbf{G}^{-1}\mathbf{W}_{2prog}\boldsymbol{\alpha}_{prog}$$

The product of $\mathbf{G}^{-1}\mathbf{W}_{2prog}$ is symmetric and only upper or lower triangular elements need to be stored. The computation of DYD is illustrated in Section 6.4.2, using the example dairy data.

6.3 Equal Design Matrices with Missing Records

When all traits in a multivariate analysis are not observed in all animals, the same methodology described in Section 6.2 can also be employed to evaluate animals, except that different residual covariance matrices have to be set up corresponding to a different combination of traits present. If the loss of traits is sequential, i.e. the presence of the ith record implies the presence of 1 to $(i - 1)$ records, then the number of residual covariance matrices is equal to the number of traits. In general, if there are n traits, there are $(2^n - 1)$ possible combinations of observed traits and therefore residual covariance matrices (Quaas, 1984).

6.3.1 An illustration

Example 6.2
For illustrative purposes, consider the data set below, obtained by modifying the data in Table 6.1.

Calf	Sex	Sire	Dam	WWG (kg)	PWG (kg)
4	Male	1	–	4.5	–
5	Female	3	2	2.9	5.0
6	Female	1	2	3.9	6.8
7	Male	4	5	3.5	6.0
8	Male	3	6	5.0	7.5
9	Female	7	–	4.0	–

The model for the analysis is the same as in Section 6.2.1 and the same genetic parameters applied in Example 6.1 are assumed. The loss of records is sequential; there are therefore two residual covariance matrices. For animals with missing records for PWG, the residual covariance matrix ($\mathbf{R}_m$) and its inverse $\left(\mathbf{R}_m^{-1}\right)$ are $\mathbf{R}_m = r_{m11} = 40$ and $\mathbf{R}_m^{-1} = r_m^{11} = \frac{1}{40} = 0.025$. For animals with records for both WWG and PWG, the residual covariance matrix ($\mathbf{R}_o$) and its inverse $\left(\mathbf{R}_o^{-1}\right)$ are:

$$\mathbf{R}_o = \begin{bmatrix} 40 & 11 \\ 11 & 30 \end{bmatrix} \quad \text{and} \quad \mathbf{R}_o^{-1} = \begin{bmatrix} 0.028 & -0.010 \\ -0.010 & 0.037 \end{bmatrix}$$

Setting up the design matrices

The $\mathbf{X}'_1$ and $\mathbf{X}'_2$ matrices, which relate sex effects for WWG and PWG, respectively, are:

$$\mathbf{X}_1' = \begin{bmatrix} 1 & 0 & 0 & 1 & 1 & 0 \\ 0 & 1 & 1 & 0 & 0 & 1 \end{bmatrix} \quad \text{and} \quad \mathbf{X}_2' = \begin{bmatrix} 0 & 0 & 0 & 1 & 1 & 0 \\ 0 & 1 & 1 & 0 & 0 & 0 \end{bmatrix}$$

$$\mathbf{X}_1'\mathbf{X}_1 = \begin{bmatrix} 3 & 0 \\ 0 & 3 \end{bmatrix} \quad \text{and} \quad \mathbf{X}_2'\mathbf{X}_2 = \begin{bmatrix} 2 & 0 \\ 0 & 2 \end{bmatrix}$$

In setting up $\mathbf{X'}_1\mathbf{R}^{11}\mathbf{X}_1$ it is necessary to account for the fact that animals (one male and one female) have missing records for PWD. Thus:

$$\mathbf{X}_1'\mathbf{R}^{11}\mathbf{X}_1 = r_m^{11}\mathbf{W'W} + r_o^{11}\mathbf{B'B} = 0.025\begin{bmatrix} 1 & 0 \\ 0 & 1 \end{bmatrix} + 0.028\begin{bmatrix} 2 & 0 \\ 0 & 2 \end{bmatrix} = \begin{bmatrix} 0.081 & 0.000 \\ 0.000 & 0.081 \end{bmatrix}$$

where the matrix $\mathbf{W}$ relates WWG records for animals 4 and 9 with missing records for PWD to sex effects and $\mathbf{B}$ relates WWG records for calves 5, 6, 7 and 8 to sex effects. The matrices $\mathbf{W}'$ and $\mathbf{B}'$ are:

$$\mathbf{W}' = \begin{bmatrix} 1 & 0 \\ 0 & 1 \end{bmatrix} \quad \text{and} \quad \mathbf{B}' = \begin{bmatrix} 0 & 0 & 1 & 1 \\ 1 & 1 & 0 & 0 \end{bmatrix}$$

However, all animals recorded for PWG also had records for WWG, therefore:

$$\mathbf{X}_1'\mathbf{R}^{22}\mathbf{X}_2 = r_o^{22}\mathbf{X}_2'\mathbf{X}_2 = 0.037\begin{bmatrix} 2 & 0 \\ 0 & 2 \end{bmatrix} = \begin{bmatrix} 0.074 & 0 \\ 0 & 0.074 \end{bmatrix}$$

and

$$\mathbf{X}_1'\mathbf{R}^{12}\mathbf{X}_2 = r_o^{12}\mathbf{X}_1'\mathbf{X}_2 = -0.010\begin{bmatrix} 2 & 0 \\ 0 & 2 \end{bmatrix} = \begin{bmatrix} -0.02 & 0.00 \\ 0.00 & -0.02 \end{bmatrix}$$

Excluding ancestors, the matrix $\mathbf{Z}_1$ is an identity matrix because every animal has a record for WWG. Therefore, $\mathbf{Z}_1'\mathbf{Z}_1 = \mathbf{I}$ and:

$$\mathbf{Z}_1'\mathbf{R}^{11}\mathbf{Z}_1 = \text{diag}(0.025, 0.028, 0.028, 0.028, 0.028, 0.025)$$

However:

$$\mathbf{Z}_2 = \text{diag}(0, 1, 1, 1, 1, 0)$$

indicating that calves 4 and 9 have no records for PWG, and:

$$\mathbf{Z}_2'\mathbf{R}^{22}\mathbf{Z}_2 = \text{diag}(0.0, 0.037, 0.037, 0.037, 0.037, 0.0)$$

To account for ancestors (animals 1 to 3), $\mathbf{Z}_1'\mathbf{R}^{11}\mathbf{Z}_1$ and $\mathbf{Z}_2'\mathbf{R}^{22}\mathbf{Z}_2$ given above augmented with three rows and columns of zeros.

The other matrices in the MME can be calculated through matrix multiplication. The matrix $\mathbf{A}^{-1}$ can be set up and $\mathbf{A}^{-1}*\mathbf{G}^{-1}$ (where * means the Kronecker product) added to the appropriate matrices, as described in Section 6.2.2, to obtain the MME. The MME are too large to be presented but solutions from solving the equations are shown below, together with solutions from the univariate analyses of WWG and PWG.

	Multivariate analysis		Univariate analysis	
Effects	WWG	PWG	WWG	PWG
Sex				
Male	4.367	6.834	4.364	6.784
Female	3.657	6.007	3.648	5.873

Continued

Continued.

Effects	Multivariate analysis		Univariate analysis	
	WWG	PWG	WWG	PWG
Animal				
1	0.130	0.266	0.077	0.273
2	−0.084	−0.075	−0.081	0.000
3	−0.098	−0.194	−0.058	−0.165
4	0.007	0.016	0.003	−0.025
5	−0.343	−0.555	−0.250	−0.463
6	0.192	0.440	0.098	0.517
7	−0.308	−0.483	−0.237	−0.460
8	0.201	0.349	0.143	0.392
9	−0.018	−0.119	0.010	−0.230

The differences for sex solutions for WWG from the multivariate and univariate analyses are very similar to those in Section 6.2 since there are no missing records in WWG. However, sex differences in the two analyses are different for PWG due to the missing records. Again, most of the benefit in terms of breeding values from the multivariate analysis was observed in WWG, as explained in Section 6.2. However, for the calves with missing records for PWG, there was a substantial change in their proofs compared with the estimates from the univariate analysis. The proofs for these calves for PWG are based on pedigree information only in the univariate analysis but include information from the records for WWG in the multivariate analysis due to the genetic and residual correlations between the two traits. Thus, the inclusion of a correlated trait in a multivariate analysis is of much benefit to animals with missing records for the other trait.

6.4 Unequal Design Matrices

Unequal design matrices for different traits arise when traits in the multivariate analysis are affected by different fixed or random effects – for instance, the multivariate analysis of yields in different lactations as different traits. Due to the fact that calving in different parities occur in different years, herd–year–season (HYS) effects associated with each lactation are different, and an appropriate model should include different HYS for yield in each parity. An example where random effects might be different for different traits is the joint analysis for weaning weight and lean per cent in beef cattle. It might be considered that random maternal effect (see Chapter 8) is only important for weaning weight and the model for the analysis will include maternal effects only for weaning weight.

6.4.1 Numerical example

Example 6.3

Using the fat yield data in Chapter 5 analysed with a repeatability model, the principles of a multivariate analysis with unequal design are illustrated below, considering yield in each parity as different traits and fitting a different HYS effect for each trait. The data with each lactational yield treated as different traits and HYS recoded for each trait are:

Cow	Sire	Dam	HYS1	HYS2	FAT1	FAT2
4	1	2	1	1	201	280
5	3	2	1	2	150	200
6	1	5	2	1	160	190
7	3	4	1	1	180	250
8	1	7	2	2	285	300

HYS1, HYS2, herd–year–season for parity 1 and 2, respectively; FAT1, FAT2, fat yield in parity 1 and 2.

The aim is to carry out a multivariate estimate of breeding values for fat yield in lactation 1 (FAT1) and 2 (FAT2) as different traits. Assume the genetic parameters are:

$$\mathbf{R} = \begin{bmatrix} 65 & 27 \\ 27 & 70 \end{bmatrix} \text{ and } \mathbf{G} = \begin{bmatrix} 35 & 28 \\ 28 & 30 \end{bmatrix} \text{ with } \mathbf{R}^{-1} = \begin{bmatrix} 0.018 & -0.007 \\ -0.007 & 0.017 \end{bmatrix}$$

$$\text{and } \mathbf{G}^{-1} = \begin{bmatrix} 0.113 & -0.105 \\ -0.105 & 0.132 \end{bmatrix}$$

The model for the analysis is the same as in Section 6.2 but the MME are different from those in Section 6.2 because HYS effects are peculiar to each trait. The MME with the equations written out separately for each trait are:

$$\begin{bmatrix} \hat{\mathbf{b}}_1 \\ \hat{\mathbf{b}}_2 \\ \hat{\mathbf{a}}_1 \\ \hat{\mathbf{a}}_2 \end{bmatrix} = \begin{bmatrix} \mathbf{X}_1'\mathbf{R}^{11}\mathbf{X}_1 & \mathbf{X}_1'\mathbf{R}^{12}\mathbf{X}_2 & \mathbf{X}_1'\mathbf{R}^{11}\mathbf{Z}_1 & \mathbf{X}_1'\mathbf{R}^{12}\mathbf{Z}_2 \\ \mathbf{X}_2'\mathbf{R}^{21}\mathbf{X}_1 & \mathbf{X}_2'\mathbf{R}^{22}\mathbf{X}_2 & \mathbf{X}_2'\mathbf{R}^{21}\mathbf{Z}_1 & \mathbf{X}_2'\mathbf{R}^{22}\mathbf{Z}_2 \\ \mathbf{Z}_1'\mathbf{R}^{11}\mathbf{X}_1 & \mathbf{Z}_1'\mathbf{R}^{12}\mathbf{X}_2 & \mathbf{Z}_1'\mathbf{R}^{11}\mathbf{Z}_1 + \mathbf{A}^{-1}g^{11} & \mathbf{Z}_1'\mathbf{R}^{12}\mathbf{Z}_2 + \mathbf{A}^{-1}g^{12} \\ \mathbf{Z}_2'\mathbf{R}^{11}\mathbf{X}_1 & \mathbf{Z}_2'\mathbf{R}^{22}\mathbf{X}_2 & \mathbf{Z}_2'\mathbf{R}^{21}\mathbf{Z}_1 + \mathbf{A}^{-1}g^{21} & \mathbf{Z}_2'\mathbf{R}^{22}\mathbf{Z}_2 + \mathbf{A}^{-1}g^{22} \end{bmatrix}^{-1}$$

$$\begin{bmatrix} \mathbf{X}_1'\mathbf{R}^{11}\mathbf{y}_1 + \mathbf{X}_1'\mathbf{R}^{12}\mathbf{y}_2 \\ \mathbf{X}_2'\mathbf{R}^{21}\mathbf{y}_1 + \mathbf{X}_2'\mathbf{R}^{22}\mathbf{y}_2 \\ \mathbf{Z}_1'\mathbf{R}^{11}\mathbf{y}_1 + \mathbf{Z}_1'\mathbf{R}^{12}\mathbf{y}_2 \\ \mathbf{Z}_2'\mathbf{R}^{21}\mathbf{y}_1 + \mathbf{Z}_2'\mathbf{R}^{22}\mathbf{y}_2 \end{bmatrix}$$

Setting up the design matrices and MME

The matrix $\mathbf{X}_1$ now relates HYS effects to FAT1 while $\mathbf{X}_2$ relates HYS effects to FAT2. The transposes of these matrices are:

$$\mathbf{X}_1' = \begin{bmatrix} 1 & 1 & 0 & 1 & 0 \\ 0 & 0 & 1 & 0 & 1 \end{bmatrix} \text{ and } \mathbf{X}_2' = \begin{bmatrix} 1 & 0 & 1 & 1 & 0 \\ 0 & 1 & 0 & 0 & 1 \end{bmatrix}$$

Matrices $\mathbf{Z}_1$ and $\mathbf{Z}_2$ are equal and they are identity matrices of order 5 by 5 considering only animals with records. The matrix $\mathbf{A}^{-1}$ has been presented in Section 5.2.2. The remaining matrices in the MME can be obtained as described in previous sections. The MME have not been presented because they are too large. The solutions to the MME are:

	Solutions			
	Multivariate analysis		Univariate analysis	
Effects	FAT1	FAT2	FAT1	FAT2
HYS				
1	175.7	243.2	175.8	237.1
2	219.6	240.6	220.4	250.0
Animal				
1	8.969	8.840	6.933	8.665
2	−2.999	−2.777	−2.59	−2.244
3	−5.970	−6.063	−4.341	−6.422
4	11.754	11.658	9.103	12.197
5	−16.253	−15.824	−12.992	−15.563
6	−17.314	−15.719	−15.197	−11.149
7	8.690	8.138	7.566	7.696
8	22.702	20.931	19.417	15.560

Similar to the results in Section 6.2.2, the largest increase in breeding value under the multivariate analysis compared with the univariate was in FAT2. This may be due to the lower heritability of FAT2 compared with FAT1, as explained earlier.

Compared with the results from the repeatability model (Section 5.2.2) on the same data with corresponding estimates of genetic parameters, the mean breeding values for FAT1 and FAT2 for animals in the multivariate analysis are similar to the breeding value estimates from the former. The ranking of animals is the same under both models. Also, the differences between solutions for corresponding levels of HYS are very similar. In general, the repeatability model on successive records of animals is very efficient compared with the multivariate model, especially when the genetic correlation among records is high. The genetic correlation used for the multivariate analysis was 0.86. Visscher (1991) reported a loss of 0–5% in efficiency in genetic gain with a repeatability model on first and second fat yield compared with the multivariate model using a selection index. Mrode and Swanson (1995) reported a rank correlation of 0.98 between breeding value estimates for milk yield in first and second lactations, from a repeatability model and multivariate analysis for bulls with 60 or more daughters. The benefit of the repeatability model compared with the multivariate is that it is less computationally demanding and fewer estimates of genetic parameters are required.

If there are missing records in addition to unequal design matrices for traits in a multivariate analysis, the analysis can be carried out using the same principles outlined in Section 6.3, defining different residual covariance matrices for each pattern of missing traits.

6.4.2 Illustrating the computation of DYD from a multivariate model

The computation of DYD from a multivariate model is illustrated using sire 1 with three daughters (cows 4, 6 and 8) in Example 6.3. As shown in Section 6.2, since each daughter has one record per each trait, $\mathbf{YD}_{ij}$ for the daughter i and trait j equals $\left(y_{ij} - x_{ij}\hat{b}\right)$. Thus:

$$\begin{pmatrix} YD_{41} \\ YD_{42} \end{pmatrix} = \begin{pmatrix} 201-175.5 \\ 280-243.2 \end{pmatrix} = \begin{pmatrix} 25.7 \\ 36.8 \end{pmatrix}; \begin{pmatrix} YD_{61} \\ YD_{62} \end{pmatrix} = \begin{pmatrix} 160-219.6 \\ 190-243.2 \end{pmatrix} = \begin{pmatrix} -59.6 \\ -53.2 \end{pmatrix}$$

and

$$\begin{pmatrix} YD_{81} \\ YD_{82} \end{pmatrix} = \begin{pmatrix} 285-219.6 \\ 300-240.6 \end{pmatrix} = \begin{pmatrix} 65.4 \\ 59.4 \end{pmatrix}$$

For all three daughters, the dams are known, therefore $\mathbf{W}_{2\text{prog}}$ in Eqn 6.13 is the same for all daughters and is:

$$\begin{aligned} \mathbf{W}_{2prog} &= \left(\mathbf{Z'R}^{-1}\mathbf{Z} + 2\mathbf{G}^{-1}\right)^{-1}\left(\mathbf{Z'R}^{-1}\mathbf{Z}\right) \\ &= \begin{pmatrix} 0.2439 & -0.2176 \\ -0.2176 & 0.2802 \end{pmatrix}^{-1} \begin{pmatrix} 0.0183 & -0.0071 \\ -0.0071 & 0.0170 \end{pmatrix} = \begin{pmatrix} 0.1713 & 0.0821 \\ 0.1078 & 0.1244 \end{pmatrix} \end{aligned}$$

The correction of the daughters' YD for the breeding values of the mates of the sire is follows:

$$\begin{pmatrix} 2YD_{41} - \hat{a}_{21} \\ 2YD_{42} - \hat{a}_{22} \end{pmatrix} = \begin{pmatrix} 51.4-(-2.999) \\ 73.6-(-2.777) \end{pmatrix} = \begin{pmatrix} 54.399 \\ 76.377 \end{pmatrix}$$

$$\begin{pmatrix} 2YD_{61} - \hat{a}_{51} \\ 2YD_{62} - \hat{a}_{52} \end{pmatrix} = \begin{pmatrix} -119.2-(-16.253) \\ -106.4-(-15.824) \end{pmatrix} = \begin{pmatrix} -102.947 \\ -90.576 \end{pmatrix}$$

$$\begin{pmatrix} 2YD_{81} - \hat{a}_{71} \\ 2YD_{82} - \hat{a}_{72} \end{pmatrix} = \begin{pmatrix} 130.8-8.690 \\ 118.8-8.138 \end{pmatrix} = \begin{pmatrix} 122.110 \\ 110.662 \end{pmatrix}$$

Since α_{prog} equals 1 for all daughters of the bull, **DYD** for sire 1, using Eqn 6.13, is:

$$\begin{aligned} \mathbf{DYD} &= \left(3\mathbf{W}_{2_{prog}}\right)^{-1}\left[\mathbf{W}_{2_{prog}}\begin{pmatrix} 53.399 \\ 76.377 \end{pmatrix} + \mathbf{W}_{2_{prog}}\begin{pmatrix} -102.947 \\ -90.576 \end{pmatrix} + \mathbf{W}_{2_{prog}}\begin{pmatrix} 122.110 \\ 110.662 \end{pmatrix}\right] \\ &= \begin{pmatrix} 24.5207 \\ 32.1543 \end{pmatrix} \end{aligned}$$

Using Eqn 6.12, the breeding value of sire 1 can be calculated as:

$$\begin{pmatrix} \hat{a}_{11} \\ \hat{a}_{12} \end{pmatrix} = \mathbf{M}_3 \begin{pmatrix} 24.5207 \\ 32.1543 \end{pmatrix} = \begin{pmatrix} 7.439 \\ 7.387 \end{pmatrix}$$

where:

$$\mathbf{M}_3 = \left(2G^{-1}0.5 + 0.5G^{-1}\sum_3 \mathbf{W}_{2_{prog}}\alpha_{prog}\right)^{-1}\left(0.5G^{-1}\sum_3 \mathbf{W}_{2_{prog}}\alpha_{prog}\right)$$

$$\mathbf{M}_3 = \begin{pmatrix} 0.1247 & -0.1110 \\ -0.1110 & 0.1432 \end{pmatrix}^{-1} \begin{pmatrix} 0.0120 & -0.0058 \\ -0.0058 & 0.0116 \end{pmatrix} = \begin{pmatrix} 0.1937 & 0.0836 \\ 0.1099 & 0.1459 \end{pmatrix}$$

The vector of breeding value calculated for sire 1 using Eqn 6.12 is slightly lower than that shown earlier in the table of results as contributions from the grand-progeny of the sire are not included.

6.5 Multivariate Models with No Environmental Covariance

In some cases, a multivariate analysis may be necessary when individual animals have records for one trait (or subset of traits) but relatives have records on a different trait (or subset of traits). For instance, in beef cattle, if selection is for dual-purpose sires, male and female calves might be reared in different environments (different feedlots) and body weight recorded in male calves and milk yield in female calves. The evaluation of the sires will be based on multivariate analysis of these two traits. A special feature of such a multivariate analysis is that there is no environmental covariance between the traits as the two traits are not observed in the same individual. In Section 6.5.1, the details of such a model are discussed and its application to example data is illustrated.

Also, when the same trait is measured on relatives in different environments such that the genetic correlation between performances in the two environments is not one, a multivariate analysis might be the optimum means to evaluate sires. For example, milk yield may be recorded on the daughters of a bull in two different environments; say, in a tropical environment and a temperate environment. Such a multivariate analysis will treat milk yield in the various environments as different traits. However, as the number of environments increases, the data might be associated with a heterogeneous fixed effects structure that might be difficult to model correctly in multivariate analysis, such that it might be useful, for practical purposes of implementation, to analyse not the original data but summaries of the data. A very good illustration of such a multivariate analysis is the multi-trait sire model used by the international bull evaluation service Interbull (Uppsala, Sweden), for the across-country evaluation of dairy sires. This multi-trait sire model, commonly referred to as MACE (multi-trait across-country evaluations), analyses de-regressed breeding values (DRB) of sires in different countries as different traits. The use of DRB could be regarded as utilizing a variable that summarizes the daughter performances of bulls in different countries. This avoids the need to model at the Interbull centre the heterogeneous fixed-effects structure, such as different herd management systems and complex national climatic conditions associated with the daughters' milk performance records in the different countries. MACE plays a very important role in the international trade of dairy cattle and in Section 6.5.2 the model for MACE is discussed and illustrated.

6.5.1 Different traits recorded on relatives

Defining the model

In this situation, with different traits recorded on relatives in different environments, the different traits are not observed on the same individual, and so there is not environmental covariance between the traits. Therefore, the residual covariance matrix **R** is diagonal. Thus, for n traits:

$$\mathbf{R} = \text{diag}\left(\sigma_{e1}^2,\ \sigma_{e2}^2,\ldots,\sigma_{en}^2\right) = \text{diag}\left(r_{11},\ r_{22},\ldots,r_{nn}\right)$$

and

$$\mathbf{R}^{-1} = \text{diag}\left(r^{11},\ r^{22},\ldots,r^{nn}\right)$$

However, var(a), where $\mathbf{a}' = [\mathbf{a}_1, \mathbf{a}_2, \ldots, \mathbf{a}_n]$, the vector of breeding values, is:

$$\text{var}(\mathbf{a}) = \mathbf{A} * \mathbf{G}$$

where * refers to the direct product, **A** is the relationship matrix and **G** the covariance matrix for additive genetic effects. Schaeffer *et al.* (1978) discussed this model in detail but from the standpoint of variance component estimation.

Assuming there are two traits, the model for the analysis is as given in Eqn 6.1 but with **R** and **G** defined as above. The MME for the BLUP of **a** and estimable functions of **b** are:

$$\begin{bmatrix} \hat{\mathbf{b}}_1 \\ \hat{\mathbf{b}}_2 \\ \hat{\mathbf{a}}_1 \\ \hat{\mathbf{a}}_2 \end{bmatrix} \begin{bmatrix} r^{11}\mathbf{X}_1'\mathbf{X}_1 & 0 & r^{11}\mathbf{X}_1'\mathbf{Z}_1 & 0 \\ 0 & r^{22}\mathbf{X}_2'\mathbf{X}_2 & 0 & r^{22}\mathbf{X}_2'\mathbf{Z}_2 \\ r^{11}\mathbf{Z}_1'\mathbf{X}_1 & 0 & r^{11}\mathbf{Z}_1'\mathbf{Z}_1 + \mathbf{A}^{-1}g^{11} & \mathbf{A}^{-1}g^{12} \\ 0 & r^{22}\mathbf{Z}_2'\mathbf{X}_2 & \mathbf{A}^{-1}g^{21} & r^{22}\mathbf{Z}_2'\mathbf{Z}_2 + \mathbf{A}^{-1}g^{22} \end{bmatrix} = \begin{bmatrix} \mathbf{X}_1'r^{11}\mathbf{y}_1 \\ \mathbf{X}_2'r^{22}\mathbf{y}_2 \\ \mathbf{Z}_1'r^{11}\mathbf{y}_1 \\ \mathbf{Z}_2'r^{22}\mathbf{y}_2 \end{bmatrix}$$

An illustration

Example 6.4

Consider the following data on the progeny of three sires born in the same herd; assuming that selection is for dual-purpose sires, such that the male and female calves are raised on different feeding regimes, with males recorded for yearling weight and females for fat yield:

Calf	Sex	Sire	Dam	HYS	Yearling weight (kg)	Fat yield (kg)
4	Female	1	Unknown	–	–	–
9	Male	1	4	1	375.0	–
10	Male	2	5	2	250.0	–
11	Male	1	6	2	300.0	–
12	Male	3	Unknown	1	450.0	–
13	Female	1	7	1	–	200.0
14	Female	3	8	2	–	160.0
15	Female	2	Unknown	3	–	150.0
16	Female	2	13	2	–	250.0
17	Female	3	15	3	–	175.0

HYS, herd–year–season.

The aim is to estimate HYS effects for both traits and predict breeding values for yearling weight and fat yield for all animals, carrying out a multivariate analysis. Note that animal 4 is just an ancestor and has no yield record for either trait. Assume that the additive genetic covariance matrix (**G**) is:

$$\mathbf{G} = \begin{bmatrix} 43 & 18 \\ 18 & 30 \end{bmatrix} \text{ and } \mathbf{R} = \text{diag}(77, 70)$$

Then $\mathbf{R}^{-1}$ = diag(1/77, 1/70) and:

$$\mathbf{G}^{-1} = \begin{bmatrix} 0.0311 & -0.0186 \\ -0.0186 & 0.0445 \end{bmatrix}$$

The MME given earlier can easily be set up using the principles discussed so far in this chapter. Solving the MME by the direct inverse of the coefficient matrix gave the following solutions:

	Solutions	
Effects	Yearling weight (kg)	Fat (kg)
HYS		
1	412.26	194.03
2	276.21	204.77
3	–	161.66
Animal		
1	−3.365	1.258
2	−1.489	3.774
3	4.237	−1.687
4	−6.940	−1.572
5	−5.012	−2.098
6	−5.012	−2.098
7	2.137	3.561
8	−4.274	−7.123
9	−12.162	−3.091
10	−8.263	−1.260
11	5.836	3.776
12	12.632	3.558
13	1.523	5.971
14	−4.292	−11.527
15	−1.870	0.011
16	4.290	11.995
17	2.684	1.663

Selection of dual-purpose sires will be based on some combination of breeding value estimates for yearling weight and fat yield. If equal weights were given to yearling weight and fat yield, sire 3 would be the best of the three sires, followed by sire 2.

6.5.2 The multi-trait across-country evaluations (MACE)

The sire model for MACE was originally proposed by Schaeffer (1994) and involved the analysis of the DYD of bulls in different countries as different traits, with the number of daughters of a bull used as a weighting factor. The genetic correlations among DYDs of bulls in different countries were incorporated. The genetic correlations accounted for genotype by environment (G × E) interactions and differences in national models for genetic evaluations among the countries. The genetic correlations among several countries used by Interbull are usually of medium to high value.

However, due to the inability of some countries to compute DYDs for bulls, the deregressed proofs (DRP) of bulls became the variable of choice (Sigurdsson and Banos, 1995) and the weighting factor became the effective daughter contributions (EDC) of bulls (Fiske and Banos, 2001). The model in matrix notation is:

$$\mathbf{y}_i = \mathbf{1}\mu_i + \mathbf{Z}_i\mathbf{Q}\mathbf{w}_i + \mathbf{Z}_i\mathbf{a}_i + \mathbf{e}_i \tag{6.14}$$

where $\mathbf{y}_i$ is the vector of DRP from country *i* for one trait such as milk yield, $\boldsymbol{\mu}_i$ is a mean effect for country *i*, which reflects the definition of the genetic base for that country, $\mathbf{w}_i$ is the vector of genetic group effects of phantom parents, $\mathbf{a}_i$ is the vector random sire proof for country *I*, and $\mathbf{e}_i$ is the vector of random mean residuals. The matrix $\mathbf{Q}_i$ relates sires to phantom groups (see Section 4.6) and $\mathbf{Z}_i$ relates DRP to sires. Given two countries, the variance–covariance matrix for **w**, **s** and **e** is:

$$\text{var}\begin{pmatrix}\mathbf{w}_1\\ \mathbf{w}_2\\ \mathbf{s}_1\\ \mathbf{s}_2\\ \mathbf{e}_1\\ \mathbf{e}_2\end{pmatrix} = \begin{pmatrix}\mathbf{A}_{pp}g_{11} & \mathbf{A}_{pp}g_{12} & \mathbf{A}_{pn}g_{11} & \mathbf{A}_{pn}g_{12} & 0 & 0\\ \mathbf{A}_{pp}g_{21} & \mathbf{A}_{pp}g_{22} & \mathbf{A}_{pn}g_{21} & \mathbf{A}_{pn}g_{22} & 0 & 0\\ \mathbf{A}_{np}g_{11} & \mathbf{A}_{np}g_{12} & \mathbf{A}_{nn}g_{11} & \mathbf{A}_{nn}g_{12} & 0 & 0\\ \mathbf{A}_{np}g_{21} & \mathbf{A}_{np}g_{22} & \mathbf{A}_{nn}g_{21} & \mathbf{A}_{nn}g_{22} & 0 & 0\\ 0 & 0 & 0 & 0 & \mathbf{D}_1\sigma^2_{e1} & 0\\ 0 & 0 & 0 & 0 & 0 & \mathbf{D}_2\sigma^2_{e2}\end{pmatrix}$$

where *n* and *p* are the number of bulls and groups, respectively, g_{ij} is the sire genetic (co)variance between countries *i* and *j*, **A** is the additive genetic relationship for all bulls and phantom parent groups based on the maternal grandsire (MGS) model (see Section 4.6), σ^2_{ei} is the residual variance for country *i*, and $\mathbf{D}_i$ is the reciprocal of the effective daughter contribution of the bull in the *i*th country.

The variable DRP, analysed in Eqn 6.14, are obtained by de-regressing the national breeding values of bulls such that they are independent of all country group effects and additive genetic relationships among bulls, their sires and paternal grandsires, which are included in the MACE analysis (Sigurdsson and Banos, 1995). DRP may therefore contain additive genetic contributions from the maternal pedigree, which are included at the national level but not in MACE. The de-regression procedure involves solving the MME associated with Eqn 6.14 for the right-hand-side details. The details of the procedure are outlined in Appendix F. The computation of the EDC of bulls used as the weighting factor for the analysis of DRP in Eqn 6.14 is dealt with in a subsequent section.

The MME for the above model, which are modified such that sire solutions have group solutions incorporated (see Section 4.6) are:

$$\begin{pmatrix}\mathbf{X}'\mathbf{R}^{-1}\mathbf{X} & \mathbf{X}'\mathbf{R}^{-1}\mathbf{Z} & 0\\ \mathbf{Z}'\mathbf{R}^{-1}\mathbf{X} & \mathbf{Z}'\mathbf{R}^{-1}\mathbf{Z}+\mathbf{A}^{-1}\otimes\mathbf{G}^{-1} & -\mathbf{A}^{-1}\mathbf{Q}\otimes\mathbf{G}^{-1}\\ 0 & -\mathbf{Q}'\mathbf{A}^{-1}\otimes\mathbf{G}^{-1} & \mathbf{Q}'\mathbf{A}^{-1}\mathbf{Q}\otimes\mathbf{G}^{-1}\end{pmatrix}\begin{pmatrix}\hat{\mathbf{c}}\\ \mathbf{Q}\hat{\mathbf{w}}+\hat{\mathbf{a}}\\ \hat{\mathbf{w}}\end{pmatrix} = \begin{pmatrix}\mathbf{X}'\mathbf{R}^{-1}\mathbf{y}\\ \mathbf{Z}'\mathbf{R}^{-1}\mathbf{y}\\ 0\end{pmatrix} \tag{6.15}$$

Genetic groups are defined for unknown sires and MGS on the basis of country of origin and year of birth of their progeny. Also, maternal granddams (MGDs) are always assumed unknown and assigned to phantom groups on the same basis.

Then $\mathbf{A}^{-1}$ can be obtained by the rules outlined in Section 3.6, which can be briefly summarized in the table below, taking into account the contribution to the groups for MGDs. Given a list of pedigrees with the *i*th line consisting of a bull, its sire or group, its MGS or group and a group for its MGD, then contributions to $\mathbf{A}^{-1}$ are as follows:

	Bull	Sire	MGS	MGD
Bull	*d*	−0.5*d*	−0.25*d*	−0.25*d*
Sire	−0.5*d*	0.25*d*	0.125*d*	0.125*d*
MGS	−0.25*d*	0.125*d*	0.0625*d*	0.0625*d*
MGD	−0.25*d*	0.125*d*	0.0626*d*	0.0625*d*

where $d = 16/(11 + m)$ and $m = 0$ if both sire and MGS are known, $m = 1$ if the sire is known but MGS is unknown, $m = 4$ if the sire is unknown and the MGS is known, and $m = 5$ if both sire and MGS are unknown.

Usually, there are dependencies among group effect equations and 1 is added to the diagonals of the phantom group effects in the inverse of the relationship matrix to overcome these dependencies. Then the group solutions sum to zero, and so the solutions for bulls are relative to the same genetic base within each country. The addition of 1 to the diagonals of the phantom groups implies that group effects are random, with expected values of zero. Since group effects represent differences in the effects of previous selection, which should not have expected values of zero, Schaeffer (1994) indicated that this approach could also be regarded as a biased estimation of the fixed effects of phantom groups. That is, a small amount of bias in the estimates of the phantom groups is accepted in exchange for the hope of getting estimates with smaller mean square errors.

Computing effective daughter contribution

The use of EDC instead of the number of daughters as a weighting factor was proposed by Fiske and Banos (2001) from a simulation study in which they demonstrated that using the numbers of daughters resulted in biased estimates of sire variances used in MACE and international reliabilities. The computation of EDC for a bull accounts for such factors as contemporary group (CG) structure for the bull's daughters, the correlation between observations on the same daughter and the reliability of the performance of the daughters' dams. Thus, the EDC provides a measure of the precision of the daughter information used to compute the DRP of the bull. The formula for the computation of EDC (Fiske and Banos, 2001), which included the performance of the dam of the daughter k of bull i is:

$$\mathrm{EDC}_i = \sum_k \frac{\lambda rel_{k(o)}}{4 - rel_{k(o)} \cdot \left(1 + rel_{dam(o)}\right)}$$

where the summation is over all the k daughters of the bull, $\lambda = (4 - h^2)/h^2$, $rel_{dam(o)}$ is the reliability of the dam's own performance, $rel_{k(o)}$ is the reliability of the animal k's own performance computed as:

$$rel_{k(o)} = \frac{n_k h^2}{1 + (n_k - 1)r}$$

with r being the reliability of the animal's records, n_k the number of lactations of the daughter k of the sire adjusted for the CG size computed as:

$$n_k = \sum_i 1 - 1 / n_{jkl}$$

where n_{jkl} is the size of the CG_j in which the daughter k of sire i made her lth lactation.

An example of MACE for two countries

Example 6.5

The data set below consists of bull breeding values (kg) and DRP for fat yield for six bulls from two countries. Two of the bulls have evaluations in both countries and, in addition, each country had two other bulls, which were the only progeny tested in that country. A MACE is implemented using the data set. Assume residual variances of 206.5 kg^2 and 148.5 kg^2 for countries 1 and 2, respectively, with corresponding sire additive genetic variances of 20.5 kg^2 and 9.5 kg^2. The sire genetic covariance between fat yield in both countries was assumed to be 12.839 kg, giving a genetic correlation of 0.92.

	Country 1			Country 2		
Sire	EDC	BV	DRP	EDC	BV	DRP
1	58	9.0	9.7229	90	13.5	14.5088
2	150	10.1	9.9717	65	7.6	7.7594
3	20	15.8	19.2651	–	–	–
4	25	–4.7	–8.5711	–	–	–
5	–	–	–	30	19.6	23.9672
6	–	–	–	55	–5.3	–9.6226

EDC = effective daughter contribution; BV = breeding value; DRP = de-regressed proof

Assume that the sires in the data set have the following pedigree structure, with unknown sires, MGS and MGD assigned to group G_i, with $i = 1, \dots 5$.

Bull	Sire	MGS	MGD
1	7	G3	G5
2	8	9	G5
3	7	2	G5
4	1	G2	G5
5	8	G3	G4
6	1	9	G4
7	G1	G2	G4
8	G1	G2	G4
9	G1	G3	G4

Computing sire breeding values

The matrix $\mathbf{G}^{-1}$ for Example 6.5 is:

$$\mathbf{G}^{-1} = \begin{pmatrix} 0.31762 & -0.42925 \\ -0.42925 & 0.68539 \end{pmatrix}$$

The inverses of the matrix of residual variances for countries 1 and 2 are:

$$\mathbf{R}_1^{-1} = \text{diag}(0.2809, 0.7264, 0.0969, 0.1211, 0, 0)$$

and

$$\mathbf{R}_2^{-1} = \text{diag}(0.6061, 0.4377, 0, 0, 0.2020, 0.3704)$$

The design matrix **X** is:

$$\mathbf{X} = \begin{pmatrix} 1 & 1 & 1 & 1 & 0 & 0 \\ 1 & 1 & 0 & 0 & 1 & 1 \end{pmatrix}$$

and

$$\mathbf{X'R}^{-1}\mathbf{X} = \begin{pmatrix} 1.2252 & 0 \\ 0 & 1.6162 \end{pmatrix}$$

The matrix **Z** is an identity matrix of order 12, considering only bulls with evaluations. The matrix $\mathbf{A}^{-1}$ is set up using the rules outlined earlier. The remaining matrices in Eqn 6.15 could be obtained through matrix multiplication and addition. The MME are of the order of 30 by 30 and have not been shown. Solutions to the MME by direct inversion gave the following results:

	Solutions			
Effects	Country 1		Country 2	
Country effect	7.268		9.036	
Animal/group	A	B	A	B
1	2.604	9.871	2.661	11.697
2	2.176	9.444	0.403	9.439
3	8.059	15.327	5.001	14.037
4	−9.865	−2.597	−5.605	3.431
5	13.634	20.902	9.728	18.764
6	−18.086	−10.818	−13.203	−4.167
7	4.310	11.578	3.071	12.106
8	7.015	14.283	4.489	13.525
9	−6.299	0.969	−5.059	3.977
G1	0.174	7.442	−0.092	8.944
G2	−0.124	7.144	0.126	9.162
G3	−0.071	7.197	0.264	9.300
G4	0.087	7.355	−0.288	8.748
G5	−0.067	7.201	−0.010	9.026

A = solutions for animals and groups from the MME; B = solutions for animals and groups expressed in each country scale

The solutions for animals and groups were expressed in each country scale by adding the solution for country effects for country *i* to the animal and group solutions of the *i*th country. As indicated earlier, the sum of the group solutions is zero. In the next section, some of the bull solutions are partitioned to contributions from various sources to gain a better understanding of MACE.

Equations for partitioning bull evaluations from MACE

The equations for sire proofs from Eqn 6.15 are:

$$\left(\mathbf{Z}'\mathbf{R}^{-1}\mathbf{Z} + \mathbf{A}^{-1} \otimes \mathbf{G}^{-1}\right)\hat{\mathbf{a}} = \left(\mathbf{A}^{-1}\mathbf{Q} \otimes \mathbf{G}^{-1}\right)\hat{\mathbf{g}} + \mathbf{Z}'\mathbf{R}^{-1}\left(\mathbf{y} - \mathbf{X}\hat{\mathbf{c}}\right) \quad (6.16)$$

where:

$$\hat{\mathbf{a}} = \mathbf{Q}\hat{\mathbf{g}} + \hat{\mathbf{s}}$$

Thus, Eqn 6.16 can be expressed as:

$$\left(\mathbf{Z}'\mathbf{R}^{-1}\mathbf{Z} + \mathbf{A}^{-1} \otimes \mathbf{G}^{-1}\right)\hat{\mathbf{a}} = \left(\mathbf{A}^{-1}\mathbf{Q} \otimes \mathbf{G}^{-1}\right)\hat{\mathbf{g}} + \mathbf{Z}'\mathbf{R}^{-1}\mathbf{Z}(\mathbf{CD}) \quad (6.17)$$

where:

$$\mathrm{CD} = \left(\mathbf{Z}'\mathbf{R}^{-1}\mathbf{Z}\right)^{-1}\left(\mathbf{Z}'\mathbf{R}^{-1}\left(\mathbf{y} - \mathbf{X}\hat{\mathbf{c}}\right)\right)$$

CD (country deviation) is simply a vector of weighted average of corrected DRP in all countries where the bull has a daughter, the weighting factor being the reciprocal of EDC multiplied by the residual variance in each country. Since $\mathbf{R}^{-1}$ is diagonal, **CD** is equal to the vector $(\mathbf{y} - \mathbf{X}\hat{\mathbf{c}})$.

For a particular bull with a direct progeny (e.g. son), Eqn 6.17 can be written as:

$$\left(\mathbf{Z}'\mathbf{R}^{-1}\mathbf{Z} + \mathbf{G}^{-1}\alpha_{bull}\right)\hat{\mathbf{a}}_{bull} = \mathbf{G}^{-1}\alpha_{par}\left(\hat{\mathbf{a}}_{sire} + 0.5\left(\hat{\mathbf{a}}_{mgs} + \hat{\mathbf{g}}\right)\right) + \mathbf{Z}'\mathbf{R}^{-1}\mathbf{Z}(\mathbf{CD}) + \mathbf{G}^{-1}\Sigma\alpha_{prog}\left(\hat{\mathbf{a}}_{prog} - 0.25\hat{\mathbf{a}}_{mate}\right) \quad (6.18)$$

where $\alpha_{par} = \frac{8}{11}, \frac{8}{15}, \frac{2}{3}$ or $\frac{1}{2}$ if both sire and MGS (maternal grandsire), only MGS, only sire or no parents are known, respectively; and $\alpha_{prog} = \frac{8}{11}$ if bull's mate (MGS of the progeny) is known or $\frac{2}{3}$ if unknown. The above values for α_{par} and α_{prog} are based on the assumption that $\mathbf{A}^{-1}$ has been calculated without accounting for inbreeding. Note that in Eqn 6.18:

$$\alpha_{bull} = 2\alpha_{par} + 0.5\alpha_{prog}$$

Equation 6.18 can be expressed as:

$$\left(\mathbf{Z}'\mathbf{R}^{-1}\mathbf{Z} + \mathbf{G}^{-1}\alpha_{bull}\right)\hat{\mathbf{a}}_{bull} = 2\mathbf{G}^{-1}\alpha_{par}(\mathbf{PA}) + \left(\mathbf{Z}'\mathbf{R}^{-1}\mathbf{Z}\right)\mathbf{CD} + 0.5\mathbf{G}^{-1}\Sigma\alpha_{prog}\left(2\hat{\mathbf{a}}_{prog} - 0.5\hat{\mathbf{a}}_{mate}\right)$$

where:

$$\mathbf{PA} = 0.5\hat{\mathbf{a}}_{sire} + 0.25\left(\hat{\mathbf{a}}_{mgs} + \hat{\mathbf{g}}\right)$$

Pre-multiplying both sides of the equation by $(\mathbf{Z}'\mathbf{R}^{-1}\mathbf{Z} + \mathbf{G}^{-1}\alpha_{bull})^{-1}$ gives:

$$\hat{\mathbf{a}}_{bull} = \mathbf{W}_1\mathbf{PA} + \mathbf{W}_2\mathbf{YD} + \mathbf{W}_3\mathbf{PC} \quad (6.19)$$

where:

$$\mathbf{PC} = \Sigma\alpha_{prog}\left(2\hat{\mathbf{a}}_{prog} - 0.5\hat{\mathbf{a}}_{mate}\right) / \Sigma\alpha_{prog} \quad \text{and} \quad \mathbf{W}_1 + \mathbf{W}_2 + \mathbf{W}_3 = \mathbf{I}.$$

The matrices $\mathbf{W}_1$, $\mathbf{W}_2$ and $\mathbf{W}_3$ are the product of $(\mathbf{Z}'\mathbf{R}^{-1}\mathbf{Z} + \mathbf{G}^{-1}\alpha_{bull})^{-1}$ and $2\mathbf{G}^{-1}\alpha_{par}$, $\mathbf{Z}'\mathbf{R}^{-1}\mathbf{Z}$, and $0.5\mathbf{G}^{-1}\Sigma\alpha_{prog}$, respectively. Using Eqn 6.19, the contributions from different sources of information from different countries to the MACE of a bull can be computed.

If the progeny in Eqn 6.19 is not a direct progeny of the bull but a maternal grandson of the bull, then α_{prog} equals $\frac{4}{11}$ if mate (sire) is known or $\frac{4}{15}$ if unknown. Then, Eqn 6.19 becomes:

$$\left(\mathbf{Z'R^{-1}Z}+\mathbf{G}^{-1}\alpha_{bull}\right)\hat{\mathbf{a}}_{bull} = \mathbf{G}^{-1}\alpha_{par}\left(\hat{\mathbf{a}}_{sire}+0.5\left(\hat{\mathbf{a}}_{mgs}+\hat{\mathbf{g}}\right)\right)+\mathbf{Z'R^{-1}Z}(\mathbf{CD})$$
$$+\mathbf{G}^{-1}\Sigma\alpha_{prog}\left(\hat{\mathbf{a}}_{prog}-0.5\hat{\mathbf{a}}_{mate}\right)$$

and α_{bull} now equals $2\alpha_{par}$ + $0.25\alpha_{prog}$ and $0.5\hat{\mathbf{a}}_{mate}$ = $0.5\hat{\mathbf{a}}_s$, the sire of the progeny. The above can be expressed as:

$$\left(\mathbf{Z'R^{-1}Z}+\mathbf{G}^{-1}\alpha_{bull}\right)\hat{\mathbf{a}}_{bull} = 2\mathbf{G}^{-1}_{\alpha_{par}}(\mathbf{PA})+\left(\mathbf{Z'R^{-1}Z}\right)\mathbf{CD}$$
$$+0.25\mathbf{G}^{-1}\Sigma\alpha_{prog}\left(4\hat{\mathbf{a}}_{prog}-2\hat{\mathbf{a}}_{mate}\right)$$

Pre-multiplying both sides by $(\mathbf{Z'R^{-1}Z} + \mathbf{G}^{-1}\alpha_{bull})^{-1}$ gives the same equation as Eqn 6.19 but with:

$$\mathbf{PC} = \Sigma\alpha_{prog}\left(4\hat{\mathbf{a}}_{prog}-2\hat{\mathbf{a}}_{mate}\right)/\Sigma\alpha_{prog} \tag{6.20}$$

and $\mathbf{W}_3$ now equals $\left(\mathbf{Z'R^{-1}Z}+\mathbf{G}^{-1}\alpha_{bull}\right)^{-1}\left(0.25\,\mathbf{G}^{-1}\Sigma\alpha_{prog}\right)$.

The use of Eqn 6.19 to partition proofs from MACE is illustrated for two bulls, one with no progeny and another with a maternal grandson. First, consider bull 3 in Example 6.5 that has DRPs only in country 1 and has no progeny. Therefore, $\mathbf{CD}_{3i}$ for bull 3 in country i is:

$$CD_{31} = y_{31}-\mu_1 = 19.2651-7.268 = 11.997 \text{ and } CD_{32} = 0.$$

Parent average for bull 3 ($\mathbf{PA}_{3i}$) in country i is:

$$PA_{31} = 0.5(\hat{a}_{71})+0.25(\hat{a}_{21}+\hat{g}_{G51}) = 0.5(4.310)+0.25(2.176+(-0.067)) = 2.68225$$

and

$$PA_{32} = 0.5(\hat{a}_{72})+0.25(\hat{a}_{22}+\hat{g}_{G52}) = 0.5(3.071)+0.25(0.403+(-0.010)) = 1.63375$$

where $\hat{a}_{ji}$ is the breeding value of animal j in country i and $\hat{g}_{Gji}$ is the solution for group j and in the ith country.

The residual variance for bull 3 in country 1 $(r_{31}) = \left(\frac{1}{20}\right)206.5 = 10.325$ and its inverse equals 0.09685. Both sire and MGS of bull 3 are known, therefore $\alpha_{bull} = \frac{16}{11}$. Then:

$$\left(\mathbf{Z'R^{-1}Z}+\mathbf{G}^{-1}\alpha_{bull}\right) = \begin{pmatrix} 0.09685 & 0 \\ 0 & 0 \end{pmatrix} + \begin{pmatrix} 0.4620 & -0.62436 \\ -0.62436 & 0.99693 \end{pmatrix}$$
$$= \begin{pmatrix} 0.55884 & -0.62436 \\ -0.62436 & 0.99693 \end{pmatrix}$$

The matrices of weights ($\mathbf{W}_i$) using Eqn 6.19 are:

$$\mathbf{W}_1 = \begin{pmatrix} 0.55884 & -0.62436 \\ -0.62436 & 0.99693 \end{pmatrix}^{-1} \begin{pmatrix} 0.4620 & -0.62436 \\ -0.62436 & 0.99693 \end{pmatrix} = \begin{pmatrix} 0.4229 & -0.3614 \\ -0.3614 & 1.0000 \end{pmatrix}$$

and

$$\mathbf{W}_2 = \begin{pmatrix} 0.55884 & -0.62436 \\ -0.62436 & 0.99693 \end{pmatrix}^{-1} \begin{pmatrix} 0.09685 & 0 \\ 0 & 0 \end{pmatrix} = \begin{pmatrix} 0.5771 & 0 \\ 0.3614 & 0 \end{pmatrix}$$

Therefore, the vector proofs of bull 3 are:

$$\begin{pmatrix} \hat{a}_{31} \\ \hat{a}_{32} \end{pmatrix} = \mathbf{W}_1 = \begin{pmatrix} 2.68225 \\ 1.63375 \end{pmatrix} + \mathbf{W}_2 = \begin{pmatrix} 11.9971 \\ 0 \end{pmatrix} = \begin{pmatrix} 1.1343 \\ 0.6644 \end{pmatrix} + \begin{pmatrix} 6.9235 \\ 4.3358 \end{pmatrix} = \begin{pmatrix} 8.058 \\ 5.000 \end{pmatrix}$$

The contribution from the DRP of bull 3 in country 1 accounts for over 85% of the MACE proof in both countries, although the bull has no DRP in country 2. Thus, with only 20 daughters, parental contribution was not very large, although, in general, parental contributions will be influenced by the heritability of the traits in both countries and the genetic correlation between them.

When a bull has a proof only in country *i* and not in *j*, its proof in country *j* can be obtained (Mrode and Swanson, 1999) as:

$$\hat{\mathbf{a}}_j = \mathbf{PA}_j - (g_{ij} / g_{ii})(\hat{\mathbf{a}}_i - \mathbf{PA}_i) \qquad (6.21)$$

where g_{ii} is the genetic variance in country *i* and g_{ij} the genetic covariance between countries *i* and *j*. Therefore, if interest was only in calculating the proof of bull 3 in country 2, it can be obtained from the above equation as:

$$\hat{a}_{32} = 1.63375 - (12.839 / 20.5)(8.059 - 2.68225) = 5.001$$

Equation 6.21 can be derived from Eqn 6.18 as follows. The equation for $\hat{a}_{32}$ from Eqn 6.18 is:

$$(g^{22}\alpha_{bull})\hat{a}_{32} = g^{22}\alpha_{par}(\hat{a}_{sire2} + 0.5(\hat{a}_{mgs2} + \hat{g}_{mgd2}) + g^{21}\alpha_{par}(\hat{a}_{sire1} + 0.5(\hat{a}_{mgs1} + \hat{g}_{mgd1}) + (g^{21}\alpha_{bull})\hat{a}_{31}$$

where $\hat{a}_{sirej}$, $\hat{a}_{mgsj}$ and $\hat{g}_{mgdj}$ are the proofs for the sire, MGS and solution for the MGD in country *j*, respectively, and g^{ii} are the inverse elements of $\mathbf{G}^{-1}$. Since $\boldsymbol{\alpha}_{bull} = 2\boldsymbol{\alpha}_{par}$ for bull 3, multiplying the above equation by $(2\alpha_{par})^{-1}$ gives:

$$g^{22}\hat{a}_{32} = g^{22}(PA_2) + g^{21}(PA_1) - g^{21}\hat{a}_{31}$$
$$g^{22}\hat{a}_{32} = g^{22}(PA_2) - g^{21}(\hat{a}_{31} - PA_1)$$
$$\hat{a}_{32} = PA_2 - g^{21} / g^{22}(\hat{a}_{31} - PA_1)$$
$$\hat{a}_{32} = PA_2 - g^{21} / g^{22}(\hat{a}_{31} - PA_1)$$

Thus, the proof of a bull in country *j* is dependent upon the parent average of the bull in country *j* and the Mendelian sampling of the bull in the *i*th country.

Partitioning the proof of bull 2 with records in both countries and a maternal grandson (bull 3) is as follows. The country deviations for bull 2 in both countries are:

$$CD_{21} = y_{21} - \mu_1 = 9.9717 - 7.268 = 2.7037$$

and

$$CD_{22} = y_{22} - \mu_2 = 7.7594 - 9.036 = -1.2766$$

Parent average for sire 2 (PA_{2i}) for country *i* is:

$$PA_{21} = 0.5(\hat{a}_{81}) + 0.25(\hat{a}_{91} + \hat{g}_{G51}) = 0.5(7.015) + 0.25(-6.299 + (-0.067)) = 1.916$$
$$PA_{22} = 0.5(\hat{a}_{82}) + 0.25(\hat{a}_{92} + \hat{g}_{G52}) = 0.5(4.489) + 0.25(-5.059 + (-0.010)) = 0.97725$$

Progeny contributions (PC) from bull 3 to sire 2 (PC_{32i}) in country i are:

$$PC_{21} = 4(\hat{a}_{31}) - 2(\hat{a}_{71}) = 4(8.059) - 2(4.310) = 23.616$$
$$PC_{22} = 4(\hat{a}_{32}) - 2(\hat{a}_{72}) = 4(5.001) - 2(3.071) = 13.862$$

The residual variance for bull 2 in country 1, $(r_{21}) = \left(\frac{1}{150}\right)206.5$ and country 2, $(r_{22}) = \left(\frac{1}{65}\right)148.5$. Corresponding inverses were 0.72639 and 0.43771, respectively. Since both sire and MGS of bull 2 are known and he has a maternal grandson, $\alpha_{bull} = 2\alpha_{par} + 0.25\alpha_{prog} = 2\left(\frac{8}{11}\right) + 0.25\left(\frac{4}{11}\right) = 1.54545$. Therefore:

$$\left(\mathbf{Z'R}^{-1}\mathbf{Z} + \mathbf{G}^{-1}\alpha_{bull}\right) = \begin{pmatrix} 0.72630 & 0 \\ 0 & 0.43771 \end{pmatrix} + \begin{pmatrix} 0.49087 & -0.66338 \\ -0.66338 & 1.05924 \end{pmatrix}$$
$$= \begin{pmatrix} 1.21726 & -0.66338 \\ -0.66338 & 1.49695 \end{pmatrix}$$

From Eqn 6.19, the matrices of weights ($\mathbf{W}_i$) are:

$$\mathbf{W}_1 = \begin{pmatrix} 1.21726 & -0.66338 \\ -0.66338 & 1.49695 \end{pmatrix}^{-1} \begin{pmatrix} 0.4620 & -0.62436 \\ -0.62436 & -0.99693 \end{pmatrix} = \begin{pmatrix} 0.2007 & -0.1977 \\ -0.3281 & 0.5783 \end{pmatrix}$$
$$\mathbf{W}_2 = \begin{pmatrix} 1.21726 & -0.66338 \\ -0.66338 & 1.49695 \end{pmatrix}^{-1} \begin{pmatrix} 0.72639 & 0 \\ 0 & 0.43771 \end{pmatrix} = \begin{pmatrix} 0.7867 & 0.2101 \\ 0.3487 & 0.3855 \end{pmatrix}$$
$$\mathbf{W}_3 = \begin{pmatrix} 1.21726 & -0.66338 \\ -0.66338 & 1.49695 \end{pmatrix}^{-1} \begin{pmatrix} 0.02887 & -0.03902 \\ -0.03902 & 0.06231 \end{pmatrix} = \begin{pmatrix} 0.0125 & -0.0124 \\ -0.02051 & 0.0361 \end{pmatrix}$$

The vector of proof for bull 2 is:

$$\begin{pmatrix} \hat{a}_{21} \\ \hat{a}_{22} \end{pmatrix} = \mathbf{W}_1 = \begin{pmatrix} 1.9160 \\ 0.9773 \end{pmatrix} + \mathbf{W}_2 = \begin{pmatrix} 2.7037 \\ -1.2766 \end{pmatrix} + \mathbf{W}_3 \begin{pmatrix} 23.616 \\ 13.862 \end{pmatrix} = \begin{pmatrix} 2.176 \\ 0.403 \end{pmatrix}$$

Again, similar to bull 3 above, the contributions from the DRPs in both countries accounted for much of the MACE proofs of the bull 2 in countries 1 and 2.

Recently, Interbull has modified the MACE systems to use sire and dam pedigree instead of sire and maternal sire pedigree. Partitioning of bull proofs can be done as in Section 6.2.3.

7 Methods to Reduce the Dimension of Multivariate Models

7.1 Introduction

One of the limitations of multivariate analysis is the large computational requirements of such high-dimensional analysis. The number of effects in multivariate analyses tends to increase linearly with the number of traits considered. In some cases, the available, or small number of, records for some of the traits can hamper the reliable estimation of a large number of covariance components simultaneously. However, developments in methodologies to model higher-dimensional data more parsimoniously (Kirkpatrick and Meyer, 2004) implies that such multivariate analysis is more feasible in terms of parameter estimation and, therefore, genetic evaluation.

Reducing the dimension of multivariate analysis includes methods such as canonical transformation and Cholesky decomposition, which involve the transformation of the vector of observations in addition to residual and genetic covariance matrices. Other approaches, such as principal component analysis and factor analysis, only involve reducing the rank of the genetic covariance matrix. Initially, methods that include the transformation of the vector of observations are discussed.

7.2 Canonical Transformation

In the example discussed in Section 6.2.2, both traits were affected by the same fixed effect and all animals were measured for both traits. Thus, the design matrices **X** and **Z** were the same for both traits; in other words, the traits are said to have equal design matrices. In addition, there was only one random effect (animal effect) for each trait apart from the residual effect. Under these circumstances, the multivariate analysis can be simplified into n (number of traits) single-trait analyses through what is called a canonical transformation (Thompson, 1977b). Canonical transformation involves using special matrices to transform the observations on several correlated traits into new variables that are uncorrelated with each other. These new variables are analysed by the usual methods for single-trait evaluation, but the results (predictions) are transformed back to the original scale of the observations. Ducrocq and Besbes (1993) have presented a methodology for applying canonical transformation when design matrices are equal for all traits but with some animals having missing traits. Details of the methodology, together with an illustration, are given in Appendix E, Section E.2.

Let **y** be vectors of observations:

$$\text{var}(\mathbf{y}) = \mathbf{G} + \mathbf{R} \tag{7.1}$$

where **G** and **R** are variance and covariance matrices for the additive genetic and residual effects, respectively. Assuming **G** and **R** are positive definite matrices, there exists a matrix Q, such that:

$$\mathbf{QRQ'} = \mathbf{I} \text{ and } \mathbf{QGQ'} = \mathbf{W}$$

 Linear Models for the Prediction of the Genetic Merit of Animals, 4th Edition (R.A. Mrode and I. Pocrnic)
DOI: 10.1079/9781800620506.0007

where **I** is an identity matrix and **W** is a diagonal matrix (Anderson, 1958). This implies that pre- and post-multiplication of **R** by the transformation matrix (**Q**) reduces it to an identity matrix and **G** to a diagonal matrix. The multiplication of **y** by **Q** yields a new vector of observations $\mathbf{y}^*$ that are uncorrelated:

$$\mathbf{y}^* = \mathbf{Q}\mathbf{y}$$

$$\text{var}(\mathbf{y}^*) = \mathbf{W} + \mathbf{I}; \text{ which is a diagonal matrix.}$$

Since there are no covariances between the transformed traits, they can be independently evaluated. The procedure for calculating the transformation matrix **Q** is given in Appendix E, Section E.1.

7.2.1 The model

A single-trait analysis is usually carried out on each of the transformed variables. The model for the *i*th transformed variable can be written as:

$$\mathbf{y}_i^* = \mathbf{X}\mathbf{b}_i^* + \mathbf{Z}\mathbf{a}_i^* + \mathbf{e}_i^* \qquad (7.2)$$

where $\mathbf{y}_i^*$ = vector of transformed variables for the *i*th transformed trait; $\mathbf{b}_i^*$ = vector of fixed effects for the *i*th transformed variable *i*; $\mathbf{a}_i^*$ = vector of random animal effects for transformed trait *i*; $\mathbf{e}_i^*$ = vector of random residual errors for the *i*th transformed trait; and **X** and **Z** are incidence matrices relating records to fixed and random effects, respectively.

The MME to be solved to obtain the BLUE of $\mathbf{a}_i^*$ and the BLUP of $\mathbf{b}_i^*$ are the same as those presented in Section 4.2 for the univariate model. These equations are:

$$\begin{bmatrix} \mathbf{X'X} & \mathbf{X'Z} \\ \mathbf{Z'X} & \mathbf{Z'Z} + \mathbf{A}^{-1}\alpha_i \end{bmatrix} \begin{bmatrix} \hat{\mathbf{b}}_i^* \\ \hat{\mathbf{a}}_i^* \end{bmatrix} = \begin{bmatrix} \mathbf{X'y}_i^* \\ \mathbf{Z'y}_i^* \end{bmatrix}$$

As explained earlier, it is assumed for the *i*th trait that:

$$\text{var}(\mathbf{a}_i^*) = \mathbf{A}w_{ii}; \ \text{var}(\mathbf{e}_i^*) = \mathbf{I} \text{ and } \text{var}(\mathbf{y}_i^*) = \mathbf{ZAZ'}w_{ii} + \mathbf{I}$$

where w_{ii} refers to the *i*th element of the diagonal matrix **W**.

The MME are solved for $\mathbf{b}_i^*$ and $\mathbf{a}_i^*$ and the transformation back to the original scale is achieved as:

$$\mathbf{b}_i = \mathbf{Q}^{-1}\mathbf{b}_i^* \qquad (7.3)$$

$$\mathbf{a}_i = \mathbf{Q}^{-1}\mathbf{a}_i^* \qquad (7.4)$$

Thus, the multivariate analysis is simplified to *i* single-trait evaluations.

7.2.2 An illustration

Example 7.1

The multivariate analysis for WWG and PWG in Section 6.2.2 is repeated below, carrying out a canonical transformation assuming the same genetic parameters.

The calculation of the transformation **Q** and the diagonal matrix **W** are given in Appendix E, Section E.1. Presented in Table 7.1 are the data for all calves in the original

Table 7.1. Weaning gain and post-weaning gain for beef calves on the original and transformed scales.

				Original scale		Transformed scale	
Calves	Sex	Sire	Dam	WWG	PWG	VAR1	VAR2
4	Male	1	–	4.5	6.8	0.208	1.269
5	Female	3	2	2.9	5.0	0.085	0.926
6	Female	1	2	3.9	6.8	0.109	1.259
7	Male	4	5	3.5	6.0	0.106	1.112
8	Male	3	6	5.0	7.5	0.236	1.400

scale and as transformed variables (VAR1 and VAR2). The observations are transformed into new uncorrelated variables using the matrix **Q**. Thus, for animal 4, the record would be transformed as:

$$\mathbf{Q}y_4 = \begin{bmatrix} 0.1659 & -0.0792 \\ 0.0168 & 0.1755 \end{bmatrix} \begin{bmatrix} 4.5 \\ 6.8 \end{bmatrix} = \begin{bmatrix} 0.208 \\ 1.269 \end{bmatrix}$$

The residual variance for each of the transformed variables is 1, thus heritability for the ith transformed variable = $w_{ii}/(1 + w_{ii})$ and $\alpha_i = 1/w_{ii}$.

Therefore $h_1^2 = 0.247$, $h_2^2 = 0.573$, $\alpha_1 = 1/0.3283 = 3.046$ and $\alpha_2 = 1/1.3436 = 0.744$. A single-trait analysis is carried out on the transformed variates for WWG and PWG using the model and the MME in Section 6.3.1 and solutions are transformed back to the original scale.

Setting up the design matrices

The matrix **X**, which relates records for either VAR1 or VAR2 to sex effects, is exactly as the matrix $\mathbf{X}_1$ in Section 6.2.2. Similarly, **Z** is the same as $\mathbf{Z}_1$ in Section 6.2.2. For animals with records, the vector of observations $\mathbf{y}_1^*$ and $\mathbf{y}_2^*$ are equal to the column of transformed variates for WWG and PWG gains, respectively, in Table 7.1. The matrices in the MME are easily obtained through matrix multiplication and the addition to the animal equations of $\mathbf{A}^{-1}\alpha_1$ for VAR1 and $\mathbf{A}^{-1}\alpha_2$ for VAR2. $\mathbf{A}^{-1}$ has been given earlier in Section 6.2.2. For instance, the MME for VAR1 only are:

$$\begin{bmatrix} \hat{b}_1^* \\ \hat{b}_2^* \\ \hat{a}_1^* \\ \hat{a}_2^* \\ \hat{a}_3^* \\ \hat{a}_4^* \\ \hat{a}_5^* \\ \hat{a}_6^* \\ \hat{a}_7^* \\ \hat{a}_8^* \end{bmatrix} = \begin{bmatrix} 3.0 & 0.0 & 0.000 & 0.000 & 0.000 & 1.000 & 0.000 & 0.000 & 1.000 & 1.000 \\ 0.0 & 2.0 & 0.000 & 0.000 & 0.000 & 0.000 & 1.000 & 1.000 & 0.000 & 0.000 \\ 0.0 & 0.0 & 5.584 & 1.523 & 0.000 & -2.031 & 0.000 & -3.046 & 0.000 & 0.000 \\ 0.0 & 0.0 & 1.523 & 6.092 & 1.523 & 0.000 & -3.046 & -3.046 & 0.000 & 0.000 \\ 0.0 & 0.0 & 0.000 & 1.523 & 6.092 & 0.000 & -3.046 & 1.523 & 0.000 & -3.046 \\ 1.0 & 0.0 & -2.031 & 0.000 & 0.000 & 6.584 & 1.523 & 0.000 & -3.046 & 0.000 \\ 0.0 & 1.0 & 0.000 & -3.046 & -3.046 & 1.523 & 8.615 & 0.000 & -3.046 & 0.000 \\ 0.0 & 1.0 & -3.046 & -3.046 & 1.523 & 0.000 & 0.000 & 8.615 & 0.000 & -3.046 \\ 1.0 & 0.0 & 0.000 & 0.000 & 0.000 & -3.046 & -3.046 & 0.000 & 7.092 & 0.000 \\ 1.0 & 0.0 & 0.000 & 0.000 & -3.046 & 0.000 & 0.000 & -3.046 & 0.000 & 7.092 \end{bmatrix}^{-1} \begin{bmatrix} 0.549 \\ 0.194 \\ 0.000 \\ 0.000 \\ 0.000 \\ 0.208 \\ 0.085 \\ 0.108 \\ 0.105 \\ 0.235 \end{bmatrix}$$

Solving the MME for each transformed trait by direct inversion of the coefficient matrix gives the following solutions on the canonical scales. Given also are solutions for WWG and PWG after transforming the solutions for the transformed variates to the original scale.

	Canonical scale		Original scale	
Effects	VAR1	VAR2	WWG	PWG
Sex				
Male	0.185	1.266	4.361	6.800
Female	0.098	1.089	3.397	5.880
Animals				
1	0.003	0.052	0.151	0.280
2	−0.002	−0.002	−0.015	−0.008
3	0.000	−0.031	−0.078	−0.170
4	−0.001	−0.002	−0.010	−0.013
5	−0.007	−0.088	−0.270	−0.478
6	0.005	0.095	0.276	0.517
7	−0.015	−0.089	−0.316	−0.479
8	0.009	0.073	0.244	0.392

The solutions are exactly the same as those obtained from the multivariate analysis in Section 6.2. The solutions are transformed to the original scale using Eqns 7.3 and 7.4. For instance, the solutions for animal 1 for both traits on the original scale are:

$$\begin{bmatrix} \hat{a}_{11} \\ \hat{a}_{12} \end{bmatrix} = \begin{bmatrix} 5.7651 & 2.6006 \\ -0.5503 & 5.4495 \end{bmatrix} \begin{bmatrix} 0.0029 \\ 0.0516 \end{bmatrix} = \begin{bmatrix} 0.151 \\ 0.280 \end{bmatrix}$$

7.3 Cholesky Transformation

When all records are measured in all animals, MBLUP may be simplified by a canonical transformation as described in Section 7.2. However, if animals have some records missing and the loss of records is sequential, then a Cholesky transformation can be applied (Quaas, 1984). Such situations can arise, for example, in dairy cattle due to sequential culling and different lactations being regarded as different traits.

7.3.1 Calculating the transformation matrix and defining the model

Cholesky transformation involves forming transformed variables (traits) that are environmentally independent of each other; i.e. there is no residual covariance among them, therefore the residual covariance matrix for the transformed traits is an identity matrix. The transformation matrix $\mathbf{T}^{-1}$ is obtained by carrying out a Cholesky decomposition of **R**, the residual covariance matrix for the traits, such that:

$$\mathbf{R} = \mathbf{T}\,\mathbf{T}'$$

where **T** is a lower triangular matrix. The transformation matrix $\mathbf{T}^{-1}$ is the inverse of **T**. The formula for calculating **T** is given in Appendix E, Section E.3.

The vector of observations $\mathbf{y}_{ki}$ for the *i*th animal is transformed as:

$$\mathbf{y}^*_{ki} = \mathbf{T}^{-1}\mathbf{y}_{ki}$$

where *k* is the number of traits recorded and $\mathbf{y}^*_{ki}$ is the transformed vector.

If traits are missing in $\mathbf{y}_{ki}$, then the corresponding rows of $\mathbf{T}^{-1}$ are set to zero when transforming the vector of observation. Thus, if $\mathbf{y}_{ki}$ is a vector of observations of *n* traits for the *i*th animal, the transformation of **y** can be illustrated as:

$$y^*_{11} = t^{11}y_{11}$$
$$y^*_{21} = t^{21}y_{11} + t^{22}y_{21}$$
$$\dots$$
$$\dots$$
$$y^*_{n1} = t^{n1}y_{11} + t^{n2}y_{21} + t^{nn}y_{n1}$$

where the t^{ij} above are the elements of $\mathbf{T}^{-1}$.

Given that the variance of $\mathbf{y}_{ki}$ is:

$$\text{var}(\mathbf{y}) = \mathbf{G} + \mathbf{R}$$

the variance of the transformed variables becomes:

$$\text{var}(\mathbf{y}^*) = \mathbf{T}^{-1}\mathbf{G}(\mathbf{T}^{-1})' + \mathbf{I} = \mathbf{G}^* + \mathbf{I} = \mathbf{M} + \mathbf{I} \tag{7.5}$$

where **G** is the covariance matrix for additive genetic effects and $\mathbf{G}^*$ is the transformed additive genetic covariance matrix. Note that $\mathbf{G}^*$ is not diagonal. Vectors of solutions ($\mathbf{b}^*_i$ and $\mathbf{a}^*_i$) are transformed back to the original scale ($\mathbf{b}^*$ and $\mathbf{a}^*$) as:

$$\mathbf{b}_i = \mathbf{T}\mathbf{b}^*_i \tag{7.6}$$

$$\mathbf{a}_i = \mathbf{T}\mathbf{a}^*_i \tag{7.7}$$

7.3.2 An illustration

Example 7.2

The methodology is illustrated using the growth data on beef calves in Section 6.3.1. The residual and additive genetic covariance matrices were:

$$\mathbf{R} = \begin{bmatrix} 40 & 11 \\ 11 & 30 \end{bmatrix} \quad \text{and} \quad \mathbf{G} = \begin{bmatrix} 20 & 18 \\ 18 & 40 \end{bmatrix}$$

Now carry out a Cholesky decomposition of **R** such that **R** = **TT′**. For the **R** above:

$$\mathbf{T} = \begin{bmatrix} 6.324555 & 0.000 \\ 1.739253 & 5.193746 \end{bmatrix} \quad \text{with} \quad \mathbf{T}^{-1} = \begin{bmatrix} 0.1581139 & 0.000 \\ -0.052948 & 0.1925393 \end{bmatrix}$$

The transformed additive genetic covariance matrix (**M**) is:

$$\mathbf{M} = \mathbf{T}^{-1}\ \mathbf{G}(\mathbf{T}^{-1})' = \begin{bmatrix} 0.5000 & 0.380539 \\ 0.380539 & 1.171972 \end{bmatrix} \quad \text{and} \quad \mathbf{M}^{-1} = \begin{bmatrix} 2.654723 & -0.862556 \\ -0.862556 & 1.133334 \end{bmatrix}$$

The transformed variables are calculated using the transforming matrix $\mathbf{T}^{-1}$. For the first two animals the transformation is as follows:

Animal 1:

$$y^*_{11} = t^{11}y_{11} = 0.1581139(4.5) = 0.712$$

Animal 2:

$$\begin{aligned} y^*_{11} &= t^{11}y_{11} = 0.1581139(2.9) = 0.459 \\ y^*_{22} &= t^{21}y_{11} + t^{22}y_{22} = -0.052948(2.9) + 0.1925393(5.0) = 0.809 \end{aligned}$$

where y_{ij} and y^*_{ij} are the original and transformed observations, respectively, for the *i*th trait and *j*th animal. The transformed variables for all calves are shown in the table below.

				Original traits		Transformed traits	
Calves	Sex	Sire	Dam	WWG	PWG	y^*_1	y^*_2
4	Male	1	–	4.5	–	0.712	–
5	Female	3	2	2.9	5.0	0.459	0.809
6	Female	1	2	3.9	6.8	0.617	1.103
7	Male	4	5	3.5	6.0	0.553	0.970
8	Male	3	6	5.0	7.5	0.791	1.179
9	Female	7	–	4.0	–	0.632	–

The model for analysis is the same as in Section 6.4.1 except that the variance of $\mathbf{y}^*$ now is:

$$\text{var}(\mathbf{y}^*) = \mathbf{T}^{-1}\mathbf{G}(\mathbf{T}^{-1})' + \mathbf{I} = \mathbf{M} + \mathbf{I}$$

The MME for the transformed variables are:

$$\begin{bmatrix} \hat{\mathbf{b}}^*_1 \\ \hat{\mathbf{b}}^*_2 \\ \hat{\mathbf{a}}^*_1 \\ \hat{\mathbf{a}}^*_2 \end{bmatrix} = \begin{bmatrix} \mathbf{X}'_1\mathbf{X}_1 & 0 & \mathbf{X}'_1\mathbf{Z}_1 & 0 \\ 0 & \mathbf{X}'_2\mathbf{X}_2 & 0 & \mathbf{X}'_2\mathbf{Z}_2 \\ \mathbf{Z}'_1\mathbf{X}_1 & 0 & \mathbf{Z}'_1\mathbf{Z}_1 + \mathbf{A}^{-1}m^{11} & \mathbf{A}^{-1}m^{12} \\ 0 & \mathbf{Z}'_2\mathbf{X}_2 & A^{-1}m^{21} & \mathbf{Z}'_2\mathbf{Z}_2 + A^{-1}m^{22} \end{bmatrix}^{-1} \begin{bmatrix} \mathbf{X}'_1\mathbf{y}^*_1 \\ \mathbf{X}'_2\mathbf{y}^*_2 \\ \mathbf{Z}'_1\mathbf{y}^*_1 \\ \mathbf{Z}'_1\mathbf{y}^*_2 \end{bmatrix}$$

The design matrices $\mathbf{X}_1$, $\mathbf{X}_2$, $\mathbf{Z}_1$ and $\mathbf{Z}_2$ and the inverse of the relationship matrix are exactly as in Section 6.4.1. The vector observations $\mathbf{y}^*$ now contain the transformed variables shown in the above table. All other matrices in the MME above can be derived from the design matrices and vector of observations through matrix multiplication and the addition of the $\mathbf{A}^{-1}m^{11}$ and $\mathbf{A}^{-1}m^{22}$ to the animal equations for trait one and two, respectively, and $\mathbf{A}^{-1}m^{12}$ to animal equations for trait one by trait two and $\mathbf{A}^{-1}m^{21}$ to equations for trait one by trait two that pertains to animals. The MME have not been shown because they are too large. However, solving the MME gives the following solutions on the transformed scale. The solutions transformed to the original scale are also shown.

	Transformed scale		Original scale	
Effects	WWG	PWG	WWG	PWG
Sex				
Male	0.691	1.085	4.367	6.834
Female	0.578	0.963	3.657	6.007
Animals				
1	0.021	0.044	0.130	0.266
2	−0.013	−0.010	−0.084	−0.075
3	−0.015	−0.032	−0.098	−0.194
4	0.001	0.003	0.007	0.016
5	−0.054	−0.089	−0.343	−0.555
6	0.030	0.075	0.192	0.440
7	−0.049	−0.077	−0.308	−0.483
8	0.032	0.056	0.201	0.349
9	−0.003	−0.022	−0.018	−0.119

These are exactly the same solutions as those obtained in Section 6.3 without any transformation. The number of non-zero elements was 188 in the analysis on the transformed variables, compared with 208 when no transformation is carried out. This difference could be substantial with large data sets and reduces storage requirements when data are transformed. The solutions were transformed to the original scale using Eqns 7.6 and 7.7. Thus, the solutions for male calves on the original scale are:

$$\begin{bmatrix} \hat{b}_{11} \\ \hat{b}_{12} \end{bmatrix} = \begin{bmatrix} 6.324555 & 0.000 \\ 1.739253 & 5.193746 \end{bmatrix} \begin{bmatrix} 0.69063 \\ 1.0846 \end{bmatrix} = \begin{bmatrix} 4.367 \\ 6.834 \end{bmatrix}$$

7.4 Factor and Principal Component Analysis

In Sections 7.2 and 7.3, the simplification of multivariate analysis using canonical transformation and Cholesky decomposition were discussed. Both approaches involved the transformation of the vector of observations as well as the residual and genetic covariance matrices. However, for multivariate analysis with a large number of traits and with high genetic correlations among the traits, a factorial or principal component analysis might be more appropriate in reducing the dimension of such analysis. Neither of these methods involve the transformation of the vector of observations. The principal component and factor analysis (FA) methods provide efficient means for reducing the rank of the genetic covariance matrix in multivariate analysis, resulting in the substantial sparsity of the MME for genetic evaluation and estimation of genetic parameters (Meyer, 2009). Therefore, both methodologies have attracted considerable attention in multivariate analysis involving many traits for parameter estimation and genetic evaluation (Kirkpatrick and Meyer, 2004; Meyer, 2005, 2007; Tyriseva *et al.*, 2011a, 2011b).

FA is mainly concerned with identifying the common factors that give rise to correlations between variables. It assumes that the traits studied are linear combinations of few latent variables, referred to as common factors. Then, any variance not explained by these common factors is modelled separately as trait specific, by fitting corresponding specific factors. Since the factors are assumed to be uncorrelated, substantial sparsity of the MME is achieved.

On the other hand, PC aims to identify factors that explain the maximum amount of variation and does not imply any underlying model. The first PC explains the maximum amount of genetic variation in the data and each successive PC explains the maximum amount of the remaining variation. Thus, for highly correlated traits, only the leading PC have a practical influence on genetic variation and those with negligible effect can be omitted without reducing the accuracy of estimation. For example, with t traits, k independent principal components ($k \leq t$) can be derived that explain a maximum proportion of the total multivariate system. Similar to the FA, the PC approach requires decomposing the genetic covariance matrix into pertaining matrices of eigenvalues and eigenvectors. The eigenvector or PC can be regarded as a linear combination of the traits and they are independent, while the corresponding eigenvalues give the variance explained.

7.4.1 Factor analysis

Assume that **w** is a vector of n variables with covariance matrix equal to **G** and that **w** can be modelled as:

$$\mathbf{w} = \boldsymbol{\mu} + \boldsymbol{\Phi}\mathbf{c} + \mathbf{s}$$

where $\boldsymbol{\mu}$ is the vector of means, **c** is a vector of common factors of length m, **s** is the vector of residuals or specific effects of length n and $\boldsymbol{\Phi}$ is the matrix of order $n \times m$ of the so-called factor loadings. In the most common form of FA, the columns of $\boldsymbol{\Phi}$ are orthogonal, i.e. $\varphi_i\varphi_j = 0$, for $i \neq j$, and, thus, the elements of **c** are uncorrelated and assumed to have unit variance, var(**c**) = **I**. The columns φ_i are determined as corresponding eigenvectors of **G**, scaled by the square root of the respective eigenvalues (Meyer, 2009).

Usually, $\boldsymbol{\Phi}$ is not unique but is often orthogonally transformed to obtain factor loadings that are more interpretable than those derived from the eigenvectors. The specific effects (**s**) are assumed to be independently distributed and therefore the variance of **s** is a diagonal matrix **S** of order n. Therefore:

$$\text{var}(\mathbf{w}) = \mathbf{G}_{\text{FA}} = \boldsymbol{\Phi}\boldsymbol{\Phi}' + \mathbf{S} \tag{7.8}$$

The above indicates that all the covariances between the levels of **w** are modelled through the common factors while the specific factors account for the additional individual variances of the elements of **w**. Thus, the $n(n + 1)/2$ elements of **G** are modelled through the n elements of the specific variances and $m(2n - m + 1)/2$ elements of $\boldsymbol{\Phi}$, and additional $m(m - 1)/2$ of $\boldsymbol{\Phi}$, which is determined by the orthogonal constraints. For example, if n is 4 and $m = 1$, then the ten elements of **G** are modelled by the four elements of **S** and the four elements of $\boldsymbol{\Phi}$. FA with a small m thus provides a parsimonious way to model the covariances among a large number of variables. When all the elements of **S** are non-zero, four traits is the minimum number of variables for which imposing an FA structure results in a reduction of the parameters (Meyer, 2009).

Mixed-model equations

Assume the following multi-trait linear mixed model in Eqn 6.1 is presented as:

$$\mathbf{y} = \mathbf{Xb} + \mathbf{Za} + \mathbf{e} \tag{7.9}$$

with terms defined as in Eqn 6.1 and MME as in Eqn 6.2. If **G** is represented by an FA structure (Eqn 7.8), an equivalent model to Eqn 7.9 is:

$$\mathbf{y} = \mathbf{Xb} + \mathbf{Z}(\mathbf{I}q \times \boldsymbol{\Phi})\mathbf{c} + \mathbf{Ws} + \mathbf{e} = \mathbf{Xb} + \mathbf{Z}^{*}\mathbf{c} + \mathbf{Ws} + \mathbf{e} \tag{7.10}$$

with q being the number of individuals, **c** is a vector of common factor effects of order m, $\mathbf{Z}^{*} = \mathbf{Z}(\mathbf{I}q \times \boldsymbol{\Phi})$, and **s** is the vector for the specific factor effects. In some contexts, application of Eqn 7.10, i.e. with elements of $\mathbf{S} \neq 0$, is referred to as the extended factor analysis (XFA) compared with models with no specific effects ($\mathbf{S} = 0$), which are simply referred to as factor analysis (FA). The MME for XFA, then, are:

$$\begin{bmatrix} \mathbf{X'R^{-1}X} & \mathbf{X'R^{-1}Z^{*}} & \mathbf{X'R^{-1}W} \\ \mathbf{Z^{*\prime}R^{-1}X} & \mathbf{Z^{*\prime}R^{-1}Z^{*}} + \mathbf{I}_m \otimes \mathbf{A}^{-1} & \mathbf{Z^{*\prime}R^{-1}W} \\ \mathbf{W'R^{-1}X} & \mathbf{W'R^{-1}Z^{*}} & \mathbf{W'R^{-1}W} + \mathbf{S}^{-1} \otimes \mathbf{A}^{-1} \end{bmatrix} \begin{pmatrix} \hat{\mathbf{b}} \\ \hat{\mathbf{c}} \\ \hat{\mathbf{s}} \end{pmatrix} = \begin{bmatrix} \mathbf{X'R^{-1}y} \\ \mathbf{Z^{*\prime}R^{-1}y} \\ \mathbf{W'R^{-1}y} \end{bmatrix} \tag{7.11}$$

The vector $\hat{\mathbf{a}}_i$ of solutions for animal i can be obtained as:

$$\hat{\mathbf{a}}_i = \boldsymbol{\Phi}\hat{\mathbf{c}}_i + \hat{\mathbf{s}}_i \tag{7.12}$$

The number of equations in the MME (Eqn 6.2) for the usual multivariate model are equal to the number of equations for **b** and **s** in Eqn 7.11. However, there are additional mq equations for the common effects and $\mathbf{Z}^{*}$, which is a vector of order m with elements φ_{ij}, and is denser than **Z** in Eqn 6.2, which contains a single element of unity in a row or column. However, the section of the coefficient matrix for random effects is much sparser as effects are genetically uncorrelated and $\mathbf{A}^{-1}$ contributes only $(m + n)$ non-zero elements compared to n^2 for Eqn 6.1. For the estimation of covariance estimates using REML, Thompson *et al.* (2003) showed that the sparsity of the MME with an XFA structure imposed dramatically reduced computational requirements compared to the standard multivariate model. Note that fitting an FA structure to **G** with no specific effects, the MME are similar to Eqn 7.11 but with the row of equations for **s** omitted, and the $\mathbf{Z}^{*}$ will be a vector of order n.

An illustration

Example 7.3

The data on pre-weaning weight gain (WWG) and post-weaning gain (PWG) in Example 6.1 are extended to include two additional traits of muscle score (MS) and backfat thickness (BFAT), and data are presented below. The objective is to undertake multi-trait analysis imposing an XFA on **G** and the results obtained compared to those from full MBLUP or FA structure on **G** with no specific factors.

Calf	Sex	WWG	PWG	MS	BFAT
4	Male	4.5	6.8	5.0	0.226
5	Female	2.9	5.0	3.0	0.573
6	Female	3.9	6.8	12.0	0.386
7	Male	3.5	6.0	8.0	0.290
8	Male	5.0	7.5	15.0	0.175

Assuming that **G** and **R**, respectively, are:

$$\mathbf{G} = \begin{pmatrix} 20 & 18 & 4 & 9 \\ 18 & 40 & 9 & 20 \\ 4 & 9 & 25 & 4.5 \\ 9 & 20 & 4.5 & 32 \end{pmatrix} \quad \text{and} \quad \mathbf{R} = \begin{pmatrix} 40 & 11 & 16 & 9 \\ 11 & 30 & 12 & 14 \\ 16 & 12 & 70 & 10 \\ 9 & 14 & 10 & 55 \end{pmatrix}$$

Applying Eqn 7.8 to **G** using the function factanal in the **R** package (The R Development Core Team, 2010) gives:

$$\boldsymbol{\Phi}' = (2.8532 \;\; 6.3056 \;\; 1.4250 \;\; 3.1678) \quad \text{and} \quad \mathbf{S}' = (11.860 \;\; 0.200 \;\; 22.975 \;\; 21.952)$$

This implies that the number of common factors, m, is equal to 1 for the example **G** above. Thus, the column vector $\mathbf{z}_i^*$ for animal i in the matrix $\mathbf{Z}^*$ in Eqn 7.11 equals $\boldsymbol{\Phi}$. Therefore, for animal i with a record $\mathbf{z}_i^{*\prime}\mathbf{r}^{-1}\mathbf{z}_i^*$ is 1.361. However, for animal i, $\mathbf{W}_i$ is a diagonal matrix and therefore $\mathbf{W}_i'\mathbf{R}^{-1}\mathbf{W}_i$ is computed as described for the MBLUP model in Section 6.2. Thus, for animal i, $\mathbf{W}_i'\mathbf{R}^{-1}\mathbf{W}_i$ is:

$$\mathbf{W}_i'\mathbf{R}^{-1}\mathbf{W}_i = \begin{pmatrix} 00297 & & & \\ -00079 & 00419 & \text{symmetric} & \\ -00052 & -00041 & 00163 & \\ -0.0019 & -0.0086 & -0.0011 & 0.0209 \end{pmatrix}$$

Although there were 48 equations in the MME defined in Eqn 7.11 for this example compared with 40 in the usual MBLUP, there were only 502 non-zero elements in the XFA compared with 620 in MBLUP, illustrating the increased sparsity of the MME with the XFA model. Solving the MME gave the following solutions. The results from the usual MBLUP gave exactly the same solutions and these have not been presented.

Solutions for sex of calf effects

	WWG	PWG	MSC	BFAT
M	4.352	6.795	9.412	0.231
F	3.487	5.959	7.095	0.535

Animal and specific solutions

		Specific effects solutions				Transformed solutions[b]			
	COM[a]	WWG	PWG	BFAT	MSC	WWG	PWG	MSC	BFAT
1	0.036	−0.008	0.095	0.000	0.005	0.095	0.227	0.340	0.010
2	−0.012	−0.001	−0.073	0.021	0.000	−0.089	−0.073	0.313	−0.050
3	−0.027	0.068	0.031	0.208	0.000	−0.086	−0.169	0.031	−0.032
4	0.021	0.046	0.113	−0.021	0.000	0.168	0.136	−0.855	0.113

Continued

Continued.

5	−0.064	−0.191	0.000	0.005	0.000	−0.191	−0.407	−0.539	−0.029
6	0.046	0.290	0.021	0.000	0.000	0.017	0.290	1.350	−0.082
7	−0.063	−0.813	0.208	0.000	0.029	−0.208	−0.399	−0.813	−0.015
8	0.028	−0.101	−0.021	0.000	0.000	−0.017	0.178	1.431	−0.101

[a]COM, solutions for common factor; [b]Transformed solutions from Eqn 7.12

Analysis with FA model

The main differences with fitting an FA model with no specific effects is that Z^* for an animal in Eqn 7.11 is of order n by n and the last row of Eqn 7.11 is omitted. Note that $\mathbf{Z}^*$ is now a product of the eigenvectors of $\mathbf{G}$ and the square root of a diagonal matrix of eigenvalues (see Section 7.4.2). Thus $\mathbf{Z}^*$ is:

$$\mathbf{Z}_i^* = \begin{pmatrix} 2.2974 & 3.2100 & -0.1259 & -2.0983 \\ -1.7761 & 5.8683 & -0.1348 & -1.5333 \\ 0.2453 & 2.0018 & 4.4348 & 1.1272 \\ 0.5878 & 4.3506 & -1.7658 & 3.0977 \end{pmatrix} \quad \text{and for animal } i:$$

$$\mathbf{Z}_i^{*\prime}\mathbf{R}^{-1}\mathbf{Z}_i^* = \begin{pmatrix} 0.372 & -0.226 & -0.042 & 0.054 \\ -0.226 & 1.236 & -0.141 & -.246 \\ -0.042 & -0.141 & 0.410 & 0.019 \\ 0.054 & -0.246 & 0.019 & 0.537 \end{pmatrix}$$

Setting the MME follows the usual rules and the MME has 40 equations for the example but with only 388 non-zero elements. The low number of non-zero elements is due to the fact that only n elements of A^{-1} are contributed compared with n^2 for the MBLUP. Solving the equations gives the following solutions:

Solutions for calf sex effects

	WWG	PWG	MSC	BFAT
M	4.352	6.795	9.412	0.231
F	3.487	5.959	7.095	0.535

Animal solutions

	Untransformed solutions				Transformed solutions[a]			
	WWG	PWG	BFAT	MSC	WWG	PWG	MSC	BFAT
1	−0.011	0.035	0.063	−0.008	0.095	0.227	0.340	0.010
2	−0.003	−0.005	0.066	0.028	−0.089	−0.073	0.313	−0.050
3	0.010	−0.020	0.010	0.021	−0.086	−0.169	0.031	−0.032
4	0.000	0.002	−0.177	−0.067	0.168	0.136	−0.855	0.113
5	0.015	−0.062	−0.099	0.019	−0.191	−0.407	−0.539	−0.029
6	−0.022	0.061	0.267	0.045	0.017	0.290	1.350	−0.082
7	0.003	−0.069	−0.153	0.005	−0.208	−0.399	−0.813	−0.015
8	−0.007	0.050	0.285	0.060	−0.017	0.178	1.431	−0.101

[a]Transformed solutions = vectors of solutions multiplied by $\mathbf{Z}^*$

7.4.2 Principal component analysis

Analysis with full PC model

The application of a full PC model with no rank reduction is similar to the FA analysis except that $\mathbf{Z}^*$ is now a matrix of eigenvectors of order n by n, and $\mathbf{Z}^{*\prime}\mathbf{R}^{-1}\mathbf{Z}^* + (\mathbf{I}_m \otimes \mathbf{A}^{-1})$ in the second row of Eqn 7.11 is replaced by $\mathbf{Z}^{*\prime}\mathbf{R}^{-1}\mathbf{Z}^* + (\mathbf{D}_n \otimes \mathbf{A}^{-1})$, where $\mathbf{D}_n$ is a diagonal matrix of eigenvalues. Again, the last row of Eqn 7.11 is omitted. It therefore involves decomposing $\mathbf{G}$ to a matrix of eigenvectors ($\mathbf{Z}^*$) and corresponding eigenvalues ($\mathbf{D}$). Thus $\mathbf{Z}^*$ and $\mathbf{D}$, respectively, are:

$$\mathbf{Z}^* = \begin{pmatrix} 0.7710 & 0.3896 & -0.02940 & -0.5029 \\ -0.5983 & 0.7139 & -0.0268 & -0.3628 \\ 0.0865 & 0.2427 & 0.9288 & 0.2664 \\ 0.2000 & 0.5288 & -0.3685 & 0.7379 \end{pmatrix} \text{ and}$$

$$\mathbf{D} = \text{diag}(8.8159 \;\; 67.6963 \;\; 22.8286 \;\; 17.6592)$$

Thus, $\mathbf{Z}_i^{*\prime}\mathbf{R}^{-1}\mathbf{Z}_i^*$ for animal i is:

$$\mathbf{Z}_i^{*\prime}\mathbf{R}^{-1}\mathbf{Z}_i^* = \begin{pmatrix} 0.042 & -0.009 & -0.003 & 0.004 \\ -0.009 & 0.018 & -0.004 & -0.007 \\ -0.003 & -0.004 & 0.018 & 0.001 \\ 0.004 & -0.007 & 0.001 & 0.030 \end{pmatrix}$$

The MME are set up as usual. Similar again to the FA model, the PC has 40 equations and 388 non-zero elements. The solutions for the various effects from solving the MME are:

Solutions for sex of calf effects

	WWG	PWG	MSC	BFAT
M	4.352	6.795	9.412	0.231
F	3.488	5.959	7.095	0.535

Animal solutions

	Untransformed solutions				Transformed solutions			
	WWG	PWG	BFAT	MSC	WWG	PWG	MSC	BFAT
1	−0.032	0.287	0.303	−0.032	0.094	0.227	0.340	0.010
2	−0.009	−0.038	0.314	0.118	−0.090	−0.073	0.313	−0.050
3	0.031	−0.163	0.047	0.089	−0.086	−0.169	0.030	−0.032
4	−0.002	0.015	−0.844	−0.279	0.170	0.136	−0.855	0.113
5	0.045	−0.511	−0.473	0.078	−0.190	−0.407	−0.539	−0.029
6	−0.062	0.496	1.276	0.186	0.014	0.290	1.350	−0.083
7	0.007	−0.571	−0.732	0.022	−0.207	−0.400	−0.812	−0.015
8	−0.018	0.413	1.362	0.252	−0.019	0.178	1.431	−0.101

[a] Transformed solutions = vectors of solutions multiplied by $\mathbf{Z}^*$

7.4.3 Analysis with reduced-rank PC model

The diagonal matrix **D** with the full PC model in Section 7.4.2 indicates that the first principal component accounts for about 8.82% of the total genetic variance. Deleting the first eigenvalue gives a diagonal $\mathbf{D}^*$ of order 3 as $\mathbf{D}^* = \text{diag}(67.6963\ 22.8286\ 17.6592)$. Then, $\mathbf{G}^*$, a new genetic covariance matrix, can be computed as $\mathbf{M'D^*M}$, where **M** is equivalent to $\mathbf{Z}^*$ in Section 7.4.2 with a full PC model fitted but with the first column deleted. Thus:

$$\mathbf{G}^* = \mathbf{M'D^*M} = \begin{pmatrix} 14.759 & 22.067 & 3.412 & 7.640 \\ 22.067 & 36.844 & 9.456 & 21.055 \\ 3.412 & 9.456 & 24.934 & 4.347 \\ 7.640 & 21.055 & 4.347 & 31.647 \end{pmatrix} \text{ with}$$

$$\mathbf{M} = \begin{pmatrix} 0.3896 & -0.02940 & -0.5029 \\ 0.7139 & -0.0268 & -0.3628 \\ 0.2427 & 0.9288 & 0.2664 \\ 0.5288 & -0.3685 & 0.7379 \end{pmatrix}$$

The application of reduced rank PC is similar to the full PC analysis with $\mathbf{Z}^*$ replaced by **M**, and **D** by $\mathbf{D}^*$. Thus, for animal *i*, $\mathbf{M}_i'\mathbf{R}^{-1}\ \mathbf{M}_i$ is:

$$\mathbf{M}_i'\mathbf{R}^{-1}\ \mathbf{M}_i = \begin{pmatrix} 0.018 & -0.004 & -0.007 \\ -0.004 & 0.018 & 0.001 \\ -0.007 & 0.001 & 0.030 \end{pmatrix}$$

The MME for the reduced PC has 32 equations and 284 non-zero elements. The solutions for the various effects from solving the MME are:

Solutions for sex of calf effects				
	WWG	PWG	MSC	BFAT
M	4.349	6.798	9.412	0.230
F	3.480	5.963	7.093	0.533

Solutions for animal effects							
	Untransformed solutions			Transformed solutions[a]			
				WWG	PWG	MSC	BFAT
1	0.295	0.305	−0.033	0.123	0.214	0.346	0.019
2	−0.037	0.314	0.118	−0.083	−0.078	0.314	−0.048
3	−0.170	0.046	0.090	−0.113	−0.156	0.025	−0.041
4	0.017	−0.844	−0.279	0.171	0.136	−0.854	0.115
5	−0.523	−0.476	0.080	−0.230	−0.390	−0.548	−0.042
6	0.511	1.279	0.185	0.069	0.263	1.362	−0.066
7	−0.576	−0.734	0.022	−0.215	−0.400	−0.816	−0.018
8	0.419	1.364	0.251	−0.003	0.171	1.435	−0.096

[a]Transformed solutions = vector of solutions multiplied by M

The deletion of the first eigenvalue in the reduced PC analysis had very little effect in terms of the EBVs of animals for traits 3 and 4. Thus there was no ranking for MSC and only two animals swapped places for BFAT compared with the result from the full PC analysis. However, only the top four and six animals were the same for WWG and PWG, respectively, compared with the full PC analysis, indicating more re-ranking was observed in WWG due to the reduction in variance. In practice, models with reduced ranks are usually applied in the analysis of many traits as in Meyer (2007), resulting in no re-ranking in the top animals, which are mainly of interest.

8 Maternal Trait Models: Animal and Reduced Animal Models

8.1 Introduction

The phenotypic expression of some traits in the progeny, such as weaning weight in beef cattle, is influenced by the ability of the dam to provide a suitable environment in the form of better nourishment. Thus, the dam contributes to the performance of the progeny in two ways: first, through her direct genetic effects passed to the progeny; and, second, through her ability to provide a suitable environment, for instance in producing milk. Traits such as birth and weaning weights in beef cattle fall into this category and are termed maternally influenced traits. The ability of the dam to provide a suitable environment for the expression of such traits in her progeny is partly genetic and partly environmental. Similar to the genetic component of an individual, the maternal genetic component can be partitioned into additive, dominance and epistatic effects (Willham, 1963). The environmental part may be partitioned into permanent and temporary environmental components. It is the maternal additive genetic component of the dam that is passed on to all her offspring, but it is expressed only when the female offspring have progeny of their own.

In the usual mixed-linear model for maternally influenced traits (Eqn 8.1), the phenotype is partitioned into:

1. additive genetic effects from the sire and the dam, usually termed the direct genetic effect;
2. additive genetic ability of the dam to provide a suitable environment, usually termed the indirect or maternal genetic effect;
3. permanent environmental effects, which include permanent environmental influences on the dam's mothering ability and the maternal non-additive genetic effects of the dam; and
4. other random environmental effects, termed residual effects.

In this chapter, the mixed-model methodology for genetic evaluation in models with maternal effects is discussed, considering a univariate situation, and the extension to multivariate analysis is also briefly presented. The application of BLUP to models with maternal effects was first presented by Quaas and Pollak (1980).

When repeated measurements for maternally influenced traits are available over a range of ages (for instance, body weight from birth to 630 days), a random regression model (see Chapter 10) might be more appropriate to analyse such a trait. A random regression model for maternally influenced traits is briefly defined in Section 10.3.6.

8.2 Animal Model for Maternal Traits

The model for maternally influenced traits in matrix notation is:

$$\mathbf{y} = \mathbf{Xb} + \mathbf{Zu} + \mathbf{Wm} + \mathbf{Spe} + \mathbf{e} \tag{8.1}$$

where $\mathbf{y}$ = vector of observations, $\mathbf{b}$ = vector of fixed effects, $\mathbf{u}$ = vector of random animal effects, $\mathbf{m}$ = vector of random maternal (indirect) genetic effects, $\mathbf{pe}$ = vector of

DOI: 10.1079/9781800620506.0008

permanent environmental effects as explained in item 3 in Section 8.1, **e** = vector of random residual effects, and **X**, **Z**, **W** and **S** are incidence matrices relating records to fixed, animal, maternal genetic and permanent environmental effects, respectively. It is assumed that:

$$\text{var}\begin{bmatrix}\mathbf{u}\\\mathbf{m}\\\mathbf{pe}\\\mathbf{e}\end{bmatrix}=\begin{bmatrix}g_{11}\mathbf{A} & g_{12}\mathbf{A} & 0 & 0\\ g_{21}\mathbf{A} & g_{22}\mathbf{A} & 0 & 0\\ 0 & 0 & \mathbf{I}\sigma^2_{pe} & 0\\ 0 & 0 & 0 & \mathbf{I}\sigma^2_e\end{bmatrix}$$

where g_{11} = additive genetic variance for direct effects, g_{22} = additive genetic variance for maternal effects, g_{12} = additive genetic covariance between direct and maternal effects, σ^2_{pe} = variance due to permanent environmental effects, and σ^2_e = residual error variance.

The variance of **y**, using the same arguments as in Section 4.2, is:

$$\text{var}(\mathbf{y})=[\mathbf{Z}\quad\mathbf{W}]\begin{bmatrix}g_{11}\mathbf{A} & g_{12}\mathbf{A}\\ g_{21}\mathbf{A} & g_{22}\mathbf{A}\end{bmatrix}\begin{bmatrix}\mathbf{Z}'\\\mathbf{W}'\end{bmatrix}+\mathbf{SI}\sigma^2_{pe}\mathbf{S}'+\mathbf{I}\sigma^2_e$$

The BLUE of estimable functions of **b** and the BLUP of **u**, **m** and **pe** in Eqn 8.1 are obtained by solving the following MME:

$$\begin{bmatrix}\hat{\mathbf{b}}\\\hat{\mathbf{u}}\\\hat{\mathbf{m}}\\\hat{\mathbf{pe}}\end{bmatrix}=\begin{bmatrix}\mathbf{X'X} & \mathbf{X'Z} & \mathbf{X'W} & \mathbf{X'S}\\ \mathbf{Z'X} & \mathbf{Z'Z}+\mathbf{A}^{-1}\alpha_1 & \mathbf{Z'W}+\mathbf{A}^{-1}\alpha_2 & \mathbf{Z'S}\\ \mathbf{W'X} & \mathbf{W'Z}+\mathbf{A}^{-1}\alpha_2 & \mathbf{W'W}+\mathbf{A}^{-1}\alpha_3 & \mathbf{W'S}\\ \mathbf{S'X} & \mathbf{S'Z} & \mathbf{S'W} & \mathbf{S'S}+\mathbf{I}\alpha_4\end{bmatrix}^{-1}\begin{bmatrix}\mathbf{X'y}\\\mathbf{Z'y}\\\mathbf{W'y}\\\mathbf{S'y}\end{bmatrix}\qquad(8.2)$$

8.2.1 An illustration

Example 8.1

Assume the data in Table 8.1 to be the birth weight for a group of beef calves. The aim is to estimate solutions for herd and pen effects and predict solutions for direct and maternal effects for all animals and permanent environmental effects for dams of

Table 8.1. Birth weight for group of beef calves.

Calf	Sire	Dam	Herds	Pen	Birth weight (kg)
5	1	2	1	1	35.0
6	3	2	1	2	20.0
7	4	6	1	2	25.0
8	3	5	1	1	40.0
9	1	6	2	1	42.0
10	3	2	2	2	22.0
11	3	7	2	2	35.0
12	8	7	3	2	34.0
13	9	2	3	1	20.0
14	3	6	3	2	40.0

progeny with records. Suppose the genetic parameters are g_{11} = 150, g_{12} = –40, g_{22} = 90, $\sigma^2_{pe} = 40$ and $\sigma^2_e = 350$ Then:

$$\mathbf{G}^{-1} = \begin{bmatrix} 0.00756 & 0.00336 \\ 0.00336 & 0.0126 \end{bmatrix} \quad \text{and} \quad \begin{bmatrix} \alpha_1 & \alpha_2 \\ \alpha_2 & \alpha_3 \end{bmatrix} = \begin{bmatrix} 2.647 & 1.176 \\ 1.176 & 4.412 \end{bmatrix}$$

and $\alpha_4 = 350 / 40 = 8.75$.

The model for the analysis is as presented in Eqn 8.1.

Setting up the design matrices

Considering only animals with records, the first three rows of matrix **X** relate records to herd effects and the last two rows to pen effects. The transpose of **X** is:

$$\mathbf{X}' = \begin{bmatrix} 1 & 1 & 1 & 1 & 0 & 0 & 0 & 0 & 0 & 0 \\ 0 & 0 & 0 & 0 & 1 & 1 & 1 & 0 & 0 & 0 \\ 0 & 0 & 0 & 0 & 0 & 0 & 0 & 1 & 1 & 1 \\ 1 & 0 & 0 & 1 & 1 & 0 & 0 & 0 & 1 & 0 \\ 0 & 1 & 1 & 0 & 0 & 1 & 1 & 1 & 0 & 1 \end{bmatrix}$$

Excluding ancestors, each animal has one record; therefore, **Z** is an identity matrix. However, **Z** is augmented with columns of zeros equal to the number of ancestors to take account of ancestors in the pedigree. The matrices **W** and **S** relate records through the dam to their effects, i.e. maternal genetic effect and permanent environmental effect, respectively. However, since maternal effect is genetic and is passed from parent to offspring, estimates of maternal effect are for all animals in the analysis while estimates of permanent environmental effects are only for dams of progeny with records. Thus, in setting up **W**, all animals are considered, while only four dams with progeny having records are taken into account for **S**. For the example data set, **W** (with rows and columns numbered by the relevant animal they relate to) is:

$$\mathbf{W} = \begin{array}{r} \\ 5 \\ 6 \\ 7 \\ 8 \\ 9 \\ 10 \\ 11 \\ 12 \\ 13 \\ 14 \end{array} \begin{array}{c} \begin{matrix} 1 & 2 & 3 & 4 & 5 & 6 & 7 & 8 & 9 & 10 & 11 & 12 & 13 & 14 \end{matrix} \\ \begin{bmatrix} 0 & 1 & 0 & 0 & 0 & 0 & 0 & 0 & 0 & 0 & 0 & 0 & 0 & 0 \\ 0 & 1 & 0 & 0 & 0 & 0 & 0 & 0 & 0 & 0 & 0 & 0 & 0 & 0 \\ 0 & 0 & 0 & 0 & 0 & 1 & 0 & 0 & 0 & 0 & 0 & 0 & 0 & 0 \\ 0 & 0 & 0 & 0 & 1 & 0 & 0 & 0 & 0 & 0 & 0 & 0 & 0 & 0 \\ 0 & 0 & 0 & 0 & 0 & 1 & 0 & 0 & 0 & 0 & 0 & 0 & 0 & 0 \\ 0 & 1 & 0 & 0 & 0 & 0 & 0 & 0 & 0 & 0 & 0 & 0 & 0 & 0 \\ 0 & 0 & 0 & 0 & 0 & 0 & 1 & 0 & 0 & 0 & 0 & 0 & 0 & 0 \\ 0 & 0 & 0 & 0 & 0 & 0 & 1 & 0 & 0 & 0 & 0 & 0 & 0 & 0 \\ 0 & 1 & 0 & 0 & 0 & 0 & 0 & 0 & 0 & 0 & 0 & 0 & 0 & 0 \\ 0 & 0 & 0 & 0 & 0 & 1 & 0 & 0 & 0 & 0 & 0 & 0 & 0 & 0 \end{bmatrix} \end{array}$$

and

$$\mathbf{S}' = \begin{array}{c} \\ 2 \\ 5 \\ 6 \\ 7 \end{array} \begin{array}{c} \begin{array}{cccccccccc} 5 & 6 & 7 & 8 & 9 & 10 & 11 & 12 & 13 & 14 \end{array} \\ \begin{bmatrix} 1 & 1 & 0 & 0 & 0 & 1 & 0 & 0 & 1 & 0 \\ 0 & 0 & 0 & 1 & 0 & 0 & 0 & 0 & 0 & 0 \\ 0 & 0 & 1 & 0 & 1 & 0 & 0 & 0 & 0 & 1 \\ 0 & 0 & 0 & 0 & 0 & 0 & 1 & 1 & 0 & 0 \end{bmatrix} \end{array}$$

The matrix **S** above implies, for instance, that animals 5, 6, 10 and 13 have the same dam (animal 2), while animals 11 and 12 are from another dam (animal 7).

The transpose of the vector of observations is:

$$y' = \begin{bmatrix} 35 & 20 & 25 & 40 & 42 & 22 & 35 & 34 & 20 & 40 \end{bmatrix}$$

The other matrices in the MME can be calculated through matrix multiplication. The inverse of the relationship matrix is calculated applying the rules in Section 3.4.1. The matrix $\mathbf{A}^{-1}\boldsymbol{\alpha}_1$ is added to animal equations, $\mathbf{A}^{-1}\boldsymbol{\alpha}_2$ to the equations for maternal genetic effects, $\mathbf{A}^{-1}\boldsymbol{\alpha}_3$ to the animal by maternal genetic equations, and α_4 to the diagonals of the equations for permanent environmental effects to obtain the MME. The MME are not presented because they are too large. There is dependency between the equations for herds and pen; thus, the row for the first herd was set to zero in solving the MME by direct inversion. Solutions to the MME are:

Effects	Solutions	
Herds		
1	0.000	
2	3.386	
3	1.434	
Pen		
1	34.540	
2	27.691	
Animals		
	Direct effects	Maternal effects
1	0.564	0.262
2	−1.244	−1.583
3	1.165	0.736
4	−0.484	0.586
5	0.630	−0.507
6	−0.859	0.841
7	−1.156	1.299
8	1.917	−0.158
9	−0.553	0.660
10	−1.055	−0.153
11	0.385	0.916
12	0.863	0.442
13	−2.980	0.093
14	1.751	0.362
Permanent environment		
2	−1.701	
5	0.415	
6	0.825	
7	0.461	

The solutions show little difference between the herds, but calves in pen 1 were heavier than those in pen 2 by about 6.85 kg at birth. The solution for level i of the fixed effect n can be calculated using Eqn 5.3, except that the sum of yields for the level of fixed effect is corrected in addition for maternal effects. That is:

$$\hat{b}_{in} = \frac{\sum_{f=1}^{\text{diag}_{in}} y_{inf} - \sum_j \hat{b}_{inj} - \sum_k \hat{a}_{ink} - \sum_l \hat{m}_{inl} - \sum_t \hat{p}e_{\text{int}}}{\text{diag}_{in}} \tag{8.3}$$

where m_{inl} is the solution for level l of genetic maternal effects within level i of the nth fixed effect and all other terms are as defined in Eqn 5.3. Thus, the solution for level 1 of pen effect is:

$$\begin{aligned}
\hat{b}_{11} &= [137 - (2\hat{h}d_1 + \hat{h}d_2 + \hat{h}d_3) - (\hat{a}_5 + \hat{a}_8 + \hat{a}_9 + \hat{a}_{13}) \\
&\quad -(2\hat{m}_2 + \hat{m}_5 + \hat{m}_6) - (2\hat{p}e_2 + \hat{p}e_5 + \hat{p}e_6)]/4 \\
&= [137 - 4.82 - (-0.986) - (-2.832) - (-2.162)]/4 \\
&= 34.540
\end{aligned}$$

where $\hat{h}d_j$ is the solution for level j of herd effect.

From the MME, the solutions for direct and maternal effects for animal i with progeny o are:

$$\begin{bmatrix} \hat{u}_i \\ \hat{m}_i \end{bmatrix} = \begin{bmatrix} n_1 + (d + k_1)\alpha_1 & (d + k_1)\alpha_2 \\ (d + k_1)\alpha_2 & n_2 + (d + k_1)\alpha_3 \end{bmatrix}^{-1} \mathbf{H}k_2 \begin{bmatrix} \hat{u}_s + \hat{u}_d \\ \hat{m}_s + \hat{m}_d \end{bmatrix}$$
$$+ \begin{bmatrix} y_i + \hat{b}_i - \hat{m}_{dam} - \hat{p}_{dam} \\ y_o - \hat{b}_o - \hat{u}_o - \hat{p}_i \end{bmatrix} + \mathbf{H}k_3 \begin{bmatrix} \hat{a}_o - 0.5(\hat{a}_{mate}) \\ \hat{m}_o - 0.5(\hat{m}_{mate}) \end{bmatrix} \tag{8.4}$$

where n_1 is the number of records for animal i; n_2 is the number of progeny records with animal i as the dam; $d = 2$ or $\frac{4}{3}$ when both, one or no parents of animal i are known; $k_2 = 1$ or $\frac{2}{3}$ when both or one parent of animal i are known; $k_1 = \frac{1}{2}$ and $k_3 = 1$ when the mate of animal i is known or $k_1 = \frac{1}{3}$ and $k_3 = \frac{2}{3}$ with the mate unknown and:

$$\mathbf{H} = \begin{bmatrix} \alpha_1 & \alpha_2 \\ \alpha_2 & \alpha_3 \end{bmatrix}$$

For instance, the solutions for direct and genetic maternal effects for animal 5 are:

$$\begin{bmatrix} \hat{u}_5 \\ \hat{m}_5 \end{bmatrix} = \begin{bmatrix} 1 + (2 + 0.5)2.647 & (2 + 0.5)1.176 \\ (2 + 0.5)1.176 & 1 + (2 + 0.5)4.412 \end{bmatrix}^{-1} \mathbf{H}k_2 \begin{bmatrix} u_1 + u_2 \\ m_2 + m_2 \end{bmatrix} + \begin{bmatrix} y_5 - \hat{b}_1 - \hat{m}_2 - \hat{p}_2 \\ y_8 - \hat{b}_1 - \hat{u}_8 - \hat{p}_5 \end{bmatrix}$$
$$+ \mathbf{H}k_3 \begin{bmatrix} \hat{a}_8 - 0.5(\hat{a}_3) \\ \hat{m}_8 - 0.5(\hat{m}_3) \end{bmatrix}$$

$$\begin{bmatrix}\hat{u}_5\\ \hat{m}_5\end{bmatrix} = \begin{bmatrix}1+(2+0.5)2.647 & (2+0.5)1.176\\ (2+0.5)1.176 & 1+(2+0.5)4.412\end{bmatrix}^{-1} \mathbf{H}(1)\begin{bmatrix}0.564+(-1.244)\\ 0.262+-(1.583)\end{bmatrix}$$
$$+\begin{bmatrix}35-0-34.54-(-1.583)-(-1.701)\\ 40-0-34.54-1.917-0.415\end{bmatrix}$$
$$+\mathbf{H}(1)\begin{bmatrix}1.917-0.5(1.165)\\ -0.158-0.5(0.736)\end{bmatrix} = \begin{bmatrix}0.630\\ -0.507\end{bmatrix}$$

The solution for the permanent environmental effect for dam j from the MME is:

$$\hat{pe}_j = (y_o - \hat{b}_o - \hat{u}_o - \hat{m}_j) / (n_2 + \alpha_3) \tag{8.5}$$

where all terms are as defined in Eqn 8.4. For animal 5, the solution for the permanent environmental effect is:

$$\hat{pe}_5 = 40 - 0 - 34.54 - 1.917 - (-0.507) / (1 + 8.75) = 0.415$$

Additive genetic maternal effects represent good mothering ability, which is passed on from dams to progeny, while permanent environment effects refer to permanent environmental and maternal non-additive genetic influences on the mothering ability of the dam. Thus, selection of dams for the next generation in a maternal line would place emphasis on good genetic maternal effects in addition to a good estimate of breeding value. If equal emphasis is placed on both effects, dams 7 and 5 would be the top two dams in the example while dam 2 ranks lowest. However, if the main interest is the performance of the future dams in the same herd, then selection of dams would be based on some combination of the solutions for direct, maternal genetic and permanent environmental effects for the dams. Again, in the example data, dam 2 ranks lowest while the best two dams are dams 6 and 7 if equal emphasis is placed on the three components.

In the case of males, the selection of sires for a maternal line, for instance, would be based on a combination of solutions for direct and maternal genetic effects. Obviously, sires 3 and 1 would be the top two bulls for such a purpose. However, if the emphasis is only on direct genetic effects, probably to breed a bull, then sire 8 in the example would be the bull of choice.

8.3 Reduced Animal Model with Maternal Effects

In Section 4.5, the use of the reduced animal model (RAM), with only one random effect apart from residual error in the model, was considered. The records of non-parents in the MME were expressed as the average of parental breeding values plus Mendelian sampling. This has the advantage of reducing the number of random animal equations in the MME. The application of RAM with multiple random effects in the model is illustrated in this section using the example data used for the full animal model in Section 8.2. The model for the analysis is the same but design matrices and the variance of non-parental animals are different. From the arguments in Section 4.5, the model for the RAM can be expressed as:

$$\begin{bmatrix}\mathbf{y}_p\\ \mathbf{y}_n\end{bmatrix} = \begin{bmatrix}\mathbf{X}_p\\ \mathbf{X}_n\end{bmatrix}\mathbf{b} + \begin{bmatrix}\mathbf{Z}_p\\ \mathbf{Z}_n\end{bmatrix}\mathbf{u}_p + \mathbf{Z}_2\mathbf{m} + \mathbf{Z}_3\mathbf{pe} + \begin{bmatrix}\mathbf{e}_p\\ \mathbf{e}_n\end{bmatrix} \tag{8.6}$$

where $\mathbf{y}_p$, $\mathbf{y}_n$ = vector of observations for parent and non-parents, respectively, $\mathbf{b}$ = vector of fixed effects, $\mathbf{u}_p$= vector of random animal effects for parents, $\mathbf{m}$ = vector of maternal genetic effects for parents, $\mathbf{pe}$ = vector of permanent environmental effects and e_p, e_n = vector of residual error for parents and non-parents, respectively.

The incidence matrices $\mathbf{Z}_2$ and $\mathbf{Z}_3$ relate records to maternal genetic and permanent environmental effect, respectively. The matrices $\mathbf{Z}_p$ and $\mathbf{X}_p$ relate records of parents to animal and fixed effects, respectively, while $\mathbf{Z}_n$ and $\mathbf{X}_n$ relate records of non-parents to parents (animal effect) and fixed effects, respectively.

It is assumed that:

$$\text{var}\begin{bmatrix}\mathbf{u}_p\\ \mathbf{m}\\ \mathbf{pe}\\ \mathbf{e}_p\\ \mathbf{e}_n\end{bmatrix} = \begin{bmatrix} g_{11}\mathbf{A} & g_{12}\mathbf{A} & 0 & 0 & 0\\ g_{21}\mathbf{A} & g_{22}\mathbf{A} & 0 & 0 & 0\\ 0 & 0 & \mathbf{I}\sigma^2_{pe} & 0 & 0\\ 0 & 0 & 0 & \mathbf{I}\sigma^2_{ep} & 0\\ 0 & 0 & 0 & 0 & \mathbf{I}\sigma^2_{en}\end{bmatrix}$$

where σ^2_{ep} is the residual variance for parents, which is equal to σ^2_e in Section 8.2, σ^2_{en} is the residual variance for non-parents and is equal to $\mathbf{I} + \mathrm{D}g_{11}$, with D being a diagonal matrix containing elements d_{jj}, which are equal to $\frac{3}{4}$ or $\frac{1}{2}$, depending on whether one or both parents are known. The matrix $\mathbf{G}$ and σ^2_{pe} are defined as in Section 8.2. Let:

$$\mathbf{X} = \begin{bmatrix}\mathbf{X}_p\\ \mathbf{X}_n\end{bmatrix},\ \mathbf{Z}_1 = \begin{bmatrix}\mathbf{Z}_p\\ \mathbf{Z}_n\end{bmatrix},\ \mathbf{R} = \begin{bmatrix}\mathbf{I}\sigma^2_{ep} & 0\\ 0 & \mathbf{I}\sigma^2_{en}\end{bmatrix} = \begin{bmatrix}\mathbf{R}_n & 0\\ 0 & \mathbf{R}_p\end{bmatrix} \text{ and } \mathbf{R}^{-1} = \begin{bmatrix}\mathbf{R}_p^{-1} & 0\\ 0 & \mathbf{R}_n^{-1}\end{bmatrix}$$

Again, the MME provide the basis of the BLUE of estimable functions of $\mathbf{b}$ and BLUP of $\mathbf{a}$, $\mathbf{m}$ and $\mathbf{pe}$ in Eqn 8.6. The relevant MME are:

$$\begin{bmatrix}\hat{\mathbf{b}}\\ \hat{\mathbf{u}}_p\\ \hat{\mathbf{m}}\\ \hat{\mathbf{pe}}\end{bmatrix}\begin{bmatrix}\mathbf{X'R^{-1}X} & \mathbf{X'R^{-1}Z_1} & \mathbf{X'R^{-1}Z_{12}} & \mathbf{X'R^{-1}Z_3}\\ \mathbf{Z_1'R^{-1}X} & \mathbf{Z_1'R^{-1}Z_1'} + \mathbf{A}_p^{-1}g^{11} & \mathbf{Z_1'R^{-1}Z_2} + \mathbf{A}_p^{-1}g^{12} & \mathbf{Z_1'R^{-1}Z_3}\\ \mathbf{Z_2'R^{-1}X} & \mathbf{Z_2'R^{-1}Z_1'} + \mathbf{A}_p^{-1}g^{21} & \mathbf{Z_2'R^{-1}Z_2} + \mathbf{A}_p^{-1}g^{22} & \mathbf{Z_2'R^{-1}Z_3}\\ \mathbf{Z_3'R^{-1}X} & \mathbf{Z_3'R^{-1}Z_1} & \mathbf{Z_3'R^{-1}Z_2} & \mathbf{Z_3'R^{-1}Z_3'} + \mathbf{I}\ 1/\sigma^2_{pe}\end{bmatrix}^{-1}$$

$$= \begin{bmatrix}\mathbf{X'R^{-1}y}\\ \mathbf{Z_1'R^{-1}y}\\ \mathbf{Z_2'R^{-1}y}\\ \mathbf{Z_3'R^{-1}y}\end{bmatrix} \qquad (8.7)$$

where g^{ii} are the elements of the inverse of $\mathbf{G}$.

As shown in Section 4.5, each block of equations in the MME above can be expressed as the sum of the contributions from parents' records and non-parents' records. Thus:

$$\mathbf{X'R^{-1}X} = \mathbf{X}_p'\ \mathbf{R}_p^{-1}\mathbf{X}_p + \mathbf{X}_n'\ \mathbf{R}_n^{-1}\mathbf{X}_n$$

Expressing Eqn 8.7 as shown for the equations for the block of fixed effects above, and multiplying by R_p, gives:

$$\begin{bmatrix} \mathbf{X}_p'\mathbf{X}_p + \mathbf{X}_n'\mathbf{R}_n^{-1}\mathbf{X}_n & \mathbf{X}_p'\mathbf{Z}_p + \mathbf{X}_n'\mathbf{R}_n^{-1}\mathbf{Z}_n & \mathbf{X}_p'\mathbf{Z}_2 + \mathbf{X}_n'\mathbf{R}_n^{-1}\mathbf{Z}_2 \\ \mathbf{Z}_p'\mathbf{X}_p + \mathbf{Z}_n'\mathbf{R}_n^{-1}\mathbf{X}_n & \mathbf{Z}_p'\mathbf{Z}_p + \mathbf{Z}_n'\mathbf{R}_n^{-1}\mathbf{Z}_n + \mathbf{A}^{-1}\alpha_1 & \mathbf{Z}_p'\mathbf{Z}_2 + \mathbf{Z}_n'\mathbf{R}_n^{-1}\mathbf{Z}_2 + \mathbf{A}^{-1}\alpha_2 \\ \mathbf{Z}_2'\mathbf{X}_p + \mathbf{Z}_2'\mathbf{R}_n^{-1}\mathbf{X}_n & \mathbf{Z}_2'\mathbf{Z}_p + \mathbf{Z}_2'\mathbf{R}_n^{-1}\mathbf{Z}_n + \mathbf{A}^{-1}\alpha_2 & \mathbf{Z}_2'\mathbf{Z}_2 + \mathbf{Z}_2'\mathbf{R}_n^{-1}\mathbf{Z}_2 + A^{-1}\alpha_3 \\ \mathbf{Z}_3'\mathbf{X}_p + \mathbf{Z}_3'\mathbf{R}_n^{-1}\mathbf{X}_n & \mathbf{Z}_3'\mathbf{Z}_p + \mathbf{Z}_3'\mathbf{R}_n^{-1}\mathbf{Z}_n & \mathbf{Z}_3'\mathbf{Z}_3 + \mathbf{Z}_3'\mathbf{R}_n^{-1}\mathbf{Z}_3 \end{bmatrix}$$

$$\begin{matrix} \mathbf{X}_p'\mathbf{Z}_3 + \mathbf{X}_n'\mathbf{R}_n^{-1}\mathbf{Z}_3 \\ \mathbf{Z}_p'\mathbf{Z}_3 + \mathbf{Z}_n'\mathbf{R}_n^{-1}\mathbf{Z}_3 \\ \mathbf{Z}_2'\mathbf{Z}_3 + \mathbf{Z}_2'\mathbf{R}_n^{-1}\mathbf{Z}_3 \\ \mathbf{Z}_3'\mathbf{Z}_3 + \mathbf{Z}_3'\mathbf{R}_n^{-1}\mathbf{Z}_3 + \mathbf{I}\alpha_4 \end{matrix}\Bigg]$$

$$\begin{bmatrix} \hat{\mathbf{b}} \\ \hat{\mathbf{a}} \\ \hat{\mathbf{m}} \\ \hat{\mathbf{pe}} \end{bmatrix} = \begin{bmatrix} \mathbf{X}_p'\mathbf{y}_p + \mathbf{X}_n'\mathbf{R}_n^{-1}\mathbf{y}_n \\ \mathbf{Z}_p'\mathbf{y}_p + \mathbf{Z}_n'\mathbf{R}_n^{-1}\mathbf{y}_n \\ \mathbf{Z}_2'\mathbf{y}_p + \mathbf{Z}_2'\mathbf{R}_n^{-1}\mathbf{y}_n \\ \mathbf{Z}_3'\mathbf{y}_p + \mathbf{Z}_3'\mathbf{R}_n^{-1}\mathbf{y}_n \end{bmatrix}$$

The α terms are as defined in Eqn 8.2 and $\mathbf{R}_n^{-1}$ now equals $1/(1+\mathbf{D}\alpha^{-1})$. The MME for the solutions of **b**, **u**, **m** and **pe** can therefore be set up as shown above or as in Eqn 8.7.

8.3.1 An illustration

Example 8.2

The same data set and genetic parameters as in Section 8.2 are used below to demonstrate the principles for setting up a RAM with maternal effects in the model using Eqn 8.5. Recollect that:

$$\mathbf{G} = \begin{bmatrix} 150 & -40 \\ -40 & 90 \end{bmatrix} \text{and } \mathbf{G}^{-1} = \begin{bmatrix} 0.00756 & 0.00336 \\ 0.00336 & 0.0126 \end{bmatrix}$$

The residual variance for parents $\sigma_{ep}^2 = 350$, and because both parents of non-parents in the data are known:

$$\sigma_{en}^2 = \sigma_e^2 + \frac{1}{2}(g_{11}) = 350 + \frac{1}{2}(150) = 425$$

with:

$$\mathbf{R} = \begin{bmatrix} \mathbf{I}\sigma_{ep}^2 & 0 \\ 0 & \mathbf{I}\sigma_{en}^2 \end{bmatrix}$$

Then:

$$\mathbf{R} = \text{diag}(350, 350, 350, 350, 350, 425, 425, 425, 425, 425)$$

and

$$\mathbf{R}^{-1} = \text{diag}(0.00286, 0.00286, 0.00286, 0.00286, 0.00286, 0.00235, 0.00235, 0.00235, 0.00235, 0.00235)$$

$$1/\sigma_{pe}^2 = \frac{1}{40} = 0.025$$

Setting up the design matrices

The matrix **X**, which relates records to fixed effects, is the same as in Section 8.2.1, considering only animals with records. The matrix $\mathbf{X'R^{-1}X}$ in the MME can be calculated through matrix multiplication from **X** and $\mathbf{R^{-1}}$ already set up. For illustrative purposes, the matrix $\mathbf{X'R^{-1}X}$, when expressed as the sum of the contributions from parents' and non-parents' records, is:

$$\mathbf{X'R^{-1}X} = r_p^{11}\mathbf{X}_p'\mathbf{X}_p + r_n^{11}\mathbf{X}_n'\mathbf{X}_n$$

$$= \begin{bmatrix} 0.0114 & 0.0 & 0.0 & 0.0057 & 0.0057 \\ 0.0 & 0.0029 & 0.0 & 0.0029 & 0.0 \\ 0.0 & 0.0 & 0.0 & 0.0 & 0.0 \\ 0.0057 & 0.0029 & 0.0 & 0.0086 & 0.0 \\ 0.0057 & 0.0 & 0.0 & 0.0 & 0.0057 \end{bmatrix} + \begin{bmatrix} 0.0 & 0.0 & 0.0 & 0.0 & 0.0 \\ 0.0 & 0.0047 & 0.0 & 0.0 & 0.0047 \\ 0.0 & 0.0 & 0.0071 & 0.0024 & 0.0047 \\ 0.0 & 0.0 & 0.0024 & 0.0024 & 0.0 \\ 0.0 & 0.0047 & 0.0047 & 0.0 & 0.0094 \end{bmatrix}$$

$$= \begin{bmatrix} 0.0114 & 0.0 & 0.0 & 0.0057 & 0.0057 \\ 0.0 & 0.0076 & 0.0 & 0.0029 & 0.0047 \\ 0.0 & 0.0 & 0.0071 & 0.0024 & 0.0047 \\ 0.0057 & 0.0029 & 0.0024 & 0.0109 & 0.0 \\ 0.0057 & 0.0047 & 0.0047 & 0.0 & 0.0151 \end{bmatrix}$$

where $\mathbf{X}_p$ and $\mathbf{X}_n$ are matrices relating parents and non-parents to fixed effects, respectively, and are:

$$\mathbf{X}_p' = \begin{bmatrix} 1 & 1 & 1 & 1 & 0 \\ 0 & 0 & 0 & 0 & 1 \\ 0 & 0 & 0 & 0 & 0 \\ 1 & 0 & 0 & 1 & 1 \\ 0 & 1 & 1 & 0 & 0 \end{bmatrix} \quad \text{and} \quad X_n' = \begin{bmatrix} 0 & 0 & 0 & 0 & 0 \\ 1 & 1 & 0 & 0 & 0 \\ 0 & 0 & 1 & 1 & 1 \\ 0 & 0 & 0 & 1 & 0 \\ 1 & 1 & 1 & 0 & 1 \end{bmatrix}$$

The matrix $\mathbf{Z}_1$, which relates records to animal effect, is: 123456789

$$\mathbf{Z}_1 = \begin{array}{c} \\ 5 \\ 6 \\ 7 \\ 8 \\ 9 \\ 10 \\ 11 \\ 12 \\ 13 \\ 14 \end{array} \begin{array}{c} \begin{array}{ccccccccc} 1 & 2 & 3 & 4 & 5 & 6 & 7 & 8 & 9 \end{array} \\ \begin{bmatrix} 0.0 & 0.0 & 0.0 & 0.0 & 1.0 & 0.0 & 0.0 & 0.0 & 0.0 \\ 0.0 & 0.0 & 0.0 & 0.0 & 0.0 & 1.0 & 0.0 & 0.0 & 0.0 \\ 0.0 & 0.0 & 0.0 & 0.0 & 0.0 & 0.0 & 1.0 & 0.0 & 0.0 \\ 0.0 & 0.0 & 0.0 & 0.0 & 0.0 & 0.0 & 0.0 & 1.0 & 0.0 \\ 0.0 & 0.0 & 0.0 & 0.0 & 0.0 & 0.0 & 0.0 & 0.0 & 1.0 \\ 0.0 & 0.5 & 0.5 & 0.0 & 0.0 & 0.0 & 0.0 & 0.0 & 0.0 \\ 0.0 & 0.0 & 0.5 & 0.0 & 0.0 & 0.0 & 0.5 & 0.0 & 0.0 \\ 0.0 & 0.0 & 0.0 & 0.0 & 0.0 & 0.0 & 0.5 & 0.5 & 0.0 \\ 0.0 & 0.5 & 0.0 & 0.0 & 0.0 & 0.0 & 0.0 & 0.0 & 0.5 \\ 0.0 & 0.0 & 0.5 & 0.0 & 0.0 & 0.5 & 0.0 & 0.0 & 0.0 \end{bmatrix} \end{array}$$

The first five rows correspond to animals 5–9, which are parents, and each has one record. The last five rows correspond to the records for animals 10–14 (non-parents), which are related to their parents. The matrices $\mathbf{Z}_2$ and $\mathbf{Z}_3$ are exactly the same as **W** and **S** in Section 8.2.1, respectively, and the vector of observation, y, is the same as in Section 8.2.1. Apart from the relationship matrix, all the matrices in the MME can easily be calculated through matrix multiplication from the design matrices and vector of observation set up above. The inverse of the relationship matrix is set up only for parents ($\mathbf{A}_p^{-1}$), i.e. for animals 1–9, using the procedure outlined in Chapter 3. The matrix $\mathbf{A}_p^{-1}g^{12}$ is added to animal equations, $\mathbf{A}_p^{-1}g^{22}$ to the equations for maternal genetic effects, $\mathbf{A}_p^{-1}g^{12}$ to the animal by maternal genetic equations, $\mathbf{A}_p^{-1}g^{21}$ to the maternal genetic by animal equations, and $1/\sigma_{pe}^2$ to the diagonals of the equations for permanent environmental effects to obtain the MME. The MME are not presented because they are too large. Solving the MME by direct inversion with the equation for the first herd set to zero gives the same solutions as from the animal model (Example 8.1). However, the number of non-zero elements in the coefficient matrix was 329 compared with 429 in the animal model, due to the reduced number of equations, indicating the advantages of the RAM.

Back-solving for non-parents

The solutions for direct animal and maternal effects for non-parents are back-solved after the MME have been solved.

Back-solving for direct effects

Solutions for direct animal effects for the non-parents are obtained from parent average and an estimate of Mendelian sampling using Eqn 4.27. Thus, the solution for the non-parent *i* is:

$$\hat{u}_i = 0.5(\hat{u}_s + \hat{u}_d) + k_i(y_i - \hat{b}_j - \hat{m}_d - \hat{pe}_d - 0.5(\hat{u}_s + \hat{u}_d)) \qquad (8.8)$$

with:

$$k_i = r^{-1}/(r^{-1} + d^{-1}g^{-1}) = 1/(1 + d^{-1}\alpha); \qquad \alpha = \sigma_e^2/\sigma_a^2$$

where d is either $\frac{1}{2}$ if both parents are known or $\frac{3}{4}$ if only one parent is known. For the example data, both parents of the non-parent individuals are known, therefore:

$$k_i = 1/(1 + 2(2.333)) = 0.17647$$

For animal 10, for instance, the breeding value is:

$$\begin{aligned}\hat{u}_{10} &= 0.5(\hat{u}_3 + \hat{u}_2) + k(y_{10} - \hat{b}_2 - \hat{b}_5 - \hat{m}_3 - \hat{p}_2 - 0.5(\hat{u}_3 + \hat{u}_2))\\ &= 0.5(1.165 + -1.244) + 0.17647(22 - 3.386 - 27.691 - (-1.583)\\ &\quad - (-1.701) - 0.5(1.165 + -1.244))\\ &= -1.055\end{aligned}$$

Back-solving for maternal effects

The equation for obtaining genetic maternal effects for non-parents can be derived as follows. From the MME, the equation for direct and genetic maternal effects for non-parent *i* is:

$$\begin{bmatrix} r^{-1}+n^{-1}g^{11} & n^{-1}g^{12} \\ n^{-1}g^{21} & n^{-1}g^{22} \end{bmatrix}\begin{bmatrix} \hat{u}_i \\ \hat{m}_i \end{bmatrix} = G^{-1}k_2\begin{bmatrix} \hat{u}_s+\hat{u}_d \\ \hat{m}_s+\hat{m}_d \end{bmatrix} + r^{-1}\begin{bmatrix} y_i-\hat{b}_i-\hat{m}_{dam}-\hat{p}e_{dam} \\ 0 \end{bmatrix} \quad (8.9)$$

where n is defined in Eqn 8.8 and other terms are as defined in Eqn 8.4.

From the above equations:

$$\begin{aligned} \hat{m}_i &= [g^{22}(\hat{m}_s+\hat{m}_d)+g^{21}(\hat{u}_s+\hat{u}_d)-n^{-1}g^{21}(\hat{u}_i)]/n^{-1}g^{22} \\ \hat{m}_i &= n(\hat{m}_s+\hat{m}_d)+[(g^{21}n(\hat{u}_s+\hat{u}_d)-g^{21}\hat{u}_i)/g^{22}] \\ \hat{m}_i &= n(\hat{m}_s+\hat{m}_d)+g^{21}/g^{22}(n(\hat{u}_s+\hat{u}_d)-\hat{u}_i) \end{aligned}$$

Note that:

$$\begin{aligned} g^{21}/g^{22} &= \{-g_{12}/(g_{11}g_{22}-g_{12}g_{21})\}\{(g_{11}g_{22}-g_{12}g_{21})/g_{11}\} \\ &= -g_{12}/g_{11} \end{aligned}$$

Therefore:

$$\hat{m}_i = n(\hat{m}_s+\hat{m}_d)+g_{12}/g_{11}(\hat{u}_i-d(\hat{u}_s+\hat{u}_d)) \quad (8.10)$$

When both parents are known:

$$\hat{m}_i = 0.5(\hat{m}_s+\hat{m}_d)+(g_{12}/g_{11})(\hat{u}_i-0.5(\hat{u}_s+\hat{u}_d))$$

For instance, for animal 10:

$$\begin{aligned} \hat{m}_{10} &= 0.5(\hat{m}_3+\hat{m}_2)+(g_{12}/g_{11})(\hat{u}_5-0.5(\hat{u}_3+\hat{u}_2)) \\ &= 0.5(0.736+(-1.583))+(-40/150)(-1.055-0.5(1.165+-1.244)) \\ &= -0.153 \end{aligned}$$

The solutions for direct and maternal effects of all non-parents in the example data (animals 10–14), applying Eqns 8.7 and 8.9, are exactly the same as those obtained for these animals in the animal model.

8.4 Sire and Maternal Grandsire Model

In some cases, due to the structure of the available data, a sire and maternal grandsire model may be fitted for traits affected by direct and maternal genetic effects. This tends to be more common for calving traits such as calving ease or stillbirth (Wiggans *et al.*, 2003). The calving event is regarded as a direct effect of the service sire (direct effect). This predicts how easily his progeny are born and is computed by fitting the service sire. The maternal effect, which predicts how easily the bull daughters calve, is computed by fitting the MGS, hence the name sire-maternal grandsire (S-MGS) model.

The model, then, is similar to Eqn 8.1 and can be written as:

$$\mathbf{y} = \mathbf{Xb}+\mathbf{Zs}+\mathbf{Wmgs}+\mathbf{Spe}+\mathbf{e} \quad (8.11)$$

where **y** = vector of observations, **s** = vector of random service sire (direct) effects, **mgs** = vector of random MGS (indirect) genetic effects and other terms defined as in Eqn 8.1, but **Z** and **W** are now incidence matrices relating records to service sire and MGS genetic effects, respectively. Note that if only first lactation data is being analysed, then the **pe** can be omitted from the model.

It is assumed that:

$$\text{var}(\mathbf{s}) = \mathbf{A}\sigma_s^2, \text{with } \sigma_s^2 = 0.25\sigma_u^2; \text{var}(\mathbf{mgs}) = \mathbf{A}\sigma_{mgs}^2 \text{ with, } \sigma_{mgs}^2 = (\frac{1}{16}\sigma_u^2 + \frac{1}{4}\sigma_m^2 + \frac{1}{4}\sigma_{u,m}),$$

where σ_u^2 and σ_m^2 are the additive genetic variance and maternal genetic variance, respectively.

$$\text{cov}(\mathbf{s}, \mathbf{mgs}) = \mathbf{A}\sigma_{s,mgs}, \quad \text{with} \quad \sigma_{s,mgs} = \frac{1}{8}\sigma_u^2 + \frac{1}{4}\sigma_{u,m}$$

$$\text{var}(\mathbf{pe}) = \sigma_{pe}^2 = (\frac{3}{16}\sigma_u^2 + \frac{3}{4}\sigma_m^2 + \frac{3}{4}\sigma_{u,m} + \sigma_{pe}^2) \text{ and}$$

$$\text{var}(\mathbf{e}) = \sigma_e^2 = (\frac{1}{2}\sigma_u^2 + \frac{1}{4}\sigma_{te}^2)$$

The same principles described in Section 8.2 can be used in the application of Eqn 8.11 to estimate breeding values and solutions for fixed effects. Note, however, that MME from such an analysis will produce predicted transmitting abilities (PTAs) (which is half of the EBV) for the service sire (direct effect). Therefore, the solutions from the MME have the following interpretation:

Solution for sire = PTA_D

Solution for mgs = ½ $\text{PTA}_\text{D} + \text{PTA}_\text{M}$

where subscripts D and M refer to direct and maternal genetic effects. Therefore, PTAs for maternal effect (PTA_M) can be computed as:

PTA_M = Solution for mgs $-$ ½PTA_D

The variance components for a S-MGS model can be converted to variances for an animal model direct and maternal effects from the details of the components of the variances defined above. Thus, the direct genetic variance component $(\sigma_u^2) = 4\sigma_s^2$, the covariance between direct and maternal component $(\sigma_{u,m}) = 4^*(\sigma_{s,mgs}) - 0.5\sigma_u^2$ and the maternal genetic variance component $(\sigma_m^2) = 4\sigma_{mgs}^2 - 0.25\sigma_u^2 - \sigma_{u,m}$. The computation of maternal genetic component (σ_m^2) can be illustrated as:

$$\sigma_m^2 = (4\sigma_{mgs}^2 - 0.25\sigma_u^2 - \sigma_{u,m}) = 4(\frac{1}{16}\sigma_u^2 + \frac{1}{4}\sigma_m^2 + \frac{1}{4}\sigma_{u,m}) - 0.25\sigma_u^2 - \sigma_{u,m} = \sigma_m^2$$

9 Social Interaction Models

9.1 Introduction

Social interaction among animals, such as competition and co-operation, can have a profound effect on the expressions of performance and welfare traits in domestic livestock populations (Muir, 2005; Bijma *et al.*, 2007a). When a group of animals relies on a limiting resource (e.g. feed) to achieve an outcome (e.g. growth), the observed phenotype of an individual (e.g. growth rate) can be influenced by both the phenotype (e.g. ability to fight for food) and the genotype (which confers this ability) of the competitors in the group. So, the growth rate of piglets, for instance, can be reduced due to competition for food. In laying-hen production systems, social interactions can result in mortality due to cannibalism when hens are housed in groups, and this poses both economic and welfare problems.

Although a major component of the social interaction among group members may appear to be environmental, there is a genetic component (Wolf *et al.*, 1998) attributable to the genes carried by others in the group, which affects how they compete; these are generally referred to as indirect genetic effects (IGE) (Cheverud and Moore, 1994; Moore *et al.*, 1997). A selection experiment to reduce mortality due to cannibalism in domestic chickens (Muir, 1996) has shown that heritable interactions (or IGE) can contribute substantially to response to selection. Selection schemes that ignore this social effect of an individual on the phenotypes of its group members could result in less optimum response or even response in the opposite direction (Griffing, 1967). This social effect or indirect genetic effect (Cheverud and Moore, 1994) is often referred to as an associative effect (Griffing, 1967). In addition, Bijma *et al.* (2007b) indicated that the existence of social interaction among individuals may increase the total heritable variance in a trait. They found that heritable variance in survival days expressed as a proportion of phenotypic variance increased from 7% to 20% due to social interactions, indicating that about two-thirds of heritable variation is due to interactions among individuals. One possible solution for improving traits affected by social interaction is to undertake group selection (Griffing, 1967). However, an optimum individual selection scheme to improve traits affected by interactions among individuals will involve the use of models that account for:

1. direct effects due to the direct effects of the genes of the individual; and
2. indirect effects due to the associative effect of the individual on its group members.

The phenotype (P_i) of an animal i for a trait influenced by social interaction belonging to a group with n members, where interaction occurs, may be modelled as:

$$P_i = A_{D,i} + Q_{D,i} + E_{D,i} = \sum_{j \neq i}^{n-1} A_{S,j} + \sum_{j \neq i}^{n-1} Q_{S,j} + E_{S,j}$$

where j is one of the $n - 1$ group mates, $A_{D,i}$ and $A_{S,j}$ are the additive direct effect and sum of the additive indirect effects of each of the $n - 1$ group mates, with corresponding non-additive components $Q_{D,i}$ and $Q_{S,i}$ and environmental components $E_{D,i}$ and $E_{S,j}$. The

DOI: 10.1079/9781800620506.0009

non-additive components may be combined with the environmental components such that the equation may be expressed:

$$P_i = A_{D,i} + E_{D,i} + \sum_{j\neq i}^{n-1} A_{S,j} + E_{S,j}$$

Therefore, the phenotypic variance can be derived as:

$$\mathrm{var}(P) = \mathrm{var}[A_{D,i} + E_{D,i} + \sum_{j\neq i}^{n-1} A_{S,j} + E_{S,j}]$$

Given that $\mathrm{cov}(E_{D,i}, E_{S,j}) = 0$ when $i \neq j$, and $\mathrm{cov}(A, E) = 0$ for all *i*, *j*, then:

$$\mathrm{var}(P) = \sigma^2_{A_D} + \sigma^2_{E_D} + \mathrm{var}(\sum_{j\neq i}^{n-1} A_{S,i}) + \mathrm{var}(\sum_{j\neq i}^{n-1} E_{S,j}) + 2\,\mathrm{cov}(A_{D,i}, \sum_{j\neq i}^{n-1} A_{S,j})$$

with $\mathrm{cov}((E_{S,j}, E_{S,j'}) = 0$
when:

$$j \neq j', \mathrm{var}\left(\sum_{j\neq i}^{n-1} E_{s,j}\right) = (n-1)\sigma^2_{E_S}$$

Also, given that $\mathrm{cov}(A_{S,j}, A_{S,j'}) = r_{jj'}\sigma^2_{A_S}$ where $r_{jj'}$ is the relatedness between animals *j* and j', then:

$$\mathrm{var}\left(\sum_{j\neq i}^{n-1} A_{s,j}\right) = (n-1)\sigma^2_{A_s} + (n-1)(n-2)r\sigma^2_{A_s}$$

with *r* equal to the mean relatedness within the groups. Finally:

$$\mathrm{cov}\left(A_{D,i}, \left(\sum_{j\neq i}^{n-1} A_{s,j}\right)\right) = (n-1)r\sigma_{A_{DS}}$$

Collecting all the terms together gives the phenotypic variance as:

$$\sigma^2_p = \sigma^2_{A_D} + \sigma^2_{E_D} + (n-1)(\sigma^2_{A_S} + \sigma^2_{E_S}) + (n-1)r[2\sigma_{A_{DS}} + (n-2)\sigma^2_{A_S}]$$

However, the total breed value (TBV; Bijma *et al.*, 2007a) for individual *i* is:

$$\mathrm{TBV}_i = A_{D,i} + (n-1)A_{s,j} \qquad (9.1)$$

Note that TBV is what the progeny of the individual *i* will inherit and is the relevant breeding value in computing response for selection for traits affected by associative effects. Therefore, the total heritable variance of the trait equals the variance of the TBVs (σ^2_{TBV}) among individuals and is:

$$\sigma^2_{TBV} = \sigma^2_{A_D} + 2(n-1)\sigma_{A_{DS}} + (n-1)^2\sigma^2_{A_S}$$

where $\sigma^2_{A_D}$, $\sigma^2_{A_S}$ and $\sigma_{A_{DS}}$ are the variance of direct breeding value (DBV), associative breeding value (SBV) and the covariance between DBV and SBV, respectively. The sign of this covariance provides a measure of the competition versus co-operation among group members. Negative values may be interpreted as 'heritable competition' in the sense that animals' positive DBV on the basis of their phenotype has a negative heritable impact on the phenotypes of their associates. On the other hand, a positive covariance may be interpreted as 'heritable co-operation' (Bijma *et al.*, 2007b).

Thus, the ratio of total heritable variance to the phenotypic variance (τ^2) for traits with associative effects (Bergsma *et al.*, 2008) can be expressed as $\tau^2 = \sigma^2_{TBV} / \sigma^2_p$. A comparison of τ^2 to the classical heritability ($h^2 = \sigma^2_A / \sigma^2_p$) indicates the proportional contribution of indirect additive effects to the total heritable variance for traits with associative effects.

Bijma *et al.* (2007a) presented this general formula for total genetic response per generation ($\Delta\bar{G}$) to selection for traits with associative effects.

$$\Delta\bar{G} = \{wt(n-1)(r+1)\sigma^2_{TBV} + (1-wt)\sigma_{p,TBV}\} \kappa / \sigma_I$$

where $\sigma_{p,TBV}$ is the covariance between the phenotype of the individual and TBV, r measures the degree of genetic relatedness, which is twice the coefficient of coancestry, κ is the selection intensity, σ_I is the standard deviation of the index (**I**) that combines individual phenotypes and phenotypes of group members, and wt defines the weights on individuals versus phenotypes of group members, such that:

$$\mathbf{I} = wtP_i + (1-wt)\sum_{j \neq i}^{n} p_j$$

Thus, for a given r, n, wt and selection intensity, response is dependent on the σ^2_{TBV} and the covariance between the phenotype of the individual and TBV. Therefore, response to selection may not necessarily follow the same direction as the selection pressure as in classical quantitative theory. The interactions among individuals affect both the direction and magnitude of selection response. Strong competition, for instance a negative $\sigma_{p,TBV}$ due to a large and negative $\sigma_{A_{DS}}$, will result in a response opposite in direction to the direction of selection.

9.2 Animal Model with Social Interaction Effects

Usually, data with associative effects tend to include animals that are full-sibs and therefore there is the need to account for the common environmental effects in the model. Thus, the MME for a trait with social interaction effects could be written as:

$$\mathbf{y} = \mathbf{Xb} + \mathbf{Z}_D\mathbf{u}_D + \mathbf{Z}_S\mathbf{u}_S + \mathbf{Wc} + \mathbf{e} \qquad (9.2)$$

where **b** is the vector of fixed effects, $\mathbf{u}_D$ and $\mathbf{u}_S$ are the vectors for direct and associative genetic effects, respectively, **c** is the vector for common environmental effects and **e** is the vector for residual error.

It is also assumed that:

$$\text{var}\begin{bmatrix} \mathbf{u}_\text{D} \\ \mathbf{u}_\text{S} \end{bmatrix} = \begin{bmatrix} g_{11}\mathbf{A} & g_{12}\mathbf{A} \\ g_{21}\mathbf{A} & \mathbf{A}_{22}\mathbf{A} \end{bmatrix}$$

and if there are n animals in a group, then for the ith animal:

$$\text{var}(e_i) = (\text{var}(E_{D,i} + E_{S,i}), j = 1, n-1 \text{ and } i \neq j) = \sigma^2_{E_D} + (n-1)\sigma^2_{E_s} \qquad (9.3)$$

Assuming that $n = 3$, with animals i, j and k in the group, then the residual covariance between animal i and j in the same group or pen is:

$$\begin{aligned} \text{cov}_{penmates} &= \text{cov}(e_i, e_j) = \text{cov}(E_{D,i} + E_{S,j} + E_{S,k}; E_{D,j} + E_{S,i} + E_{S,k}) \\ &= \text{cov}\left(E_{D,i}, E_{S,i}\right) + \text{cov}\left(E_{S,j}, E_{D,j}\right) + \text{cov}\left(E_{S,k}, E_{S,k}\right) = 2\sigma_{E_{DS}} + (n-2)\sigma^2_{E_s} \qquad (9.4) \end{aligned}$$

Therefore, the correlation among animals in the same group (ρ) can be defined as:

$$\rho = \text{cov}(e_i, e_j)/\text{var}(e) = [2\sigma_{E_{DS}} + (n-2)\sigma^2_{E_s}]/[\sigma^2_{E_D} + (n-1)\sigma^2_{E_s}]$$

Assuming that residual covariance among different groups is zero, the residual variance structure can then be defined as $\text{var(e)} = \mathbf{R}$, with $r_{ii} = \sigma^2_e$, $r_{ij} = \rho(\sigma^2_e)$ for animals i and j in the same group and $r_{ij} = 0$ for animals i and j in different groups. Thus, **R** is block diagonal, and with $n = 3$, the block diagonal for one group is:

$$\mathbf{R} = \begin{bmatrix} 1 & \rho & \rho \\ \rho & 1 & \rho \\ \rho & \rho & 1 \end{bmatrix} \sigma^2_e$$

All elements between the various block diagonals are zero. However, Bergsma *et al.* (2008) indicated that the residual covariance within groups ($\text{cov}_{penmates}$) equals the variance among group means (σ^2_g). Thus, when $\text{cov}_{penmates}$ or ρ is > 0, instead of fitting the correlated residual structure described above, a random group effect can be fitted as an equivalent model, with:

$$\sigma^2_g = 2\sigma_{E_{DS}} + (n-2)\sigma^2_{E_s}$$

and residual variance now defined as:

$$\sigma^2_{e^*} = \sigma^2_e - \sigma^2_g$$

Therefore, the equivalent model to Eqn 9.2 is:

$$\mathbf{y} = \mathbf{Xb} + \mathbf{Z}_D\mathbf{u}_D + \mathbf{Z}_s\mathbf{u}_s + \mathbf{Vg} + \mathbf{Wc} + \mathbf{e} \tag{9.5}$$

where **g** is the vector of random group effects with $\mathbf{g} \sim N(0, \mathbf{I}_g\sigma^2_g)$. The MME to be solved, then, are:

$$\begin{pmatrix} \mathbf{X'X} & \mathbf{X'Z_D} & \mathbf{X'Z_S} & \mathbf{X'V} & \mathbf{X'W} \\ \mathbf{Z'_DX} & \mathbf{Z'_D Z_D} + \mathbf{A}^{-1}\alpha_1 & \mathbf{Z'_D Z_S} + \mathbf{A}^{-1}\alpha_2 & \mathbf{Z'_DV} & \mathbf{Z'_DW} \\ \mathbf{Z'_SX} & \mathbf{Z'_S Z_D} + \mathbf{A}^{-1}\alpha_2 & \mathbf{Z'_S Z_S} + \mathbf{A}^{-1}\alpha_3 & \mathbf{Z'_SV} & \mathbf{Z'_SW} \\ \mathbf{V'X} & \mathbf{V'Z_D} & \mathbf{V'Z_S} & \mathbf{V'V} + \mathbf{I}\alpha_4 & \mathbf{V'W} \\ \mathbf{W'X} & \mathbf{W'Z_D} & \mathbf{W'Z_S} & \mathbf{W'V} & \mathbf{W'W} + \mathbf{I}\alpha_5 \end{pmatrix} \begin{pmatrix} \hat{\mathbf{b}} \\ \hat{\mathbf{u}}_D \\ \hat{\mathbf{u}}_S \\ \hat{\mathbf{g}} \\ \hat{\mathbf{c}} \end{pmatrix} = \begin{pmatrix} \mathbf{X'y} \\ \mathbf{Z'_Dy} \\ \mathbf{Z'_Sy} \\ \mathbf{V'y} \\ \mathbf{W'y} \end{pmatrix} \tag{9.6}$$

If $\mathbf{G}^{-1} = \begin{bmatrix} g^{11} & g^{12} \\ g^{21} & g^{22} \end{bmatrix}$ then $\begin{bmatrix} \alpha_1 & \alpha_2 \\ \alpha_2 & \alpha_3 \end{bmatrix} = \sigma^2_{e^*} \begin{bmatrix} g^{11} & g^{12} \\ g^{21} & g^{22} \end{bmatrix}$, $\alpha_4 = \sigma^2_{e^*}/\sigma^2_g$ and $\alpha_5 = \sigma^2_{e^*}/\sigma^2_c$.

However, when $\text{cov}_{penmates}$ is ≤ 0, the MME to be solved are:

$$\begin{bmatrix} \mathbf{X'R^{-1}X} & \mathbf{X'R^{-1}Z_D} & \mathbf{X'R^{-1}Z_S} & \mathbf{X'R^{-1}W} \\ \mathbf{Z_D'R^{-1}X} & \mathbf{Z_D'R^{-1}Z_D} + \mathbf{A}^{-1}g^{11} & \mathbf{Z_D'R^{-1}Z_S} + \mathbf{A}^{-1}g^{12} & \mathbf{Z_D'R^{-1}W} \\ \mathbf{Z_S'R^{-1}X} & \mathbf{Z_S'R^{-1}Z_D} + \mathbf{A}^{-1}g^{21} & \mathbf{Z_S'R^{-1}Z_S} + \mathbf{A}^{-1}g^{22} & \mathbf{Z_S'R^{-1}W} \\ \mathbf{W'R^{-1}X} & \mathbf{W'R^{-1}Z_D} & \mathbf{W'R^{-1}Z_S} & \mathbf{W'R^{-1}W} + \mathbf{I}\sigma_c^2 \end{bmatrix}_{ss}$$

$$\begin{bmatrix} \hat{\mathbf{b}} \\ \hat{\mathbf{u}}_D \\ \hat{\mathbf{u}}_S \\ \hat{\mathbf{c}} \end{bmatrix} = \begin{bmatrix} \mathbf{X'R^{-1}y} \\ \mathbf{Z_D'R^{-1}y} \\ \mathbf{Z_S'R^{-1}y} \\ \mathbf{W'R^{-1}y} \end{bmatrix} \qquad (9.7)$$

Although the number of equations to be fitted for Eqn 9.6 is usually more than for Eqn 9.7, the systems of equations for Eqn 9.7 are denser and more difficult to set up.

9.2.1 Illustration of a model with social interaction

Example 9.1

Table 9.1 contains the growth rate data of nine pigs housed in three pens during the finishing period in groups of three. The pigs are from three different litters and the aim is to estimate the direct and associative breeding values for all pigs, estimate sex effect and common environment effect as some of the pigs are full-sibs. It is assumed that genetic variances for direct and associative effects are 25.70 g^2 and 3.60 g^2, respectively, with a covariance of 2.25 g between them. Also, it is assumed that the variance for common environmental variance (σ_c^2) is 12.5 g^2 and residual variances for direct ($\sigma_{E_D}^2$) and associative ($\sigma_{E_s}^2$) effects are 40.6 g^2 and 10.0 g^2, respectively, and the correlation among pigs in the same pen (ρ) is 0.2.

The MME in Eqn 9.6 are initially used to analyse the data. Based on the given genetic parameters:

$$\text{var(e)} = \sigma_{E_D}^2 + (n-1)\sigma_{E_S}^2 = 40.6 + (3-1)10 = 60.6$$

Since $\rho = \text{cov}(e_i, e_j)/\text{var(e)} = 0.2$ and $\text{cov}(e_i, e_j) = \sigma_g^2$ in Eqn 9.6, then $\sigma_g^2 = \rho\,\text{var(e)} =$ 0.2*60.6 = 12.12.

Therefore, the residual variance relevant to the analysis using Eqn 9.6 with groups fitted is $\text{var}(e^*) = \text{var}(e) - \sigma_g^2 = 60.6 - 12.12 = 48.48$

and:

$$\begin{bmatrix} \alpha_1 & \alpha_2 \\ \alpha_2 & \alpha_3 \end{bmatrix} \text{in[8.6]} = \mathbf{G}^{-1}\sigma_{e^*}^2 = \begin{bmatrix} 1.9956 & -1.2472 \\ -1.2472 & 14.2462 \end{bmatrix}$$

Setting up the incidence matrices **X**, **V**, **W** and $\mathbf{Z}_D$ in Eqn 9.6 follows the pattern already described for other models in previous chapters, with $\mathbf{Z}_D$ being a diagonal matrix for animals with records and

$$\mathbf{V} = \begin{pmatrix} 1 & 1 & 1 & 0 & 0 & 0 & 0 & 0 & 0 \\ 0 & 0 & 0 & 1 & 1 & 1 & 0 & 0 & 0 \\ 0 & 0 & 0 & 0 & 0 & 0 & 1 & 1 & 1 \end{pmatrix}$$

relating records to pen (groups).

Table 9.1. The growth rate of a set of finishing pigs.

Animal	Sire	Dam	Pen	Sex	Growth rate (g/day)*10
7	1	4	1	Male	5.50
8	1	4	1	Female	9.80
9	2	5	1	Female	4.90
10	1	4	2	Male	8.23
11	2	5	2	Female	7.50
12	3	6	2	Female	10.00
13	2	5	3	Male	4.50
14	3	6	3	Female	8.40
15	3	6	3	Male	6.40

The matrix $\mathbf{Z}_S$ that relates an individual to other members of the same group is:

$$\mathbf{Z}_S = \begin{pmatrix} 0 & 1 & 1 & 0 & 0 & 0 & 0 & 0 & 0 \\ 1 & 0 & 1 & 0 & 0 & 0 & 0 & 0 & 0 \\ 1 & 1 & 0 & 0 & 0 & 0 & 0 & 0 & 0 \\ 0 & 0 & 0 & 0 & 1 & 1 & 0 & 0 & 0 \\ 0 & 0 & 0 & 1 & 0 & 1 & 0 & 0 & 0 \\ 0 & 0 & 0 & 1 & 1 & 0 & 0 & 0 & 0 \\ 0 & 0 & 0 & 0 & 0 & 0 & 0 & 1 & 1 \\ 0 & 0 & 0 & 0 & 0 & 0 & 1 & 0 & 1 \\ 0 & 0 & 0 & 0 & 0 & 0 & 1 & 1 & 0 \end{pmatrix}$$

Setting up the MME therefore follows a similar pattern to that described in previous chapters. Solving the MME (Eqn 9.6) for this example gives the following set of solutions. The results from an analysis that ignored associative effects but fitted random animal and common environmental effects and fixed effects of sex of pigs and pen effects are also presented.

Model with associative effects				Model with no associative effects[a]
Sex of pig effects				
Male	6.004			0.000
Female	8.243			2.169
Animal effects				
	DBV	SBV	TBV	
1	0.296	–0.044	0.207	0.336
2	–0.483	0.028	–0.428	–0.478
3	0.188	0.017	0.221	0.142
4	0.296	–0.044	0.207	0.336
5	–0.483	0.028	–0.428	–0.478
6	0.188	0.017	0.221	0.142
7	0.125	–0.076	–0.027	0.279
8	0.522	–0.099	0.324	0.652

Continued

Continued.

Model with associative effects				Model with no associative effects[a]
9	−0.874	0.009	−0.856	−0.738
10	0.536	−0.003	0.530	0.412
11	−0.488	0.083	−0.321	−0.628
12	0.399	0.060	0.519	0.216
13	−0.572	0.019	−0.534	−0.547
14	0.153	0.005	0.163	0.162
15	0.199	0.002	0.203	0.192
Common environment effects				
1	0.333			0.327
2	−0.515			−0.465
3	0.183			0.139
Group effects				
1	−0.269			
2	0.359			
3	−0.090			

DBV, direct EBV; SBV, associative EBV; TBV, total EBV = (DBV + (n – 1)SBV).
[a]Model also fitted pen effects, and solutions were 5.160, 7.131 and 5.838 for pens 1, 2 and 3, respectively.

Although solutions for sex of pig and common environmental effects were generally in the same direction in models with or without associative effects, there was a major re-ranking of animals based on the EBVs. Griffing (1967) has indicated that selection schemes that ignore this social effect of an individual on the phenotypes of its group members could result in less optimum response, while Bijma *et al.* (2007a) observed that the presence of social interaction among individuals may increase the total heritable variance in a trait.

9.3 Partitioning Evaluations from Associative Models

The equations for DBV and SBV for animal i can be written as:

$$\begin{bmatrix} \mathbf{Z}'_{iD}\ \mathbf{Z}_{iD} + \mathbf{A}^{1}_{\alpha_1} & \mathbf{Z}'_{iD}\ \mathbf{Z}_{iS} + \mathbf{A}^{-1}_{\alpha_2} \\ \mathbf{Z}'_{iS}\ \mathbf{Z}_{iD} + \mathbf{A}^{-1}_{\alpha_2} & \mathbf{Z}'_{iS}\ \mathbf{Z}_{iS} + \mathbf{A}^{-1}_{\alpha_3} \end{bmatrix} \begin{bmatrix} \hat{\mathbf{u}}_{D} \\ \hat{\mathbf{u}}_{S} \end{bmatrix} = \begin{bmatrix} \mathbf{Z}'_{iD}(y_i - \mathbf{X}\hat{\mathbf{b}} - \mathbf{Z}_{jS}\hat{\mathbf{u}}_{jS}) \\ \mathbf{Z}'_{jS}(y_j - \mathbf{X}\hat{\mathbf{b}} - \mathbf{Z}_{jS}\hat{\mathbf{u}}_{jS} - \mathbf{Z}_{jD}\hat{\mathbf{u}}_{jD}) \end{bmatrix}$$

$$= \begin{pmatrix} \mathbf{Z}'_{iD}\ \mathbf{Z}_{iD} & 0 \\ 0 & \mathbf{Z}'_{iS}\ \mathbf{Z}_{iS} \end{pmatrix} \begin{pmatrix} \mathbf{yd}_1 \\ \mathbf{yd}_2 \end{pmatrix} \tag{9.8}$$

with $i \neq j$ and $j = (1,\ n-1)$, where n is the number of animals in the same group and:

$$\mathbf{yd}_1 = \left(\mathbf{Z}'_{iD}\ \mathbf{Z}_{iD}\right)^{-1} \mathbf{Z}'_{iD}\left(y_i - \mathbf{X}\hat{\mathbf{b}} - \mathbf{Z}_{jS}\ \hat{\mathbf{u}}_{jS}\right) \text{ and}$$

$$\mathbf{yd}_2 = \left(\mathbf{Z}'_{iS}\ \mathbf{Z}_{iS}\right)^{-1} \mathbf{Z}'_{iS}\left(y_i - \mathbf{X}\mathbf{b} - \mathbf{Z}_{jS}\ \hat{\mathbf{u}}_{jS} - Z_{jD}\ \hat{\mathbf{u}}_{jD}\right)$$

Thus $\mathbf{yd}_1$ is the yield record of animal *i* corrected for all fixed effects and the SBVs of all other members in the same group, and $\mathbf{yd}_2$ is the average of the yield records of all animals in the same group apart from animal *i* corrected for all fixed effects, the DBVs and SBVs of the members of the group. Transferring the left non-diagonal terms of $\mathbf{A}^{-1}$ in Eqn 9.8 to the right side of the equation gives:

$$\begin{bmatrix} \mathbf{Z}'_{iD}\,\mathbf{Z}_{iD} + a^{ii}\alpha_1 & \mathbf{Z}'_{iD}\,\mathbf{Z}_{iS} + a^{ii}\alpha_2 \\ \mathbf{Z}'_{iS}\,\mathbf{Z}_{iD} + a^{ii}\alpha_2 & \mathbf{Z}'_{iS}\,\mathbf{Z}_{iS} + a^{ii}\alpha_3 \end{bmatrix} \begin{bmatrix} \hat{\mathbf{u}}_D \\ \hat{\mathbf{u}}_S \end{bmatrix} = 2a_{par}\begin{pmatrix} \alpha_1 & \alpha_2 \\ \alpha_2 & \alpha_3 \end{pmatrix}\begin{pmatrix} \mathbf{PA}_1 \\ \mathbf{PA}_2 \end{pmatrix}$$

$$+\begin{pmatrix} \mathbf{Z}'_{iD}\,\mathbf{Z}_{iD} & 0 \\ 0 & \mathbf{Z}'_{iS}\,\mathbf{Z}_{iS} \end{pmatrix}\begin{pmatrix} \mathbf{yd}_1 \\ \mathbf{yd}_2 \end{pmatrix} + 0.5a_{prog}\begin{pmatrix} \alpha_1 & \alpha_2 \\ \alpha_2 & \alpha_3 \end{pmatrix}\begin{pmatrix} 2\hat{\mathbf{u}}_{Dprog} - \hat{\mathbf{u}}_{Dmate} \\ 2\hat{\mathbf{u}}_{Sprog} - \hat{\mathbf{u}}_{Smate} \end{pmatrix}$$

where $\mathbf{PA}_1$ and $\mathbf{PA}_2$ are the parent averages for DBV and SBV for animal i; $a_{par} = 1$, $\frac{2}{3}$ or $\frac{1}{2}$ if both, one or neither parents are known, respectively; and $a_{prog} = 1$ if the animal's mate is known and $\frac{2}{3}$ if unknown. Note that $a^{ii} = 2a_{par} + 0.5a_{prog}$, therefore premultiplying both sides of the above equation by the inverse of **DIAG**, with:

$$\mathbf{DIAG} = \begin{bmatrix} \mathbf{Z}'_{iD}\mathbf{Z}_{iD} + a^{ii}\alpha_1 & \mathbf{Z}'_{iD}\mathbf{Z}_{iS} + a^{ii}\alpha_2 \\ \mathbf{Z}'_{iS}\mathbf{Z}_{iD} + a^{ii}\alpha_2 & \mathbf{Z}'_{iS}\mathbf{Z}_{iS} + a^{ii}\alpha_3 \end{bmatrix}$$

gives:

$$\begin{bmatrix} \hat{\mathbf{u}}_D \\ \hat{\mathbf{u}}_S \end{bmatrix} = \mathbf{WT}_1\begin{pmatrix} \mathbf{PA}_1 \\ \mathbf{PA}_2 \end{pmatrix} + \mathbf{WT}_2\begin{pmatrix} \mathbf{yd}_1 \\ \mathbf{yd}_2 \end{pmatrix} + \mathbf{WT}_3\begin{pmatrix} \mathbf{PC1} \\ \mathbf{PC2} \end{pmatrix} \qquad (9.9)$$

where:

$$\begin{pmatrix} \mathbf{PC1} \\ \mathbf{PC2} \end{pmatrix} = \Sigma a_{prog}\begin{pmatrix} 2\hat{\mathbf{u}}_{Dprog} - \hat{\mathbf{u}}_{Dmate} \\ 2\hat{\mathbf{u}}_{Sprog} - \hat{\mathbf{u}}_{Smate} \end{pmatrix} / \Sigma a_{prog}$$

The weights $\mathbf{W}_1$, $\mathbf{W}_2$ and $\mathbf{W}_3 = \mathbf{I}$, with:

$$\mathbf{W}_1 = (\mathbf{DIAG})^{-1}\left(2a_{par}\begin{pmatrix} \alpha_1 & \alpha_2 \\ \alpha_2 & \alpha_3 \end{pmatrix}\right),\ \mathbf{W}_2 = (\mathbf{DIAG})^{-1}\begin{pmatrix} \mathbf{Z}'_{iD}\,\mathbf{Z}_{iD} & 0 \\ 0 & \mathbf{Z}'_{iS}\,\mathbf{Z}_{iS} \end{pmatrix}$$

$$\mathbf{W}_3 = (\mathbf{DIAG})^{-1}0.5\begin{pmatrix} \alpha_1 & \alpha_2 \\ \alpha_2 & \alpha_3 \end{pmatrix}\Sigma a_{prog}$$

Equation 9.9 is illustrated below using pig 7 in Example 9.1. For pig 7:

$$\begin{aligned} \mathrm{yd}_1 &= \left(y_7 - \hat{b}_1 - \hat{u}_{S8} - \hat{u}_{S9} - \hat{c}_1 - \hat{g}_1\right) \\ &= \left(5.50 - 6.004 - (-0.099) - 0.009 - 0.333 - (-0.269)\right) = -0.478 \end{aligned}$$

and

$$\begin{aligned} \mathrm{yd}_2 &= 1/(n-1)\left((y_8 - y_9) - 2\hat{b}_2 - \hat{u}_{D8} - \hat{u}_{D9} - \hat{u}_{S8} - \hat{u}_{S9} - \hat{c}_1 - \hat{c}_2 - 2\hat{g}_1\right) \\ &= ((9.8 + 4.9) - 2(8.243) - 0.522 - (-0.874) - (-0.099) - 0.009 \\ &\quad - 0.333 - (-0.515) - 2(-0.269)) = -0.312 \end{aligned}$$

Since both parents are known:

$$\mathbf{DIAG} = 2\begin{pmatrix} \alpha_1 & \alpha_2 \\ \alpha_2 & \alpha_3 \end{pmatrix} + \begin{pmatrix} 1 & 0 \\ 0 & 2 \end{pmatrix} = \begin{pmatrix} 4.991 & -2.494 \\ -2.494 & 30.492 \end{pmatrix}$$

Therefore:

$$\mathbf{WT}_1 = (\mathbf{DIAG})^{-1} 2\begin{pmatrix} \alpha_1 & \alpha_2 \\ \alpha_2 & \alpha_3 \end{pmatrix} = \begin{pmatrix} 0.791 & -0.034 \\ -0.017 & 0.932 \end{pmatrix} \text{ and}$$

$$\mathbf{WT}_2 = (\mathbf{DIAG})^{-1}\begin{pmatrix} 1 & 0 \\ 0 & 2 \end{pmatrix} = \begin{pmatrix} 0.209 & 0.034 \\ 0.017 & 0.068 \end{pmatrix}$$

From Eqn 9.9:

$$\begin{pmatrix} \hat{u}_{D7} \\ \hat{u}_{S7} \end{pmatrix} = \mathbf{WT}_1\begin{pmatrix} PA_1 \\ PA_2 \end{pmatrix} + \mathbf{WT}_2\begin{pmatrix} yd_1 \\ yd_2 \end{pmatrix} = \mathbf{WT}_1\begin{pmatrix} 0.296 \\ -0.044 \end{pmatrix} + \mathbf{WT}_2\begin{pmatrix} -0.478 \\ -0.312 \end{pmatrix} = \begin{pmatrix} 0.125 \\ -0.075 \end{pmatrix}$$

The weights indicate that the relative emphasis on parent contribution was higher for the SBV compared to the DBV. This might be due to the lower genetic variance for associative effects in the model.

9.4 Analysis Using Correlated Error Structure

The analysis of the same data using Eqn 9.7 gave the same solutions obtained from Eqn 9.6. Since the major difference is the structure of the residual covariance, R, this section has only focused on illustrating the structure of R for this example. Although the number of equations using Eqn 9.7 were three less compared to Eqn 9.6, the number of non-zero elements was higher (481 compared with 462 for Eqn 9.6). This is due to the correlated residual variance structure in Eqn 9.7.

As mentioned earlier, residual error structure is block diagonal with all elements between the various block diagonals being zero. Thus, for the example data in Table 9.1, with $n = 3$, the **R** block diagonal structure for one group is:

$$\mathbf{R} = \begin{bmatrix} 1 & \rho & \rho \\ \rho & 1 & \rho \\ \rho & \rho & 1 \end{bmatrix}\sigma_e^2 = \begin{pmatrix} 1 & 0.2 & 0.2 \\ 0.2 & 1 & 0.2 \\ 0.2 & 0.2 & 1 \end{pmatrix} 60.6 \text{ and}$$

$$\mathbf{R}^{-1} = \begin{pmatrix} 0.01768 & -0.00295 & -0.00295 \\ -0.00295 & 0.01768 & -0.0029 \\ -0.00295 & -0.00295 & 0.01768 \end{pmatrix}$$

The MME can then easily be set following the usual principles.

10 Analysis of Longitudinal Data

10.1 Introduction

In Chapter 5, the use of a repeatability model to analyse repeated measurements on individuals was discussed and illustrated. The basic assumption of the model was that repeated measurements were regarded as expression of the same trait over time. In other words, a genetic correlation of unity was assumed between repeated measurements. The model has been employed mostly in the genetic evaluation of milk production traits of dairy cattle in most countries up to 1999 (Interbull, 2000). The main advantages of this model are its simplicity, fewer computation requirements and fewer parameters compared to a multivariate model (see Chapter 6). However, the model has some drawbacks. First, test-day records within lactation are assumed to measure the same trait during the whole lactation length and are used to compute 305-day yields. These test-day records are actually repeated observations measured along a trajectory (days in milk), and the mean and covariance between measurements change gradually along the trajectory. Several studies have reported that heritability of daily milk yields varied with days in milk. In addition, genetic correlations between repeated measurements usually tended to decrease as the time between them increases (Meyer, 1989; Pander *et al.*, 1992). The extension of test records to compute 305-day yields is unable to account for these changes in the covariance structure. Second, the assumption that 305-day yields across parities measure the same trait suffers from the same limitations.

However, in beef cattle, repeated measurements of growth have been analysed somewhat differently, with the assumption that measurements are genetically different but correlated traits. Usually, a multivariate model has been employed in the genetic evaluation of these traits. While the multivariate model is an improvement on the repeatability model by accounting for the genetic correlations among different records, it would be highly over-parameterized if records were available at many ages or time periods. For instance, a multivariate model for daily body weight up to yearly weight in beef cattle as different traits will not only be over-parameterized, but it will also be difficult to obtain accurate estimates of the necessary genetic parameters.

An appropriate model for the analysis of repeated measurements over time or age (also termed longitudinal data) should account for the mean and covariance structure that changes with time or age and should be feasible in terms of estimating the required genetic parameters. In 1994, Schaeffer and Dekkers introduced the concept of the random regression (RR) model for the analysis of test-day records in dairy cattle as a means of accounting for the covariance structure of repeated records over time or age. Almost at the same time, Kirkpatrick *et al.* (1990, 1994) introduced covariance functions (CFs) to handle the analysis of longitudinal data, illustrating their methodology with growth data. The application of RR models in animal breeding for the analysis of various types of data has been comprehensively reviewed by Schaeffer (2004). Prior to the development of the RR model for genetic evaluation, milk yield test-day records were analysed by Ptak and Schaefer (1993)

DOI: 10.1079/9781800620506.0010

using a fixed regression model. The details of this model are discussed and illustrated in the next section, followed by its extension to a RR model. This is then followed by a brief presentation of CF, and the equivalence of the RR model and CF is demonstrated.

10.2 Fixed Regression Model

The theoretical framework for the fixed regression model and its application for the analysis of longitudinal data such as test-day milk production traits were presented by Ptak and Schaefer in 1993. On a national scale, a fixed regression model was implemented for the genetic evaluation of test-day records of milk production traits and somatic cell counts in Germany from 1995 until 2002. The model involved the use of individual test-day records, thereby avoiding the problem of explicitly extending test-day yields into 305-day yield, and accounted for the effects peculiar to all cows on the same test day within herds (herd–test–day (HTD) effect). Therefore, corrections for temporary environmental effects on the day of test are more precise compared to evaluations based on 305-day yields. The model also accounted for the general shape of the lactation curve of groups of similar age, and calving in the same season and region. The latter was accomplished by regressing lactation curve parameters on days in milk (hence the name of the model) within the groupings for cows. Inclusion of the curve therefore allows for correction of the means of test-day yields at different stages of lactation. Fitting residual variances relevant to the appropriate stage of lactation could also account for the variation of test-day yields with days in milk. The only major disadvantage is that the volume of data to be analysed is much larger, especially in the dairy situation, as ten or more test-day observations are stored relative to a single 305-day yield.

Similar to the repeatability model, at the genetic level, the fixed regression model assumes that test-day records within a lactation are repeated measurements of the same trait, i.e. a genetic correlation of unity among test-day observations. Usually, the permanent environmental effect is included in the model to account for environmental factors with permanent effects on all test-day yields within lactation.

The fixed regression model is of the form:

$$y_{tij} = htd_i + \sum_{k=0}^{nf} \phi_{tjk}\beta_k + u_j + pe_j + e_{tij}$$

where y_{tij} is the test-day record of cow j made on day t within HTD subclass i; β_k are fixed regression coefficients; u_j and pe_j are vectors of animal additive genetic and permanent environmental effects, respectively, for animal j; ϕ_{tjk} is the vector of the kth Legendre polynomials or any other curve parameter, for the test-day record of cow j made on day t; nf is the order of fit for Legendre polynomials used to model the fixed regressions (fixed lactation curves) and e_{tij} is the random residual. In matrix notation, the model may be written as:

$$\mathbf{y} = \mathbf{Xb} + \mathbf{Qu} + \mathbf{Zpe} + \mathbf{e} \qquad (10.1)$$

where **y** is the vector of TD yields, **b** is a vector of solutions for HTD and fixed regressions, and **u** and **pe** are vectors of animal additive genetic and permanent environmental effects, respectively. The variances of **u** and **pe** are as defined in Eqn 5.1. The matrices **X**, **Q** and **Z** are incidence matrices and are described in detail in the next section, which illustrates the application of the model. It is assumed that $\text{var}(\mathbf{u}) = \mathbf{A}\sigma_u^2$, and $\text{var}(\mathbf{pe}) = \mathbf{I}\sigma_p^2$, and $\text{var}(\mathbf{e}) = \mathbf{I}\sigma_e^2 = \mathbf{R}$. The MME for Eqn 10.1 are:

$$\begin{pmatrix} \mathbf{X'X} & \mathbf{X'Q} & \mathbf{X'Z} \\ \mathbf{Q'X} & \mathbf{Q'Q} + \mathbf{A}^{-1}\alpha_1 & \mathbf{Q'Z} \\ \mathbf{Z'X} & \mathbf{Z'Q} & \mathbf{Z'Z} + \alpha_2 \end{pmatrix} \begin{pmatrix} \hat{\mathbf{b}} \\ \hat{\mathbf{u}} \\ \hat{\mathbf{pe}} \end{pmatrix} = \begin{pmatrix} \mathbf{X'y} \\ \mathbf{Q'y} \\ \mathbf{Z'y} \end{pmatrix}$$

with $\alpha_1 = \sigma_e^2 / \sigma_u^2$ and $\alpha_2 = \sigma_e^2 / \sigma_p^2$.

10.2.1 An illustration

Example 10.1

Given in Table 10.1 are the test-day fat yields of five cows in a herd with details of HTD and days in milk (DIM). The aim is to estimate solutions for HTD effects, regression coefficients for a fixed lactation curve fitting Legendre polynomials of order 4, solutions for permanent environmental effects and breeding values for animal effects using Eqn 10.1. Assume that the estimated variances for additive genetic effects, permanent environmental effects and residual variances were 5.521 kg^2, 8.470 kg^2 and 3.710 kg^2, respectively. Then:

$$\alpha_1 = \sigma_e^2 / \sigma_u^2 = 3.710 / 5.521 = 0.672$$

and:

$$\alpha_2 = \sigma_e^2 / \sigma_p^2 = 3.710 / 8.470 = 0.438$$

The modelling of the fixed lactation curve by means of Legendre polynomials implies the need to compute $\mathbf{\Phi}$, which is the matrix of Legendre polynomials evaluated at the different DIM. The matrix $\mathbf{\Phi}$ is of order t (the number of DIM) by k (where k is the order of fit) with element $\phi_{ij} = \phi_j(a_t)$, which is the jth Legendre polynomial evaluated at the standardized DIM t (a_t). Therefore $\mathbf{\Phi} = \mathbf{M\Lambda}$, where $\mathbf{M}$ is the matrix containing the polynomials of the standardized DIM values and Λ is a matrix of order k containing the coefficients of Legendre polynomials. The calculation of $\mathbf{\Phi}$ is outlined in Appendix G and matrix $\mathbf{\Phi}$ for Example 10.1 is shown in Eqn g.1.

Table 10.1. Test-day fat yields (TDY) for some cows in a herd.

	Animals									
	4		5		6		7		8	
DIM	HTD	TDY	HTD	TDY	HTD	TDY	HTD	TDY	HTD	TDY
4	1	17.0	1	23.0	6	10.4	4	22.8	1	22.2
38	2	18.6	2	21.0	7	12.3	5	22.4	2	20.0
72	3	24.0	3	18.0	8	13.2	6	21.4	3	21.0
106	4	20.0	4	17.0	9	11.6	7	18.8	4	23.0
140	5	20.0	5	16.2	10	8.4	8	18.3	5	16.8
174	6	15.6	6	14.0			9	16.2	6	11.0
208	7	16.0	7	14.2			10	15.0	7	13.0
242	8	13.0	8	13.4					8	17.0
276	9	8.2	9	11.8					9	13.0
310	10	8.0	10	11.4					10	12.6

DIM, days in milk; HTD, herd–test-day.

Setting up the incidence matrices for the MME

In Eqn 10.1, let $\mathbf{Xb} = \mathbf{X}_1\mathbf{b}_1 + \mathbf{X}_2\mathbf{b}_2$, then in Example 10.1, the matrix $\mathbf{X}_1$, which relates records to HTD effects, is of order n_{td} (number of TD records) and is too large to be presented. However, $\mathbf{X}_1'\mathbf{X}_1$ is diagonal and is:

$$\mathbf{X}_1'\mathbf{X}_1 = \text{diagonal}[3,3,3,4,4,5,5,5,5,5]$$

The matrix $\mathbf{X}_2$ of order n_{td} by nf contains Legendre polynomials (covariables) corresponding to the DIM of the *i*th TD yield. Thus the *i*th row of $\mathbf{X}_2$ contains elements of the row of $\mathbf{\Phi}$ corresponding to the DIM for the *i*th record. The matrix $\mathbf{X}_2$, with rows for the first three TD records of cow 4 and the last three TD records of cow 8 is:

$$\begin{bmatrix} 0.7071 & -1.2247 & 1.5811 & -1.8704 & 2.1213 \\ 0.7071 & -0.9525 & 0.6441 & -0.0176 & -0.6205 \\ 0.7071 & -0.6804 & -0.0586 & 0.7573 & -0.7757 \\ \vdots & \vdots & \vdots & \vdots & \vdots \\ 0.7071 & 0.6804 & -0.0586 & -0.7573 & -0.7757 \\ 0.7071 & 0.9525 & 0.6441 & 0.0176 & -0.6205 \\ 0.7071 & 1.2247 & 1.5811 & 1.8704 & 2.1213 \end{bmatrix}$$

and $\mathbf{X}_2'\mathbf{X}_2$ is:

$$\mathbf{X}_2'\mathbf{X}_2 = \begin{bmatrix} 20.9996 & -4.4261 & 4.0568 & -0.8441 & 8.7149 \\ -4.4261 & 24.6271 & -4.7012 & 11.1628 & -3.0641 \\ 4.0568 & -4.7012 & 31.0621 & -6.6603 & 19.0867 \\ -0.8441 & 11.1628 & -6.6603 & 38.6470 & -8.8550 \\ 8.7149 & -3.0641 & 19.0867 & -8.8550 & 48.2930 \end{bmatrix}$$

Considering only animals with records, $\mathbf{Q} = \mathbf{Z}$ and is a matrix of order 5 (number of animals) by n_{td}. The matrix $\mathbf{Q}'$ could be represented as:

$$\mathbf{Q}' = \begin{bmatrix} \mathbf{q}_4' & 0 & 0 & 0 & 0 \\ 0 & \mathbf{q}_5' & 0 & 0 & 0 \\ 0 & 0 & \mathbf{q}_6' & 0 & 0 \\ 0 & 0 & 0 & \mathbf{q}_7' & 0 \\ 0 & 0 & 0 & 0 & \mathbf{q}_8' \end{bmatrix}$$

where $\mathbf{q}_i'$ is a vector of ones with size equal to the number of TD records for the *i*th cow. The matrices $\mathbf{Q}'\mathbf{Q}$ and $\mathbf{Z}'\mathbf{Z}$ are both diagonal and equal. Thus:

$$\mathbf{Q}'\mathbf{Q} = \mathbf{Z}'\mathbf{Z} = \text{diag}[10,10,5,7,10]$$

The matrix $\mathbf{A}^{-1}$ has been given in Example 5.1. The remaining matrices in the MME could be obtained as outlined in earlier chapters. Solving the MME, with the solution for the tenth level of HTD effects constrained to zero, give the following results:

Effects	Solutions	
HTD		
1	10.9783	
2	7.9951	
3	8.7031	
4	8.2806	
5	6.3813	
6	3.1893	
7	3.3099	
8	3.3897	
9	0.6751	
10	0.0000	
Fixed regression coefficients		
1	16.3082	
2	−0.5227	
3	−0.1245	
4	0.5355	
5	−0.4195	
Animal effect		
	EBV for daily yield	EBV for 305-day yield
1	−0.3300	−100.6476
2	−0.1604	−48.9242
3	0.4904	149.5718
4	0.0043	1.3203
5	−0.2449	−74.7065
6	−0.8367	−255.2063
7	1.1477	350.0481
8	0.3786	115.4757
Permanent environmental effects		
Cow	Solutions for daily yield	Solutions for daily yield
4	−0.6156	−187.7634
5	−0.4151	−126.6150
6	−1.6853	−514.0274
7	2.8089	856.7092
8	−0.0928	−28.3035

EBV, estimated breeding value.

The solutions for the fixed regressions are regression coefficients from which plots of lactation curves can be obtained. In practice, the fixed regressions are usually fitted within group of cows calving in the same season in the same parity and of similar age. Thus the curves obtained for various groups of cows are useful for examining the influence of different environmental factors on lactation curves. In Example 10.1, one fixed lactation curve was fitted for all cows and a vector (**v**) of actual daily fat yield (kg) from days 4 to 310 can be obtained as:

$$\mathbf{v} = \mathbf{\Phi}\hat{\mathbf{b}} = \sum_{t=4}^{310}\sum_{j=1}^{nf}\phi_{ij}\ \hat{b}_{2j}$$

where $\mathbf{\Phi}$ is a matrix of Legendre polynomials evaluated from 4 to 310 DIM, as described in Appendix G. From the above equation, $\mathbf{v}^{38}$, for instance, is:

$$\mathbf{v}_{38} = [0.7071 \quad -0.9525 \quad 0.6441 \quad -0.0176 \quad -0.6205]\hat{\mathbf{b}}_2 = 12.2001$$

For the DIM in the example data set, **v** is:

(DIM)	4	38	72	106	140	174	208	242	276	310
v =	[10.0835	12.2001	12.6254	12.2077	11.5679	11.0407	10.9156	11.1111	11.2500	10.8297]

A graph of the fixed lactation curve can be obtained by plotting the elements of **v** against DIM.

The EBV for animals and solutions for permanent environmental effect obtained by solving the MME are those for daily fat yield. To obtain EBV or solutions for **pe** effects on the nth DIM, these solutions are multiplied by n. This is implicit from the assumptions stated earlier of genetic correlations of unity among TD records. Thus EBVs for 305 days, shown in the table of results above, were obtained by multiplying the solutions for daily fat yield by 305.

Partitioning breeding values and solutions for permanent environmental effects

Similar to the repeatability model, EBVs of animals can be partitioned in terms of contributions from various sources, using Eqn 4.8. The YD for an animal is now calculated as the average of corrected TD records. The correction is for effects of HTD, fixed regressions and **pe**. Thus for cow 6, with five TD records, YD_6 is:

$$YD_6 = (\mathbf{Q'Q})^{-1}\mathbf{Q'}(\mathbf{y}_6 - \mathbf{X}_1\hat{\mathbf{b}}_1 - \mathbf{X}_2\hat{\mathbf{b}}_2 - \hat{\mathbf{pe}})$$

with $\mathbf{y}_6 - \mathbf{X}_1\hat{\mathbf{b}}_1 - \mathbf{X}_2\hat{\mathbf{b}}_2 - \hat{\mathbf{pe}} = \mathbf{y}_c$

$$= \begin{bmatrix} 10.4 \\ 12.3 \\ 13.2 \\ 11.6 \\ 8.4 \end{bmatrix} - \begin{bmatrix} 3.1893 \\ 3.3099 \\ 3.3897 \\ 0.6751 \\ 0.0000 \end{bmatrix} - \begin{bmatrix} 10.0835 \\ 12.2001 \\ 12.6254 \\ 12.2077 \\ 11.5679 \end{bmatrix} - \begin{bmatrix} -1.6853 \\ -1.6853 \\ -1.6853 \\ -1.6853 \\ -1.6853 \end{bmatrix} = \begin{bmatrix} -1.1875 \\ -1.5247 \\ -1.1298 \\ 0.4025 \\ -1.4826 \end{bmatrix}$$

and:

$$YD_6 = (\mathbf{Q'Q})^{-1}\mathbf{Q'}(\mathbf{y}_c) = \frac{1}{5}\left(\mathbf{q}'_6 \begin{bmatrix} -1.1875 \\ -1.5247 \\ -1.1298 \\ 0.4025 \\ -1.4826 \end{bmatrix}\right) = -4.9221/5 = -0.9844$$

Then the solution for additive genetic effect for animal 6 using Eqn 4.8 is:

$$\begin{aligned} \hat{u}_6 &= w_1((\hat{u}_1 + \hat{u}_5)/2) + w_2(YD_6) \\ &= w_1((-0.3300 + -0.2449)/2) + w_2(-0.9844) = -0.8367 \end{aligned}$$

with $w_1 = 2(0.672)/6.344$, $w_2 = 5/6.344$ and 6.344 = the sum of the numerators of w_1 and w_2.

For animal 8 with ten TD records, the solution for additive genetic effect is:

$$\begin{aligned}\hat{u}_8 &= w_1((\hat{u}_1 + \hat{u}_7)/2) + w_2(YD_8) \\ &= w_1((-0.3300 + 1.1477)/2) + w_2(0.3746) = 0.3786\end{aligned}$$

with w_1 = 2(0.672)/11.344, w_2 = 10/11.344 and 11.344 = the sum of the numerators of w_1 and w_2. The weights on YDs were 0.7882 and 0.8815 for animals 6 and 8, respectively. This illustrates the fact that as the number of TD increases, more emphasis is placed on performance records of the animal. Considering animal 4 with ten TD records and a progeny, her breeding value can be calculated as:

$$\begin{aligned}\hat{u}_4 &= w_1((\hat{u}_1 + \hat{u}_2)/2) + w_2(YD_4) + w_3(\hat{u}_7 - 0.5\hat{u}_3) \\ &= w_1((-0.3300 + -0.1604)/2) + w_2(-0.0226) \\ &\quad + w_3(2(1.1477) - 0.4934) = 0.0043\end{aligned}$$

where w_1 = 2(0.672)/11.68, w_2 = 10/11.68 and w_3 = 0.5(0.672)/11.68 and 11.68 is the sum of the numerators of w_1, w_2 and w_3. There was a slight reduction to the weight given to parent average from 0.1185 (animal 8) to 0.1151 (animal 4) due to the additional information from progeny.

The solution for pe of an animal can be calculated as in Section 5.2.2, using Eqn 5.4. Here, the correction of the TD records is for the estimates for HTD effects and fixed regressions and animal effect. Thus, for cow 6, $\hat{\mathbf{pe}}_6$ can be calculated as:

$$\begin{aligned}\hat{\mathbf{pe}}_6 &= \left(\mathbf{t}'\begin{bmatrix}10.4\\12.3\\13.2\\11.6\\8.4\end{bmatrix} - \begin{bmatrix}3.1893\\3.3099\\3.3897\\0.6751\\0.0000\end{bmatrix} - \begin{bmatrix}10.0835\\12.2001\\12.6254\\12.2077\\11.5679\end{bmatrix} - \begin{bmatrix}-0.8367\\-0.8367\\-0.8367\\-0.8367\\-0.8367\end{bmatrix}\right) \Big/ 5.4380 \\ &= -9.1650/5.4380 = -1.6853\end{aligned}$$

where **t** is a column vector of order 5 (number of TD records for the animal), with all elements equal to one. However, in contrast to pe estimates in Example 5.1, these pe estimates represent permanent environmental factors affecting TD records within lactation.

10.3 Random Regression Model

In Section 10.2, the advantage of including fixed regressions on days in milk in the model was to account for the shape of the lactation curve for different groups of cows. However, the breeding values estimated represented genetic differences between animals at the height of the curves. Although different residual variances associated with different stages of lactation could be fitted with the fixed regression model, the model did not account for the covariance structure at the genetic level. Schaeffer and Dekkers (1994) extended the fixed regression model for genetic evaluation by considering the regression coefficients on the same covariables as random, therefore allowing for between-animal variation in the shape of the curve. Thus the genetic differences among animals could be modelled as deviations from the fixed lactation curves by means of random parametric curves (see Guo and Swalve, 1997) or orthogonal polynomials such as Legendre polynomials (Brotherstone *et al.*, 2000), or even non-parametric curves such as natural cubic splines (White *et al.*, 1999). Most studies have used Legendre polynomials as they make no assumption about the shape of the curve and are easy to apply. The RR model has also been employed for the analysis of growth data in pigs (Andersen and Pedersen, 1996) and beef cattle (Meyer, 1999). An additional benefit of the RR model in dairy cattle is that it provides the possibility of genetic evaluation for persistence of the lactation. A typical random regression model (RRM), especially for the analysis of dairy cattle test-day records, is of the form:

$$y_{tijk} = htd_i + \sum_{k=0}^{nf} \phi_{jtk} \boldsymbol{\beta}_k + \sum_{k=0}^{nr} \phi_{jtk} \mathbf{u}_{jk} + \sum_{k=0}^{nr} \phi_{jtk} \mathbf{pe}_{jk} + e_{tijk}$$

where y_{tijk} is the test-day record of cow *j* made on day *t* within HTD subclass *i*; $\boldsymbol{\beta}_k$ are fixed regression coefficients; $\mathbf{u}_{jk}$ and $\mathbf{pe}_{jk}$ are vectors of the *k*th random regression for animal and permanent environmental effects, respectively, for animal *j*; φ_{jtk} is the vector of the *k*th Legendre polynomials for the test-day record of cow *j* made on day *t*; *nf* is the order of polynomials fitted as fixed regressions; *nr* is the order of polynomials for animal and **pe** effects; and e_{tljk} is the random residual. The model in matrix notation is:

$$\mathbf{y} = \mathbf{Xb} + \mathbf{Qu} + \mathbf{Zpe} + \mathbf{e}$$

The vectors **y**, **b** and the matrix **X** are as described in Example 10.1. However, **u** and **pe** are now vectors of random regressions for animal additive genetic and pe effects. The matrices Q and **Z** are covariable matrices and, if only animals with records are considered, the *i*th row of these matrices contains the orthogonal polynomials (covariables) corresponding to the DIM of the *i*th TD yield. If the order of fit is the same for animal and pe effects, Q = Z, considering only animals with records. This would not be the case if the order of fit is different for animal and pe effects. In general, considering animals with records, the order of either **Q** or **Z** is n_{td} (number of TD records) by *nk*, where *nk* equals *nr* times the number of animals with records. It is assumed that var(**u**) = **A*****G**, var(**pe**) = **I*****P** and $\text{var}(\mathbf{e}) = I\sigma_e^2 = \mathbf{R}$, where **A** is the numerator relationship matrix, * is the Kronecker product and **G** and **P** are of the order of polynomial fitted for animal and pe effects. The MME are:

$$\begin{pmatrix} \mathbf{X'R^{-1}X} & \mathbf{X'R^{-1}Q} & \mathbf{X'R^{-1}Z} \\ \mathbf{Q'R^{-1}X} & \mathbf{Q'R^{-1}Q + A^{-1} \otimes G} & \mathbf{Q'R^{-1}Z} \\ \mathbf{Z'R^{-1}X} & \mathbf{Z'R^{-1}Q} & \mathbf{Z'R^{-1}Z + P} \end{pmatrix} \begin{pmatrix} \hat{\mathbf{b}} \\ \hat{\mathbf{u}} \\ \hat{\mathbf{pe}} \end{pmatrix} = \begin{pmatrix} \mathbf{X'R^{-1}y} \\ \mathbf{Q'R^{-1}y} \\ \mathbf{Z'R^{-1}y} \end{pmatrix}$$

10.3.1 Numerical application

Example 10.2

Analysis of the data in Table 10.1 is undertaken fitting an RR model with Legendre polynomials of order 4 fitted for the fixed lactation curve and Legendre polynomials of order 2 fitted for both random animal and **pe** effects. The covariance matrices for the random regression coefficients for animal effect and **pe** effects are:

$$\mathbf{G} = \begin{bmatrix} 3.297 & 0.594 & -1.381 \\ 0.594 & 0.921 & -0.289 \\ -1.381 & 0.289 & 1.005 \end{bmatrix}; \quad \mathbf{P} = \begin{bmatrix} 6.872 & -0.254 & -1.101 \\ -0.254 & 3.171 & 0.167 \\ -1.101 & 0.167 & 2.457 \end{bmatrix}$$

and the residual variance equals 3.710 for all stages of lactation.

As indicated earlier, the above **G** or **P** matrix models the genetic or permanent environment covariance structure of fat yields over the whole lactation length. Thus the genetic covariance between DIM *i* and *j* along the trajectory can be calculated from **G**. For instance, the genetic variance for DIM *i*, (v_{ii}) can be calculated as:

$$v_{ii} = \mathbf{t}_i \mathbf{G} \mathbf{t}'_t$$

where $\mathbf{t}_i = \phi_{ik}$, the ith row vector of $\mathbf{\Phi}$, for day i, and k is the order of fit. The genetic covariance between DIM i and j (v_{ij}) therefore is:

$$v_{ij} = \mathbf{t}_i \mathbf{G}'_i$$

Using the **G** matrix in Example 10.1, the genetic variance for DIM 106 equals 2.6433 kg^2, with $\mathbf{t}^{106} = [0.7071 \; -0.4082 \; -0.5271]$, and the genetic covariance between DIM 106 and 140 equals 3.0219 kg, with $\mathbf{t}^{140} = [0.7071 \; -0.1361 \; -0.7613]$. The plots of daily genetic and permanent environmental variances against DIM are shown in Fig. 10.1, indicating how these variances change through the lactation length.

Setting up the matrices for the MME

The setting of the matrix **X** has been described in Example 10.1. The matrix $\mathbf{X'R^{-1}X}$ can easily be obtained by matrix multiplication. Considering only animals with records, Q′ can be represented as:

$$\mathbf{Q}' = \begin{bmatrix} \mathbf{Q}'_4 & 0 & 0 & 0 & 0 \\ 0 & \mathbf{Q}'_5 & 0 & 0 & 0 \\ 0 & 0 & \mathbf{Q}'_6 & 0 & 0 \\ 0 & 0 & 0 & \mathbf{Q}'_7 & 0 \\ 0 & 0 & 0 & 0 & \mathbf{Q}'_8 \end{bmatrix}$$

where $\mathbf{Q}'_i$ is the matrix of order nr by k (number of TD records for animal i). Thus for animal 6, $\mathbf{Q}'^6$ is:

$$\mathbf{Q}'_6 = \begin{bmatrix} 0.7071 & 0.7071 & 0.7071 & 0.7071 & 0.7071 \\ -1.2247 & -0.9525 & 0.6804 & -0.4082 & -0.1361 \\ 1.5811 & 0.6441 & -0.0586 & 0.5271 & -0.7613 \end{bmatrix}$$

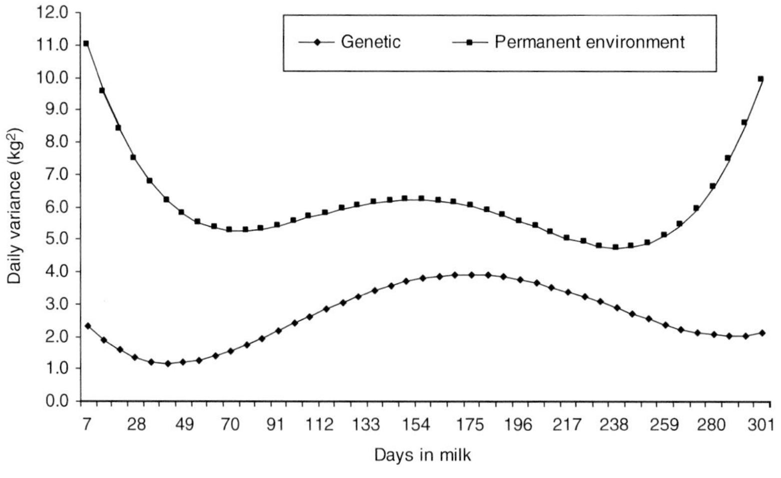

Fig. 10.1. The estimates of daily genetic and permanent environmental variances by days in milk.

For all animals with records, $\mathbf{Q'R^{-1}Q} = \mathbf{Z'R^{-1}Z}$ and are block diagonal. For instance, $\mathbf{Q'R^{-1}Q}$ for the first three cows (cows 4, 5 and 6) with records is:

$$\begin{bmatrix} 1.348 & 0.000 & 0.335 & 0 & 0 & 0 & 0 & 0 & 0 \\ 0.000 & 1.647 & 0.000 & 0 & 0 & 0 & 0 & 0 & 0 \\ 0.335 & 0.000 & 2.035 & 0 & 0 & 0 & 0 & 0 & 0 \\ 0 & 0 & 0 & 1.348 & 0.000 & 0.335 & 0 & 0 & 0 \\ 0 & 0 & 0 & 0.000 & 1.647 & 0.000 & 0 & 0 & 0 \\ 0 & 0 & 0 & 0.335 & 0.000 & 2.035 & 0 & 0 & 0 \\ 0 & 0 & 0 & 0 & 0 & 0 & -0.674 & -0.648 & 0.167 \\ 0 & 0 & 0 & 0 & 0 & 0 & 0.648 & 0.824 & -0.591 \\ 0 & 00 & 0 & 0 & 0 & 0 & 0.167 & -0.591 & 1.018 \end{bmatrix}$$

When all animals are considered, $\mathbf{Q'R^{-1}Q}$ is augmented by *nr* columns and rows per ancestor without records (i.e. animals 1–3). The matrix $\mathbf{G}^{-1}$ is then added to $\mathbf{Q'R^{-1}Q}$ and $\mathbf{P}^{-1}$ added to $\mathbf{Z'R^{-1}Z}$ to obtain the MME. Solving the MME by direct inversion with the solution for level 10 of HTD effects constrained to zero gave the following results:

Effects	Solutions			
HTD				
1	10.0862			
2	7.5908			
3	8.5601			
4	8.2430			
Effects	Solutions			
5	6.3161			
6	3.0101			
7	3.1085			
8	3.1718			
9	0.5044			
10	0.0000			
Fixed regression				
1	16.6384			
2	−0.6253			
3	−0.1346			
4	0.3479			
5	−0.4218			
Animal		Regression coefficients		305-day breeding value
1	−0.0583	0.0552	−0.0442	−12.3731
2	−0.0728	−0.0305	−0.0244	−15.7347
3	0.1311	−0.0247	0.0686	28.1078
4	0.3445	0.0063	−0.3164	74.8132
5	−0.4537	−0.0520	0.2798	−98.4153
6	−0.5485	0.0730	0.1946	−118.4265
7	0.8518	−0.0095	−0.3131	184.1701
8	0.2209	0.0127	−0.0174	47.6907

Continued

Continued.

Effects	Solutions			
Permanent environmental effects				
		Regression coefficients		
Cow				305-day solutions
4	–0.6487	–0.3601	–1.4718	–138.4887
5	–0.7761	0.1370	0.9688	–168.5531
6	–1.9927	0.9851	–0.0693	–427.2378
7	3.5188	–1.0510	–0.4048	756.9415
8	–0.1013	0.2889	0.9771	–22.6619

The solutions for HTD and fixed regression for the RRM are similar to those from the fixed regression model. Lactation curves can be constructed from the fixed regression, as described in Section 10.2.1, and influences of different environmental factors on the curves can be evaluated. Each animal has *nr* regression coefficients as solutions for animal and permanent environmental effects. These are not useful for ranking animals and need to be converted to breeding values for any particular day of interest. Usually, in dairy cattle, values are calculated for 305-day yields and these have been shown above in the table of results. The EBV from days 6 to m for animal k (EBV_{km}) is calculated as:

$$EBV_{km} = \mathbf{t}\hat{\mathbf{u}}_k; \qquad with \quad \mathbf{t} = t_i = \sum_{i=6}^{m}\sum_{j=0}^{nr}\phi_{ij} \tag{10.2}$$

where $\mathbf{t}$ is a row vector of order nr, with the jth elements equal to the sum of the jth orthogonal polynomial from days 6 to m and $\hat{\mathbf{u}}_k$ is the vector for the regression coefficient of animal k. For Example 10.2, the matrix $\mathbf{\Phi}$ for days 4–310 has not been shown because of the size but can be generated as described in Appendix G. Assuming 305-day breeding values are computed from days 6 to 310, then the vector $\mathbf{t}$ for Example 10.2 calculated from days 6 to 310 is:

$$\mathbf{t} = [215.6655 \quad 2.4414 \quad 1.5561]$$

The breeding value for 305-day yield for animal 4, for instance, can be calculated as:

$$\mathbf{t}\hat{\mathbf{u}}_4 = [215.6655 \quad 2.4414 \quad -1.5561]\begin{pmatrix} 0.3445 \\ 0.0063 \\ -0.3164 \end{pmatrix} \approx 74.81$$

Over the lactation length, daily breeding values can be computed for each animal from the random regression coefficients. Genetic lactation curves can be obtained for each animal by plotting these daily breeding values against DIM and differences between curves for different animals can then be studied. Let $\mathbf{v}$ be a vector containing daily breeding values for days 6–310, then $\mathbf{v}$ can be calculated as:

$$\mathbf{v} = \mathbf{T}\hat{\mathbf{u}}_k; \qquad \text{with} \quad \mathrm{T} = t_{ij} = \sum_{i=6}^{310}\sum_{j=0}^{nr}\phi_{ij}$$

The plots of the daily breeding values for animals 2, 3 and 8 are shown in Fig. 10.2. The plots indicate that the animal with the highest 305-day breeding value for fat yield also had the highest daily breeding values along the lactation length.

If the trait being analysed is milk yield, persistence breeding values can be calculated from the daily breeding values. For instance, persistence predicted transmitting ability (PS_{PTA}) for milk yield can be calculated (Schaeffer *et al.*, 2000) as:

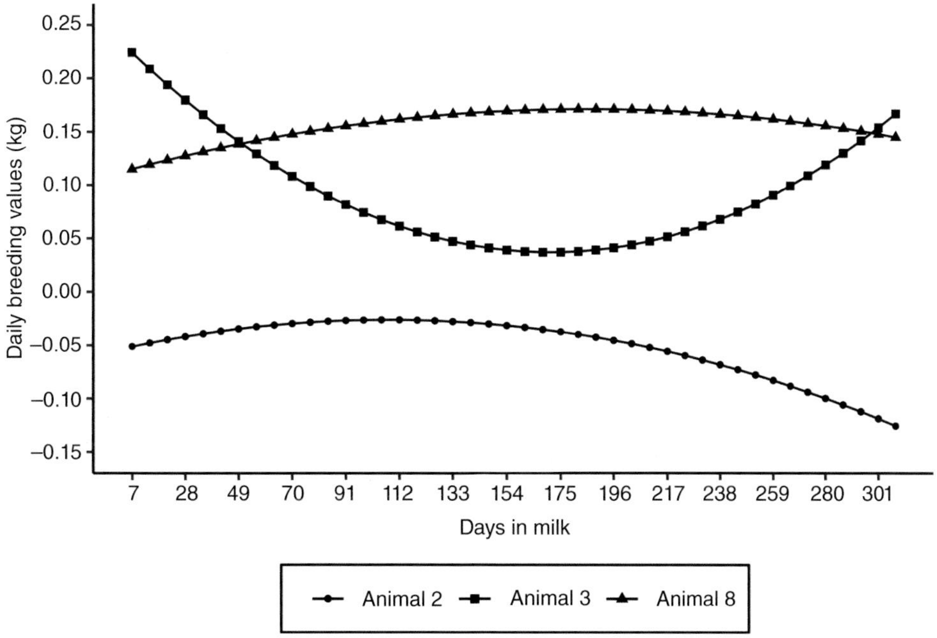

Fig. 10.2. The estimates of daily breeding values for some animals by days in milk.

$$PS_{PTA} = \frac{PTA_{280} - PTA_{60} + \bar{y}_{280}}{\bar{y}_{60}}(100)$$

where PTA_{60} and PTA_{280} are predicted transmitting abilities for day milk yield for an animal at days 60 and 280, respectively, and y_{60} and y_{280} are the average milk yields of cows in the genetic base at days 60 and 280, respectively.

10.3.2 Partitioning animal solutions from random regression model

Equations for calculating the contribution of information from various sources to the solutions (random regression coefficients) of an animal from an RRM were presented by Mrode and Swanson (2004). These equations are the same as those presented in Section 6.2.3 for the multivariate model. Test-day records of cows contribute to random regressions for the animal effect through the yield deviations. The calculation of the vector of yield deviations (**YD**) is first examined. Using the same argument for deriving Eqn 6.7, the equation for **YD** for an RRM is:

$$\mathbf{YD} = (\mathbf{Q'R}^{-1}\mathbf{Q})^{-1}(\mathbf{Q'R}^{-1}(\mathbf{y} - \mathbf{Xb} - \mathbf{Z}\mathbf{p}\hat{\mathbf{e}})) \qquad (10.3)$$

While this equation is similar to Eqn 6.6 for yield deviation under a multivariate model, here **YD** is a vector of weighted regressions of the animal's TD yields adjusted for all effects other than additive genetic effect, on orthogonal polynomials for DIM. Since **YD** is a vector of regressions, it can be used to generate actual yield deviations for any DIM using Eqn 10.2. Thus actual yield deviation (yd*) for day *m*, for instance, equals **v′YD**, where **v** is a vector of order *nr* with $v_m = \varphi_{mj}$ and $j = 1,nr$. The actual yield deviation for 305-day yield can be calculated using Eqn 10.2 but with **u** replaced with **YD**.

The calculation of **YD** for cow 6 in Example 10.2 is illustrated below. First, the vector of TD records for cow 6, corrected for all effects ($\mathbf{y}_c$) other than the additive genetic effect, is:

$$\mathbf{y}_c = \mathbf{y}_6 - \mathbf{X}_1\hat{\mathbf{b}}_1 - \mathbf{X}_2\hat{\mathbf{b}}_2 - \hat{\mathbf{pe}}$$

$$\mathbf{y}_c = \begin{bmatrix} 10.4 \\ 12.3 \\ 13.2 \\ 11.6 \\ 8.4 \end{bmatrix} - \begin{bmatrix} 3.0101 \\ 3.1085 \\ 3.1718 \\ 0.5044 \\ 0.0000 \end{bmatrix} - \begin{bmatrix} 10.7725 \\ 12.5295 \\ 12.7890 \\ 12.3454 \\ 11.7541 \end{bmatrix} - \begin{bmatrix} -2.7251 \\ -2.3920 \\ -2.0752 \\ -1.7746 \\ -1.4904 \end{bmatrix} - \begin{bmatrix} -0.6576 \\ -0.9460 \\ -0.6856 \\ 0.5249 \\ -1.8738 \end{bmatrix}$$

where $\hat{\mathbf{b}}_1$ and $\hat{\mathbf{b}}_2$ are vectors of solutions for HTD and fixed regression coefficients. The matrices $\mathbf{Q'R}^{-1}\mathbf{Q}$ and $\mathbf{Q'R}^{-1}\mathbf{y}_c$ are:

$$\mathbf{Q'R}^{-1}\mathbf{Q} = \begin{bmatrix} 0.6738 & -0.6484 & 0.1674 \\ -0.6484 & 0.8235 & -0.5906 \\ 0.1674 & -0.5906 & 1.0177 \end{bmatrix} \quad \text{and} \quad \mathbf{Q'R}^{-1}\mathbf{y}_c = \begin{bmatrix} -0.6934 \\ 0.5967 \\ -0.1237 \end{bmatrix}$$

Using Eqn 10.3, yield deviation for cow 6 ($\mathbf{YD}_6$) is:

$$\mathbf{YD}_6 = (\mathbf{Q'R}^{-1}\mathbf{Q})^{-1}\mathbf{Q'R}^{-1}\mathbf{y}_c = \begin{bmatrix} -5.0004 \\ -4.6419 \\ -1.9931 \end{bmatrix}$$

The actual yield deviation at 305 DIM for cow 6 using Eqn 10.2 with $\hat{\mathbf{u}}$ replaced with $\mathbf{YD}_6$ is –1086.6450.

The equation for the partitioning of random regression coefficients for animals to contributions for parent average, yield deviations and progeny is:

$$\hat{\mathbf{u}}_{anim} = \mathbf{W}_1\mathbf{PA} + \mathbf{W}_2(\mathbf{YD}) + \mathbf{W}_3\mathbf{PC} \tag{10.4}$$

with:

$$\mathbf{PC} = \Sigma\alpha_{prog}(2\,\hat{\mathbf{u}}_{prog} - \hat{\mathbf{u}}_{mate}) / \Sigma\alpha_{prog} \text{ and } \mathrm{W}_1 + \mathrm{W}_2 + \mathrm{W}_3 = 1$$

This is the same equation as Eqn 5.8, which partitioned breeding values under the multivariate model. The weights $\mathbf{W}_1$, $\mathbf{W}_2$ and $\mathbf{W}_3$ are as defined in Eqn 5.8, but here $\mathbf{W}_i$ is of the order of orthogonal polynomials for animal effects. Illustrating with cow 6, the weights on parent average ($\mathbf{W}_1$) and yield deviation ($\mathbf{W}_2$) can be calculated as:

$$\mathbf{W}_1 = \underbrace{\begin{bmatrix} 2.1520 & -0.9957 & 2.0986 \\ -0.9957 & 3.2921 & -0.3580 \\ 2.0986 & -0.3580 & 5.7284 \end{bmatrix}^{-1}}_{\mathbf{Q'R}^{-1}\mathbf{Q}+\mathbf{G}^{-1}} \underbrace{\begin{bmatrix} 1.4781 & -0.3473 & 1.9313 \\ -0.3473 & 2.4685 & 0.2326 \\ 1.9313 & 0.2326 & 4.7107 \end{bmatrix}}_{(2\mathbf{G}^{-1}\alpha_{par})}$$

$$= \underbrace{\begin{bmatrix} 0.6156 & 0.1940 & 0.2987 \\ 0.0935 & 0.8107 & 0.2402 \\ 0.1175 & 0.0202 & 0.7279 \end{bmatrix}}_{\mathbf{W}_1}$$

and

$$\mathbf{W}_2 = \underbrace{\begin{bmatrix} 2.1520 & -0.9957 & 2.0986 \\ -0.9957 & 3.2921 & -0.3580 \\ 2.0986 & -0.3580 & 5.7284 \end{bmatrix}^{-1}}_{(\mathbf{Q'R^{-1}Q + G^{-1}}\alpha_6)^{-1}} \underbrace{\begin{bmatrix} 0.6738 & -0.6484 & 0.1674 \\ -0.6484 & 0.8235 & -0.5906 \\ 0.1674 & -0.5906 & 1.0177 \end{bmatrix}}_{(\mathbf{Q'R^{-1}Q})}$$

$$= \underbrace{\begin{bmatrix} 0.3844 & -0.1940 & -0.2987 \\ -0.0935 & 0.1893 & -0.2402 \\ -0.1175 & -0.0202 & 0.2721 \end{bmatrix}}_{\mathbf{W}_2}$$

The contributions from **PA** and **YD** to the random regression coefficients for cow 6 are:

$$\begin{bmatrix} \hat{u}_0 \\ \hat{u}_1 \\ \hat{u}_2 \end{bmatrix} = \mathbf{W}_1 \begin{bmatrix} -0.2560 \\ 0.0016 \\ 0.1178 \end{bmatrix} + \mathbf{W}_2 \begin{bmatrix} -5.0004 \\ -4.6419 \\ -1.9931 \end{bmatrix} = \begin{bmatrix} -0.1221 \\ 0.0057 \\ 0.0557 \end{bmatrix} + \begin{bmatrix} -0.4265 \\ 0.0674 \\ 0.1389 \end{bmatrix} = \begin{bmatrix} -0.5485 \\ 0.0730 \\ 0.1946 \end{bmatrix}$$

For cow 8 with ten TD records and no progeny, Eqn 10.4 is:

$$\begin{bmatrix} \hat{u}_0 \\ \hat{u}_1 \\ \hat{u}_2 \end{bmatrix} = \underbrace{\begin{bmatrix} 0.3893 & -0.0844 & 0.1763 \\ -0.0604 & 0.5903 & 0.0353 \\ 0.1576 & 0.0425 & 0.6379 \end{bmatrix}}_{(\mathbf{W}_1)} \underbrace{\begin{bmatrix} 0.3967 \\ 0.0228 \\ -0.1787 \end{bmatrix}}_{(\mathbf{PA})}$$

$$+ \underbrace{\begin{bmatrix} 0.6107 & 0.0844 & 0.1763 \\ 0.0604 & 0.4097 & -0.0353 \\ -0.1576 & -0.0425 & 0.3621 \end{bmatrix}}_{(\mathbf{W}_2)} \underbrace{\begin{bmatrix} 0.2102 \\ 0.0574 \\ 0.1893 \end{bmatrix}}_{(\mathbf{YD}_8)}$$

and

$$\begin{bmatrix} \hat{u}_0 \\ \hat{u}_1 \\ \hat{u}_2 \end{bmatrix} = \begin{bmatrix} 0.1210 \\ -0.0168 \\ -0.0505 \end{bmatrix} + \begin{bmatrix} 0.0998 \\ 0.0295 \\ 0.0330 \end{bmatrix} = \begin{bmatrix} 0.2208 \\ 0.0127 \\ -0.0175 \end{bmatrix}$$

Considering cow 4, with ten TD records and a progeny:

$$\begin{bmatrix} \hat{u}_0 \\ \hat{u}_1 \\ \hat{u}_2 \end{bmatrix} = \underbrace{\begin{bmatrix} 0.3488 & -0.0684 & 0.1393 \\ -0.0490 & 0.5132 & 0.0284 \\ 0.1245 & 0.0343 & 0.5451 \end{bmatrix}}_{(\mathbf{W}_1)} \underbrace{\begin{bmatrix} -0.0655 \\ 0.0123 \\ -0.0343 \end{bmatrix}}_{(\mathbf{PA})} + \underbrace{\begin{bmatrix} 0.5640 & 0.0856 & -0.1741 \\ 0.0613 & 0.3585 & -0.0355 \\ -0.1557 & -0.0428 & 0.3186 \end{bmatrix}}_{(\mathbf{W}_2)}$$

$$\underbrace{\begin{bmatrix} 0.2711 \\ -0.0508 \\ -0.6412 \end{bmatrix}}_{(\mathbf{YD})} + \underbrace{\begin{bmatrix} 0.0872 & -0.0171 & 0.0348 \\ -0.0123 & 0.1283 & 0.0071 \\ 0.0311 & 0.0086 & 0.1363 \end{bmatrix}}_{(\mathbf{W}_3)} \underbrace{\begin{bmatrix} 1.5725 \\ 0.0057 \\ -0.6948 \end{bmatrix}}_{(\mathbf{PC})}$$

$$\begin{bmatrix} \hat{u}_0 \\ \hat{u}_1 \\ \hat{u}_2 \end{bmatrix} = \begin{bmatrix} -0.0285 \\ 0.0086 \\ -0.0264 \end{bmatrix} + \begin{bmatrix} 0.2602 \\ 0.0212 \\ -0.2443 \end{bmatrix} + \begin{bmatrix} 0.1128 \\ -0.0235 \\ -0.0457 \end{bmatrix} = \begin{bmatrix} 0.3445 \\ 0.0063 \\ -0.3164 \end{bmatrix}$$

Equation 10.4 is useful in explaining the evaluations for animals in terms of contributions from different sources of information, and how these contributions vary with different DIM could also be examined. However, Eqn 10.4 relates to random regression coefficients. Usually, the EBV at a particular stage of the longitudinal scale, such as 305 days for milk yield or body weight at one year of age, is published. Therefore, the interest might be in calculating the contributions from the various sources of information to the published EBV. Using milk yield as an example, the contribution to 305-day estimated BV from various sources of information can be calculated as:

$$\begin{aligned}\hat{\mathbf{u}}_{(305)anim} &= \mathbf{V}_1\mathbf{PA} + \mathbf{V}_2\mathbf{YD} + \mathbf{V}_3\mathbf{PC} \\ \hat{\mathbf{u}}_{(305)anim} &= \mathbf{PA}^* + \mathbf{YD}^* + \mathbf{PC}^*\end{aligned} \qquad (10.5)$$

where $\mathbf{V}_i = \mathbf{DW}_i$, with $\mathbf{D}$ being a diagonal matrix such that $d_{ii} = t_i$, with t_i being the element of the row vector $\mathbf{t}$ in Eqn 10.2, $\mathbf{PA}^* = \mathbf{V}_1\mathbf{PA}$, $\mathbf{YD}^* = \mathbf{V}_2\mathbf{YD}$ and $\mathbf{PC}^* = \mathbf{V}_3\mathbf{PC}$. However, $\mathbf{V}_1 + \mathbf{V}_2 + \mathbf{V}_3 \neq \mathbf{I}$. Thus, the estimated BV at 305 days ($\mathrm{BV}_{(305)anim}$) from Eqn 10.5 is:

$$BV_{(305)anim} = \sum_{i=1}^{nr}\hat{u}_{(305)anim} = \sum_{i=1}^{nr}PA_i^* + \sum_{i=1}^{nr}YD_i^* + \sum_{i=1}^{nr}PC_i^*$$

where the contributions to the EBV at 305 days from PA, YD and PC are:

$$\sum_{i=1}^{nr}PA_i^* + \sum_{i=1}^{nr}YD_i^*, \text{ and} \sum_{i=1}^{nr}PC_i^*, \text{ respectively.}$$

Using Eqn 10.5, the contributions from various sources of information can be calculated for EBV at days or ages *j* to *n* along the longitudinal scale, and this could be plotted to examine how the contributions vary with days or age.

Using cow 6 in Example 10.2, the matrix $\mathbf{D}$ used in calculating the $\mathbf{V}$ terms in Eqn 10.5 is:

$$\mathbf{D} = \text{diag}(215.6655, 2.4414, -1.5561)$$

Using the $\mathbf{W}_1$ and $\mathbf{W}_2$ calculated earlier for cow 6:

$$\mathbf{V}_1 = \mathbf{DW}_1 = \begin{pmatrix} 132.7637 & 41.8391 & 64.4193 \\ 0.2283 & 1.9792 & 0.5864 \\ 0.1828 & -0.0314 & -1.1327 \end{pmatrix}$$

$$\mathbf{V}_2 = \mathbf{DW}_2 = \begin{pmatrix} 82.9018 & -41.8391 & -64.4193 \\ -0.2283 & 0.4622 & -0.5864 \\ 0.1828 & 0.0314 & -0.4234 \end{pmatrix} \text{ and}$$

$$\hat{\mathbf{u}}_{(305)6} = \mathbf{V}_1\mathbf{PA} + \mathbf{V}_2\mathbf{YD} = \begin{pmatrix} -26.3320 \\ 0.0138 \\ 0.1178 \end{pmatrix} + \begin{pmatrix} -91.9351 \\ 0.1649 \\ -0.2160 \end{pmatrix}$$

Therefore, contributions from **PA** and **YD** are –26.4049 and –91.9862, respectively, and:

$$BV_{(305)6} = -26.4049 + 91.9862 = -118.3911$$

Thus contribution from parent average is about 22% of the EBV at 305 days. The EBV at 305 days calculated above is slightly different from the value of –118.4265 shown earlier, due to rounding.

10.3.3 Calculating daughter yield deviations

The equation for calculating daughter yield deviation under an RRM is the same as Eqn 6.12 presented for the multivariate models. However, with the RRM, **DYD** in Eqn 6.12 is a vector of random regression coefficients, and the weights $\mathbf{M}_1$, $\mathbf{M}_2$ and $\mathbf{M}_3$ are of the order *nr*. Actual daughter yield deviation for any DIM can be generated using Eqn 10.2.

As indicated in Section 6.2, for ease of computation, $\mathbf{W}_{2prog}$ in Eqn 6.12 is pre-multiplied with $\mathbf{G}^{-1}$, such that the equation for **DYD** becomes:

$$\mathbf{DYD} = \sum \mathbf{G}^{-1}\mathbf{W}_{2prog}\alpha_{prog}(2\mathbf{YD} - \hat{\mathbf{u}}_{mate}) / \mathbf{G}^{-1}\mathbf{W}_{2prog}\alpha_{prog}$$

10.3.4 Reliability of breeding values

The reliability of an EBV depends on its prediction error variance (PEV) relative to the genetic variance. It can therefore be regarded as a statistic summarizing the value of information available in calculating the EBV. The published EBV from an RR model is usually a linear function of the random regression coefficients obtained by solving the MME. The principles for calculating PEV and reliability under this situation are presented using the diagonal elements of the inverse of the coefficient matrix of the MME for Example 10.2.

Let $\mathbf{k'u}$ define the EBV for the trait of interest for animal *i* from the RR model. The vector $\mathbf{k} = \mathbf{w}_i\mathbf{t}$, where $\mathbf{w}_i$ might be the weighting factor for the *i*th age or lactation if the study was on body weight at several ages or fat yield in different lactations analysed as different traits. For instance, if fat yields in lactations 1 and 2 were analysed as different traits, w_i' might be [0.70 0.3], indicating a weight of 0.7 and 0.3, respectively, for first and second lactation EBV. The vector **t** defines how within-lactation EBV was calculated and is the same as in Eqn 10.2. For Example 10.2, *k* is a scalar with a value of 1. Given that **G** is the additive genetic covariance matrix for random regression effect for animal effects and **P** is the covariance matrix for pe effects, then the additive genetic variance of $\mathbf{k'u} = g = \mathbf{k'Gk}$ and the variance for the pe effect for the trait of interest $= p = \mathbf{k'Pk}$. The heritability of $\mathbf{k'u}$ can therefore be calculated as $(g/(g + p + e)$ and $a = (4 - h^2)/h^2$.

Let $\mathbf{C}_{ii}$ be the subset of the inverse of MME corresponding to the genetic effect for the *i*th animal. Then for animal *i*, prediction error variance $(\text{PEV}_i) = \mathbf{k'C}_{ii}\mathbf{k}$. The reliability of $\mathbf{k'u}$ can therefore be calculated as $1 - \text{PEV}_i/g$. As an illustration, in Example 10.2, $\mathbf{k'} = \mathbf{wT} = [215.66552.4414 - 1.5561]$, $g = \mathbf{k'Gk} = 154896.766 \text{ kg}^2$, $p = \mathbf{k'Pk} = 323462.969 \text{ kg}^2$ and $h^2 = 0.32$. For animal 1, the matrix $\mathbf{C}^{11}$ is:

$$\mathbf{C}^{11} = \begin{bmatrix} 2.9911 & 0.5159 & -1.2295 \\ 0.5159 & 0.8683 & -0.2480 \\ -0.2295 & -0.2480 & -0.9183 \end{bmatrix}$$

and

$$\text{PEV}_1 = \mathbf{k'C}^{11}\mathbf{k} = 140499.97$$

Therefore, reliability for animal 1 equals $1 - 140499.97 / 154896.766 = 0.09$. The reliabilities for the animals in Example 10.2 are:

Animal	Reliability
1	0.09
2	0.04
3	0.07
4	0.12
5	0.15
6	0.06
7	0.10
8	0.05

In practice, calculating the inverse of the MME is not feasible for large populations and PEV has to be approximated. As indicated earlier, EBV from RR models are linear functions of the random regressions; therefore, methods to approximate reliabilities should simultaneously approximate PEV and the prediction error covariance (PEC) among the individual random regressions (Liu *et al.*, 2002; Meyer and Tier, 2003). Such an approximation method presented by Meyer and Tier (2003) is outlined in Appendix D, Section D.2.

10.3.5 Random regression models with spline function

Random regression models with Legendre polynomials have been considered to have better convergence properties as the regressions are orthogonal. However, some studies have reported high genetic variances at the extremes of the lactation and negative correlations between the most distant test days. In order to overcome this limitation, some workers have fitted RRM using splines (Misztal, 2006; Bohmanova *et al.*, 2008). Splines are piecewise functions consisting of independent segments that are connected in knots. The segments are described by lower-order polynomials. Linear splines are the simplest spline function where the segments are fitted by linear polynomials between two knots adjacent to the record and zero between all other knots. Thus, the system of equations is sparse as only two coefficients are non-zero for a given record. The use of cubic splines for the modelling of the lactation curve has also been presented by White *et al.* (1999). However, the linear spline is considered in this section.

Let **T** be a vector of n knots, then the covariables of the linear spline for DIM t ($\boldsymbol{\Phi}_i(t)$) located between knots T_i and T_{i+1} can be calculated as:

$$\begin{aligned}\boldsymbol{\Phi}_i(t) &= (t - T_i)/(T_{i+1} - T_i)\\ \boldsymbol{\Phi}_{i+1}(t) &= (T_{i+1} - t)/(T_{i+1} - T_i)\\ &= 1 - \boldsymbol{\Phi}_i(t) \text{ and } \boldsymbol{\Phi}_{1\ldots i-1,i+2\ldots n} = 0\end{aligned}$$

If $t = T_i$, $\boldsymbol{\Phi}_i(t) = 1$ and $\boldsymbol{\Phi}_{1\ldots i-1,i+1\ldots n} = 0$.

Thus the vector $\boldsymbol{\Phi}$ for DIM t has, at most, two non-zero elements, which sum up to one. The above formula assumes that $T_i \le t < T_n$. If, however, $t < T_i$ or $t > T_n$, the following can be used and the sum of the elements of the vector will not sum up to one:

$$\begin{aligned}&\text{if } t < T_1, \boldsymbol{\Phi}_1(t) = t / T_1 \text{ and } \boldsymbol{\Phi}_{1+i\ldots n} = 0\\ &\text{if } t > T_n, \boldsymbol{\Phi}_n(t) = T_5 / t \text{ and } \boldsymbol{\Phi}_{1\ldots n-1} = 0\end{aligned}$$

Using the data in Example 10.1, assume that the four knots are fitted for the fixed lactation curve and knots are placed at days 4, 106, 208 and 310, then the covariables for the spine function for particular DIM are as follows:

DIM	Φ_0	Φ_1	Φ_2	Φ_3
4	1.000	0.000	0.000	0.000
38	0.333	0.667	0.000	0.000
72	0.667	0.333	0.000	0.000
106	0.000	1.000	0.000	0.000
140	0.000	0.333	0.667	0.000
174	0.000	0.667	0.333	0.000
208	0.000	0.000	1.000	0.000
242	0.000	0.000	0.333	0.667
276	0.000	0.000	0.667	0.333
310	0.000	0.000	0.000	1.000

As an illustration, the covariates for DIM 38 can be computed as:
$\boldsymbol{\Phi}_1(38) = (38-4)/106-4) = 0.333$ and $\boldsymbol{\Phi}_2(38) = 1-0.333 = 0.667$

Thus $\boldsymbol{\Phi}(38) = [0.333, 0.667, 00]$
A random regression model can therefore be fitted as:

$$y_{tijk} = htd_i + \sum_{k=0}^{nf}\phi_{jtk}\beta_k + \sum_{k=0}^{nr}\phi_{jtk}\mathbf{u}_{jk} + \sum_{k=0}^{nr}\phi_{jtk}\mathbf{pe}_{jk} + e_{tij}$$

where all terms are as defined in Section 10.3 but the φ_{jtk} is the vector of the kth spline function for the test-day record of cow j made on day t. The same procedure described in Section 10.3.1 can be used in the application of the model for the analysis of data and interpretation of results.

10.3.6 Random regression models for maternal traits

Maternal genetic effects are important in growth traits in beef cattle, and models that account for these effects have been discussed in Chapter 8. However, the RR model could also be augmented to include random regressions for maternal genetic and maternal permanent environmental effects. Albuquerque and Meyer (2001) examined different orders of fit for the random regressions for both effects. One of the favoured models was the one in which the order of Legendre polynomials for direct genetic, maternal genetic, animal pe and maternal pe effects were 5, 5, 5 and 3, respectively.

Such a model, excluding all fixed effects, could be written as:

$$y_{ijktd} = \sum_{i=0}^{k_1-1}\phi_{jti}\mathbf{u}_{ji} + \sum_{i=0}^{k_2-1}\phi_{jti}\mathbf{m}_{ji} + \sum_{i=0}^{k_3-1}\phi_{jti}\mathbf{pe}_{ji}\sum_{i=0}^{k_4-1}\phi_{dti}\mathbf{pp}_{di} + e_{ijktd}$$

where y_{ijktd} is the body weight of cow j taken at age t that has a dam d; $\mathbf{u}_{ji}$, $\mathbf{m}_{ji}$ and $\mathbf{pe}_{ji}$ are the random regressions for direct, maternal genetic and animal pe effects for animal j, respectively; $\mathbf{pp}_{di}$ is the random regression for dam pe effects and e_{ijktd} is random error; φ_{jti} and φ_{dti} are the vector of the ith Legendre polynomial for body weight at age t for cow j and dam d, respectively. They assumed a zero covariance between direct and maternal genetic effects to simplify the computation. The variance for direct effects increased from birth to 365 days while maternal genetic variance increased from birth to about 115 days and decreased thereafter.

10.4 Covariance Functions

Kirkpatrick *et al.* (1990, 1994) introduced the concept of analysing repeated records taken along a trajectory such as time or age by means of covariance functions. In view of the fact that such a trait can take on a value at each of an infinite number of ages and its value at each age can be regarded as a distinct trait, the trajectory for such a trait could be regarded as an infinite-dimensional trait. Thus the growth trajectory or milk yield trajectory of an individual could be represented by a continuous function. Covariance function describes the covariance structure of an infinite-dimensional character as a function of time. Therefore, the covariance function is the infinite-dimensional equivalent of a covariance matrix for a given number of records taken over time at different ages. The value of the phenotypic covariance function, $þ(t_i, t_j)$, gives the phenotypic covariance between the value of the trait at ages t_i and t_j. Similarly, the value of the additive genetic covariance function, $f(t_i, t_j)$, gives the additive genetic covariance between the value of the trait at ages t_{i} and t_j. In mathematical terms, given t ages, the covariance between breeding values u_l and u_m on an animal at ages a_l and a_m could be written as:

$$\operatorname{cov}(u_l, u_m) = f(a_l, a_m) = \sum_{i=0}^{k-1}\sum_{i=0}^{k-1}\phi_i(a_l)\phi_j(a_m)\mathrm{C}_{ij} \tag{10.6}$$

$$= \sum_{i=0}^{k-1}\sum_{i=0}^{k-1}\tau_{ij}a_l^i a_m^j \tag{10.7}$$

where f with factors τ_{ij} is the covariance function (CF), **C** is the coefficient matrix associated with the CF with elements C_{ij}, a_{l} is the lth age standardized to the intervals for which the polynomials are defined and k is the order of fit. Kirkpatrick *et al.* (1990, 1994) used Legendre polynomials that span the interval –1 to + 1. The ages can be standardized as described in Appendix G.

Given that **G** is the observed genetic covariance matrix of order t, and assuming a full- order polynomial fit ($k = t$), Eqn 10.6 can be written in matrix notation as:

$$\hat{\mathrm{G}} = \mathbf{\Phi}\,\hat{\mathrm{C}}\,\mathbf{\Phi}' \tag{10.8}$$

and $\hat{\mathrm{C}}$ can be estimated as:

$$\hat{\mathrm{C}} = \mathbf{\Phi}^{-1}\,\hat{\mathrm{G}}(\mathbf{\Phi}^{-1}) \tag{10.9}$$

where $\mathbf{\Phi}$ is the matrix of Legendre polynomials of order t by k with element $\phi_{ij} = \phi_j(a_t)$ = the jth polynomial evaluated at standardized age t.

As an illustration, assume body weight measurements in beef cattle have been taken at three different ages – 90, 160 and 240 months old – and that the genetic covariance matrix ($\hat{\mathrm{G}}$) estimated was:

$$\hat{\mathrm{G}} = \begin{bmatrix} 132.3 & 127.0 & 136.6 \\ 127.0 & 172.8 & 200.8 \\ 136.6 & 200.8 & 288.0 \end{bmatrix}$$

Using the method described in Appendix G, the vector of standardized ages is:

$\mathbf{a}' = [-1.0 \; -0.0667 \; 1.000]$

and **M** becomes:

$$\mathbf{M} = \begin{bmatrix} 1.0000 & -1.0000 & 1.0000 \\ 1.0000 & -0.0667 & 0.0044 \\ 1.0000 & 1.0000 & 1.0000 \end{bmatrix}$$

Thus, for t = 3, Λ is:

$$\Lambda = \begin{bmatrix} 0.7071 & 0.0000 & -0.7906 \\ 0.0000 & 1.2247 & 0.0000 \\ 0.0000 & 0.0000 & 2.3717 \end{bmatrix}$$

and **Φ** is:

$$\mathbf{\Phi} = \begin{bmatrix} 0.7071 & -1.2247 & 1.5811 \\ 0.7071 & -0.0816 & -0.7801 \\ 0.7071 & 1.2247 & 1.5811 \end{bmatrix}$$

and from Eqn 10.9, the coefficient matrix $\hat{\mathrm{C}}$ is:

$$\hat{\mathrm{C}} = \begin{bmatrix} 344.7117 & 45.2787 & -3.2062 \\ 45.2787 & 24.5185 & -0.1475 \\ -3.2062 & -0.1475 & 3.2768 \end{bmatrix}$$

The covariance between two different ages can be calculated using Eqn 10.8. For instance, the variances at days 90 and 200 of body weight and the covariance between body weight on both days are $\mathbf{\Phi}_{90}\hat{\mathrm{C}}\,\mathbf{\Phi}'_{90} = 132.30$, $\mathbf{\Phi}_{200}\hat{\mathrm{C}}\,\mathbf{\Phi}'_{200} = 218.50$, $\mathbf{\Phi}_{90}\hat{\mathrm{C}}\,\mathbf{\Phi}'_{200} = 129.71$, respectively, with:

$\mathbf{\Phi}_{90} = \mathbf{m}_{90}\mathbf{\Lambda} = [0.7071 - 1.22471.5811]$

and

$\mathbf{\Phi}_{200} = \mathbf{m}_{200}\mathbf{\Lambda} = [0.70710.5716 - 0.2740]$

where $\mathbf{m}_i$ are the appropriate row vectors of the matrix **M**.

Also, from Eqn 10.8 and Appendix G, $\hat{\mathrm{G}}$ can be written as:

$\hat{\mathrm{G}} = \mathbf{M\Lambda\hat{C}\Lambda'M'}$

Therefore, $\hat{\mathrm{G}} = \mathbf{MTM'}$ with $\mathbf{T} = \mathbf{\Lambda\hat{C}\Lambda}$ or calculated as $\mathbf{T} = \mathbf{M}^{-1}\hat{\mathrm{G}}(\mathbf{M}^{-1})'$, where **T** is the matrix with elements τ_{ij} in Eqn 10.7. Substituting **T** in Eqn 10.7 the full estimate of the CF, $f(a_l, a_m)$, can be obtained. Using the example data:

$$\mathbf{T} = \begin{pmatrix} 177.99 & 39.35 & -11.52 \\ 39.35 & 36.78 & -0.43 \\ -11.52 & -0.43 & 18.43 \end{pmatrix}$$

Therefore, the full estimate of the covariance function, $f(a_l, a_m)$, is:

$$\begin{aligned} f(a_l, a_m) &= 177.99 + 39.35(a_l + a_m) + 36.78a_la_m - 11.52(a_l^2 + a_m^2) \\ &\quad -0.43(a_l^2a_m + a_la_m^2) + 18.43a_l^2a_m^2 \end{aligned}$$

The application of CF in genetic evaluation involves defining an equivalent model using Eqn 10.8. For instance, using the example of the body weight of beef cattle, assume that the multivariate model for observations measured on one animal is:

$$\mathrm{y} = \mathbf{Xb} + \mathbf{a} + \mathbf{e}$$

where **y**, **X**, **b**, **a** and **e** are vectors defined as in Eqn 5.1 with $i = t$, with var(a)= $\breve{\mathbf{G}}$ and var(**e**) = **R**. Assuming a CF has also been fitted for the covariance matrix for environmental effects with a term included to account for measurement error, then:

$$\mathbf{R} = \mathbf{\Phi C}_p\mathbf{\Phi}' + \mathbf{I}\sigma^2\varepsilon$$

where $\mathbf{C}_p$ contains the coefficient matrix associated with the CF for pe and variance ϵ is $\mathrm{I}\sigma^2\epsilon$. Using this equation and Eqn 10.8, an equivalent model to the multivariate model can be written as:

$$\mathrm{y} = \mathbf{Xb} + \mathbf{\Phi u} + \mathbf{\Phi pe} + \varepsilon$$

where **u** and **pe** are now vectors of random regression coefficients for random animal and pe effects. Then var(**u**) = $\mathbf{\Phi C \Phi}'$ and var(**pe**) = $\mathbf{\Phi C}_p\mathbf{\Phi}'$. The application of the above model in genetic evaluation is illustrated in Example 10.2. Thus, the breeding value a_n for any time n can be calculated as:

$$a_n = \sum_{i=0}^{k-1}\phi_i(t_n)\mathbf{u}_i$$

where $\varphi(t_n)$ is the vector of Legendre polynomial coefficients evaluated at age t_n. Thus, with a full-order fit, the covariance function model is exactly equivalent to the multivariate model. However, in practice, the order of fit is chosen such that the estimated covariance matrix can be appropriately fitted with as few parameters as possible. In the next section, the fitting of a reduced-order CF is discussed.

10.4.1 Fitting a reduced-order covariance function

Equation 10.8 and the illustration given in the above section assumed a full-order polynomial fit of **G** ($k = t$). Therefore, it was possible to get an inverse of **Φ** and hence estimate C. However, for a reduced-order ($k < t$) fit, **Φ** has only k columns and a direct inverse may not be possible. With the reduced fit, the number of coefficients to be estimated are reduced to $k(k + 1)/2$. This is particularly important for large Λ, such as test-day milk yield within a lactation with t equal to 10 or 305, assuming monthly or daily sampling and requiring $t(t + 1)/2$ coefficients to be estimated. Thus a reduced-order fit with k substantially lower than t could be very beneficial.

Kirkpatrick *et al.* (1990) proposed weighted least squares as an efficient method of obtaining an estimate of the reduced coefficient matrix ($\breve{\mathbf{C}}$) from the linear function of the elements of $\breve{\mathbf{G}}$. They outlined the following steps for the weighted least-square procedure. The procedure is illustrated using the example $\breve{\mathbf{G}}$ for the body weight in beef cattle given earlier, fitting polynomials of order 1, i.e. only the first two Legendre polynomials are fitted, thus $k = 2$. Initially, a vector $\breve{\mathbf{g}}$ of order t^2 is formed by stacking the successive columns of $\breve{\mathbf{G}}$. Thus:

$$\breve{\mathbf{g}}' = \breve{G}_{11},\ldots,\breve{G}_{n1},\breve{G}_{12},\ldots,\breve{G}_{n2,}\breve{G}_{1n},\ldots,\breve{G}_{nn}$$

Thus, for the example $\breve{\mathbf{G}}$:

$\breve{\mathbf{g}}' = [132.3 \quad 127.0 \quad 136.6 \quad 172.8 \quad 200.8 \quad 288.0]$

Define Φ_r of order t by k, obtained by deleting $(t - k)$ columns of $\mathbf{\Phi}$ corresponding to those ϕ_j not in the reduced-order fit. The relationship between the observed covariance matrix, $\breve{\mathbf{g}}$, and the coefficient matrix of the reduced fit to be estimated is given by the following regression equation:

$$\breve{\mathbf{g}} = \mathbf{X}_s \, \breve{\mathbf{c}} + \mathbf{e} \qquad (10.10)$$

where $\mathbf{e}$ is the vector of the difference between observed covariances and those predicted by the covariance function, $\breve{\mathbf{c}} = (\breve{C}^{00},..., \breve{C}_k{}^0,..., \breve{C}_k{}^1,..., \breve{C}_{kk})$ $\breve{\mathbf{c}}$ is a vector of dimension k^2, containing the elements of the coefficient matrix of the reduced fit $\breve{\mathbf{C}}$. The order of elements of $\breve{\mathbf{C}}$ in $\breve{\mathbf{c}}$ is the same as in $\breve{\mathbf{g}}$: that is, $\breve{\mathbf{c}} = (\breve{C}_{00},..., \breve{C}_{k0},..., \breve{C}_{k1},..., \breve{C}_{kk})$. $\mathbf{X}_s$ is the Kronecker product of $\mathbf{\Phi}_r$ with itself ($\mathbf{X}_s = \mathbf{\Phi}_r \times \mathbf{\Phi}_r$) and is of the order t^2 by k^2. Since only the first two polynomials are fitted, the matrix $\mathbf{\Phi}_r$ can be derived by deleting $\mathbf{\Phi}$ **from** the third column, corresponding to the missing second-degree polynomial. Thus, for the beef cattle example:

$$\mathbf{\Phi}_r = \begin{bmatrix} 0.7071 & -1.2247 \\ 0.7071 & -0.0816 \\ 0.7071 & 1.2247 \end{bmatrix}$$

and $\mathbf{X}_s$ is:

$$\mathbf{X}_s = \begin{bmatrix} 0.5000 & -0.8660 & -0.8660 & 1.4999 \\ 0.5000 & -0.0577 & -0.8660 & 0.0999 \\ 0.5000 & 0.8660 & -0.8660 & -1.4999 \\ 0.5000 & -0.8660 & -0.0577 & 0.0999 \\ 0.5000 & -0.0577 & -0.0577 & 0.0067 \\ 0.5000 & 0.8660 & -0.0577 & -0.0999 \\ 0.5000 & -0.8660 & 0.8660 & -1.4999 \\ 0.5000 & -0.0577 & 0.8660 & -0.0999 \\ 0.5000 & 0.8660 & 0.8660 & 1.4999 \end{bmatrix}$$

The application of weighted least squares to obtain solutions for $\breve{\mathbf{c}}$ in Eqn 10.10 requires the covariance matrix (**V**) of sampling errors of $\breve{\mathbf{g}}$. Kirkpatrick *et al.* (1990) presented several methods for estimating **V**, examining three different experimental designs. However, in animal breeding, most estimates of $\breve{\mathbf{G}}$ are from field data and may not fit strictly to the designs they described, but estimates of sampling variances from REML analysis could be used. For the example $\breve{\mathbf{G}}$ for the beef cattle data, **V** has been estimated using the formula given by Kirkpatrick *et al.* (1990) for a half-sib design, assuming that 60 sires were each mated to 20 dams. The mean cross-product for the residual effect ($\hat{W}_e$) was estimated as $\hat{W}_{e,ij} = P_{ij} - 0.25\,\breve{G}_{ij}$ and that among sires ($\hat{W}_a$) as $\hat{W}_{a,ij} = (n - 1/4)\breve{G}_{ij} + P_{ij}$, where P_{ij} is the phenotypic variance and n is the number of dams. Sampling variance for $\breve{\mathbf{g}}$ was then calculated as: $\mathbf{V} = (16/n^2)[\mathrm{cov}(\hat{W}_{a,ij}, \hat{W}_{a,kl}) + \mathrm{cov}(\hat{W}_{e,ij}, \hat{W}_{e,kl})]$, where $\mathrm{cov}\,(\hat{W}_{ij}, \hat{W}_{kl}) = (\hat{W}_{ik}\hat{W}_{jl} + \hat{W}_{il}\hat{W}_{jk})/\,\mathrm{df}$, with df = number of degrees of freedom plus 2. In estimating $\mathrm{cov}(\hat{W}_{a,ij}, \hat{W}_{a,kl})$ and $\mathrm{cov}(\hat{W}_{e,ij}, \hat{W}_{e,kl})$, df = $(s - 1) + 2$ and $s(n - 1) + 2$, respectively. The estimated **V** therefore is:

$$\hat{V} = \begin{bmatrix} 3450.0 & 2256.4 & 2184.6 & 2256.4 & 1480.3 & 1434.7 & 2184.6 & 1434.7 & 1390.9 \\ 2256.4 & 2959.6 & 2430.9 & 2959.6 & 2903.5 & 2530.2 & 2430.9 & 2530.2 & 2180.1 \\ 2184.6 & 2430.9 & 3889.7 & 2430.9 & 2249.1 & 3181.6 & 3889.7 & 3181.6 & 4051.5 \\ 2256.4 & 2959.6 & 2430.9 & 2959.6 & 2903.5 & 2530.2 & 2430.9 & 2530.2 & 2180.1 \\ 1480.3 & 2903.5 & 2249.1 & 2903.5 & 5711.4 & 4410.0 & 2249.1 & 4410.0 & 3417.5 \\ 1434.7 & 2530.2 & 3181.6 & 2530.2 & 4410.0 & 5818.8 & 3181.6 & 5818.8 & 6354.3 \\ 2184.6 & 2430.9 & 3889.7 & 2430.9 & 2249.1 & 3181.6 & 3889.7 & 3181.6 & 4051.5 \\ 1434.7 & 2530.2 & 3181.6 & 2530.2 & 4410.0 & 5818.8 & 3181.6 & 5818.8 & 6354.3 \\ 1390.9 & 2180.1 & 4051.5 & 2180.1 & 3417.5 & 6354.3 & 4051.5 & 6354.3 & 11835.0 \end{bmatrix}$$

However, the symmetry of $\breve{\mathbf{G}}$ resulted in redundancies in the vector $\breve{\mathbf{g}}$ such that **V** is singular. The vector $\breve{\mathbf{g}}$ can be redefined to be of the order s by 1, which contains only the elements in the lower half of $\breve{\mathbf{G}}$, where $s = t(t + 1)/2$. Therefore, delete from $\breve{\mathbf{g}}$ the elements $\breve{\mathbf{G}}_{ij}$ for which $i < j$. Thus, for the example $\breve{\mathbf{G}}$, the vector $\breve{\mathbf{g}}$ becomes:

$$\breve{\mathbf{g}} = [132.3 \quad 127.0 \quad 136.6 \quad 172.8 \quad 200.8 \quad 288.0]$$

Then delete from **V** those columns and rows corresponding to elements $\breve{\mathbf{G}}_{ij}$ with $i < j$. This involves deleting rows and columns 4, 7 and 8 from the matrix **V** given above. The **V** of reduced order (s by s) is:

$$\hat{V} = \begin{bmatrix} 3450.0 & 2256.4 & 2184.6 & 1480.3 & 1434.7 & 1390.9 \\ 2256.4 & 2959.6 & 2430.9 & 2903.5 & 2530.2 & 2180.1 \\ 2184.6 & 2430.9 & 3889.7 & 2249.1 & 3181.6 & 4051.5 \\ 1480.3 & 2903.5 & 2249.1 & 5711.4 & 4410.0 & 3417.5 \\ 1434.7 & 2530.2 & 3181.6 & 4410.0 & 5818.8 & 6354.3 \\ 1390.9 & 2180.1 & 4051.5 & 3417.5 & 6354.3 & 11835.0 \end{bmatrix}$$

Similarly, the rows corresponding to those elements of $\breve{\mathbf{g}}$ for which $\breve{\mathbf{G}}_{ij}$ has $i < j$ are deleted from $\mathbf{X}_s$. In the example $\mathbf{X}_s$, rows 4, 7 and 8 are deleted. Thus $\mathbf{X}_s$ becomes:

$$\mathbf{X}_s = \begin{bmatrix} 0.5000 & -0.8660 & -0.8660 & 1.4999 \\ 0.5000 & -0.0577 & -0.8660 & 0.0999 \\ 0.5000 & 0.8660 & -0.8660 & -1.4999 \\ 0.5000 & -0.0577 & -0.0577 & 0.0067 \\ 0.5000 & 0.8660 & -0.0577 & -0.0999 \\ 0.5000 & 0.8660 & 0.8660 & 1.4999 \end{bmatrix}$$

Also, for each element of $\breve{\mathbf{c}}$ for which $\breve{C}_{ij}$ has $i < j$, add the corresponding column of $\mathbf{X}_s$ to the column corresponding to $\breve{C}_{ji}$, then delete the former column. For the beef cattle example, the vector of coefficients, $\breve{\mathbf{c}}' = [\breve{C}_{00}\ \breve{C}_{10}\ \breve{C}_{01}\ \breve{C}_{11}]$. Therefore, the third column of $\mathbf{X}_s$ corresponding to $\breve{C}_{01}$,is added to the second column and the third column is deleted. The matrix $\mathbf{X}_s$ then becomes:

$$\mathbf{X}_s = \begin{bmatrix} 0.5000 & -1.7320 & 1.4999 \\ 0.5000 & -0.9237 & 0.0999 \\ 0.5000 & 0.0000 & -1.4999 \\ 0.5000 & -0.1154 & 0.0067 \\ 0.5000 & 0.8083 & -0.0999 \\ 0.5000 & 1.7320 & 1.4999 \end{bmatrix}$$

Finally, delete from $\check{\mathbf{c}}$ the elements for which $\check{\mathbf{C}}$, has elements $i < j$. The matrix $\check{\mathbf{c}}$ is now of the order $k(k + 1)/2$ by 1. For the example data, $\check{\mathbf{c}}' = [\check{C}_{00}\ \check{C}_{10}\ \check{C}_{11}]$. The vector $\check{\mathbf{c}}$ can now be calculated by a weighted least-square procedure as:

$$\check{\mathbf{c}} = (\mathbf{X}_s'\hat{\mathbf{V}}^{-1}\mathbf{X}_s)^{-1}\mathbf{X}_s'\hat{\mathbf{V}}^{-1}\breve{\mathbf{g}}$$

For the example data, $\check{\mathbf{c}}$ calculated using the above equation is:

$$\check{\mathbf{c}}\ [341.8512 \quad 45.0421 \quad 24.5405]$$

The reduced coefficient matrix $\check{\mathbf{C}}$ is then constructed from the calculated $\check{\mathbf{c}}$. Then a row and column of zeros are inserted in positions corresponding to those polynomials not included to obtain $\check{\mathbf{C}}$. For the example data, $\check{\mathbf{C}}$ is now:

$$\check{\mathbf{C}} = \begin{bmatrix} 341.8512 & 45.0421 & 0.0 \\ 45.0421 & 24.5405 & 0.0 \\ 0.0 & 0.0 & 0.0 \end{bmatrix}$$

Kirkpatrick *et al.* (1990) presented the following chi-square statistic to test the goodness of fit of the reduced covariance function to $\breve{\mathbf{G}}$:

$$\chi^2_{(m-p)}\ (\breve{\mathbf{g}} - \mathbf{X}_s\,\check{\mathbf{c}})'\mathbf{V}^{-1}\ (\breve{\mathbf{g}} - \mathbf{X}_s\,\check{\mathbf{c}})$$

where $m = t(t + 1)/2$ is the number of degrees of freedom in $\breve{\mathbf{G}}$ and $p = (k(k + 1)/2$ is the number of parameters being fitted. A significant result indicates that the model is inconsistent with the data, and a higher order of fit may be needed. For the beef cattle example, the value of χ^2 was 0.2231 with $m = 6$ and $p = 3$. This value of χ^2 was not significant with three degrees of freedom and thus the reduced covariance function was not significantly different from $\breve{\mathbf{G}}$.

Another method of fitting a reduced-order CF, proposed by Mantysaari (1999), involved eigenvalue decomposition of the coefficient matrix. The largest k eigenvalues of $\hat{\mathbf{C}}$ in Eqn 10.9, for instance, are kept in a diagonal matrix ($\mathbf{D}_a$) and the matrix $\boldsymbol{\Phi}$ replaced by the k corresponding eigenfunctions. Thus, $\breve{\mathbf{G}}$ in Eqn 10.7 can be approximated as:

$$\hat{\mathbf{G}} \approx \Phi[\mathrm{v}_1\ \mathrm{v}_2 \cdots \mathrm{v}_k]\mathbf{D}_a[\mathrm{v}_1\mathrm{v}_2 \cdots \mathrm{v}_k]'\Phi' = T\mathbf{D}_a\mathbf{T}'$$

where the $\mathbf{v}_i$ are the eigenvectors of $\hat{\mathbf{C}}$ corresponding to eigenvalues in $\mathbf{D}_a$.

Similarly, if CF has been fitted to the environmental covariance matrix, a similar reduction can be carried as follows:

$$\begin{aligned} \mathbf{R} &= \Phi C_p \Phi + \mathbf{I}\sigma^2\varepsilon \\ &= \Phi[\mathrm{v}_1\ \mathrm{v}_2 \cdots \mathrm{v}_k]\mathbf{D}_p[\mathrm{v}_1\ \mathrm{v}_2 \cdots \mathrm{v}_k]'\Phi' + \mathbf{I}\sigma^2\varepsilon = \mathbf{Q}\mathbf{D}_p\mathbf{Q}' + \mathbf{I}\sigma^2\varepsilon \end{aligned} \qquad (10.11)$$

where $\mathbf{D}_p$ contains the k largest eigenvalues of $\mathbf{C}_p$. However, Mantysaari (1999) indicated that with several biological traits, Eqn 10.11 could easily lead to a non-positive definite $\mathbf{C}_p$ and the decomposition may not be possible. He used an expectation maximization (EM) algorithm to fit the CF to the environmental covariance matrix. However, if $\mathbf{C}_p$ has been estimated directly using REML (Meyer and Hill, 1997), the EM algorithm would not be necessary and the covariance matrix for pe can be approximated as $\mathbf{QD}_p\mathbf{Q}'$. In addition to reducing the number of equations to k per animal in the MME with this method, the system of equations is very sparse since $\mathbf{D}_a$ or $\mathbf{D}_p$ are diagonal.

10.5 Equivalence of the Random Regression Model to the Covariance Function

Meyer and Hill (1997) indicated that the RR model is equivalent to a covariance function model. The equivalence of the RR model fitting either a parametric curve or Legendre polynomials to the CF model is presented below. Similar to the model in Section 10.3, the RR model with a parametric curve can be represented as:

$$y_{jt} = F_{jt} + \sum_{m=0}^{f-1} z_m(t)\beta_m + \sum_{m=0}^{k-1} z_m(t)\alpha_{jm} + \sum_{m=0}^{k-1} z_m(t)\lambda_{jm} + e_{jt} \tag{10.12}$$

where y_{jt} is the test day record of cow j made on day t; β_m are fixed regressions coefficients; α_{jm} and λ_{jm} are the additive genetic and permanent environmental random regressions for cow j; F_{jt} represents the remaining fixed effects in the model; $z_m(t)$ is the mth parameter of a parametric function of days in milk; and e_{jt} is the random error term. For example, in the model of Jamrozik *et al.* (1997), z was a function of days in milk with five parameters: $z = (1\ c\ c^2\ d\ d^2)$, where $c = t/305$ and $d = \ln(1/c)$, with ln being the natural logarithm. Then the covariance between breeding values u_i and u_l on an animal recorded at DIM t_i and t_1 is:

$$\text{cov}(u_i, u_l) = f(t_i, t_l) = \sum_{m=0}^{k-1}\sum_{r=0}^{k-1} z_m(t_i) z_r(t_l)\ \text{cov}(\alpha_m, \alpha_r) \tag{10.13}$$

However, instead of a parametric curve, assume that orthogonal polynomials such as Legendre polynomials were fitted in an RR model as described in Section 10.3. Let a_i and a_l represent TD records on days t_i and t_l of animal j standardized to the interval −1 to 1 as outlined in Appendix G. Furthermore, assume that the mth Legendre polynomial of a_i be $\varphi_m(a_l)$, for $m = 0, \ldots, k - 1$. The covariance between breeding values u_i and u_1 on an animal recorded at DIM a_i and a_1 could then be represented as:

$$\text{cov}(u_i, u_l) = f(a_i, a_l) = \sum_{m=0}^{k-1}\sum_{r=0}^{k-1} \phi_m(a_i)\phi_r(a_l)\ \text{cov}(\alpha_m, \alpha_r) \tag{10.14}$$

The right-hand sides of Eqns 10.13 and 10.14 are clearly equivalent to the right-hand side of Eqn 10.6, with $\text{cov}(\alpha_m, \alpha_r)$ equal to $\mathbf{C}_{ij}$, the ijth element of the coefficient matrix of the covariance function. This equivalence of the RR model with the covariance function is useful when analysing data observed at many ages or time periods, as only k regression coefficients and their $k(k + 1)/2$ covariances need to be estimated for each source of variation in a univariate model.

11 Genomic Prediction and Selection

11.1 Introduction

As introduced in Chapter 1, a genetic marker is a fragment of DNA that is associated with a certain location (chromosome) within the genome. In the 1990s, most genetic markers used in livestock studies were microsatellites. DNA microsatellites, which are also referred as simple sequence repeats (SSR), consist of a specific sequence of DNA bases or nucleotides which contain mono, di, tri or tetra tandem repeats. For example, AAAAAAAAAAA or CTGCTGCTGCTG, which may be referred to as (A)11 or (CTG)4, respectively. Alleles at a specific location (locus) can differ in the number of repeats (polymorphic) and hence they are used as genetic markers. Microsatellites are inherited in a Mendelian fashion and are typically co-dominant, i.e. the heterozygote genotype could be distinguished from either homozygote.

Genetic markers are useful in identifying portions of the chromosomes that are associated with particular quantitative traits. The incorporation of information on marker loci that are linked to quantitative trait loci (QTL), together with phenotypic information in a genetic evaluation procedure, would increase the accuracy of evaluations and therefore of selection. The use of breeding values with marker information incorporated in the selection of animals in a breeding programme is termed marker-assisted selection (MAS). The gains from MAS depend on the amount of genetic variation explained by the marker information and are larger for traits with low heritabilities, and therefore estimated breeding values from phenotype are of low accuracy (Goddard and Hayes, 2002). Similarly, MAS should result in larger increases in accuracies for traits, which are sex-limited such as milk yield or measured only in culled animals, for instance carcass traits. In the 1990s, methodologies for the use of microsatellites as markers for the purpose of marker-assisted selection (linkage analysis) were presented by Fernando and Grossman (1989). A detailed description and application of marker-assisted selection in animal breeding is presented in Mrode (2014) and it is therefore not covered in this section.

In outbreeding populations, the incorporation of molecular information in breeding programmes on the basis of the linkage analysis was limited, as the marker maps are rather sparse and linkage between the markers and QTL may not be sufficiently close enough to persist across the population. Thus, the linkage phase between marker and QTL must be established for every family in which the marker is intended to be used for selection.

However, a huge amount of variation has been discovered in the genome at the DNA level as a result of sequencing the genomes of most livestock species. The most abundant form of variation are the single nucleotide polymorphisms (SNPs). An SNP is a DNA sequence variation occurring when a single nucleotide (A, T, C or G) in the genome differs between paired chromosomes in an individual. For example, two sequenced DNA fragments from different individuals, AAGCCTA to AAGC*T*TA, contain a difference in a single nucleotide. In this case we say that there are two alleles: C and T. Generally, SNPs

DOI: 10.1079/9781800620506.0011

are diallelic. In view of the high frequency of SNPs in the genome, and developments in genotyping technology that mean many thousands of SNPs can be genotyped very cheaply, they have been proposed as markers for use in QTL analysis and in association studies in place of microsatellites.

The main emphasis of this chapter is the use of SNPs to directly compute EBVs of animals, which are often called direct genomic breeding values (DGV). This is usually combined with some measure of the traditional EBV, say parent index, from an animal model to produce what is termed genomic breeding values (GEBV), which are officially published and used for the selection of animals.

The use of GEBV in the selection of animals has been referred to as genomic selection. Genomic selection requires that markers (SNPs) are in linkage disequilibrium (LD) with the QTLs across the whole population. LD can be defined as the non-random association between the alleles of two loci (e.g. between alleles of a marker and a QTL). Given a marker locus, A (with alleles A_1 and A_2), and a QTL locus, B (with alleles B_1 and B_2), on the same chromosome, LD can be measured as the squared correlation (r^2) between the marker and the QTL as:

$$D = \text{freq}(A_1B_1) * \text{freq}(A_2B_2) - \text{freq}(A_1B_2) * \text{freq}(A_2B_1)$$
$$r^2 = D^2 / \left[\text{freq}(A_1) * \text{freq}(A_2) * \text{freq}(B_1) * \text{freq}(B_2)\right]$$

The r^2 between the marker and the QTL indicates the proportion of the variance for the QTL that can be explained at the marker.

The basic assumption is that the use of SNPs as markers enables all QTL in the genome to be traced through the tracing of chromosome segments defined by adjacent SNPs. It is assumed that the effects of the chromosome segments will be the same across the population as a result of the LD between the SNPs and QTL. Thus, it is important that marker density is high enough to ensure that all QTL are in LD with at least a marker.

The main advantages of genomic selection are similar to those associated with MAS. Briefly, it results in a reduction of the generation interval, as young animals can be genotyped early in life and their GEBV computed for the purposes of selection. In the dairy cattle situation, GEBV computed early in life can be used to select young bulls, thereby reducing the cost of progeny testing, provided the GEBV are accurate enough. In addition, higher accuracy of GEBV, about 20–30% above that from a parent average, has been reported for young bulls. The computation of GEBV for an individual on the basis of the SNPs it has inherited, means that the differences in the genomic merit of full-sibs can be captured.

The implementation of genomic selection involves three steps: (i) estimating the SNP effects in a reference or training population that consists of individuals with phenotypic records and genotypes; (ii) validation in another data set (validation candidates or population) but with records excluded to determine accuracy of prediction; and (iii) the prediction of GEBV for selection candidates that do not yet have phenotypes of their own for the purposes of selecting individuals of superior genetic merit.

The efficiency of genomic selection firstly depends on the accuracy with which SNP effect are estimated. The higher the accuracy, the more effective is genomic selection. The accuracy is, in turn, influenced by factors such as the size of reference population, heritability of the trait and statistical methodology employed. In general, larger reference populations and traits with higher heritabilities are associated with higher accuracy. Goddard (2009) showed that to achieve a GEBV accuracy of 0.5, about 5000 individuals with phenotypic records in the reference population are needed when heritability is 0.2 compared to 3500 phenotypic records when heritability is 0.3. Secondly, the higher the linkage disequilibrium (LD) between markers and QTL is, the more effective is genomic selection. The LD further depends on the marker density (chip size), the effective size of the population (the number of independent chromosome segments segregating) and the relationship between selection and reference candidates. In general, the higher the marker density, the smaller the effective population size, and the closer the relationship between the selection and reference candidates, the higher are LD and accuracy of GEBVs.

The general linear model underlying genomic evaluation is of the form:

$$\mathbf{y} = \mathbf{Xb} + \sum_{m}^{i=1} \mathbf{M}_i\mathbf{g}_i + \mathbf{e} \tag{11.1}$$

where m is the number of SNPs or markers across the genome, $\mathbf{y}$ is the data vector, $\mathbf{b}$ the vector for mean or fixed effects, $\mathbf{g}_i$ the genetic effect of the ith SNP genotype and $\mathbf{e}$ is the error. The matrices $\mathbf{X}$ and $\mathbf{M}_i$ are design matrices for the mean (or fixed effects) and the ith SNP, respectively. The matrix $\mathbf{M}$ is of dimension n (number of animals) by m. The assumption is that all the additive genetic variance is explained by all the marker's effects such that the estimate of an animal's total genetic merit or breeding value ($\mathbf{a}$) is:

$$\mathbf{a} = \sum_{i=1}^{m} \mathbf{M}_i\mathbf{g}_i$$

However, if it is assumed that a certain proportion of the additive genetic variance is not explained by markers, then the model can be extended to include a residual polygenic effect ($\mathbf{u}$), which is the proportion of the additive genetic variance not captured by markers. The model can then be written as:

$$\mathbf{y} = \mathbf{Xb} + \sum_{i=1}^{m} \mathbf{M}_i\mathbf{g}_i + \mathbf{Wu} + \mathbf{e} \tag{11.2}$$

where $\mathbf{W}$ is the design matrix linking records to random animal or sire effects if an animal or sire model has been fitted.

11.2 Coding and Scaling Genotypes

As explained in Eqn 11.1, $\mathbf{M}$ is the genotypic matrix that contains which marker alleles each individual inherited. The genotypes of animals are commonly coded as 2 and 0 for the two homozygotes (AA and BB) and 1 for the heterozygotes (AB or BA). If alleles are expressed in terms of nucleotides, and the reference allele at a locus is G and the alternative allele is C, then the code is 0 = GG, 1 = GC and 2 = CC. The diagonal elements of $\mathbf{MM}'$ then indicate the individual relationship with itself (inbreeding) and the off-diagonals indicate the number of alleles shared by relatives (VanRaden, 2007).

Commonly, in genomic evaluations (VanRaden, 2008), the elements of $\mathbf{M}$ are scaled to set the mean values of the allele effects to zero and account for differences in allele frequencies of the various SNPs. Let the frequency of the second or alternative allele at locus j be p_j and then elements of $\mathbf{M}$ can be scaled by subtracting $2p_j$. Let the element for column j of a matrix $\mathbf{P}$ equal $2p_j$, then the matrix $\mathbf{Z}$, which contained the scaled elements of $\mathbf{M}$, can be computed as $\mathbf{Z} = \mathbf{M} - \mathbf{P}$. Note that the sum of the elements of each column of $\mathbf{Z}$ equals zero. Furthermore, the elements of $\mathbf{Z}$ can be normalized by dividing the column for marker j by its standard deviation, which is assumed to be $\sqrt{2p_j(1-p_j)}$. This is assuming that the locus is at Hardy-Weinberg equilibrium. However, in this chapter **Z** computed as **M** – **P** has been used.

11.3 Fixed-effect Model for SNP Effects

Several methods for genomic selection were presented by Meuwissen *et al.* (2001), and one such method includes the least squares approach with chromosome segments or SNPs considered as fixed. There is no assumption made about the distribution of the SNP effects and it usually involves two steps:

1. analysis of each SNP using the simple model in Eqn 11.1, with $\mathbf{g}_i$ defined as the vector of fixed *i*th SNP effect; and
2. selection of the *k* most significant SNPs and estimating their effects simultaneously (in the same data) using a multiple regression with the term for SNP effects in Eqn 11.1 equal to:

$$\sum_{i=1}^{k} \mathbf{M}_i \mathbf{g}_i$$

This approach suffers from two major limitations. First, the estimation of effects based on an SNP selected by single SNP analysis will result in overestimation of the SNP effects, as the large amount of multiple testing ensures the selected SNPs are those with positive error terms. Second, determining the level of significance for the choice of SNPs to include in the final analysis is far from straightforward.

In an animal breeding context, assuming the few SNPs that have significant effects on a trait have been identified, then these SNPs can be fitted as fixed effects in a model that includes the polygenic effect as a random effect. Thus, the genomic breeding value for the *i*-th animal ($GEBV_i$) can be computed as a sum of the direct genomic breeding value (DGV_i) calculated from the marker (SNP) effects as $\mathbf{M}_i\hat{\mathbf{g}}_i$ and the polygenic effects $(\hat{\mathbf{u}}_i)$. Such a linear model could be written as:

$$\mathbf{y} = \mathbf{Xb} + \mathbf{Zg} + \mathbf{Wu} + \mathbf{e} \tag{11.3}$$

where **g** represents the fixed marker or SNP effects, **Z** is the scaled matrix of genotypes defined in Section 11.2, which relates SNPs to phenotypes, and other terms are defined as in Eqn 11.2.

The equations for obtaining the solutions for SNP and polygenic effects are:

$$\begin{bmatrix} \mathbf{X'X} & \mathbf{X'Z} & \mathbf{X'W} \\ \mathbf{Z'X} & \mathbf{Z'Z} & \mathbf{Z'W} \\ \mathbf{W'X} & \mathbf{W'Z} & \mathbf{W'W} + \mathbf{A}^{-1}\alpha \end{bmatrix} \begin{bmatrix} \hat{\mathbf{b}} \\ \hat{\mathbf{g}} \\ \hat{\mathbf{u}} \end{bmatrix} = \begin{bmatrix} \mathbf{X'y} \\ \mathbf{Z'y} \\ \mathbf{W'y} \end{bmatrix}$$

where:

$$\alpha = \sigma_e^2 / \sigma_u^2$$

If the vector of observations, **y** in Eqn 11.3, are de-regressed breeding values of bulls (see Section 6.5.2), then each observation may be associated with differing reliabilities. Thus, a weighted analysis may be required to account for these differences in bull reliabilities. The weight (wt_i) for each observation could be the reciprocal of the effective daughter contribution (see Section 6.5.2) or $wt_i = (1/rel_{dtr}) - 1$, where rel_{dtr} is the bull's reliability from daughters with parent information excluded (VanRaden, 2008). Then the MME are:

$$\begin{bmatrix} X'R^{-1}X & X'R^{-1}Z & X'R^{-1}W \\ Z'R^{-1}X & Z'R^{-1}Z & Z'R^{-1}W \\ W'R^{-1}X & W'R^{-1}Z & W'R^{-1}W + A^{-1}\alpha \end{bmatrix} \begin{bmatrix} \hat{b} \\ \hat{g} \\ \hat{u} \end{bmatrix} = \begin{bmatrix} X'R^{-1}y \\ Z'R^{-1}y \\ W'R^{-1}y \end{bmatrix} \tag{11.5}$$

where **R** = **D** and **D** is a diagonal matrix with diagonal element $d_{ii} = wt_i$.

Example 11.1

Given below is the real genotype for the first ten SNPs of a popular dairy bull and those of his sons and some other unrelated bulls genotyped using the 50K Illumina chip. The genotypes of animals are coded as described in Section 11.2. The observations are the DYDs for fat yield, and the effective daughter contribution (EDC) for each bull is also given. The EDCs can be used as weights in the analysis. It is assumed that the genetic variance for fat yield is 35.241 kg^2 and residual variance of 245 kg^2. Animals 13–20 are assumed as the reference population and 21–26 as selection candidates. Assuming that the first three SNPs have been identified as having the most significant effect, the aim is to fit Eqn 11.3 with and without weights using these three SNPs:

Animal	Sire	Dam	Mean	EDC	Fat DYD	SNP Genotype									
13	0	0	1	558	9.0	2	0	1	1	0	0	0	2	1	2
14	0	0	1	722	13.4	1	0	0	0	0	2	0	2	1	0
15	13	4	1	300	12.7	1	1	2	1	1	0	0	2	1	2
16	15	2	1	73	15.4	0	0	2	1	0	1	0	2	2	1
17	15	5	1	52	5.9	0	1	1	2	0	0	0	2	1	2
18	14	6	1	87	7.7	1	1	0	1	0	2	0	2	2	1
19	14	9	1	64	10.2	0	0	1	1	0	2	0	2	2	0
20	14	9	1	103	4.8	0	1	1	0	0	1	0	2	2	0
21	1	3	1	13	7.6	2	0	0	0	0	1	2	2	1	2
22	14	8	1	125	8.8	0	0	0	1	1	2	0	2	0	0
23	14	11	1	93	9.8	0	1	1	0	0	1	0	2	2	1
24	14	10	1	66	9.2	1	0	0	0	1	1	0	2	0	0
25	14	7	1	75	11.5	0	0	0	1	1	2	0	2	1	0
26	14	12	1	33	13.3	1	0	1	1	0	2	0	1	0	0

EDC, effective daughter contribution; DYD, daughter yield deviation

The prediction of marker effects and polygenic effects for the reference population and selection candidates can be done simultaneously by including $\mathbf{A}^{-1}$ for all animals but using only the fat yield records for the reference animals. Thus **y**′ = (9.0 13.4 12.7 15.4 5.9 7.7 10.2 4.8). The incidence matrix $\mathbf{X} = \mathbf{1}_q$, with q = 8 (the number of animals in the reference population).

Computing the matrix Z

The computation of **Z** requires calculating the allele frequency for each SNP. The allele frequency for the *i*th SNP was computed as:

$$\frac{\sum_{j=1}^{n} m_{ij}}{2n}$$

where n = 14, the number of animals with genotypes, and m_{ij} are elements of **M**. The allele frequencies for the ten SNPs were 0.321, 0.179, 0.357, 0.357, 0.143, 0.607, 0.071, 0.964, 0.571 and 0.393, respectively. However, only the first three SNPs are needed for this example, therefore **Z** is of order 8 by 3, with elements $z_{i,j} = m_{i,j} - p_{i,j}$, with j =1, 2, and 3. Thus:

$$\mathbf{Z} = \begin{bmatrix} 1.357 & -0.357 & 0.286 \\ 0.357 & -0.357 & -0.714 \\ 0.357 & 0.643 & 1.286 \\ -0.643 & -0.357 & 1.286 \\ -0.643 & 0.643 & 0.286 \\ 0.357 & 0.643 & -0.714 \\ -0.643 & -0.357 & 0.286 \\ -0.643 & 0.643 & 0.286 \end{bmatrix}$$

The **W** matrix is a diagonal matrix for the eight reference animals with records. This is augmented with 12 columns of zeros to account for ancestors 1–12. For the weighted analysis, the **R** was a diagonal matrix with the diagonal elements equal to the EDCs of the first eight animals in the data set. The matrix $\mathbf{A}^{-1}$ is computed using the usual rules for all 26 animals and α = 245/35.241 = 6.952. Solving the system of equations gives the following results:

	Unweighted analysis		Weighted analysis	
Mean effect				
	9.895		9.196	
SNP effect				
1	0.607		1.158	
2	−4.080		−3.956	
3	1.934		2.527	
Reference animals				
	DGV	Polygenic	DGV	Polygenic
13	2.834	−0.299	3.708	−3.821
14	0.293	0.256	0.022	4.116
15	0.081	0.142	1.120	2.248
16	3.554	0.254	3.918	2.158
17	−2.460	−0.085	−2.565	−0.450
18	−3.787	0.271	−3.934	2.402
19	1.620	−0.092	1.391	−0.212
20	−2.460	−0.181	−2.565	−1.542
	−2.460	−0.181	−2.565	−1.542
Selection animals				
	DGV	Polygenic	DGV	Polygenic
25	0.900	0.000	1.181	0.000
26	−0.314	0.128	−1.136	2.058
27	−2.460	0.128	−2.565	2.058
28	0.293	0.128	0.022	2.058
29	−0.314	0.128	−1.136	2.058
30	2.227	0.128	2.549	2.058

With this small amount of data, it seems that when records are properly weighted, polygenic effects were very close to zero. The GEBVs for reference and selection animals equal $\mathbf{Z\hat{g}} + \mathbf{\hat{u}}$. This would be equal to 2.535 for animal 13, for instance. The **Z** has been given for reference animals, and for the selection candidates the corresponding matrix $\mathbf{Z}_2$ is:

$$\mathbf{Z}_2 = \begin{bmatrix} 1.357 & -0.357 & -0.714 \\ -0.643 & -0.357 & -0.714 \\ -0.643 & 0.643 & 0.286 \\ 0.357 & -0.357 & -0.714 \\ -0.643 & -0.357 & -0.714 \\ 0.357 & -0.357 & 0.286 \end{bmatrix}$$

11.4 Mixed Linear Model for Computing SNP Effects

Several methods that fit SNP effects as random have been presented by various researchers (Meuwissen *et al.*, 2001; VanRaden *et al.*, 2008; Habier *et al.*, 2011). The most common random model used in the national evaluation centres for genomic evaluation, especially of dairy animals, assumes that the effect of the SNP is normally distributed and all SNP are from a common normal distribution (e.g. the same genetic variance for all SNPs). There are two equivalent models with these assumptions:

1. A model fitting individual SNP effects simultaneously. In this model (SNP-BLUP), DGVs for selection candidates are calculated as $\text{DGV} = \mathbf{Z\hat{g}}$, where $\mathbf{\hat{g}}$ are the estimates of random SNP effects. This method involves knowing σ_g^2, but this may not be the case in practice, and σ_g^2 may have to be approximated from σ_a^2, the additive genetic variance. In such situations, this method is also referred to as ridge regression.
2. A model estimating breeding values directly (GBLUP), with the (co)variance among breeding values $\mathbf{G}\sigma_a^2$ fitted, where **G** is the genomic relationship matrix. The matrix **G** represents the realized proportion of the genome that animals share in common and is estimated from the SNPs.

These models will now be described in more detail.

11.4.1 SNP-BLUP model

In matrix form, the mixed linear model for estimating SNP effects can be written as (Meuwissen *et al.*, 2001; VanRaden, 2008):

$$\mathbf{y} = \mathbf{Xb} + \mathbf{Zg} + \mathbf{e} \qquad (11.6)$$

where **g** is a vector of additive genetic effects corresponding to allele substitution effects for each SNP and all other terms are defined as in Eqn 11.3. The matrix **Z** relates SNP effects to the phenotypes. The sum of **g** over all marker loci is assumed equal to the vector of breeding values (**a**), i.e. DGV = **a** = **Zg**. The MME for Eqn 11.6 are:

$$\begin{bmatrix} \mathbf{X'R^{-1}X} & \mathbf{X'R^{-1}Z} \\ \mathbf{Z'R^{-1}X} & \mathbf{Z'R^{-1}Z} + \mathbf{I}\alpha \end{bmatrix} \begin{bmatrix} \mathbf{\hat{b}} \\ \mathbf{\hat{g}} \end{bmatrix} = \begin{bmatrix} \mathbf{X'R^{-1}y} \\ \mathbf{Z'R^{-1}y} \end{bmatrix} \qquad (11.7)$$

where $\alpha = \sigma_e^2 / \sigma_g^2$ and **R** is a diagonal matrix of weights (see Eqn 11.5). The MME in Eqn 11.7 can easily be set up and solutions obtained for each SNP and the fixed effects. However, in practice, the value of σ_g^2 may not be known but could be obtained either as $\sigma_g^2 = \sigma_a^2 / m$, with m = the number of markers, or as $\sigma_g^2 = \sigma_a^2 / 2\Sigma p_j(1-p_j)$. The latter is preferred as it takes into account the differences in allele frequencies. With the latter, $\alpha = 2\Sigma p_j(1-p_j) * [\sigma_e^2 / \sigma_a^2]$, with σ_a^2 being the additive genetic variance for the trait and p_j is as defined in Section 11.2.

Hayes and Daetwyler (2013) indicated that there is a potential problem with this estimate as it assumes the LD between SNP and QTL is perfect and all genetic variance is captured by the SNP. This may not be the case in practice, and they recommended the method described by Moser *et al.* (2010) for estimating α through cross-validation. The method involves estimating SNP effects with different values of α and predicting DGV in validation data sets that have not contributed to the estimation of SNP effects. The value of α that minimizes the mean square error between the DGV and **y** is taken as the appropriate estimate. This process can be repeated, dropping out different subsets of the data and obtaining an estimate of α by averaging across data sets.

Example 11.2

Using the data and genetic parameters given in Example 11.1, SNP effects are predicted using Eqn 11.6 and all ten SNPs. Then DGVs are computed for the reference and validation animals. Initially, analyses are carried out without weights, thus $\mathbf{R} = \mathbf{I}\sigma_e^2$. Then the data were re-analysed using EDCs as weights, with **R** in Eqn 11.7 being a diagonal matrix containing elements (1/EDC) for reference bulls.

Computing the required matrices and α

The allele frequencies for the ten SNPs have been calculated in Example 11.1. Using those frequencies, $2\Sigma p_j(1 - p_j) = 3.5383$. Thus $\alpha = 3.5383*(245/35.242) = 24.598$.

The matrix **X** in Eqn 11.7 is the same as **X** in Example 11.1 and **Z** computed as **Z** = **M** − **P** is:

$$\mathbf{Z} = \begin{pmatrix} 1.357 & -0.357 & 0.286 & 0.286 & -0.286 & -1.214 & -0.143 & 0.071 & -0.143 & 1.214 \\ 0.357 & -0.357 & -0.714 & -0.714 & -0.286 & 0.786 & -0.143 & 0.071 & -0.143 & -0.786 \\ 0.357 & 0.643 & 1.286 & 0.286 & 0.714 & -1.214 & -0.143 & 0.071 & -0.143 & 1.214 \\ -0.643 & -0.357 & 1.286 & 0.286 & -0.286 & -0.214 & -0.143 & 0.071 & 0.857 & 0.214 \\ -0.643 & 0.643 & 0.286 & 1.286 & -0.286 & -1.214 & -0.143 & 0.071 & -0.143 & 1.214 \\ 0.357 & 0.643 & -0.714 & 0.286 & -0.286 & 0.786 & -0.143 & 0.071 & 0.857 & 0.214 \\ -0.643 & -0.357 & 0.286 & 0.286 & -0.286 & 0.786 & -0.143 & 0.071 & 0.857 & -0.786 \\ -0.643 & 0.643 & 0.286 & -0.714 & -0.286 & -0.214 & 0.143 & 0.071 & 0.857 & -0.786 \end{pmatrix}$$

The MME in Eqn 11.7 can then be easily set up. The solutions for the mean and SNP effects from solving the MME, either using weights or no weights, are shown in Table 11.1. The DGVs for the reference animals are then computed as $\mathbf{Z}\hat{\mathbf{g}}$. The results are shown in Table 11.2.

Table 11.1. Solutions for mean and SNP effects from various models.

	Unweighted	Weighted
Mean effect		
	9.944	11.876
SNP effects solutions		
1	0.087	−0.633
2	−0.311	−3.041
3	0.262	3.069
4	−0.080	−1.267
5	0.110	2.600
6	0.139	4.447
7	0.000	0.000
8	0.000	0.000
9	−0.061	−3.240
10	−0.016	1.883

Similarly, the DGV of the validation animals are computed as $\mathbf{Z}_2\hat{\mathbf{g}}$, where $\mathbf{Z}_2$ contains the centralized genotypes for the selection candidates. Thus, for the unweighted analysis:

$$\begin{bmatrix}\hat{a}_{21}\\\hat{a}_{22}\\\hat{a}_{23}\\\hat{a}_{24}\\\hat{a}_{25}\\\hat{a}_{26}\end{bmatrix} = \begin{pmatrix} 1.357 & -0.357 & -0.714 & -0.714 & -0.286 & -0.214 & 1.857 & 0.071 & -0.143 & 1.214 \\ -0.643 & -0.357 & -0.714 & 0.286 & 0.714 & 0.786 & -0.143 & 0.071 & -1.143 & -0.786 \\ -0.643 & 0.643 & 0.286 & -0.714 & -0.286 & -0.214 & -0.143 & 0.071 & 0.857 & 0.214 \\ 0.357 & -0.357 & -0.714 & -0.714 & 0.714 & -0.214 & -0.143 & 0.071 & -1.143 & -0.786 \\ -0.643 & -0.357 & -0.714 & 0.286 & 0.714 & 0.786 & -0.143 & 0.071 & -0.143 & -0.786 \\ 0.357 & -0.357 & 0.286 & 0.286 & -0.286 & 0.786 & -0.143 & -0.929 & -1.143 & -0.786 \end{pmatrix} \begin{pmatrix} 0.087 \\ -0.311 \\ 0.262 \\ -0.080 \\ 0.110 \\ 0.139 \\ 0.000 \\ 0.000 \\ -0.061 \\ -0.016 \end{pmatrix}$$

$$= \begin{pmatrix} 0.027 \\ 0.114 \\ -0.240 \\ 0.143 \\ 0.054 \\ 0.354 \end{pmatrix}$$

11.4.2 Equivalent models: GBLUP

An equivalent model to Eqn 11.6 is the application of the usual BLUP MME but with the inverse of the numerator relationship matrix ($\mathbf{A}^{-1}$) replaced by the inverse of the genomic relationship matrix ($\mathbf{G}^{-1}$) (Habier *et al.*, 2007; Hayes *et al.*, 2009). This tends to be referred to generally as GBLUP. The DGVs are computed directly from the MME as the sum of the SNP effects ($\mathbf{a} = \mathbf{Z}\hat{\mathbf{g}}$), with the assumption that SNP effects are normally distributed. Assume the following mixed linear model:

$$\mathbf{y} = \mathbf{Xb} + \mathbf{Wa} + \mathbf{e} \qquad (11.8)$$

Table 11.2. Direct genomic breeding (DGV) values from various models.

	SNP-BLUP	GBLUP	Selection index	SNP-BLUP (weighted)
Reference animals				
13	0.070	0.069	0.070	–2.651
14	0.111	0.116	0.111	1.307
15	0.045	0.049	0.045	0.611
16	0.253	0.260	0.253	1.007
17	–0.495	–0.500	–0.495	–5.693
18	–0.357	–0.359	–0.357	–4.358
19	0.145	0.146	0.146	0.502
20	–0.224	–0.231	–0.225	–5.718
Selection candidates				
21	0.027	0.028	0.028	–0.006
22	0.114	0.115	0.115	6.513
23	–0.240	–0.240	–0.240	–3.835
24	0.143	0.143	0.143	2.701
25	0.054	0.054	0.054	3.273
26	0.354	0.353	0.353	6.350

where **y** is the vector of observations, **a** is the vector of DGVs and **W** is the design matrix linking records to breeding value (random animal or sire effect if an animal or sire model has been fitted) and **e** is random residual effect. Given that $\mathbf{a} = \mathbf{Z\hat{g}}$, then:

$$\mathrm{Var}(\mathbf{a}) = \mathbf{ZZ}'\sigma_g^2$$

Noting that:

$$\sigma_g^2 = \frac{\sigma_a^2}{2\sum_{j=1}^{m} p_j(1-p_j)}$$

the matrix **ZZ′** can be scaled such that:

$$\mathbf{G} = \frac{\mathbf{ZZ}'}{2\sum_{j=1}^{m} p_j(1-p_j)}$$

and $\mathrm{Var}(\mathbf{a}) = \mathbf{G}\sigma_a^2$. The above division scales **G** to be analogous to the numerator relationship matrix (**A**). The genomic inbreeding coefficient for individual i is $G_{ii} - 1$, and the genomic relationship between individuals i and j, which are analogous to the relationship coefficients (Wright, 1922), can be obtained by dividing the elements G_{ij} by the square roots of the diagonals of G_{ii} and G_{jj}. The matrix **G** is generally positive semi-definite but can be singular if two individuals have identical genotypes or the number of markers (m) is less than genotyped individuals (n). If the number of markers is limited ($m < n$), an improved non-singular matrix $\mathbf{G}_{wt}$ can be obtained as wt**G** + (1 – wt)**A**. VanRaden (2008) indicated that wt = 0.90, 0.95 and 0.98 gave good results.

Another method for computing **G** involves scaling **ZZ′** by the reciprocals of the expected variance of marker loci (VanRaden, 2008). Thus, **G** = **ZDZ′**, where **D** is diagonal with:

$$d_{ii} = \frac{1}{m\left[2p_j(1-p_j)\right]}$$

The MME for Eqn 11.8 are:

$$\begin{bmatrix} \mathbf{X}'\mathbf{R}^{-1}\mathbf{X} & \mathbf{X}'\mathbf{R}^{-1}\mathbf{W} \\ \mathbf{W}'\mathbf{R}^{-1}\mathbf{X} & \mathbf{W}'\mathbf{R}^{-1}\mathbf{W}+\mathbf{G}^{-1}\alpha \end{bmatrix} \begin{bmatrix} \hat{\mathbf{b}} \\ \hat{\mathbf{a}} \end{bmatrix} = \begin{bmatrix} \mathbf{X}'\mathbf{R}^{-1}\mathbf{y} \\ \mathbf{W}'\mathbf{R}^{-1}\mathbf{y} \end{bmatrix} \tag{11.9}$$

where α now equals σ_e^2 / σ_a^2.

This approach for genomic evaluation has the advantage that existing software for genetic evaluation can be used by replacing **A** with **G** and the systems of equations are of the size of animals, which tend to be fewer than the number of SNPs. In pedigree populations, **G** discriminates among sibs and other relatives, allowing us to say whether these sibs are more or less alike than expected, so we can capture information on Mendelian sampling. Also, the method is attractive for populations without good pedigree, as **G** will capture this information among the genotyped individuals (Hayes and Daetwyler, 2013).

Note that Eqn 11.9 assumes all the additive genetic variance (σ_a^2) is captured by the SNP, but this may not be the case if the linkage disequilibrium between SNP and QTL is not perfect. Later, in Section 11.5, a model is discussed that might capture any residual polygenic variance not captured by the SNPs. Another possible limitation is that there are no direct rules for computing $\mathbf{G}^{-1}$ and in large populations the computation may not be feasible.

Example 11.3

The data in Example 11.1 are analysed using Eqns 11.8 and 11.9 and the same genetic parameters to compute DGVs for both the reference and validation animals without using weights.

The matrix **X** in Eqn 11.9 is the same as **X** in Example 11.1, **W** is a diagonal matrix for the eight reference animals with records, and α = 245/35.241 = 6.952. The **G** matrix constructed from **Z** for the ten SNPs as:

$$\frac{\mathbf{ZZ}'}{2\sum_{j=1}^{m} p_j(1-p_j)}$$

with $2\Sigma p_j(1 - p_j)$ = 3.5383 is:

13	1.472													
14	-0.446	0.746												
15	0.988	-0.930	1.634											
16	0.059	-0.446	0.422	0.907										
17	0.685	-0.950	1.048	0.402	1.593					symmetric				
18	-0.163	0.180	-0.365	-0.163	-0.102	0.746								
19	-0.708	0.201	-0.627	0.423	-0.365	0.201	0.786							
20	-0.547	0.079	-0.183	0.301	-0.203	0.079	0.382	0.826						
21	0.887	0.100	0.120	-0.526	-0.183	0.100	-0.728	-0.567	2.280					
22	-0.789	0.402	-0.708	-0.506	-0.446	-0.163	0.140	-0.264	-0.526	1.190				
23	-0.203	-0.143	0.160	0.362	0.140	0.140	0.160	0.604	-0.224	-0.486	0.665			
24	-0.143	0.483	-0.345	-0.708	-0.648	-0.365	-0.345	-0.183	0.120	0.705	-0.405	1.068		
25	-0.829	0.362	-0.748	-0.264	-0.486	0.079	0.382	-0.022	-0.567	0.867	-0.244	0.382	0.826	
26	-0.264	0.362	-0.466	-0.264	-0.486	-0.203	0.100	-0.304	-0.284	0.584	-0.526	0.382	0.261	1.109

For the purposes of comparison, the **G** matrix computed from 41866 SNPs ($\mathbf{G}_{all}$) with $2\Sigma p_j(1 - p_j)$ = 15555.80 and the **A** computed from a five-generation pedigree are shown below:

G_{all} =

13	0.957													
14	-0.108	0.973												
15	0.452	-0.116	1.182											
16	0.209	-0.058	0.424	1.025										
17	0.234	-0.083	0.425	0.312	1.037									
18	-0.040	0.438	0.097	-0.047	-0.043	1.151		symmetric						
19	-0.089	0.458	0.039	-0.067	-0.070	0.426	1.175							
20	-0.093	0.460	0.053	-0.058	-0.063	0.432	0.707	1.183						
21	0.077	-0.082	0.064	0.104	0.082	-0.071	-0.069	-0.069	1.031					
22	-0.056	0.418	0.093	-0.046	-0.038	0.408	0.355	0.342	-0.044	1.139				
23	-0.005	0.464	-0.038	-0.035	-0.038	0.206	0.223	0.215	0.011	0.280	0.993			
24	-0.070	0.468	0.075	-0.027	-0.053	0.403	0.521	0.550	-0.079	0.424	0.260	1.198		
25	-0.052	0.416	0.098	-0.009	-0.031	0.386	0.363	0.342	-0.038	0.370	0.219	0.419	1.125	
26	-0.070	0.493	-0.084	-0.039	-0.044	0.258	0.241	0.270	-0.072	0.253	0.178	0.259	0.214	1.009

A =

13	1.008													
14	0.033	1.037												
15	0.545	0.021	1.041											
16	0.288	0.021	0.536	1.016										
17	0.285	0.031	0.541	0.293	1.020									
18	0.047	0.580	0.036	0.028	0.032	1.062								
19	0.033	0.613	0.021	0.021	0.031	0.365	1.095		symmetric					
20	0.033	0.613	0.021	0.021	0.031	0.365	0.613	1.095						
21	0.099	0.031	0.082	0.118	0.074	0.028	0.031	0.031	1.021					
22	0.046	0.586	0.032	0.031	0.039	0.351	0.373	0.373	0.044	1.068				
23	0.096	0.569	0.067	0.043	0.047	0.329	0.357	0.357	0.042	0.338	1.050			
24	0.041	0.574	0.027	0.019	0.026	0.331	0.406	0.406	0.028	0.335	0.335	1.056		
25	0.033	0.548	0.035	0.039	0.039	0.315	0.336	0.336	0.037	0.321	0.310	0.310	1.029	
26	0.035	0.588	0.023	0.024	0.039	0.337	0.376	0.376	0.036	0.347	0.341	0.348	0.325	1.070

The matrix **A** is more similar to $\mathbf{G}_{all}$ than to **G**, thus with more SNPs, the genomic relationship matrix captures more relationships. The $\mathbf{G}^{-1}$ used in this example was computed after adding 0.01 to the diagonals of **G** obtained from the ten SNPs.

The matrices required for Eqn 11.9 have been described. Solving Eqn 11.9 gives the DGVs directly for both the reference and selection animals and these are shown in Table 11.2. The solution for the mean effects was 9.944. Thus, the model gave the same results as the SNP model.

11.4.3 Females in reference

In dairy cattle, when the number of proven bulls is limited, such as in Australia, females have been included in the reference population for genomic prediction. In addition, for some cases, the recording of the more difficult-to-measure traits only started in recent years; therefore the reliabilities of bull evaluations are limited; thus cows with observations are included in the reference population. Some adjustments are necessary to avoid double-counting of information when cows and their sires are included in the reference population when using SNP-BLUP or GBLUP models. When a bull and his daughters are both present in the reference population the de-regressed proofs (or response variable) and associated weights such as effective daughter contributions of the bull are to be adjusted for the contribution from those daughters. The information needed to compute these adjustments can be found in VanRaden and Wiggans (1991). If yield deviations (YDs) are used as the response variable for cows in the reference population, it may be necessary to ensure the variances of YDs are similar to that of de-regressed proofs for the bull through some adjustment. However, the wide use of single-step GBLUP (ssGBLUP, see Chapter 12) for genomic prediction, more recently, allows for easy conclusion of both males and females in the reference population without any adjustments.

11.4.4 Computing SNP solutions from GBLUP

Since SNP-BLUP and GBLUP are equivalent models, SNP effects ($\hat{\mathbf{g}}$) can easily be obtained from the GBLUP solutions, i.e. the DGVs (**a**). This process involves back-solving using known matrices **G** and **Z**, and the vector of DGVs **a** (Stranden and Garrick, 2009). As defined previously, the DGVs are sum of the SNP effects (**a** = **Zg**) with the corresponding variance $\mathrm{Var}(\mathbf{a}) = ZZ'\sigma_g^2$, and if we assume SNP effects are normally distributed with $\mathrm{Var}(\mathbf{g}) = \mathbf{D} = \mathbf{I}\sigma_g^2 = \mathbf{I}\frac{\sigma_a^2}{2\sum p_j(1-p_j)}$, then SNP effects can extracted as:

$$\mathbf{g} = \mathbf{D}\mathbf{Z}'(\mathbf{Z}\mathbf{D}\mathbf{Z}')^{-1}\hat{\mathbf{a}} = \frac{1}{2\sum p_j(1-p_j)}\mathbf{Z}'\mathbf{G}^{-1}\hat{\mathbf{a}}$$

It is also possible to show this equivalence following the derivations based on the selection index approach in Section 11.4.5. Furthermore, different weights can easily be assigned to SNP effects mimicking Bayesian methods (Section 11.7), e.g. via the iterative method (Wang *et al.*, 2012) by defining **D** as diagonal matrix with weights for each SNP $d_i = g_i^2 2p_j(1\text{-}p_j)$.

Example 11.4

We will take the vector of DGVs for both the reference and selection animals obtained by GBLUP model from Example 11.3, as presented in the Table 11.2, and using equation from Section 11.4.4 to obtain the SNP solutions. Same as in the previous example, we assume $\frac{1}{2\sum pj(1-pj)} = 0.2826$, **Z** is now a complete (14 by 10) matrix, and $\mathbf{G}^{-1}$ is obtained after adding 0.01 to the diagonals of **G** as in the Example 11.3. SNP solutions obtained by back-solving from DGVs are $\hat{\mathbf{g}}$ = (0.087, –0.311, 0.262, –0.080, 0.110, 0.139, 0.000, 0.001, –0.061, –0.016), thus the same as the SNP effects solutions obtained directly by SNP-BLUP model demonstrated in Table 11.1 (unweighted example). This, once again, shows equivalence between SNP-BLUP and GBLUP models.

11.4.5 Equivalent models: selection index approach

VanRaden (2008) presented a selection index approach which is equivalent to Eqn 11.9. The method is of limited use in practice as it is assumed that the solutions of the vector of fixed ($\hat{\mathbf{b}}$) effects are known. It does, however, demonstrate the equivalence of the selection index approach to GBLUP.

Selection index equations to predict DGV ($\hat{\mathbf{a}}$) are constructed as the covariance between **y** and **a** multiplied by the inverse of the variance of **y** and the deviation of **y** from fixed-effects solutions. Thus:

$$\hat{\mathbf{a}} = \mathbf{G}\left(\mathbf{G} + \mathbf{R}\left(\frac{\sigma_e^2}{\sigma_a^2}\right)\right)^{-1}\left(\mathbf{y} - \mathbf{X}\hat{\mathbf{b}}\right) \qquad (11.10)$$

The vector of estimates of SNP effects $(\hat{\mathbf{g}})$ can be obtained from Eqn 11.10 as:

$$\hat{\mathbf{g}} = \left(\frac{1}{2\sum p_j\left(1-p_j\right)}\right)\mathbf{Z}'\left(\mathbf{G} + \mathbf{R}\left(\frac{\sigma_e^2}{\sigma_a^2}\right)\right)^{-1}\left(\mathbf{y} - \mathbf{X}\hat{\mathbf{b}}\right) \qquad (11.11)$$

The DGV of validation candidates without records can then be computed with the selection index approach as:

$$\hat{\mathbf{a}} = \mathbf{C}\left(\mathbf{G} + \mathbf{R}\left(\frac{\sigma_e^2}{\sigma_a^2}\right)\right)^{-1}\left(\mathbf{y} - \mathbf{X}\hat{\mathbf{b}}\right) \qquad (11.12)$$

where $\mathbf{C}$ is the genomic covariance between animals with and without records computed as:

$$\frac{\mathbf{Z}_2\mathbf{Z}'}{2\sum p_j\left(1-p_j\right)}$$

with $\mathbf{Z}_2$ being the matrix of centralized genotypes for the validation animals (see Example 11.3).

Example 11.5

The data in Example 11.1 are again analysed using Eqn 11.10 and the same genetic parameters to compute DGVs for the reference animals without using weights. The solution of 9.994 has been assumed for the mean.

The $\mathbf{X}$ matrix in Eqn 11.10 equals $\mathbf{X}$ in Example 11.1, the $\mathbf{G}$ matrix is of order 8 for the reference animals only and corresponds to the first eight rows and columns of $\mathbf{G}$ computed in Example 11.3 and $\mathbf{R} = \mathbf{I}\sigma_e^2$, assuming no weights are used in the analysis.

Solutions from solving Eqn 11.10 are shown in Table 11.2. Similarly, the DGV of the selection candidates were obtained by Eqn 11.12 and these are also shown in Table 11.2. The same solutions were obtained for both reference and validation animals as obtained from the SNP or GBLUP models.

11.4.6 Computing base population allele frequencies

Most commonly, $\mathbf{G}$ is constructed based on the allele frequencies calculated in the recent genotyped population, while, in theory, those should come from the base population. This assumption can potentially raise issues with the bias and accuracy of predictions, especially in the context of the single-step GBLUP as further described in Chapter 12. Since the genotypes of the base population are unknown, pedigree information can be used to calculate expected genotypes for ungenotyped animals. The classical method proposed by Gengler *et al.* (2007) is based on the linear method, where each genotype is treated as a phenotype in a standard BLUP model, with corresponding heritability of 0.99. Within this method, the imputed predicted genotypes are BLUPs. The proposed model can be summarized as:

$$\mathrm{q} = 1\mu + \mathbf{M}d + e$$

where **q** is a vector of observed genotypes, μ is an overall mean, **M** is the incidence matrix indicating genotyped and ungenotyped individuals, and **e** is the residual error. The corresponding mixed-model equation is:

$$\begin{bmatrix} 1'1 & 1'M \\ M'1 & M'M + A^{-1}\varepsilon \end{bmatrix} \begin{bmatrix} \hat{\mu} \\ \hat{d} \end{bmatrix} = \begin{bmatrix} 1'q \\ M'q \end{bmatrix}$$

where **A** is the relationship matrix that can be partitioned as $A = \begin{bmatrix} A_{gg} & A_{gn} \\ A_{ng} & A_{nn} \end{bmatrix}$, $\hat{d}$ can be partitioned to $\hat{d} = \begin{bmatrix} \hat{d}_g \\ \hat{d}_n \end{bmatrix}$, $M = \begin{bmatrix} I_g \\ 0_n \end{bmatrix}$, and $\varepsilon = \frac{\sigma_e^2}{\sigma_d^2} = 0.01$ (or even smaller number). Therefore, $\hat{d}$ represents estimates of gene content (genotypes) for genotyped (g) and ungenotyped (n) animals and $\hat{\mu}$ is equal to $2\hat{p}_j$ where p_j is the allele frequency for the *j*-th SNP.

Example 11.6

In this example, we illustrate imputation by applying the linear method described in Section 11.4.6 on the same dataset as in Example 11.1. We assume that ungenotyped animals 1–12 form a base of unrelated animals and the goal of the example is to impute genotypes for those animals. We will show how to calculate it for the SNP #1. Let **q** be vector of known SNP #1 genotype: $q' = (2\ 1\ 1\ 0\ 0\ 1\ 0\ 0\ 2\ 0\ 0\ 1\ 0\ 1)$, **M** is design matrix connecting animals to the genotypes in **q**, $\mathbf{A}^{-1}$ is inverse of pedigree-based relationship matrix among 26 animals, partitioned into ungenotyped and genotyped blocks, and ε is set to 0.01. By solving MME described in Section 11.4.6, we obtain a mean of 0.597 and predicted genotypes for SNP #1:

Animal	Predicted BLUPs	Predicted genotypes (Mean + BLUPs)	Predicted genotypes (Rounded)	True SNP
Ungenotyped animals				
1	0.695	1.292	1	–
2	−0.523	0.074	0	–
3	0.695	1.292	1	–
4	−0.198	0.399	0	–
5	−0.523	0.074	0	–
6	0.140	0.737	1	–
7	−0.518	0.079	0	–
8	−0.518	0.079	0	–
9	−0.779	−0.182	0	–
10	0.140	0.737	1	–
11	−0.518	0.079	0	–
12	0.140	0.737	1	–
Genotyped animals				
13	1.387	1.984	2	2
14	0.380	0.977	1	1
15	0.397	0.993	1	1
16	−0.586	0.010	0	0
17	−0.586	0.010	0	0
18	0.400	0.997	1	1
19	−0.589	0.008	0	0
20	−0.589	0.008	0	0
21	1.389	1.986	2	2

Continued

Continued.

Animal	Predicted BLUPs	Predicted genotypes (Mean + BLUPs)	Predicted genotypes (Rounded)	True SNP
22	−0.587	0.010	0	0
23	−0.587	0.010	0	0
24	0.400	0.997	1	1
25	−0.587	0.010	0	0
26	0.400	0.997	1	1

The solutions are BLUPs and are represented as deviation from the base population mean. Therefore, to get integer genotypes we add the base population mean to each solution and round it to the nearest integer, as shown in the table. The mean of 0.597 corresponds to $2\hat{p}_j$ for SNP #1, so the base allele frequency for SNP #1 is 0.298. Alternatively, this could be obtained by a generalized least squares (GLS) method (e.g. Garcia-Baccino *et al.*, 2017) as:

$$\hat{\mu} = (\mathbf{1}'A_{22}^{-1}\mathbf{1})^{-1}\mathbf{1}'A_{22}^{-1}q = 0.575$$

where A_{22}^{-1} is the block of the inverse of pedigree relationship matrix corresponding to genotyped animals, and $\mathbf{q}$ is as previously defined.

11.5 Mixed Linear Models with Polygenic Effects

The genomic BLUP model used to estimate SNP effects in most livestock populations is based on chips with densities of about 60K, and it is usually assumed that these SNPs explain all the genetic variation for the traits analysed. However, fitting a residual polygenic effect (RP) may account for the fact that SNPs may not explain all the genetic variance and it has also been found to render SNP effects less biased (Solberg *et al.*, 2009). Liu *et al.* (2011) have demonstrated that the optimum level of RP may differ for traits of different heritabilities but tends to vary between 10% and 20% of the genetic variance.

A mixed linear model with polygenic effects included is of this form:

$$\mathbf{y} = \mathbf{Xb} + \mathbf{Wu} + \mathbf{Zg} + \mathbf{e} \qquad (11.13)$$

where $\mathbf{u}$ is the vector of random residual polygenic effects, $\mathbf{W}$ is the design matrix that relates records to animals and other terms are defined as in Eqn 11.6. If a SNP-BLUP model is fitted, the MME to be solved are:

$$\begin{bmatrix} \mathbf{X'R^{-1}X} & \mathbf{X'R^{-1}W} & \mathbf{X'R^{-1}Z} \\ \mathbf{W'R^{-1}X} & \mathbf{W'R^{-1}W} + \mathbf{A}^{-1}\alpha_1 & \mathbf{W'R^{-1}Z} \\ \mathbf{Z'R^{-1}X} & \mathbf{Z'R^{-1}W} & \mathbf{Z'R^{-1}Z} + \mathbf{I}\alpha_2 \end{bmatrix} \begin{bmatrix} \hat{\mathbf{b}} \\ \hat{\mathbf{u}} \\ \hat{\mathbf{g}} \end{bmatrix} = \begin{bmatrix} \mathbf{X'R^{-1}y} \\ \mathbf{W'R^{-1}y} \\ \mathbf{Z'R^{-1}y} \end{bmatrix} \qquad (11.14)$$

where $\alpha_1 = \sigma_e^2 / \sigma_u^2$, with σ_u^2 equal to the chosen percentage of the additive genetic variance fitted as polygenic effect and $\alpha_2 = \sigma_e^2 / \sigma_g^2$, with σ_g^2 calculated to account for the percentage of additive genetic variance attributed to the polygenic effect. Thus, $\alpha_2 = \left(\sigma_e^2 - \sigma_u^2\right)/m$ with m = number of markers or $2\sum p_j\left(1-p_j\right)*\left[\sigma_e^2 / \left(\sigma_a^2 - \sigma_u^2\right)\right]$.

However, if a GBLUP model is to be fitted, then the mixed linear model is:

$$\mathbf{y} = \mathbf{Xb} + \mathbf{Wu} + \mathbf{Wa} + \mathbf{e} \qquad (11.15)$$

where **a** is the vector of DGVs and all other terms are as defined in Eqn 11.8. The MME to be solved are:

$$\begin{bmatrix} \mathbf{X'R^{-1}X} & \mathbf{X'R^{-1}W} & \mathbf{X'R^{-1}W} \\ \mathbf{W'R^{-1}X} & \mathbf{W'R^{-1}W} + \mathbf{A}^{-1}\alpha_1 & \mathbf{W'R^{-1}W} \\ \mathbf{W'R^{-1}X} & \mathbf{W'R^{-1}W} & \mathbf{W'R^{-1}W} + \mathbf{G}^{-1}\alpha_2 \end{bmatrix} \begin{bmatrix} \hat{\mathbf{b}} \\ \hat{\mathbf{u}} \\ \hat{\mathbf{a}} \end{bmatrix} = \begin{bmatrix} \mathbf{X'R^{-1}y} \\ \mathbf{W'R^{-1}y} \\ \mathbf{W'R^{-1}y} \end{bmatrix} \tag{11.16}$$

where:

$$\alpha_1 = \sigma_e^2 / \sigma_u^2 \text{ and } \alpha_2 = \sigma_e^2 / \left(\sigma_a^2 - \sigma_u^2\right)$$

Example 11.7

The data in Example 11.1 are analysed assuming the same genetic parameters to compute DGVs for the reference animals without using weights. It is also assumed that 10% of the additive genetic variance is due to residual polygenic effect in the model. The analysis has been carried out using both Eqns 11.14 and 11.16 without any weights.

Given that $\sigma_a^2 = 35.241$, then $\sigma_u^2 = 0.1 * 35.241 = 3.5241$. Therefore, for both Eqns 11.14 and 11.16, $\alpha_1 = \sigma_e^2 / \sigma_u^2 = 245 / 3.5241 = 69.521$. However, for Eqn 11.14, $\alpha_2 = \sigma_e^2 / \sigma_g^2$ and now equals $2\sum p_j\left(1-p_j\right) * \left[\sigma_e^2\left(\sigma_a^2 - \sigma_u^2\right)\right] = 3.5383 * \left(245/\left(35.241 - 3.5241\right) = 27.332\right.$, while in Eqn 11.16, $\alpha_2 = \sigma_e^2 / \left(\sigma_a^2 - \sigma_u^2\right) = 7.725$.

The matrix **Z** in Eqn 11.14 is as defined in Example 11.2, while **W** in Eqns 11.14 and 11.16 have been set up in Example 11.3. The matrix $\mathbf{A}^{-1}$ is for the eight reference animals. All matrices for Eqns 11.14 and 11.16 have therefore been defined. The mean and SNP solutions from solving the MME in Eqn 11.14 are given in Table 11.3.

The mean solution from solving Eqn 11.16 was 9.940. The DGVs for the reference and validation populations from both sets of MME are given in Table 11.4. As expected, Eqns 11.14 and 11.16 gave similar results, but for this example, the inclusion of 10% polygenic effects decreased the range of SNP solutions slightly but increased the range for DGVs.

Table 11.3. Mean and SNP effects from SNP-BLUP model with polygenic effects.

Mean effects	
	9.940
SNP effects	
1	0.078
2	−0.280
3	0.234
4	−0.075
5	0.098
6	0.128
7	0.000
8	0.000
9	−0.054
10	−0.018

Table 11.4. Direct genomic breeding values from models with polygenic effects.

	SNP-BLUP model		GBLUP	
	Polygenic	DGV	Polygenic	DGV
Reference animals				
13	0.011	0.066	0.011	0.064
14	−0.007	0.102	−0.007	0.106
15	0.043	0.071	0.043	0.074
16	0.076	0.299	0.076	0.305
17	−0.015	−0.473	−0.015	−0.477
18	−0.025	−0.343	−0.025	−0.345
19	−0.021	0.115	−0.021	0.115
Selection candidates				
20	−0.056	−0.254	−0.056	−0.260
21	0.005	0.028	0.005	0.029
22	−0.006	0.102	−0.006	0.102
23	−0.004	−0.220	−0.004	−0.220
24	−0.008	0.125	−0.008	0.125
25	−0.003	0.051	−0.003	0.051
26	−0.006	0.316	−0.006	0.315

11.6 Haplotype Models

An alternative to SNPs-based model is the haplotype model, where instead of fitting individual SNPs either directly (SNP-BLUP) or via **G** matrix (GBLUP), one fits blocks of SNP markers, i.e. the haplotypes. The main motivation for applying haplotype-based models in genomic predictions is the assumption that such models have a higher probability to capture QTL effects in a strong LD with the markers, thus potentially raising the accuracy of predictions. The first and crucial step in the application of the haplotype-based model is construction of the haplotype blocks, which can be achieved in several ways and largely varies across the literature. Most commonly, the haplotype blocks are constructed by predefining a certain LD threshold, so that each block consists of SNP markers with reasonably high LD, or by predefining certain length of a genome segment that can be in the base pairs, in the centimorgans, or in the number of adjacent SNP markers. It is worth noting, that construction of haplotypes involves an additional step where the SNP marker genotypes are phased, that is alleles inherited from sire or dam at a particular locus are known. The model for analysing is the same as for the SNP-BLUP (Eqn 11.6) but replacing individual SNP effects (**g**) for haplotype effects (**h**). The matrix **Z** is now matrix of haplotype counts. After solving the MME as in Eqn 11.7, the sum of **h** over all haplotypes is assumed equal to the vector of breeding values, i.e. $\mathbf{Z}\hat{\mathbf{h}}$. The haplotype model can also be extended by including pedigree-based polygenic components as explained in Section 11.5. Haplotype genomic relationship matrix ($\mathbf{G}_h$) can be constructed following the same logic as **G** based on the SNP-markers (see Section 11.4.2) and fit within the GBLUP model by substituting **G** with $\mathbf{G}_h$ matrix. Commonly, instead of using haplotype blocks directly, genotypes are recoded according to haplotype alleles, sometimes termed as pseudo-SNPs, representing the number of copies of each haplotype allele in the haploblock (e.g. Teissier *et al.*, 2020).

Example 11.8

To illustrate the construction and application of the haplotype-based model, we used stochastic simulation via AlphaSimR (Gaynor *et al.*, 2021) to generate five animals with information on nine phased SNP genotypes, i.e. with known maternal (f) and paternal (m) haplotypes. In this example, we defined the haplotypes by a fixed length of three adjacent SNP markers. The first haplotype (Hap 1) has six haplotype alleles (101, 100, 011, 010, 111, 000), the second (Hap 2) has five haplotype alleles (010, 000, 111, 011, 001), and the third (Hap 3) has five haplotype alleles (110, 010, 011, 001, 000). Based on the number of copies of each haplotype allele, pseudo-SNPs are formed. For example, animal 1 has one copy of haplotype 1 allele 101 and 1 copy of allele 100, therefore pseudo-SNPs based on haplotype 1 are 110000. The same animal has two copies of haplotype 3 allele 110, and therefore pseudo-SNPs based on haplotype 3 are 20000.

											Haplotypes			Pseudo-SNPs		
Animal*		Phased SNP									Hap 1	Hap 2	Hap 3	Hap 1	Hap 2	Hap 3
1	f	1	0	1	0	1	0	1	1	0	101	010	110	1 1 0 0 0 0	1 1 0 0 0	2 0 0 0 0
	m	1	0	0	0	0	0	1	1	0	100	000	110			
2	f	0	1	1	1	1	1	0	1	0	011	111	010	0 0 1 1 0 0	0 1 1 0 0	0 1 1 0 0
	m	0	1	0	0	0	0	0	1	1	010	000	011			
3	f	1	1	1	0	1	1	0	0	1	111	011	001	0 1 0 0 1 0	0 0 0 1 1	0 0 0 1 1
	m	1	0	0	0	0	1	0	0	0	100	001	000			
4	f	0	1	1	1	1	1	0	1	0	011	111	010	0 1 1 0 0 0	0 1 1 0 0	1 1 0 0 0
	m	1	0	0	0	0	0	1	1	0	100	000	110			
5	f	1	0	0	0	0	1	0	0	1	100	001	001	0 1 0 0 0 1	0 1 0 0 0	0 0 1 1 0
	m	0	0	0	0	0	0	0	1	1	000	000	011			

*Animal ID (1–5), maternal (f) and paternal (m) haplotypes

Pseudo-SNPs can thereafter be fitted in a same manner as regular SNP genotypes would (see Section 11.4.2). Assuming observed allele frequency of $2\sum p_j(1-p_j) = 4.06$, the regular SNP-based **G** matrix is:

$$\mathbf{G} = \begin{bmatrix} 1.350 & -0.670 & -0.522 & 0.266 & -0.424 \\ -0.670 & 1.005 & -0.325 & 0.217 & -0.227 \\ -0.522 & -0.325 & 1.054 & -0.374 & 0.414 \\ 0.266 & 0.217 & -0.374 & 0.414 & -0.522 \\ -0.424 & -0.227 & 0.167 & -0.522 & 1.005 \end{bmatrix}$$

While haplotypes based $\mathbf{G_h}$ with $2\sum p_j(1-p_j) = 4.42$ is:

$$\mathbf{G_h} = \begin{bmatrix} 0.986 & -0.462 & -0.326 & 0.036 & -0.235 \\ -0.462 & 0.805 & -0.416 & 0.172 & -0.100 \\ -0.326 & -0.416 & 1.077 & -0.371 & 0.036 \\ 0.036 & 0.172 & -0.371 & 0.443 & -0.281 \\ -0.235 & -0.100 & 0.036 & -0.281 & 0.579 \end{bmatrix}$$

For illustration, we will assign phenotypes to animals following WWG in Table 4.1. We solve the simple GBLUP model $y = 1\mu + Wa + e$, using **G** and $\mathbf{G}_h$ matrices such that **a**~N(0, **G** or **Gh** σ_a^2), and variance ratios $\alpha = \sigma_e^2 / \sigma_a^2 = 40/20$. Solutions from SNP-based GBLUP are 3.960 for overall mean, and 0.253, −0.474, 0.057, −0.224, and 0.388 for DGVs. Solutions from the haplotype-based GBLUP are 3.960 for overall mean, and 0.235, −0.380, 0.079, −0.196, and 0.262 for DGVs. Therefore, both models result in equal ranking of animals, and correlations between their respective DGVs are 0.999.

11.7 Bayesian Methods for Computing SNP Effects

The assumption of equal variance explained by all loci in the SNP-BLUP or GBLUP model has the advantage that only one variance has to be estimated. However, this may be unrealistic across all traits, which may have different genetic architecture. Also, one of the problems with GBLUP is that it does not allow for moderate to large QTL effects; if these are actually present, they will be severely reduced. The other problem is that with GBLUP, SNP effects cannot be zero, they always have (often very small) effects. Meuwissen *et al.* (2001) presented a Bayesian method that assumes *t* distributions at the level of the SNP effect, modelled using different genetic variances for each SNP (the so-called BayesA method) and another method in which some SNPs are assumed to have effects following a *t*-distribution, and others have zero effects (BayesB). Other variations of the Bayesian methods such as BayesC and BayesCπ (where some SNPs are assumed to have zero effects, and others are assumed to follow a normal distribution) have been published by Habier *et al.* (2011). This section presents some of these methods.

11.7.1 BayesA

Instead of the assumption of a normal distribution for SNP effects as in the SNP-BLUP model, another possible assumption is that the distribution follows a student's *t*-distribution. This allows for a higher probability of moderate to large SNP effects than a normal distribution. However, the *t*-distribution is not easy to incorporate into prediction of marker effects, so a mathematically tractable way of achieving this is to assume that each SNP effect comes from a normal distribution but σ_g^2 can be varied among the SNPs. Thus if σ_g^2 is large then $\hat{g}$ will be large and if σ_g^2 is small, then $\hat{g}$ will likely be small as it will regress towards zero (Hayes and Daetwyler, 2013). This leads to modelling the data at two levels: first, at the level of the data that is similar to SNP-BLUP to estimate the SNP effects; and second, at the variances of the chromosome segments or SNPs, which are assumed to be different at every segment or locus. The procedure uses a Gibbs sampling approach, which involves sampling from the posterior distributions conditioned on other effects. If the reader is not familiar with Gibbs sampling, they may want to read Chapter 16, where application of the Gibbs sampling for the estimation of genetic parameters is discussed.

Thus, given the linear model in Eqn 11.6, the conditional distribution that generates the data, **y**, is:

$$\mathbf{y} \mid \mathbf{b}, \mathbf{g}, \sigma_e^2 \sim \mathrm{N}\left(\mathbf{Xb} + \mathbf{Zg} + \mathbf{R}\sigma_e^2\right)$$

Prior distributions

Specification of the Bayesian model involves defining the prior distributions. Usually, an improper or 'flat' prior distribution is assigned to **b**. Thus $P(\mathbf{b}) \sim$ constant.

The overall mean effect (**b**) is then sampled from the following conditional distribution as:

$$\mathbf{X'Xb} \mid \mathbf{g},\ \sigma^2_{gi},\ \sigma^2_e, \mathbf{y} \sim N\left(\mathbf{X'}(\mathbf{y}-\mathbf{Zg}),\ \mathbf{X'X}\sigma^2_e\right)$$

Therefore:

$$\mathbf{b} \mid \mathbf{g},\ \sigma^2_{gi},\ \ \sigma^2_e, \mathbf{y} \sim N\left(\hat{\mathbf{b}}, (\mathbf{X'X})^{-1}\ \sigma^2_e\right) \qquad (11.20)$$

where $\hat{\mathbf{b}} = (\mathbf{X'X})^{-1}\mathbf{X'}(\mathbf{y}-\mathbf{Zg})$.

A scaled inverted chi-square distribution, $\chi^{-2}(\upsilon, S)$ is usually used as prior for the variance components, with υ being the degrees of freedom and S the scaled parameter (Wang *et al.*, 1993). Thus, for the residual variance, prior uniform distribution ($\chi^{-2}(-2, 0)$) or flat prior can be assumed. Sampling is, then, from the following conditional posterior distribution:

$$\sigma^2_e \mid e_i \sim \chi^{-2}\left(n-2,\ e_i'e_i\right) \qquad (11.21)$$

where $e_i = (y_i - \mathbf{x}_i\mathbf{b} - \mathbf{z}_i\mathbf{g})$; $i = 1, n$ with n equal to the number of records or animals.

Similarly, σ^2_{gi} is sampled from the following conditional posterior distribution:

$$\sigma^2_{gi} \mid \mathbf{g}_i \sim \chi^{-2}\left(\upsilon+k,\ S+\mathbf{g}_i'\mathbf{g}_i\right) \qquad (11.22)$$

with $\upsilon = 4.012$ and S derived as:

$$\frac{\tilde{\sigma}^2(\upsilon-2)}{\upsilon}$$

where $\tilde{\sigma}^2$ is the a prior value of σ^2_{gi} and k_i equals 1 for the *i*th SNP.

Other researchers (Xu, 2003; Ter Braak *et al.*, 2005) have published similar approaches with different priors for estimating σ^2_{gi}.

Finally, $\hat{\mathbf{g}}_i$ for the *i*th SNP is sampled from the following distribution as:

$$\mathbf{g}_i \mid \mathbf{b}, \mathbf{g}_j,\ \sigma^2_{gi},\ \sigma^2_e,\ \mathbf{y} \sim N\left(\hat{\mathbf{g}}_i, \left(\mathbf{z}_i'\mathbf{z}_i+\alpha\right)^{-1}\sigma^2_e\right);\ i \neq j \qquad (11.23)$$

with:

$$\hat{\mathbf{g}}_i = \left(\mathbf{z}_i'\mathbf{z}_i+\alpha\right)^{-1}\mathbf{z}_i'\left(\mathbf{y}-\mathbf{Xb}-\mathbf{z}_j\mathbf{g}_j\right) \text{ and } \alpha = \sigma^2_e / \sigma^2_{gi}$$

The Gibbs sampling procedure, then, consists of setting initial values for **b**, **g**, σ^2_e and σ^2_g, and iteratively sampling successively from Eqns 11.20 to 11.23, using updated values of the parameters from the i round in the $i + 1$ round. Assuming that p rounds of iteration were performed, then p is called the length of the chain. The first j samples are usually discarded as the burn-in period. This is to ensure that samples saved are not influenced by the priors but are drawn from the posterior distribution. Posterior means are then computed from the saved samples.

Example 11.9

Using the data in Example 11.1, the application of BayesA is illustrated using residual updating (Legarra and Misztal, 2008). The data for the reference animals are analysed by fitting the model in Eqn 11.6. Thus, n, the number of records, is 8 and a flat prior has been assumed for **b**. It is also assumed that v = 4.012 and S is derived as: $\frac{\tilde{\sigma}^2(v-2)}{v} = 0.352$ where $\tilde{\sigma}^2 = 0.702$. Note that the matrix of genotypes **Z** used in the computation below has not been centralized and there **Z** equals **M** in Section 11.2.

The starting value for $\hat{\mathbf{b}}$ was computed as $\hat{\mathbf{b}} = (\mathbf{X'X})^{-1}\mathbf{X'y} = 79.1/8 = 9.888$ and those for $\hat{\mathbf{g}}$ and σ^2_{gi} were 0.05 and 0.702, respectively, for all SNPs. The starting value for σ^2_a was set as 2.484, thus $\sigma^2_{gi} = \sigma^2_a / 2\Sigma p_j(1-pj) = 0.702$. The starting values for DGV for animals in the reference population were computed as $\mathbf{a} = \mathbf{Z}\hat{\mathbf{g}}$. Thus:

$$\mathbf{a}' = \begin{pmatrix}0.45 & 0.30 & 0.55 & 0.45 & 0.45 & 0.50 & 0.40 & 0.35\end{pmatrix}$$

Initially, a vector of residuals $\hat{\mathbf{e}}$ was computed as $\hat{\mathbf{e}} = \mathbf{y} - \mathbf{X}\hat{\mathbf{b}} - \mathbf{Z}\hat{\mathbf{g}}$. Thus:

$$\begin{pmatrix}\hat{e}_1\\ \hat{e}_2\\ \hat{e}_3\\ \hat{e}_4\\ \hat{e}_5\\ \hat{e}_6\\ \hat{e}_7\\ \hat{e}_8\end{pmatrix} = \begin{pmatrix}9.0\\ 13.4\\ 12.7\\ 15.4\\ 5.9\\ 7.7\\ 10.2\\ 4.8\end{pmatrix} - \begin{pmatrix}9.888\\ 9.888\\ 9.888\\ 9.888\\ 9.888\\ 9.888\\ 9.888\\ 9.888\end{pmatrix} - \begin{pmatrix}0.45\\ 0.30\\ 0.55\\ 0.45\\ 0.45\\ 0.50\\ 0.40\\ 0.35\end{pmatrix} = \begin{pmatrix}-1.388\\ 3.213\\ 2.263\\ 5.063\\ -4.438\\ -2.688\\ -0.088\\ -5.437\end{pmatrix}$$

From the above, $\hat{\mathbf{e}}'\hat{\mathbf{e}} = 99.345$, and thus given the value of 8.131 sampled from the inverted χ^2 distribution with n – 2 degrees of freedom, $\sigma^2_e[1] = 99.345/8.131 = 12.218$, using Eqn 11.21. The superscript in brackets denotes the iteration number.

Then, sample **b**[1] using Eqn 11.20, with $\hat{\mathbf{b}}$ calculated as $(\mathbf{x}_j'\mathbf{x})^{-1}\mathbf{1}'\hat{\mathbf{e}} = 9.456$ after initially updating $\hat{\mathbf{e}}$, the vector of residuals to include information on **b** as:

$$\hat{\mathbf{e}}_i = \hat{\mathbf{e}}_i + \mathbf{X}\hat{\mathbf{b}} \text{ with } i = 1,\ n$$

Assuming the random number generated from a normal distribution is 0.873 and $(\mathbf{x}_j'\mathbf{x})^{-1}\sigma^2_e = 12.218/8 = 1.527$, then $\mathbf{b}_1^{[1]} = (9.456 + 0.873\sqrt{(1.527)} = 10.535$.

After sampling for **b**, the $\hat{\mathbf{e}}$ is updated to exclude the information on **b** as:

$$\hat{\mathbf{e}}_i = \hat{\mathbf{e}}_i - \mathbf{X}\hat{\mathbf{b}} \text{ with } i = 1,\ n$$

Using Eqn 11.22, σ^2_{gi} for the ith SNP effect is sampled from the inverted χ^2 distribution with degrees of freedom 5.012 and S = 0.352 computed earlier. For the first SNP, $\hat{\mathbf{g}}_1^2 = 0.003$, thus given the value of 11.422 sampled from the inverted χ^2 distribution $\sigma^2_{g1}[1] = (S + \hat{\mathbf{g}}^2)/11.422 = 0.031$. The variance estimates for other SNPs in the first iteration are shown in Table 11.5.

Finally, estimates of $\hat{\mathbf{g}}$ are sampled from the normal distribution using Eqn 11.23. First update the vector of residuals to include information on the jth SNP. Thus, for the jth SNP effect:

$$\hat{\mathbf{e}}_i = \hat{\mathbf{e}}_i + \mathbf{z}_{ij}\hat{\mathbf{g}}_j \text{ with } i = 1,\ n$$

Table 11.5. SNP solutions and variances from BayesA and BayesB.

	BayesA				BayesB			
	First iteration		Posterior means		First iteration		Posterior means	
SNP	Effects	Var	Effects	Var	Effects	Var	Effects	Var
1	0.289	0.031	0.018	0.170	2.187	1.105	0.038	0.316
2	0.279	0.049	−0.064	0.179	−1.565	0.516	−0.107	0.319
3	−0.010	0.070	0.058	0.179	−0.156	0.124	0.067	0.293
4	0.023	0.097	−0.023	0.176	−0.309	0.118	−0.034	0.300
5	0.045	0.052	0.022	0.167	0.413	0.363	0.047	0.328
6	−0.321	0.050	0.025	0.171	−0.521	0.161	0.031	0.283
7	0.411	0.256	−0.006	0.186	0.000	0.000	0.009	0.335
8	0.408	0.056	−0.008	0.168	−0.010	0.431	0.008	0.261
9	0.115	0.034	−0.003	0.162	0.000	0.000	−0.006	0.294
10	−0.578	0.152	−0.008	0.165	0.000	0.000	−0.017	0.286

Var, SNP variances

Thus, for the first SNP effect, $\hat{\mathbf{g}}_1 = (\mathbf{z}'_{i1}\mathbf{z}_{i1} + \alpha)^{-1}\mathbf{z}'_{i1}\hat{\mathbf{e}}_i = (7 + 393.201)^{-1}(-2.775) = -0.007$. Assuming the random number generated from a normal distribution is 1.692, then $\mathbf{g}_j^{[1]}$ can be sampled as $\mathbf{g}_1^{[1]} = -0.007 + 1.692\sqrt{(12.218 / 400.201)} = 0.289$. After computing $\mathbf{g}_1^{[1]}$, the residual vector is updated as $(\hat{\mathbf{e}}_i = \hat{\mathbf{e}}_i - \mathbf{z}_{i1}\,\mathbf{g}_1^{[1]},\ i = 1, n)$ before computing the next SNP effect. The estimates of $\mathbf{g}_2^{[1]}$ to $\mathbf{g}_8^{[1]}$ are given in Table 11.5. The next cycle of sampling then begins again with sampling residual variance without setting up of the vector of residuals.

For this example, the Gibbs sampling chain was ran 10,000 times, with the first 3000 considered as the burn-in period. The posterior means computed from the remaining 7000 samples for $\hat{\mathbf{b}}$ and σ_e^2 were 9.890 kg and 33.119 kg², respectively. The estimates for $\hat{\mathbf{g}}$ and σ_{gi}^2 are given in Table 11.4.

The DGV of animals in the validation set can then be predicted using the solutions for the SNP effects in Table 11.5 as $\mathbf{Z}_2\hat{\mathbf{g}}$, where $\mathbf{Z}_2$ is a matrix of genotypes for the validation of test animals given in Example 11.2.

11.7.2 BayesB

The basic assumption in BayesA is that there is genetic variance at every loci or chromosome segment. It is possible that some SNPs will have zero effects as they are in genomic regions with no QTL. The prior density of BayesA does not account for such SNPs with zero effects as BayesA density peak at $\sigma_{gi}^2 = 0$; in fact, its probability of $\sigma_{gi}^2 = 0$ is infinitesimal (Meuwissen *et al.*, 2001). It is possible that genetic variance may be observed in relatively few marker loci containing QTL. Meuwissen *et al.* (2001) introduced BayesB to address this situation. Thus, the prior distribution of BayesB is a mixture distribution with some SNPs with zero effects and the rest with a *t*-distribution (Hayes and Daetwyler, 2013). BayesB, therefore, uses a prior that has a high density, π, at $\sigma_{gi}^2 = 0$ and has an inverted chi-squared distribution for $\sigma_{gi}^2 > 0$. Thus, the prior distribution for BayesB is:

$$\sigma_{gi}^2 = 0 \text{ with probability } \pi$$
$$\sigma_{gi}^2 \sim \chi^{-2}(v, S) \text{ with probability } (1-\pi) \quad (11.24)$$

where S is the scaling parameter, v the degrees of freedom and π is assumed known. They set S to be to 0.0429 and computed it as in Eqn 11.22 while v was set to 4.234.

While the Gibbs sampling algorithm used for BayesA can also be used for BayesB, it will not, however, move through the entire sampling space, as the sampling of $\sigma_{gi}^2 = 0$ is not possible if $(\mathbf{g}_i'\mathbf{g}_i)$ is greater than zero. Also, if $\sigma_{gi}^2 = 0$, the sampling of $\mathbf{g}_i$ has an infinitesimal probability. This problem is overcome by sampling σ_{gi}^2 and $\mathbf{g}_i$ simultaneously from the distribution:

$$p\left(\sigma_{gi}^2, \mathbf{g}_i \mid \mathbf{y}^*\right) = p\left(\sigma_{gi}^2 \mid \mathbf{y}^*\right) \times p\left(\mathbf{g}_i \mid \sigma_{gi}^2, \ \mathbf{y}^*\right) \quad (11.25)$$

where $\mathbf{y}^*$ is the data vector $\mathbf{y}$ corrected for the mean and all genetic effects apart from $\mathbf{g}_i$. The first term in Eqn 11.25 implies sampling σ_{gi}^2 without conditioning on $\mathbf{g}_i$ and then sampling from the second term of Eqn 11.25 for $\mathbf{g}_i$ conditional on σ_{gi}^2 and $\mathbf{y}^*$ as in BayesA. The distribution $p\left(\sigma_{gi}^2 \mid \mathbf{y}^*\right)$ cannot be expressed in the form of a known distribution, therefore Meuwissen *et al.* (2001) used the Metropolis–Hastings (MH) algorithm to sample from $p(\sigma_{gi}^2 \mid \mathbf{y}^*)$ using the prior distribution, $p\left(\sigma_{gi}^2\right)$, as the driver distribution to suggest updates for the MH chain as follows:

1. Sample $\sigma_{gi(\text{new})}^2$ from the prior distribution $p\left(\sigma_{gi}^2\right)$.
2. Replace the current σ_{gi}^2 by $\sigma_{gi(\text{new})}^2$ with a probability of k:

$$k = \text{minimize}\left\{p\left(\mathbf{y}^* \mid \sigma_{gi(\text{new})}^2\right) / p\left(\mathbf{y}^* \mid \sigma_{gi}^2\right); 1\right\}$$

and then go to step 1: where $p\left(\mathbf{y}^* \mid \sigma_{gi}^2\right)$ is the likelihood of the data given σ_{gi}^2. The likelihood can be calculated as:

$$L\left(\mathbf{y}^* \mid \sigma_{gi}^2\right) = \frac{1}{2\pi^{1/2n}\sqrt{|\mathbf{V}|}} e^{-1/2\left(\mathbf{y}^{*\prime}\mathbf{V}^{-1}\mathbf{y}\right)} \quad (11.26)$$

where $\mathbf{V} = \mathbf{z}_i\left(\mathbf{I}\sigma_{gi}^2\right)\mathbf{z}_i' + \mathbf{I}\sigma_e^2$ and |V| is the determinant of V. Note that if σ_{gi}^2 is zero, as will happen in the course of the MH sampling, then $\mathbf{V} = \mathbf{I}\sigma_e^2$.

The computation of the required likelihood is easier to implement in a log-likelihood form. Fernando (2010) presented the following algorithm for the log-likelihood:

$$\log \text{LH} = -0.5\left(\log(\mathbf{V})\right) + \left(\left(\left(\mathbf{z}_i'\mathbf{y}^*\right)' \mathbf{V}^{-1}\right)\mathbf{z}_i'\mathbf{y}^*\right) \quad (11.27)$$

with:

$$\mathbf{V} = \left(\mathbf{z}_i'\mathbf{z}_i\mathbf{I}\sigma_{gi}^2, \mathbf{z}_i'\mathbf{z}_i\right) + \mathbf{z}_i'\mathbf{z}_i * \sigma_e^2 \text{ or } \mathbf{V} = \mathbf{z}_i'\mathbf{z}_i * \sigma_e^2 \text{ when } \sigma_{gi}^2 \text{ is zero}$$

In practice, a required number of MH cycles are implemented per cycle of Gibbs sampling. The implementation of each MH cycle involves:

1. Using Eqn 11.26 or 11.27 compute an initial likelihood (LH1) using the current σ_{gi}^2. Note that the current σ_{gi}^2 could be zero and LH1 is also computed but with **V** appropriately defined.
2. Then commence the MH cycle, by drawing r from a uniform distribution. Set $\sigma_{gi(\text{new})}^2$ to be zero. If $r < (1 - \pi)$, sample a $\sigma_{gi(\text{new})}^2$ from the driver distribution using Eqn 11.25. Compute likelihood (LH2) using $\sigma_{gi(\text{new})}^2$ and calculate k as k = minimize (LH2/LH1; 1). Note that if log-likelihood Eqn 11.27 is used, then k = exp(LH2 – LH1). The value of k is

compared with a number s drawn from a uniform distribution. If s is less than k, then accept $\sigma^2_{gi(\text{new})}$ and then set LH1 = LH2. Go to step 1 and begin another MH cycle until required MH cycles are complete.

After the required number of MH cycles, if $\sigma^2_{gi(\text{new})}$ is > 0, then $\mathbf{g}_i$ is sampled as in BayesA, otherwise $\mathbf{g}_i = 0$. Similarly, the sampling of $\mathbf{b}$ and σ^2_e is implemented as described in BayesA.

Example 11.10

The application of BayesB is illustrated using the data in Example 11.1 with residual updating. The data for the reference animals are analysed with the model in Eqn 11.6. The initial parameters are the same as outlined for BayesA in Example 11.9 and the starting value of π was set at 0.30.

The starting values for $\hat{\mathbf{b}}$, $\hat{\mathbf{g}}$, σ^2_{gi} and $\hat{\mathbf{a}}$ were the same as for BayesA. The sampling procedure for parameters is the same as for BayesA apart from sampling for σ^2_{gi}.

Initially, the vector of residuals, $\hat{\mathbf{e}}$, is set up and this has been given in Example 11.9. Therefore, in the first iteration $\sigma_e^{2[1]} = 99.345/8.131 = 12.218$.

Similarly, $\mathbf{b}_1^{[1]} = (9.456 + 0.873\sqrt{(1.527)} = 10.535$, as in Example 11.9.

Using the steps outlined for the MH cycle for BayesB, σ^2_{gi} for the ith SNP effect is then sampled, which could result in either $\sigma^2_{gi} = 0$ or $\sigma^2_{gi} > 0$. In this example, 20 MH samples were evaluated per each round of Gibbs sampling, and for the first SNP, the estimate of $\sigma^2_{g1} = 1.105$. Therefore, $\hat{\mathbf{g}}_1$ was sampled from the normal distribution using Eqn 11.23 as described in Example 11.9 but with $\boldsymbol{\alpha}_1 = 12.218/1.105 = 11.057$. In this example, σ^2_{gi} and $\hat{\mathbf{g}}_i$ for SNP_i, with i = 7, 9 and 10 were zero in the first round of iteration. The solutions for σ^2_{gi} and $\hat{\mathbf{g}}_i$ for the first round of iteration are presented in Table 11.5.

The Gibbs sampling was run for 10,000 cycles, with the first 3000 regarded as the burn-in period. The posterior means computed from the remaining 7000 samples for $\hat{\mathbf{b}}$ and σ^2_e were 9.792 kg and 34.930 kg^2, respectively. The estimates for $\hat{\mathbf{g}}$ and σ^2_{gi} are given in Table 11.5. The DGV of animals in the validation set can then be predicted using the solutions for the SNP effects in Table 11.2 as $\mathbf{Z}_2\hat{\mathbf{g}}$, where $\mathbf{Z}_2$ is defined as in Example 11.9.

11.7.3 BayesC

Habier *et al.* (2011) indicated the estimation of individual SNP variances in BayesA and BayesB has only one additional degree of freedom compared with its prior, and so the shrinkage of SNP effects is largely dependent on the scale parameter, S. To overcome this limitation, they proposed BayesC, which involves estimating a single variance that is common to all SNPs, thereby reducing the influence of the scale parameter. Similar to BayesB, BayesC allows for some SNPs to have zero effects with probability π while the remaining SNPs have non-zero effect with probability (1 – π). Habier *et al.* (2011) indicated that since the priors of all SNP effects have a common variance, the effect of an SNP fitted with probability (1 – π) comes from a mixture of multivariate student's t-distributions.

In BayesC, it is assumed that π is known and the decision to include SNP_i depends on the full conditional posterior of an indicator variable δ_i. This indicator variable equals 1 if SNP_i is fitted, otherwise it is zero. Thus, the decision to include the ith SNP involves computing the probability k of $\delta_i = 1$ as $k = 1/\left\{1 + \left(p(\mathbf{y}^* \mid \delta_i = 0, \theta)/p(\mathbf{y}^* \mid \delta_i = 1,\ \sigma^2_g, \theta)\right)\right\}$, where $\left(p(\mathbf{y}^* \mid \delta_i = 1, \theta)\right)$ denotes the likelihood of the data given that SNP_i is fitted with common variance σ_g^2, θ refers to accepted values for all other parameters, $\left(p\left(\mathbf{y}^* \mid \delta_i = 0, \theta\right)\right)$ denotes the likelihood of the data model without the *i*th SNP and where $\mathbf{y}^*$ is the data vector **y** corrected for the mean and all genetic effects apart from $\mathbf{g}_i$.

The computation of the required likelihood is easier to implement in a log-likelihood form. Fernando (2010) presented such an algorithm based on the log-likelihood.

Given current estimates of σ_g^2, and σ_e^2, logLH1 with $\delta_i = 1$ is computed as:

$$\text{logLH1} = -0.5(\log(\mathbf{V})) + (\mathbf{z}_i'\mathbf{y}^*)'\mathbf{V}^{-1}\mathbf{z}_i'\mathbf{y}^* + \log(1-\pi) \text{ with}$$
$$\mathbf{V} = (\mathbf{z}_i'\mathbf{z}_i\mathbf{I}\sigma_g^2, \mathbf{z}_i'\mathbf{z}_i) + \mathbf{z}_i'\mathbf{z}_i * \sigma_e^2$$

Similarly, the log-likelihood when $\delta_i = 0$ is computed as logLH0 = $-0.5(\log(\mathrm{V})) + (\mathrm{z}_i'\mathrm{y}^*)'\mathrm{V}^{-1}\mathrm{z}_i'\mathrm{y}^* + \log(\pi)$ but with $\mathbf{V} = \mathbf{z}_i'\mathbf{z}_i * \sigma_e^2$.

Then compute probability *k* of $\delta_i = 1$ as $k = 1/(1 + \exp(\text{logLH0} - \text{logLH1}))$.

If *k* is greater than *r*, where *r* is a random drawn from a uniform distribution, then SNP*i* is fitted and $\mathbf{g}_i^{[j]}$ is sampled from the normal distribution using Eqn 11.23, otherwise $\mathbf{g}_i^{[j]} = 0$.

After sampling the vector **g**, σ_g^2 is sampled from the following conditional posterior distribution as:

$$\sigma_g^2 \mid \mathbf{g} \sim \chi^{-2}\left(v + k^{[j]}, S + \mathbf{g}_i'\mathbf{g}_i\right) \qquad (11.28)$$

with terms defined as in Eqn 11.22 but with degrees of freedom equal to $v + k^{[j]}$, where $k^{[j]}$ is the number of SNPs with non-zero effects fitted in the *j*th iteration.

Example 11.11

The data in Example 11.1 is used to illustrate BayesC by applying the model in Eqn 11.6. The assumptions and the starting values for $\hat{\mathbf{b}}$, $\hat{\mathbf{g}}$ and **a** were the same as outlined for BayesA in Example 11.9. The starting value of π was assumed at 0.30 while the starting value of σ_g^2 was set at 0.702.

The sampling procedure for σ_e^2 and **b** were as outlined in BayesA and therefore with the same solutions in the first iteration. Then, for the *i*th SNP, the probability of $\hat{\mathbf{g}}_i$ having a zero effect or otherwise was computed as described earlier in this section. In the first iteration, the first SNP has a non-zero effect; therefore, $\hat{\mathbf{g}}_1 = \left(\mathbf{z}_{i1}'\mathbf{z}_{i1} + \alpha\right)^{-1}\mathbf{z}_{i1}'\hat{\mathbf{e}}_i = (7 + 17.045)^{-1}(-2.775) = -0.115$, with α = 12.218/0.702. Assuming the random number generated from a normal distribution is 0.748, $\mathbf{g}_j^{[1]}$ was sampled using Eqn 11.23 as $\mathbf{g}_1^{[1]} = -0.115 + 0.748\sqrt{(12.218/24.045)} = 0.418$. In the first round of iteration, two SNPs (5 and 10) had zero effects. The solutions for $\hat{\mathbf{g}}_i$ in the first iteration are presented in Table 11.6.

The sampling of common variance was done using Eqn 11.28. For this example, eight SNPs had non-zero effects in the first iteration; therefore, σ_g^2 in the first iteration was sampled from the inverted χ^2 distribution with degrees of freedom now equal to 8 + 4.012 = 12.012, *S* = 0.352 and $\sum\hat{\mathbf{g}}_i^2 = 1.435$. Thus, given the value of 16.294 sampled from the inverted χ^2 distribution, then in the first iteration $\sigma_g^{2[1]} = \left(S + \sum\hat{\mathbf{g}}_i^2\right)/16.294 = 0.110$.

The Gibbs sampling was run for 10,000 cycles, with the first 3000 regarded as the burn-in period. The posterior means computed from the remaining 7000 samples for $\hat{\mathbf{b}}$, σ_e^2 and σ_g^2 were 9.828 kg, 32.377 kg^2 and 0.184 kg^2, respectively. The estimates for $\hat{\mathbf{g}}$ are given in Table 11.6.

Table 11.6. Solutions for SNP effects from BayesC and BayesCπ.

	BayesC		BayesCπ
SNP	First iteration	Posterior means	Posterior means
1	0.416	0.015	0.010
2	−0.360	−0.045	−0.029
3	−0.590	0.044	0.028
4	0.465	−0.014	−0.018
5	0.000	0.014	0.013
6	0.360	0.025	0.010
7	−0.586	−0.002	0.004
8	−0.307	0.009	0.003
9	−0.041	−0.013	−0.011
10	0.000	−0.002	−0.006

11.7.4 BayesCπ

In BayesC, there is the implicit assumption that the probability, $\pi > 0$, i.e. a SNP has zero effect, is regarded as known. Habier *et al.* (2011) argued that the shrinkage of SNP effects is affected by π and should be estimated from the data and proposed BayesCπ, which incorporates this estimation step. Thus, compared to BayesC, the additional feature of BayesCπ is estimating π from the data. The sampling procedure for parameters in BayesCπ are therefore the same as BayesC apart from the additional step of sampling for π. Thus, only the procedure for sampling π is described.

The parameter π is sampled from a beta distribution, with shape parameters $(m - k^{[j]} + 1)$ and $(k^{[j]} + 1)$, with m equal to the total number of SNPs in the analysis and $k^{[j]}$ is the number of SNPs with non-zero effects fitted in the jth iteration.

Example 11.12

The application of BayesCπ is illustrated using the data in Example 11.1. The reference animals are analysed by applying the model in Eqn 11.6 using residual updating. The initial parameters are the same as outlined for BayesA in Example 11.9. The starting values of π and σ_g^2 were set at 0.30 and 0.702, respectively.

The sampling procedure for σ_e^2 and **b** were as outlined in BayesA and therefore with the same solutions in the first iteration. Then, for the ith SNP, the probability of $\hat{\mathbf{g}}_i$ having a zero effect or otherwise was computed as described earlier in this section. In the first iteration, the first SNP has a non-zero effect; therefore, $\hat{\mathbf{g}}_1 = \left(\mathbf{z}_{i1}'\mathbf{z}_{i1} + \alpha\right)^{-1} \mathbf{z}_{i1}'\hat{\mathbf{e}}_i = \left(7 + 17.045\right)^{-1}$ $(-2.775) = -0.115$, with $\boldsymbol{\alpha} = 12.218/0.702$. Assuming the random number generated from a normal distribution is 0.748, $\mathbf{g}_j^{[1]}$ was sampled using Eqn 11.23 as $\mathbf{g}_1^{[1]} = -0.115 + 0.748(12.218/24.045) = 0.418$. In the first round of iteration, two SNPs (5 and 10) had zero effects and the solutions for **g** were the same as obtained for BayesC (Table 11.6).

The sampling of common variance follows the same procedure for BayesC, again with the degrees of freedom equal to the number of SNPs with non-zero effects. For this example, eight SNPs had non-zero effects in the first iteration; therefore, σ_g^2 in the first iteration was sampled from the inverted χ^2 distribution with degrees of freedom now equal to 8 + 4.012 = 12.012, $S = 0.352$ and $\sum \hat{g}_i^2 = 1.435$. Thus, given the value of 16.294 sampled from the inverted χ^2 distribution, then in the first iteration, $\sigma_i^{2[1]} = \left(S + \sum \hat{g}_i^2\right) / 16.294 = 0.110$.

Then $\pi^{[1]}$ was sampled from the beta distribution with shape parameters $((m - k^{[1]} + 1) = 3)$ and $((k^{[1]} + 1) = 9)$, given eight SNPs had non-zero effects. A value of 0.339 was sampled for π.

A total of 10,000 cycles was implemented for the Gibbs sampling and the first 3000 were discarded as the burn-in period. The posterior means computed from the remaining 7000 samples for $\hat{\mathbf{b}}$, σ_e^2, σ_g^2 and π were 9.898 kg, 32.343 kg^2, 0.162 kg^2 and 0.51, respectively. The estimates for $\hat{\mathbf{g}}$ were given in Table 11.6.

The estimates for $\hat{\mathbf{b}}$ and σ_e^2 were very consistent for the Bayesian models considered. Similarly, BayesC and BayesCπ gave very similar estimates of σ_g^2, which were consistent with estimates for BayesA but SNP solutions were different from the different models. The estimates of σ_{gi}^2 for BayesB were almost double those from the other models.

11.8 Multivariate Genomic Models

The extension of univariate GBLUP to multiple trait analyses (MGBLUP) follows the same logic and model construction as previously explained for the pedigree-based multivariate model (MBLUP) in Chapter 6. If we assume the univariate GBLUP model is defined with Eqn 11.8 and MME as in Eqn 11.9, extending to multiple traits would involve introduction of additive genetic covariance ($\mathbf{G}_a$) and residual covariance ($\mathbf{R}$) matrices as follows:

$$\begin{bmatrix} \mathbf{X}'(\mathbf{R}^{-1} \otimes \mathbf{I})\mathbf{X} & \mathbf{X}'(\mathbf{R}^{-1} \otimes \mathbf{I})\mathbf{W} \\ \mathbf{W}'(\mathbf{R}^{-1} \otimes \mathbf{I})\mathbf{X} & \mathbf{W}'(\mathbf{R}^{-1} \otimes \mathbf{I})\mathbf{W} + \mathbf{G}_a^{-1} \otimes \mathbf{G}^{-1} \end{bmatrix} \begin{bmatrix} \hat{\mathbf{b}} \\ \hat{\mathbf{a}} \end{bmatrix} = \begin{bmatrix} \mathbf{X}'(\mathbf{R}^{-1} \otimes \mathbf{I})\mathbf{y} \\ \mathbf{W}'(\mathbf{R}^{-1} \otimes \mathbf{I})\mathbf{y} \end{bmatrix}.$$

One can easily identify that the only difference between previously described MBLUP and MGBLUP models is substitution of pedigree-based relationship matrix among animals (**A**) with genomic relationship matrix (**G**) among genotyped animals, and all design matrices are as previously defined.

Example 11.13

For illustrative purposes, assume the same data in Table 6.1 is now extended in a way that each calf has information on genotypes for the ten SNPs obtained by stochastic simulation via AlphaSimR (Gaynor *et al.*, 2021):

Calves	Sex	Sire	Dam	WWG	PWG	SNP Genotype									
4	Male	1	–	4.5	6.8	0	2	1	1	1	2	0	1	1	1
5	Female	3	2	2.9	5.0	2	1	2	1	1	0	1	0	2	0
6	Female	1	2	3.9	6.8	1	2	1	0	0	1	1	1	2	0
7	Male	4	5	3.5	6.0	1	2	2	1	0	1	1	1	2	1
8	Male	3	6	5.0	7.5	1	1	1	1	1	0	1	1	2	0

Assume that the additive genetic covariance ($\mathbf{G}_a$) and residual covariance ($\mathbf{R}$) matrices are the same as in the MBLUP example:

$$\mathbf{G}_a = \begin{bmatrix} 20 & 18 \\ 18 & 40 \end{bmatrix} \text{ and}$$

$$\mathbf{R} = \begin{bmatrix} 40 & 11 \\ 11 & 30 \end{bmatrix}, \text{ with corresponding inverses :}$$

$$\mathbf{G}_a^{-1} = \begin{bmatrix} 0.084 & -0.038 \\ -0.038 & 0.042 \end{bmatrix} \text{ and}$$

$$\mathbf{R}^{-1} = \begin{bmatrix} 0.028 & -0.010 \\ -0.010 & 0.037 \end{bmatrix}.$$

The **G** matrix based on the ten SNPs is constructed as in Section 11.4.2 and assuming observed allele frequency of $2\sum pj(1-pj) = 4.08$, is:

$$\mathbf{G} = \begin{bmatrix} 1.137 & -0.725 & -0.088 & 0.010 & -0.333 \\ -0.725 & 0.843 & -0.235 & -0.137 & 0.255 \\ -0.088 & -0.235 & 0.402 & 0.010 & -0.088 \\ 0.010 & -0.137 & 0.010 & 0.353 & -0.235 \\ -0.333 & 0.255 & -0.088 & -0.235 & 0.402 \end{bmatrix}$$

Note that a comparision with **A** from Chapter 6 is not plausible as **G** in this example is based on only 10 SNPs. The $\mathbf{G}^{-1}$ used in this example was computed after adding 0.01 to the diagonals of **G**, to ensure invertibility. Due to these differences, it is hard to directly compare solutions between MBLUP and MGBLUP models, as it can be seen from the table:

	MBLUP		MGBLUP	
Effects	WWG	PWG	WWG	PWG
Sex				
Male	4.361	6.800	4.323	6.753
Female	3.397	5.880	3.416	5.921
Animals				
1	0.151	0.280	–	–
2	–0.015	–0.008	–	–
3	–0.078	–0.170	–	–
4	–0.010	–0.013	0.081	0.108
5	–0.270	–0.478	–0.212	–0.419
6	0.276	0.517	0.181	0.377
7	–0.316	–0.479	–0.209	–0.302
8	0.244	0.392	0.159	0.235

Nevertheless, the ranking among the animals with genotypes is similar for both models, but not the same. We can explain this re-ranking as the average relationship between these animals and the rest of the population in **A** matrix and in the **G** matrix is different. With this result, we can also illustrate the main flaw of the GBLUP model, which is the inability to produce the solutions for ungenotyped individuals (in this example animals 1–3). Therefore, in the next chapter we introduce the single-step method that can provide solutions for ungenotyped animals as well.

11.9 Cross-validation and Genomic Reliabilities

As described in previous sections, the computation of SNP effects is usually in a reference population using animals with observations. In the case of the dairy industry, the estimation of SNP effects has been carried out using mostly bulls with high reliability as the reference population with de-regressed breeding values (DRP) used as observations. Recently, some countries have started including cows in the reference populations, which require weighting the cow records appropriately. Ideally, it is necessary that the estimates of SNP effects are validated in another data set, which has not contributed any information to the reference population to assess accuracy of prediction. In general, animals born after the animals in the reference population are used as the validation data set and this is termed forward validation, as we are assessing the accuracy of prediction in younger animals. However, in some cases, when the number of genotyped animals is limited such that adequate reference and validation data sets cannot be created, then a cross-validation approach can be implemented. This involves randomly sampling animals from all genotyped candidates to be used as the validation data set and their records are excluded in the estimation of SNP effects.

The DGV computed for the validation data sets are compared with their DRP. An estimate of the correlation between the DGV and the DRP in the validation animals provides an estimate of the accuracy of genomic predictions, although this does not take into account the accuracy of the DRP themselves. For the purposes of illustration, the correlation between the DGV from the SNP or GBLUP models with the DRP for the validation animals in the data for Example 11.1 is 0.49, which gives a reliability of 0.24. The accuracies or reliabilities from the cross-validation studies are usually referred to as realized reliabilities.

We typically assess the realized reliability and the dispersion bias for validation set by regressing the animal's pseudo-phenotypes (e.g. DYD or phenotypes adjusted for the fixed effects) on their GEBV as presented in Fig. 11.1. In this sense, R^2 serves as a measure of

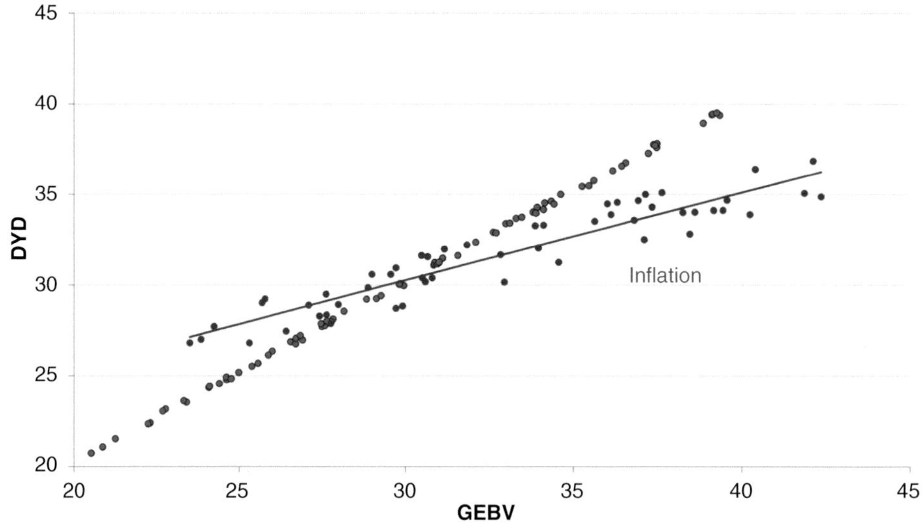

Fig. 11.1 Regression of DYD on GEBV. (Blue dots represent the training set and red dots represent the validation set.)

realized reliability, while a regression coefficient (the slope of the regression) smaller or greater than 1 indicates the GEBV are either inflated or deflated, respectively. In the example in Fig. 11.1, the regression coefficient is smaller than 1, meaning GEBV are inflated, that is overestimated. It is worth noticing, that for the training set the slope is 1, as expected.

Theoretical reliabilities, as calculated in traditional BLUP, can also be computed from the inverse of equations similar to those used to compute DGV. For individuals with observations, reliabilities for the DGV can be computed (VanRaden, 2008) by first computing **B** as follows:

$$\mathbf{B} = \mathbf{G}\left(\mathbf{G} + \mathbf{R}\left(\frac{\sigma_e^2}{\sigma_a^2}\right)\right)^{-1}\mathbf{G}$$

Then, reliability for animal $i = \text{rel}_i = 1 - \left(b_{ii}^{*}\ \sigma_e^2 / \sigma_a^2\right)$, where b_{ii} is the diagonal element of **B** for the animal. Similarly, for validation candidates with no records, **B** is:

$$\mathbf{B} = \mathbf{C}\left(\mathbf{G} + \mathbf{R}\left(\frac{\sigma_e^2}{\sigma_a^2}\right)\right)^{-1}\mathbf{C}'$$

Then reliability is computed from the diagonal elements of **B** as described for the reference animals.

However, these theoretical reliability estimates tend to be too high. These can be scaled by the realized reliabilities from the validation study. In addition, with a large data set, the inversion required for the computation of the reliabilities could be a source of limitation to the use of the methodology. However, for ssGBLUP, Misztal *et al.* (2013a) presented two methods of approximating reliability based on the decomposition of a function of reliability into contributions from records, pedigrees and genotypes, expressed in terms of records or daughter equivalents. The first approximation method involved inversion of a matrix that contains inverses of the genomic relationship matrix and the pedigree relationship matrix for genotyped animals. The second approximation method involved only the diagonal elements of those inverses. More recently, Zaabza *et al.* (2021) presented a full Monte Carlo sampling-based method for approximating reliability in the SNP-BLUP model that avoids the inverse of the MME. An advantage of the method is its low computational demand but it tends to overestimate reliability for animals with low reliability, especially when the weight of the residual polygenic effects is high. Increasing the number of MC samples can reduce the overestimation problem. Zaabza *et al.* (2020) published the software for the implementation of this approach.

12 Single-step Approaches to Genomics

12.1 Basic Principle

As discussed in Chapter 11, SNP-BLUP and GBLUP became popular models for incorporating genomic information (SNPs) in animal evaluations, where the resulting DGVs must be combined with the conventional BLUP results to formulate GEBVs used for ranking the animals. Due to several separate steps involved, such a procedure was named a multi-step genomic evaluation in which the combination of DGVs and the conventional evaluations is based on some sort of selection index approach. The selection index presented by VanRaden *et al.* (2009) was:

$$\text{GEBV} = wt_1\text{DGV} + wt_2PTA_1 + wt_3PTA_2$$

for animals in the reference population. Similarly, for the selection candidates with no daughter information:

$$\text{GEBV} = wt_1\text{DGV} + wt_2PA_1 + wt_3PA_2$$

where PTA_1 and PTA_2 are predicted transmitting abilities from the official evaluations based on all records and the evaluations of only the bulls in the reference population using the **A** matrix, respectively. Correspondingly, PA_1 and PA_2 are parent averages from the respective evaluations. The weights (wt_i) were computed as $\mathbf{c'V^{-1}}$. The matrix **V** is of order 3 × 3 with diagonal elements equal to the reliabilities for DGV, PTA_1 (PA_1) and PTA_2 (PA_2), respectively. The off-diagonal elements were calculated as $v_{12} = v_{22}$, $v_{23} = v_{22}$ and $v_{13} = v_{22} + (v_{11} - v_{22})(v_{33} - v_{22})/(1 - v_{22})$. The vector **c′** has elements v_{11}, v_{22} and v_{33}. The main weaknesses of multi-step genomic evaluation are its restriction to genotyped animals exclusively, the need for pseudo-phenotypes (e.g. daughter yield deviations or de-regressed proofs), and inability to properly account for genomic preselection, which altogether is said to generate bias and inaccuracy in genomic evaluations (Patry and Ducrocq, 2011; Legarra *et al.*, 2014). While these issues are still present today, they were especially expressed at the advent of genomic prediction when genotyping was expensive and only a handful of animals in population were genotyped.

To overcome these issues, Misztal *et al.* (2009) proposed the single-step method in which all the available information (pedigree, phenotypes and genomic information) would be fitted into a single MME jointly, resulting in the direct (G)EBV prediction for both genotyped and ungenotyped animals. Assume the following mixed linear model:

$$\mathbf{y} = \mathbf{Xb} + \mathbf{Wa} + \mathbf{e} \qquad (12.1)$$

where **y** is a vector of phenotypes or de-regressed breeding values, **a** is a vector of breeding values, and **W** is a design matrix that relates records to all animals (including genotyped and ungenotyped animals). Initial developments supposed **a** is portioned as $\mathbf{a}_1$ for ungenotyped animals and $\mathbf{a}_2$ for genotyped animals, and then:

$$\text{Var}\begin{bmatrix}\mathbf{a}_1\\ \mathbf{a}_2\end{bmatrix} = \begin{bmatrix}\mathbf{A}_{11} & \mathbf{A}_{12}\\ \mathbf{A}_{21} & \mathbf{G}\end{bmatrix}\sigma_a^2 = \mathbf{A} + \begin{bmatrix}0 & 0\\ 0 & \mathbf{G}-\mathbf{A}_{22}\end{bmatrix}\sigma_a^2$$

DOI: 10.1079/9781800620506.0012

where $\mathbf{A}_{22}$ is the relationship matrix only for the genotyped animals.

It has already been shown in Section 11.4.2 that $\mathbf{a}_2 = \mathbf{Zg}$ and $\text{var}(\mathbf{a}_2) = \mathbf{G}\sigma_a^2$. Based on selection index theory, $\mathbf{a}_1$ can be predicted from the genotyped animals (Legarra *et al.*, 2009) as:

$$\mathbf{a}_1 = \mathbf{A}_{12}\mathbf{A}_{22}^{-1}\mathbf{Zg} + \boldsymbol{\omega}$$

where $\boldsymbol{\omega}$ is the residual term, such that:

$$\text{Var}(\mathbf{a}_1) = \mathbf{A}_{12}\mathbf{A}_{22}^{-1}\mathbf{G}\mathbf{A}_{22}^{-1}\mathbf{A}_{21} + \mathbf{A}_{11} - \mathbf{A}_{12}\mathbf{A}_{22}^{-1}\mathbf{A}_{12}$$

and this reduces to

$$\text{Var}(\mathbf{a}_1) = \mathbf{A}_{11} + \mathbf{A}_{12}\mathbf{A}_{22}^{-1}(\mathbf{G} - \mathbf{A}_{22})\mathbf{A}_{22}^{-1}\mathbf{A}_{21}$$

Finally, $\text{Cov}(\mathbf{a}_1, \mathbf{a}_2) = \mathbf{A}_{12}\mathbf{A}_{22}^{-1}\mathbf{G}$

Putting all terms together into a matrix $\mathbf{H}$, a covariance matrix of breeding values including genomics information (Legarra *et al.*, 2009; Christensen and Lund, 2010) is:

$$\mathbf{H} = \begin{bmatrix} \mathbf{H}_{11} & \mathbf{H}_{12} \\ \mathbf{H}_{21} & \mathbf{H}_{22} \end{bmatrix} = \begin{bmatrix} \mathbf{A}_{11} + \mathbf{A}_{12}\mathbf{A}_{22}^{-1}(\mathbf{G} - \mathbf{A}_{22})\mathbf{A}_{22}^{-1}\mathbf{A}_{21} & \mathbf{A}_{12}\mathbf{A}_{22}^{-1}\mathbf{G} \\ \mathbf{G}\mathbf{A}_{22}^{-1}\mathbf{A}_{21} & \mathbf{G} \end{bmatrix}$$

The matrix $\mathbf{H}$ could be regarded as a matrix that combines pedigree and genomic relationships.

The single-step methodology involves the use of matrix $\mathbf{H}$, and Aguilar *et al.* (2010) and Christensen and Lund (2010) found the inverse of $\mathbf{H}$ has the following simple form:

$$\mathbf{H}^{-1} = \mathbf{A}^{-1} + \begin{bmatrix} 0 & 0 \\ 0 & \mathbf{G}^{-1} - \mathbf{A}_{22}^{-1} \end{bmatrix} \tag{12.2}$$

where $\mathbf{A}_{22}^{-1}$ is the inverse of the relationship matrix for genotyped animals. This implies that by replacing $\mathbf{A}^{-1}$ with $\mathbf{H}^{-1}$ in the usual MME, direct prediction of (G)EBVs and genomic evaluations can be obtained for ungenotyped and genotyped animals. Therefore, the MME for the single-step procedure (Eqn 12.1) are:

$$\begin{bmatrix} \mathbf{X'R}^{-1}\mathbf{X} & \mathbf{X'R}^{-1}\mathbf{W} \\ \mathbf{W'R}^{-1}\mathbf{X} & \mathbf{W'R}^{-1}\mathbf{W} + \mathbf{H}^{-1}\alpha \end{bmatrix} \begin{bmatrix} \hat{\mathbf{b}} \\ \hat{\mathbf{a}} \end{bmatrix} = \begin{bmatrix} \mathbf{X'R}^{-1}\mathbf{y} \\ \mathbf{W'R}^{-1}\mathbf{y} \end{bmatrix} \tag{12.3}$$

where:

$\alpha = \sigma_e^2 / \sigma_a^2$ and $\mathbf{R}^{-1}$ is the inverse of the diagonal elements of weights. If the weights are removed, MME simplifies to:

$$\begin{bmatrix} \mathbf{X'X} & \mathbf{X'W} \\ \mathbf{W'X} & \mathbf{W'W} + \mathbf{H}^{-1}\alpha \end{bmatrix} \begin{bmatrix} \hat{\mathbf{b}} \\ \hat{\mathbf{a}} \end{bmatrix} = \begin{bmatrix} \mathbf{X'y} \\ \mathbf{W'y} \end{bmatrix}.$$

The main advantage of the single-step approach is that existing software for genetic predictions can easily be modified to implement this method. The core of the ssGBLUP is the $\mathbf{H}$ matrix and the core of the $\mathbf{H}$ matrix is the $\mathbf{G} - \mathbf{A}_{22}$ block. This raises important complications in the single-step approach, as $\mathbf{G}$ must be on exactly the same scale (e.g. scaled to the same base animals) as $\mathbf{A}$, otherwise animals with genotypes will have biased GEBVs. Since $\mathbf{G}$ is usually constructed based on the observed allele frequencies of the recent genotyped population and not based on the allele frequencies of the base population, expectation of GEBVs of genotyped animals is assumed to be zero. This assumption is commonly false since domesticated animal populations have undergone a process of selection, which should consequently reduce the genetic variance and introduce changes to the average breeding values (Legarra *et al.*, 2014). It is worth mentioning that the base allele frequencies can be inferred by a method of Gengler *et al.* (2007) as presented in Chapter 11.4.6, but with limited accuracy (e.g. estimates out of bounds) and varying success depending on the data set. As a solution to this compatibility issue, Vitezica *et al.* (2011) proposed an adjustment factor (a) to match (or 'tune') the averages of **G** to the averages of $\mathbf{A}_{22}$. This adjustment factor can be represented as:

$$\mathrm{a} = \overline{\mathbf{A}}_{22} - \overline{\mathbf{G}}$$

and it is used to create a 'tuned' **G** matrix as:

$$\mathbf{G}_{\mathrm{t}} = \mathrm{b}\mathbf{G} + \mathbf{11}'a \text{ where } \mathrm{b} = 1 - \mathrm{a}/2.$$

In such a form, $\mathbf{G}_{\mathrm{t}}$ accounts for both changes in the mean due to genetic trend and reduction in genetic variance (Legarra *et al.*, 2014). Christensen *et al.* (2012) proposed to rather estimate these adjustment factors (a and b) from the system of equations:

$$\frac{\mathrm{trace}(\mathbf{G})}{n}\mathrm{b} + \mathrm{a} = \frac{\mathrm{trace}(A_{22})}{n}$$

and

$$\mathrm{a} + \mathrm{b}\overline{\mathbf{G}} = \overline{\mathbf{A}}_{22}$$

When Hardy-Weinberg equilibrium approximately holds, $\mathrm{b} \approx 1 - \mathrm{a}/2$, equivalently to Vitezica *et al.* (2011). While these commonly used tuning approaches are based on scaling **G** to $\mathbf{A}_{22}$, there is also an opposite approach that advocates scaling the $\mathbf{A}_{22}$ to match the **G** (Christensen, 2012).

The compatibility between the bases, i.e. the 'tuning', should be clearly differentiated from what is usually called a 'blending' step in the single-step procedure. In this step the original **G** matrix is 'blended' with the $\mathbf{A}_{22}$ matrix with two purposes. First, if it is assumed that a certain proportion of the genetic variance is not explained by the SNPs; this will add the residual polygenic effect. Following Legarra *et al.* (2014), GEBVs can be decomposed into a part explained by the SNPs ($\mathrm{a}_{\mathrm{m},2}$) and a residual part explained by the pedigree ($\mathrm{a}_{\mathrm{p},2}$) as $\mathrm{a}_2 = \mathrm{a}_{\mathrm{m},2} + \mathrm{a}_{\mathrm{p},2,}$ with corresponding variances $\sigma_a^2 = \sigma_{a,m}^2 + \sigma_{a,p}^2$. Based on that, the 'blended' **G** matrix can be constructed as: $\mathbf{G}_{\mathrm{wt}} = wt\mathbf{G} + (1 - wt)\mathbf{A}_{22}$ (VanRaden, 2008; Aguilar *et al.*, 2010; Christensen and Lund, 2010).

Secondly, this 'blending' will make the initial **G** invertible. The *wt* parameter varies across the studies and populations, with 0.95 being the most common choice (VanRaden, 2008). It is worth noting that adding a small number to the diagonal, for example as $\mathbf{G}_{\mathrm{wt}} = \mathbf{G} + 0.01\mathbf{I}$, would also make the matrix invertible while at the same time keeping $\mathbf{G}_{\mathrm{wt}}$ more similar to **G**.

Example 12.1

We used data from Example 11.1 and extended them to ungenotyped animals (1–12). The model is like the GBLUP model presented in the example, except a combined pedigree and genomic relationship matrix (**H**) is used. In practice, **H** matrix is usually not created, but rather $\mathbf{H}^{-1}$ is directly created due to its simple structure. In the example, we show a general procedure for setting up and solving a single-step GBLUP model. In this example, we first created the **G** matrix following Example 11.3, using $2\Sigma p_j(1 - p_j) = 3.5383$ and **Z** matrix for 14 animals and 10 SNPs as described in Example 11.2. That initial **G** was blended with $\mathbf{A}_{22}$ as $\mathbf{G}_b = 0.95\mathbf{G} + 0.05\mathbf{A}_{22}$, and afterwards tuned as $\mathbf{G}_t = 0.202 + 0.709\mathbf{G}_b$. The tuning parameters a and b were obtained as in Chen *et al.* (2011):

$$b = \frac{\bar{\mathbf{A}}_{22,diag} - \bar{\mathbf{A}}_{22}}{\bar{\mathbf{G}}_{diag} - \bar{\mathbf{G}}}$$

and

$$a = \bar{\mathbf{A}}_{22} - b\bar{\mathbf{G}}.$$

The reader is advised that various software implementations will have a slightly different order of presented steps. For example, some implementations might blend **G** before tuning and others might do the opposite. Also, as previously discussed, tuning and blending parameters may vary across the data sets. Tuned matrix $\mathbf{G}_t$ was then inverted, and the $\mathbf{G}_t - \mathbf{A}_{22}^{-1}$ block of the $\mathbf{H}^{-1}$ was created, following Eqn 12.2. The matrices in Eqn 12.3 have all been previously defined and solving these equations with $\alpha = 245/35.241 = 6.952$ gives the following solutions:

	SNP-BLUP	GBLUP	Single-step GBLUP
1	–	–	0.009
2	–	–	0.121
3	–	–	0.009
4	–	–	0.010
5	–	–	–0.250
6	–	–	–0.196
7	–	–	–0.001
8	–	–	0.026
9	–	–	–0.076
10	–	–	0.040
11	–	–	–0.137
12	–	–	0.136
13	0.070	0.069	0.048
14	0.111	0.116	0.082
15	0.045	0.049	0.039
16	0.253	0.260	0.202
17	0.495	–0.500	–0.356
18	–0.357	–0.359	–0.254
19	0.145	0.146	0.098
20	–0.224	–0.231	–0.169
21	0.027	0.028	0.018
22	0.114	0.115	0.079
23	–0.240	–0.240	–0.164
24	0.143	0.143	0.101
25	0.054	0.054	0.039
26	0.354	0.353	0.245

The solutions for animals with genotypes (13–26) are following a very similar ranking between the methods. An important distinction is that with single-step GBLUP, solutions for ungenotyped animals are available as well. The reader should note that due to differences in data structure and relationship matrices used, it is not possible to directly compare single-step GBLUP and GBLUP solutions.

12.2 Alternative Approaches

Equivalent to the single-step GBLUP, there is an alternative single-step method based on the direct estimation of the SNP effects. Those two models are sometimes considered as a breeding value model (BVM) and a marker effects model (MEM). Single-step MEM was proposed by Fernando *et al.* (2014) and named single-step Bayesian regression model (ssBR). The main premise of ssBR model is that marker effects are fitted directly for the genotyped animals, while for the ungenotyped animals they are imputed by including the 'imputation residual effect' to account for deviations between the true (unobserved) and imputed marker genotypes. Following Fernando *et al.* (2014), the ssBR model can be derived from a general form:

$$\begin{bmatrix} \mathbf{y}_n \\ \mathbf{y}_g \end{bmatrix} = \begin{bmatrix} \mathbf{X}_n \\ \mathbf{X}_g \end{bmatrix} \mathbf{b} + \begin{bmatrix} \mathbf{Z}_n & 0 \\ 0 & \mathbf{Z}_g \end{bmatrix} \begin{bmatrix} \mathbf{M}_n \boldsymbol{\alpha} + \boldsymbol{\epsilon} \\ \mathbf{M}_g \boldsymbol{\alpha} \end{bmatrix} + \mathbf{e}$$

where subscripts *n* and *g* denote ungenotyped and genotyped animals, respectively, **y** are vectors of phenotypes, **X** are the incidence matrices for the fixed effects **b**, **Z** are incidence matrices that relate breeding values of animals to the phenotypic values, and **e** is the vector of residuals. Furthermore, the breeding values for ungenotyped and genotyped animals are:

$$\mathbf{u}_n = \mathbf{M}_n \boldsymbol{\alpha} + \boldsymbol{\epsilon}$$

and

$$\mathbf{u}_g = \mathbf{M}_g \boldsymbol{\alpha}$$

where $\mathbf{M}_g$ is matrix of centred marker genotypes, $\mathbf{M}_n$ is matrix of imputed marker genotypes, $\boldsymbol{\alpha}$ is the vector of marker effects, and $\boldsymbol{\epsilon}$ is a vector of imputation residual effects. The purpose of this imputation residual term is to account for deviations between the true (unobserved) and imputed marker genotypes, and it is assumed to be normally distributed as:

$$\boldsymbol{\varepsilon} \sim N[0, (\mathbf{A}_{nn} - \mathbf{A}_{ng} \mathbf{A}_{gg}^{-1} \mathbf{A}_{gn}) \sigma_g^2]$$

Here, the pedigree relationship matrix is partitioned into ungenotyped and genotyped individuals:

$$\mathbf{A} = \begin{bmatrix} \mathbf{A}_{nn} & \mathbf{A}_{ng} \\ \mathbf{A}_{gn} & \mathbf{A}_{gg} \end{bmatrix}$$

and has the corresponding inverse:

$$\mathbf{A}^{-1} = \begin{bmatrix} \mathbf{A}^{nn} & \mathbf{A}^{ng} \\ \mathbf{A}^{gn} & \mathbf{A}^{gg} \end{bmatrix}.$$

An important detail to notice is that $\mathbf{M}_n$ is empty or rather unknown and must be inferred and can therefore be considered a matrix of imputed marker genotypes. It can be represented as:

$$\hat{\mathbf{M}}_n = \mathbf{A}_{ng} \mathbf{A}_{gg}^{-1} \mathbf{M}_g,$$

and it is obtained by iteratively solving the system of equations:

$$\mathbf{A}^{nn} \hat{\mathbf{M}}_n = -\mathbf{A}^{ng} \mathbf{M}_g$$

Finally, the complete MME for the ssBR model are:

$$\begin{bmatrix} X'X & X'ZM & X_n'Z_n \\ X'ZM & M'Z'ZM + I\frac{\sigma_e^2}{\sigma_\alpha^2} & M_n'Z_n'Z_n \\ X_n'Z_n & M_n'Z_n'Z_n & Z_n'Z_n + A^{nn}\frac{\sigma_e^2}{\sigma_g^2} \end{bmatrix} \begin{bmatrix} \hat{b} \\ \hat{\alpha} \\ \hat{\varepsilon} \end{bmatrix} = \begin{bmatrix} X'y \\ M'Z'y \\ Z_n'y_n \end{bmatrix}$$

where σ_α^2 is marker effects variance, σ_g^2 is the additive genetic variance, and σ_e^2 is the residual variance, $y = \begin{bmatrix} y_n \\ y_g \end{bmatrix}$, $X = \begin{bmatrix} X_n \\ X_g \end{bmatrix}$, $Z = \begin{bmatrix} Z_n & 0 \\ 0 & Z_g \end{bmatrix}$, and $M = \begin{bmatrix} M_n \\ M_g \end{bmatrix}$.

Eventually, Fernando *et al.* (2016) also proposed an optimized ssBR model usually referred to as a 'hybrid' model with somewhat better computational properties, where genotyped animals are processed by MEM and ungenotyped animals by BVM, thus the name 'hybrid' model. The difference between the hybrid model and the ssBR model is that for ungenotyped individuals, breeding values are fitted directly instead of their imputation residual term. Computational variation of MEM is an 'on-the-fly' genotypes imputation approach (Taskinen *et al.*, 2017) that eases computations as only the true genotypes (genotyped animals) are stored in the memory while for the ungenotyped animals they are imputed implicitly as needed during each round of iterative solving without storing them.

Like the developments of Fernando *et al.* (2014), an alternative single-step MEM with incorporated polygenic residual effect was presented by Liu *et al.* (2014, 2016) and was named single-step SNP-BLUP (ssSNP-BLUP). Rather than imputing the marker genotypes for ungenotyped animals, this model simultaneously estimates breeding values and marker effects. Such a model can be expressed by MME as:

$$\begin{bmatrix} X'X & X'W_n & X'W_g & 0 \\ W_n'X & W_n'W_n + A^{nn}\frac{\sigma_e^2}{\sigma_g^2} & A^{ng}\frac{\sigma_e^2}{\sigma_g^2} & 0 \\ W_g'X & A^{gn}\frac{\sigma_e^2}{\sigma_g^2} & W_g'W_g + (A^{gg} + (\frac{1}{k} - 1)A_{gg}^{-1}\frac{\sigma_e^2}{\sigma_g^2} & -\frac{1}{k}A_{gg}^{-1}Z\frac{\sigma_e^2}{\sigma_g^2} \\ 0 & 0 & -\frac{1}{k}Z'A_{gg}^{-1}\frac{\sigma_e^2}{\sigma_g^2} & (B^{-1} + \frac{1}{k}Z'A_{gg}^{-1}Z)\frac{\sigma_e^2}{\sigma_g^2} \end{bmatrix} \begin{bmatrix} \hat{b} \\ \hat{u}_n \\ \hat{u}_g \\ \hat{g} \end{bmatrix}$$

$$= \begin{bmatrix} X'y \\ W_n'y \\ W_g'y \\ 0 \end{bmatrix}$$

where **X** is the incidence matrix for fixed effects, **W** are the incidence matrices for additive genetic effects for ungenotyped (n) and genotyped (g) individuals, **B** is the covariance matrix for marker effects (size is m × m, where m is the number of markers), **Z** is the matrix of marker genotypes, **b** is the vector of fixed effects, **u** are the vectors of additive genetic effects, **g** is the vector of marker effects, σ_g^2 is the additive genetic variance, and σ_e^2 is the residual variance. In the case when the marker effects are assumed to be normally distributed and uncorrelated (i.e. SNP-BLUP), **B** is diagonal matrix, and reduces to $\left(\frac{1-k}{m}\right)\mathbf{I}$ (Liu *et al.*, 2014). The parameter k [0, 1] is used to control the residual polygenic effect and can be considered as a proportion of additive genetic variance not explained by the markers like the *wt* parameter described in Section 12.1.

There are two main benefits of single-step MEM models compared to single-step BVM. The MEM can bypass the construction and inverse of **G** matrix and can fit various priors for the marker effects such as Bayesian mixture models (BayesB, BayesC etc.). On the other hand, the widespread implementation is somewhat limited due to various convergence issues and high computing costs related to the Markov chain Monte Carlo (MCMC) techniques.

12.3 Groups and Metafounders in Single-step Procedures

A pedigree-based animal model with groups was presented in Chapter 4, together with its importance to avoid the bias in the prediction of the breeding values. The recent move towards the single-step GBLUP evaluations has raised the issue of how to define and model unknown parent groups (UPG) in such a system of equations and is still an active area of research. The approaches that have been examined can roughly be divided into two groups: (i) approach based on expanding the traditional UPG modelling to novel genomic models (Misztal *et al.*, 2013b); and (ii) an alternative 'metafounders' approach (Legarra *et al.*, 2015). Both methods are recently reviewed by Masuda *et al.* (2022).

Misztal *et al.* (2013b) presented an approach to incorporate UPG into ssGBLUP by extending the traditional UPG model by substituting the **A** matrix with the **H** matrix. If we consider the same model presented in Chapter 4, MME for groups in ssGBLUP are:

$$\begin{bmatrix} X'X & X'Z & X'ZQ \\ Z'X & Z'Z + H^{-1}\alpha & Z'ZQ \\ Q'Z'X & Q'Z'Z & Q'Z'ZQ \end{bmatrix} \begin{bmatrix} \hat{b} \\ \hat{a} \\ \hat{g} \end{bmatrix} = \begin{bmatrix} X'y \\ Z'y \\ Q'Z'y \end{bmatrix}$$

Then the modified MME by the QP-transformation (Quaas and Pollak, 1981) are:

$$\begin{bmatrix} X'X & X'Z & 0 \\ Z'X & Z'Z + H^{-1}\alpha & -H^{-1}Q\alpha \\ 0 & -Q'H^{-1}\alpha & Q'H^{-1}Q\alpha \end{bmatrix} \begin{bmatrix} \hat{b} \\ \hat{a} + Q\hat{g} \\ \hat{g} \end{bmatrix} = \begin{bmatrix} X'y \\ Z'y \\ 0 \end{bmatrix}$$

While in the case of the pedigree-based group model **Q**, the matrix does not need to be directly constructed, but rather a set of rules can be used (see Chapter 4), this is not the case with the ssGBLUP. In the case of ssGBLUP, the LHS of the coefficient matrix of the MME (ignoring the fixed effects) becomes:

$$\begin{bmatrix} Z_1'Z_1 & Z_1'Z_2 & 0 \\ Z_2'Z_1 & Z_2'Z_2 & 0 \\ 0 & 0 & 0 \end{bmatrix} + \begin{bmatrix} A^{11} & A^{12} & -\begin{bmatrix} A^{11} & A^{12} \end{bmatrix} Q \\ A^{21} & A^{22} & -\begin{bmatrix} A^{21} & A^{22} \end{bmatrix} Q \\ -Q'\begin{bmatrix} A^{11} \\ A^{21} \end{bmatrix} & -Q'\begin{bmatrix} A^{12} \\ A^{22} \end{bmatrix} & Q'A^{-1}Q \end{bmatrix} +$$

$$\begin{bmatrix} 0 & 0 & 0 \\ 0 & G^{-1} - A_{22}^{-1} & -\left(G^{-1} - A_{22}^{-1}\right)Q_2 \\ 0 & -Q_2'\left(G^{-1} - A_{22}^{-1}\right) & Q_2'\left(G^{-1} - A_{22}^{-1}\right)Q_2 \end{bmatrix} = C1 + C2 + C3$$

where Q_2 is the matrix assigning genotyped animals to the groups. The 1 denotes ungenotyped animals, and 2 genotyped animals. Here, the UPG are fitted to **A** as presented in Chapter 4 and as illustrated with block **C2** in MME, and then in **C3** extended to genomic information. In that sense, $C2 + C3$ components constitute the $\mathbf{H}^{-1}$, following principles as in Eqn 12.2. The other matrices are the same as defined before. This approach will yield the same results as fitting the UPGs as covariates directly in the ssGBLUP model.

There is also a modified version, where the purpose of UPG is to compensate for missing the pedigree relationships only and not for the genomic relationships, so the UPG are modelled for **A** component (**C1**) and $\mathbf{A}_{22}$ component (within **C3**) only. This is the currently recommended version in use for national evaluations. The evaluations are the same as before, but the G^{-1} is removed from the presented equations, and $C3$ becomes:

$$C3 = \begin{bmatrix} 0 & 0 & 0 \\ 0 & G^{-1} - A_{22}^{-1} & -\left(-A_{22}^{-1}\right)\mathbf{Q}_2 \\ 0 & -\mathbf{Q'}_2\left(-A_{22}^{-1}\right) & \mathbf{Q'}_2\left(-A_{22}^{-1}\right)\mathbf{Q}_2 \end{bmatrix}$$

Like before, $C2 + C3$ components constitute the $\mathbf{H}^{-1}$, following principles as in Eqn 12.2.

An alternative approach to model the groups is the metafounders approach proposed by Legarra *et al.* (2015). A metafounder can be considered as pseudo-individual who is the father and mother of all base animals, and details of this approach are provided in Chapter 3, Section 8. Such an approach can be considered a generalization of the UPG model in which related and possibly inbreed founder animals, i.e. the metafounders, are added to the model as the substitute for the UPG. The relationship coefficients between those pseudo-animals are commonly denoted by the covariance matrix Γ, or ancestral relationship matrix, whose elements are a function of covariance between allele frequencies across all loci (p) in metafounder populations i and j, such that elements are $\Gamma_{i,j} = 8cov(p_i, p_j)$; see section 3.8.4 for the derivation. Those covariances among metafounders can be accurately estimated by a generalized least-squares or maximum-likelihood approach (Garcia-Baccino *et al.*, 2017). In the context of single-step GBLUP, metafounders are implemented by modification of the H matrix, whose inverse is then:

$$H^{\Gamma-1} = A^{\Gamma-1} + \begin{bmatrix} 0 & 0 \\ 0 & G^{-1} - A_{22}^{\Gamma-1} \end{bmatrix}$$

The metafounders theory assumes that the **G** is constructed using fixed-allele frequency of 0.5, rather than the observed allele frequencies from the recent genotyped population.

Masuda *et al.* (2021) provided a novel derivation of the $\mathbf{H}^{-1}$ matrix incorporating UPG as the random effects, named 'UPG-encapsulated $\mathbf{H}^{-1}$' (EUPG). In the EUPG model, the LHS of the coefficient matrix of the MME (ignoring the fixed effects) is the same as before, but with modification in the genomic part ($\mathbf{C3}$):

$$C3 = \begin{bmatrix} 0 & 0 & 0 \\ 0 & G^{-1} - A_{22}^{*} & 0 \\ 0 & 0 & 0 \end{bmatrix}$$

where

$$A_{22}^{*} = A^{22} - \begin{bmatrix} A^{21} & \left(-\begin{bmatrix} A^{21} & A^{22} \end{bmatrix} Q\right) \end{bmatrix} \begin{bmatrix} A^{11} & \left(-\begin{bmatrix} A^{11} & A^{12} \end{bmatrix} Q\right) \\ \left(Q' \begin{bmatrix} A^{11} \\ A^{21} \end{bmatrix}\right) & \left(Q'A^{-1}Q + \Sigma^{-1}\right) \end{bmatrix} \begin{bmatrix} A^{12} \\ \left(Q' \begin{bmatrix} A^{12} \\ A^{22} \end{bmatrix}\right) \end{bmatrix}$$

Here the groups are assumed to be a random effect and following a multivariate normal distribution with mean zero and covariance $\Sigma\sigma_u^2$. Inbreeding coefficients and additive relationships between the UPG that form the Σ are estimated by a recursive method (VanRaden, 1992; Aguilar and Misztal, 2008). If UPG are assumed to be unrelated (or non-inbreed), $\Sigma = I$ can be used. Masuda *et al.* (2021) also presented the proof that such inverse has the same structure as the H^{-1} obtained with the metafounders approach.

12.4 APY Approach

The simplicity of the ssGBLUP method is that, 'under the hood', it is technically a conventional BLUP with A^{-1} replaced by the H^{-1}. However, the computation of H^{-1} requires efficient computation of G^{-1}. Thus, this is a major limitation, with large numbers of animals genotyped, since there are no simple rules for computing the inverse of G directly. The computation cost for G^{-1} is cubic in the number of genotyped animals and the computational limit for the current systems is between 100,000 and 150,000 genotyped animals. Nowadays, some organizations and commercial companies have millions of genotyped animals in their databases, with an ever-growing genotyping trend from year to year. Running ssGBLUP evaluations with such big data became unfeasible so Misztal *et al.* (2014) proposed to approximate inverse of G by Algorithm for Proven and Young (APY). The base of the APY approach is the recursion and division of the genotyped animals into 'proven' and 'young' subsets. In the current implementation, those categories are named as 'core' (c) and 'non-core' (n) animals with the G matrix partitioned as:

$$\mathbf{G} = \begin{bmatrix} \mathbf{G}_{cc} & \mathbf{G}_{cn} \\ \mathbf{G}_{nc} & \mathbf{G}_{nn} \end{bmatrix} \tag{12.4}$$

A theoretical framework and detailed derivations were further developed by Misztal (2016) via the concept of limited dimensionality of genomic information, where the breeding values of the non-core animals are expressed as a linear function of the breeding values of core animals:

$$\mathbf{u}_n = \mathbf{P}_{nc}\mathbf{u}_c + \Phi_n,$$

where $\mathbf{P}_{nc}$ is a matrix relating breeding values of non-core to core animals and Φ_n is the estimation error term. Subsequently, in the matrix notation:

$$\begin{bmatrix} \mathbf{u}_c \\ \mathbf{u}_n \end{bmatrix} = \begin{bmatrix} \mathbf{I} & 0 \\ \mathbf{P}_{nc} & \mathbf{I} \end{bmatrix} \begin{bmatrix} \mathbf{u}_c \\ \mathbf{\Phi}_n \end{bmatrix}$$

Based on this, the G can be written as:

$$\operatorname{var}(\mathbf{u}) = \sigma_a^2 \mathbf{G}_{APY} = \begin{bmatrix} \mathbf{I} & 0 \\ \mathbf{P}_{nc} & \mathbf{I} \end{bmatrix} \begin{bmatrix} \mathbf{G}_{cc} & 0 \\ 0 & \mathbf{M}_{nn} \end{bmatrix} \begin{bmatrix} \mathbf{I} & \mathbf{P}_{cn} \\ 0 & \mathbf{I} \end{bmatrix}$$

where $\mathbf{M}_{nn} = \text{var}(\mathbf{\Phi}_n)$ and its inverse is:

$$\mathbf{G}_{APY}^{-1} = \begin{bmatrix} \mathbf{I} & -\mathbf{P}_{cn} \\ 0 & \mathbf{I} \end{bmatrix} \begin{bmatrix} \mathbf{G}_{cc}^{-1} & 0 \\ 0 & \mathbf{M}_{nn}^{-1} \end{bmatrix} \begin{bmatrix} \mathbf{I} & 0 \\ -\mathbf{P}_{nc} & \mathbf{I} \end{bmatrix} / \sigma_a^2$$

Using conditional distributions, $\mathbf{P}_{nc} = \mathbf{G}_{nc}\mathbf{G}_{cc}^{-1}$ and $\mathbf{M}_{nn,ii} = \mathbf{g}_{ii} - \mathbf{g}_{ic}\mathbf{G}_{cc}^{-1}\mathbf{g}_{ci}$ (12.5)

where g_{ii} is the ith diagonal element of $\mathbf{G}_{nn}$ and $\mathbf{g}_{ci} = \mathbf{g}'_{ci}$ is the ith column of $\mathbf{G}_{cn}$ (or the ith row of $\mathbf{G}_{nc}$), the final APY inverse equation is as originally defined in Misztal *et al.* (2014):

$$\mathbf{G}_{APY}^{-1} = \begin{bmatrix} \mathbf{G}_{cc}^{-1} & 0 \\ 0 & 0 \end{bmatrix} + \begin{bmatrix} -\mathbf{G}_{cc}^{-1}\mathbf{G}_{cn} \\ \mathbf{I} \end{bmatrix} \mathbf{M}_{nn}^{-1} \begin{bmatrix} -\mathbf{G}_{nc}\mathbf{G}_{cc}^{-1} & \mathbf{I} \end{bmatrix} \quad (12.6)$$

In this equation, the direct inversion is needed only for the $\mathbf{G}_{cc}$ block and the diagonal matrix M_{nn}. Therefore, the inversion by APY has a cubic computational cost for the core animals and the linear cost for the non-core animals.

There are two major questions with the application of APY method: (i) how many; and (ii) which animals to choose as a core animals. One of the proposed theories links the number of core individuals with the dimensionality of genomic information that can be assessed by the eigenvalue decomposition of **G** or equivalently by the singular value decomposition of **Z** (Misztal, 2016; Pocrnic *et al.*, 2016). For example, by setting the number of core individuals to the number of eigenvalues that explain more than 98% of the variance in **G**, GEBV obtained by the APY inverse were comparable to those obtained by the standard inverse (Pocrnic *et al.*, 2016). Empirical testing suggested that the definition of core subset is quite robust, and that random selection of core individuals gives satisfactory results (e.g. Fragomeni *et al.*, 2015; Bradford *et al.*, 2017). As the random sampling of core animals could generate fluctuations between subsequent evaluations, more optimal core selection methods are becoming the subject of current research (e.g. Pocrnic *et al.*, 2022).

Example 12.2

In this example, we will illustrate how to invert the **G** matrix from Example 11.3 using the APY approach. The first step is to set the number of core animals. For that, we randomly selected four animals in a core subset (animals 15, 20, 22 and 26), while another ten animals were considered a non-core subset. Then, following Eqn 12.4, the **G** matrix from Example 11.3 is partitioned into the core block (animals 15, 20, 22 and 26):

$$\mathbf{G}_{cc} = \begin{bmatrix} 1.634 & & & \\ -0.183 & 0.826 & \textit{symmetric} & \\ -0.708 & -0.264 & 1.190 & \\ -0.466 & -0.304 & 0.584 & 1.109 \end{bmatrix}$$

and the non-core block (animals 13, 14, 16, 17, 18, 19, 21, 23, 24 and 25):

$$\mathbf{G}_{cn} = \begin{bmatrix} 0.988 & -0.930 & 0.422 & 1.048 & -0.365 & -0.627 & 0.120 & 0.160 & -0.345 & -0.748 \\ -0.547 & 0.079 & 0.301 & -0.203 & 0.079 & 0.382 & -0.567 & 0.604 & -0.183 & -0.022 \\ -0.789 & 0.402 & -0.506 & -0.446 & -0.163 & 0.140 & -0.526 & -0.486 & 0.705 & 0.867 \\ -0.264 & 0.362 & -0.264 & -0.486 & -0.203 & 0.099 & -0.284 & -0.526 & 0.382 & 0.261 \end{bmatrix}$$

The $\mathbf{G}_{nc}$ is simply the transpose of $\mathbf{G}_{cn}$, so we do not show it. It is easy to see that a non-core block of **G** is connecting non-core animals (13, 14, 16, 17, 18, 19, 21, 23, 24 and 25) to core animals (15, 20, 22 and 26). The last component needed is the diagonal matrix **M** for non-core animals following Eqn 12.5, where diag(**M**) = (0.234, 0.204, 0.600, 0.848, 0.485, 0.420, 1.107, 0.104, 0.644, 0.102). Direct inversion is need for the core block

$$\mathbf{G}_{cc}^{-1} = \begin{bmatrix} 0.987 & & & \\ 0.494 & 1.616 & \textit{symmetric} & \\ 0.576 & 0.449 & 1.489 & \\ 0.247 & 0.414 & -0.419 & 1.340 \end{bmatrix}$$

and for the diagonal matrix **M**, which has a trivial inverse ($1/\mathbf{M}_{nn}$).

Finally, we use the components ($\mathbf{G}_{cc}^{-1}, \mathbf{G}_{cn}, \mathbf{G}_{nc}, \mathbf{M}_{nn}^{-1}$) inside the Eqn 12.6 to construct the APY inverse:

$$\mathrm{G}_{\mathrm{APY}}^{-1} = \begin{bmatrix}
4.425 & 0.373 & -0.527 & 0.412 & -0.795 & 2.731 & -0.349 & -0.658 & 0.959 & 0.776 & 0.483 & -0.447 & -0.108 & 1.828 \\
0.373 & 0.412 & 3.893 & -0.299 & 3.668 & 0.003 & -0.599 & 0.250 & 0.431 & -0.981 & 1.094 & -5.949 & -0.014 & -0.910 \\
-0.527 & 3.893 & 10.984 & -1.202 & 3.161 & 0.259 & 0.441 & -0.062 & 0.684 & 0.055 & 0.768 & 1.341 & -0.946 & -7.269 \\
0.412 & -0.299 & -1.202 & 2.655 & 0.024 & -0.585 & -0.146 & 0.342 & 0.539 & -0.186 & 0.330 & 2.035 & -0.086 & 2.034 \\
-0.795 & 3.668 & 3.161 & 0.024 & 4.272 & \square & \square & \square & \square & \square & \square & \square & \square & \square \\
2.731 & 0.003 & 0.259 & -0.585 & \square & 4.893 & \square & \square & \square & \square & \square & \square & \square & \square \\
-0.349 & -0.599 & 0.441 & -0.146 & \square & \square & 1.667 & \square & \square & \square & \square & \square & \square & \square \\
-0.658 & 0.250 & -0.062 & 0.342 & \square & \square & \square & 1.180 & \square & \square & \square & \square & \square & \square \\
0.959 & 0.431 & 0.684 & 0.539 & \square & \square & \square & \square & 2.061 & \square & \square & \square & \square & \square \\
0.776 & -0.981 & 0.055 & -0.186 & \square & \square & \square & \square & \square & 2.382 & \square & \square & \square & \square \\
0.483 & 1.094 & 0.768 & 0.330 & \square & \square & \square & \square & \square & \square & 0.904 & \square & \square & \square \\
-0.447 & -5.949 & 1.341 & 2.035 & \square & \square & \square & \square & \square & \square & \square & 9.611 & \square & \square \\
-0.108 & -0.014 & -0.946 & -0.086 & \square & \square & \square & \square & \square & \square & \square & \square & 1.554 & \square \\
1.828 & -0.910 & -7.269 & 2.034 & \square & \square & \square & \square & \square & \square & \square & \square & \square & 9.824
\end{bmatrix}$$

Here, shaded in red, we can directly recognize $\mathrm{M}_{\mathrm{nn}}^{-1}$, while shaded in blue and grey are the product of matrix multiplications from the Eqn 12.6. For example, the block shaded in blue is a result of $\mathrm{G}_{\mathrm{cc}}^{-1} + \mathrm{G}_{\mathrm{cc}}^{-1}\mathrm{G}_{\mathrm{cn}}\mathrm{M}_{\mathrm{nn}}^{-1}\mathrm{G}_{\mathrm{nc}}\mathrm{G}_{\mathrm{cc}}^{-1}$, and the upper block shaded in grey is a result of $-\mathrm{G}_{\mathrm{cc}}^{-1}\mathrm{G}_{\mathrm{cn}}\mathrm{M}_{\mathrm{nn}}^{-1}$.

The DGVs obtained using the regular **G** inverse are the same as previously reported in Example 12.1, Example 11.3 and Table 11.2. The DGVs obtained using the APY inverse for animals 13–26 are 0.065, 0.119, 0.030, 0.365, 0.471, –0.244, –0.055, –0.220, 0.010, 0.127, –0.229, 0.097, –0.007 and 0.363, respectively. In this example, the correlations between DGVs obtained from the full and APY inverse are 0.95, and there was re-ranking of animals. The reader should be advised that this example is of purely illustrative nature due to the small size of the **G** matrix (14 animals genotyped for 10 SNPs). In the real world, we construct the core subset such that the resulting correlations between DGVs with the full and APY inverse pass 0.99 and no significant re-ranking of animals is observed. Nevertheless, even from this small example, it is possible to recognize the computational benefits arising from the sparse structure of the APY inverse.

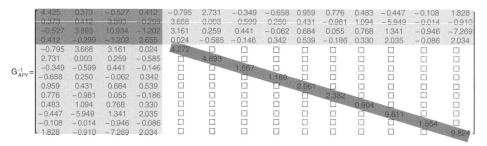

13 Non-additive Animal Models

13.1 Introduction

The models considered in the previous chapters have dealt with only additive genetic effects. However, the total genetic value of an animal is a function of both additive and non-additive effects and both effects could provide a better prediction of future phenotypes of individual animals. The availability of estimates of non-additive genetic effects for individuals could be useful in mate allocation and the performance of livestock can be improved through assortative mating when either dominance or epistasis, or both, are present (Toro and Varona, 2010; Aliloo *et al.*, 2017).

Henderson (1985) provided a statistical framework for modelling additive and non-additive genetic effects (dominance and epistatic effects) using the mixed-model methodology with pedigree information when there is no inbreeding.

Prior to the availability and use of genomic information, the application of non-additive models in livestock breeding programmes has been limited due to difficulties of precisely estimating genetic parameters due to limited pedigree structure. Usually, large full-sib families to achieve accurate estimation of non-additive genetic effects are needed and these effects tend to be highly confounded with others, such as common maternal environment.

However, the advent of the genomic era has created renewed interest in the estimation of variances for non-additive effects and their utilization in breeding programmes, especially in pigs and poultry (Su *et al.*, 2012; Wellmann and Bennewitz, 2012). Genomic information makes it easier to determine which animals are heterozygotes at each locus evaluated but the prediction of the genotypic value of future matings is also more straightforward (Toro and Varona, 2010).

Initially, the application of mixed-model methodology for the utilization of non-additive effects is outlined and methodologies based on genomic information are subsequently presented. Reviews at the time of writing are Varona *et al.* (2018, 2022). Extension to F1 crosses is in González-Diéguez *et al.* (2021) for hybrid crops and Christensen *et al.* (2015) for hybrid animals.

13.2 Dominance Relationship Matrix Using Pedigree Information

Dominance genetic effects result from the action of pairs of alleles at a locus on a trait. If two animals have the same set of parents or grandparents, it is possible that they possess the pair of alleles in common. The dominance relationship between two such animals represents the probability that they have the same pair of alleles in common. Thus, for a group of animals, the dominance genetic relationship matrix (**D**) among them can be set

DOI: 10.1079/9781800620506.0013

up. The dominance relationship between an individual x with parents s and d, and y with parents f and m in a non-inbred population can be calculated (Cockerham, 1954) as:

$$d_{xy} = 0.25(u_{sf}\ u_{dm} + u_{sd}\ u_{fm}) \quad (13.1)$$

where u_{ij} represents the additive genetic relationship between i and j. For instance, for two full-sibs with both parents unrelated to each other:

$$d = 0.25[(1)(1) + (0)(0)] = 0.25$$

with the assumption that there is no common environmental variance.

Thus, **D** can be generated from the additive genetic relationship in non-inbred pedigrees. However, the prediction of dominance effects requires the inverse of **D**. This could be obtained by calculating **D** by Eqn 13.1 and inverting it: this is not computationally feasible with large data sets. Hoeschele and VanRaden (1991) developed a methodology for obtaining a rapid inversion of **D** and more information on this method is given in Section 13.3.2. Initially, the principles involved in using $\mathbf{D}^{-1}$ from Eqn 13.1 for the prediction of dominance effects are discussed.

13.3 Animal Model with Dominance Effect

The model with dominance included is:

$$\mathbf{y} = \mathbf{Xb} + \mathbf{Za} + \mathbf{Wd} + \mathbf{e} \quad (13.2)$$

where **y** = vector of observations, **b** = vector of fixed effects representing the mean, **a** and **d** = vectors for genetic additive and dominance deviation effects, and **e** = random residual error.

It is assumed that:

$$\text{var}(\mathbf{a}) = \mathbf{A}\sigma_a^2,\ \text{var}(\mathbf{d}) = \mathbf{D}\sigma_d^2 \quad \text{and} \quad \text{var}(\mathbf{e}) = \sigma_e^2$$
$$\text{var}(\mathbf{y}) = \mathbf{ZAZ}' + \mathbf{WDW}' + \mathbf{I}\sigma_e^2$$

The MME to be solved for the BLUP of **a** and **d** and the BLUE of **b** are:

$$\begin{bmatrix} X'X & X'Z & X'W \\ Z'X & Z'Z + A^{-1}\alpha_1 & Z'W \\ W'X & W'Z & W'W + \mathrm{D}^{-1}\alpha_2 \end{bmatrix} \begin{bmatrix} \hat{b} \\ \hat{a} \\ \hat{\mathrm{d}} \end{bmatrix} = \begin{bmatrix} X'y \\ Z'y \\ W'y \end{bmatrix} \quad (13.3)$$

with $\alpha_1 = \sigma_e^2 / \sigma_a^2$ and $\alpha_2 = \sigma_e^2 / \sigma_d^2$. When considering only animals with records, **Z** = **W** in Eqn 13.3. However, we are interested in the total genetic merit (**g**) of the animal, which is **g** = a + **d**. The MME could be modified such that the total genetic merit is solved directly. Since **g** = a + **d**, then:

$$var(g) = \mathbf{M} = \mathbf{A}\sigma_a^2 + \mathbf{D}\sigma_d^2$$

The MME become:

$$\begin{bmatrix} X'X & X'Z \\ Z'X & Z'Z + \mathrm{M}^{-1}\sigma_e^2 \end{bmatrix} \begin{bmatrix} \hat{b} \\ \hat{g} \end{bmatrix} = \begin{bmatrix} X'y \\ Z'y \end{bmatrix} \quad (13.4)$$

The individual components of **g** can be obtained as:

$$\hat{a} = \sigma_a^2 \mathbf{AM}^{-1}\hat{g} \quad \text{and} \quad \hat{d} = \sigma_d^2 \mathbf{DM}^{-1}\hat{g}$$

13.3.1 Solving for additive and dominance genetic effects separately

Example 13.1
Suppose the data below are the weaning weights for some piglets in a herd.

Pig	Sire	Dam	Pen	Weaning weight (kg)
5	1	2	1	17.0
6	3	4	1	20.0
7	6	5	1	18.0
8	0	5	1	13.5
9	3	8	2	20.0
10	3	8	2	15.0
11	6	8	2	25.0
12	6	8	2	19.5

The aim is to estimate pen effects and predict solutions for genetic additive (or breeding, BV) and dominance deviation (DV) effects, assuming that $\sigma_e^2 = 120$, $\sigma_a^2 = 90$ and $\sigma_d^2 = 80$. This has been illustrated below, solving for additive and dominance effects separately (Eqn 13.3). From the above parameters, α_1 = 1.333 and α_2 = 1.5.

Setting up the MME

The matrix **X** relates records to pen effects. Its transpose, considering only animals with records, is:

$$\mathbf{X}' = \begin{bmatrix} 1 & 1 & 1 & 1 & 0 & 0 & 0 & 0 \\ 0 & 0 & 0 & 0 & 1 & 1 & 1 & 1 \end{bmatrix}$$

The matrices **Z** and **W** are both identity matrices since each animal has one record. The transpose of the vector of observations **y′** = [17 20 18 13.5 20 15 25 19.5].

The other matrices in the MME, apart from $\mathbf{A}^{-1}$ and $\mathbf{D}^{-1}$, can be obtained through matrix multiplication from the matrices already calculated. The inverse of the additive relationship matrix is set up using rules outlined in Section 3.5.1. Using Eqn 13.1, the dominance relationship matrix is:

$$\mathbf{D} = \begin{bmatrix} 1.000 & 0.000 & 0.000 & 0.000 & 0.000 & 0.000 & 0.000 & 0.000 & 0.000 & 0.000 & 0.000 & 0.000 \\ 0.000 & 1.000 & 0.000 & 0.000 & 0.000 & 0.000 & 0.000 & 0.000 & 0.000 & 0.000 & 0.000 & 0.000 \\ 0.000 & 0.000 & 1.000 & 0.000 & 0.000 & 0.000 & 0.000 & 0.000 & 0.000 & 0.000 & 0.000 & 0.000 \\ 0.000 & 0.000 & 0.000 & 1.000 & 0.000 & 0.000 & 0.000 & 0.000 & 0.000 & 0.000 & 0.000 & 0.000 \\ 0.000 & 0.000 & 0.000 & 0.000 & 1.000 & 0.000 & 0.000 & 0.000 & 0.000 & 0.000 & 0.000 & 0.000 \\ 0.000 & 0.000 & 0.000 & 0.000 & 0.000 & 1.000 & 0.000 & 0.000 & 0.000 & 0.000 & 0.000 & 0.000 \\ 0.000 & 0.000 & 0.000 & 0.000 & 0.000 & 0.000 & 1.000 & 0.000 & 0.062 & 0.062 & 0.125 & 0.125 \\ 0.000 & 0.000 & 0.000 & 0.000 & 0.000 & 0.000 & 0.000 & 1.000 & 0.000 & 0.000 & 0.000 & 0.000 \\ 0.000 & 0.000 & 0.000 & 0.000 & 0.000 & 0.000 & 0.062 & 0.000 & 1.000 & 0.250 & 0.125 & 0.125 \\ 0.000 & 0.000 & 0.000 & 0.000 & 0.000 & 0.000 & 0.062 & 0.000 & 0.250 & 1.000 & 0.125 & 0.125 \\ 0.000 & 0.000 & 0.000 & 0.000 & 0.000 & 0.000 & 0.125 & 0.000 & 0.125 & 0.125 & 1.000 & 0.250 \\ 0.000 & 0.000 & 0.000 & 0.000 & 0.000 & 0.000 & 0.125 & 0.000 & 0.125 & 0.125 & 0.250 & 1.000 \end{bmatrix}$$

and its inverse is:

$$\mathbf{D}^{-1} = \begin{bmatrix} 1.000 & 0.000 & 0.000 & 0.000 & 0.000 & 0.000 & 0.000 & 0.000 & 0.000 & 0.000 & 0.000 & 0.000 \\ 0.000 & 1.000 & 0.000 & 0.000 & 0.000 & 0.000 & 0.000 & 0.000 & 0.000 & 0.000 & 0.000 & 0.000 \\ 0.000 & 0.000 & 1.000 & 0.000 & 0.000 & 0.000 & 0.000 & 0.000 & 0.000 & 0.000 & 0.000 & 0.000 \\ 0.000 & 0.000 & 0.000 & 1.000 & 0.000 & 0.000 & 0.000 & 0.000 & 0.000 & 0.000 & 0.000 & 0.000 \\ 0.000 & 0.000 & 0.000 & 0.000 & 1.000 & 0.000 & 0.000 & 0.000 & 0.000 & 0.000 & 0.000 & 0.000 \\ 0.000 & 0.000 & 0.000 & 0.000 & 0.000 & 1.000 & 0.000 & 0.000 & 0.000 & 0.000 & 0.000 & 0.000 \\ 0.000 & 0.000 & 0.000 & 0.000 & 0.000 & 0.000 & 1.028 & 0.000 & -0.032 & -0.032 & -0.096 & -0.096 \\ 0.000 & 0.000 & 0.000 & 0.000 & 0.000 & 0.000 & 0.000 & 1.000 & 0.000 & 0.000 & 0.000 & 0.000 \\ 0.000 & 0.000 & 0.000 & 0.000 & 0.000 & 0.000 & -0.032 & 0.000 & 1.084 & -0.249 & -0.080 & -0.080 \\ 0.000 & 0.000 & 0.000 & 0.000 & 0.000 & 0.000 & -0.032 & 0.000 & -0.249 & 1.084 & -0.080 & -0.080 \\ 0.000 & 0.000 & 0.000 & 0.000 & 0.000 & 0.000 & -0.096 & 0.000 & -0.080 & -0.080 & 1.092 & -0.241 \\ 0.000 & 0.000 & 0.000 & 0.000 & 0.000 & 0.000 & -0.096 & 0.000 & -0.080 & -0.080 & -0.241 & 1.0921 \end{bmatrix}$$

The matrices $\mathbf{A}^{-1}\alpha_1$ and $\mathbf{D}^{-1}a_2$ are added to $\mathbf{Z'Z}$ and $\mathbf{W'W}$ in the MME. The MME are of the order 26 × 26 and are too large to be presented. However, the solutions to the MME by direct inversion of the coefficient matrix are:

Effects	Solutions	
Pen		
1	16.980	
2	20.030	
Animal	BV[a]	DV[a]
1	−0.160	0.000
2	−0.160	0.000
3	0.059	0.000
4	0.819	0.000
5	−0.320	0.136
6	1.259	0.705
7	0.555	0.237
8	−0.998	−0.993
9	−0.350	0.000
10	−1.350	−1.333
11	1.061	1.428
12	−0.039	−0.038

[a]BV, DV, solutions for random additive and dominance effects, respectively.

The results indicate that pigs in the second pen were heavier than those in the first pen by about 3.05 kg at weaning. The breeding value for animal i, $\hat{a}_i$ from the MME can be calculated using Eqn 3.8, except that yield deviation is corrected not only for fixed effects but also for dominance effect. Thus, the solution for animal 6 can be calculated as:

$$\begin{aligned} \hat{a}_6 &= n_1((\hat{a}_3 + \hat{a}_4)/2) + n_2(y_6 - \hat{b}_1 - \hat{d}_6) + n_3(2\hat{a}_{12} - \hat{a}_8) + n_3(2\hat{a}_{11} - \hat{a}_8) + n_3(\hat{a}_7 - \hat{a}_5) \\ &= n_1(0.059 + 0.819)/2 + n_2(20 - 16.980 - 0.705) + n_3(2(-0.039) - (-0.998)) \\ &\quad + n_3(2(1.061) - (-0.998)) + n_3(2(0.555) - (-0.320)) \\ &= 1.259 \end{aligned}$$

where $n_1 = 2\alpha_1/wt$, $n_2 = 1/wt$, $n_3 = 0.5\alpha_1/wt$, with wt equal to the sum of the numerator of n_1, n_2 and $3(n_3)$.

The solution for the dominance effect of animal i from the MME is:

$$\hat{d}_i = \left[-\alpha_2\left(\sum_j c_{ij}\hat{d}_j\right) + \left(y_i - \hat{b}_k - \hat{a}_i\right)\right] \Big/ \left(n + c_{ij}\alpha_2\right)$$

where c_{ij} is the inverse element of $\mathbf{D}$ between animal i and j, and n is the number of records. For instance, the dominance effect of animal 6 is:

$$\hat{d}_6 = (0 + (20 - 16.980 - 1.259)) / (1 + 1.5) = 0.705$$

The dominance effect for an individual represents interactions of pairs of genes from both parents and Mendelian sampling; it therefore gives an indication of how well the genes from two parents combine. This could be used in the selection of mates.

13.3.2 Solving for total genetic merit directly

Example 13.2

Using the same data and genetic parameters as in Example 13.1, solving directly for total genetic merit $(\hat{\mathbf{a}} + \mathbf{d})$ applying Eqn 13.4 is illustrated.

Setting up the MME

The design matrices $\mathbf{X}$ and $\mathbf{Z}$ are exactly the same as in Eqn 13.3. However, in Eqn 13.4, $\mathbf{M} = \mathbf{A}\sigma_a^2 + \mathbf{D}\sigma_d^2$. The matrix $\mathbf{D}$ has been given earlier and $\mathbf{A}$ can be calculated as outlined in Section 3.3. Then, $\mathbf{M}^{-1}\sigma_e^2$ is added to $\mathbf{Z'Z}$ to obtain the MME (Eqn 13.4). Solving the MME by direct inversion of the coefficient matrix gives the following solutions:

Effects		Solutions	
Pen			
1	16.980		
2	20.030		
Animal + dominance		*Animal + dominance*	
1	−0.160	7	0.792
2	−0.160	8	−1.991
3	0.059	9	−0.349
4	0.819	10	−2.683
5	−0.184	11	2.489
6	1.963	12	−0.078

The vector of solutions for additive genetic effects can then be calculated as $\hat{a} = \sigma_a^2 \mathbf{A}\mathbf{M}^{-1}\overset{\frown}{g}$ and as $\overset{\frown}{\mathrm{d}} = \sigma_d^2 \mathbf{D}\mathbf{M}^{-1}\overset{\frown}{g}$ for dominance effects, as mentioned earlier. It should be noted that the sum of $\hat{a}_i$ and $\overset{\frown}{d}_i$ for animal i in Example 13.1 equals the solution for animal i above, indicating that the two sets of results are equivalent. The advantage of using Eqn 13.4 is the reduction in the number of equations to be solved.

The major limitation to the application of models that include dominance effects was computing the inverse of the dominance matrix for real data of meaningful size. The theory for computing the dominance relationship in inbred populations was developed by De Boer and Hoeschele (1993). Hoeschele and VanRaden (1991) developed a method for computing directly the inverse of the dominance relationship matrix for populations that are not inbred, by including sire and dam or sire and maternal grandsire subclass effects in the model. With the availability of genomic data, the method is likely to be of limited use and readers interested should see Chapter 12 of the third edition of this text (*Linear Models for the Prediction of Animal Breeding Values* (Mrode, 2014)).

13.4 Genomic Models for Dominance Effects

Vitezica *et al.* (2013) compared two different genomic models for the estimation of dominance effects: (i) the classical model based on directly estimating breeding values and dominance deviations, which is compatible with pedigree-based estimations described in Section 13.2; and (ii) the model proposed by Su *et al.* (2012) for estimating additive and dominant genotypic effects (biological values at the gene level) which can also be used in the prediction of future phenotypes (or total genetic values) but not to predict breeding values or variance components directly.

13.4.1. Estimating breeding values and dominance deviation effects

The total genetic variance can be statistically partitioned in classical terms as variances of breeding values, dominant deviations, epistatic deviations, and so on (Hill *et al.*, 2008) and this is very useful in terms of prediction of genetic merit and selection. In addition, it is an orthogonal partition, which means that increasing levels of complexity in the model (adding higher-order interactions) do not change the meaning (or, if data were infinitely informative, the estimates) of the effects. However, the total genetic variance may also be partioned in terms of additive and dominance biological effects of the markers at individual loci with genotypic values computed for these effects. The formulation of the model on the basis of biological effects (Su *et al.*, 2012) is different from the statistical partitioning which is usually used in the estimation for the variances of breeding values and dominant deviations. The estimates of variances and effects from the statistical model are comparable with pedigree-based estimates, but those from the biological estimates are not directly comparable. However, if distribution of allelic frequencies are available, translation of estimates from one model to the other is straightforward and is illustrated later. This section will focus on the statistical models for the estimation of breeding values and dominant deviations, while the biological effects are dealt with in a subsequent section.

Similar to (13.2), the model including statistical additive (substitution effect) and statistical dominant effects of SNP markers can be written as:

$$\mathbf{y} = \mathbf{Xb} + \mathbf{Tu}_a + \mathbf{Vu}_d + \mathbf{e} \tag{13.5}$$

where **y** and **e** are as defined in Eqn 13.2, **b** = vector of fixed effect consisting of the mean, $\mathbf{u}_a$ and $\mathbf{u}_d$ are vectors for random SNP additive genetic and dominance effects, respectively; the t_{ij} element of **T** for the ith animal and jth SNP genotype equals $2 - 2p$, $1 - 2p$, $-2p$

for SNP genotypes AA, AB, and BB, respectively, and p is the allele frequency of the minor allele. Correspondingly, the v_{ij} element of **V**, the dominant component, for the ith animal and jth SNP genotype equals $-2q^2$, $2pq$ and $-2p^2$ for SNP genotypes AA, AB and BB, respectively. Equation 13.5 will yield SNP solutions for additive genetic and dominant effects but most studies have implemented the equivalent GBLUP dominance model.

Equivalent GBLUP dominance model

The equivalent GBLUP dominance model to Eqn 13.5 can be written as:

$$\mathbf{y} = \mathbf{Xb} + \mathbf{Za} + \mathbf{Wd} + \mathbf{e} \tag{13.6}$$

where terms are as defined in Eqn 13.2; therefore, **a** and **d** are vectors of random additive and dominance deviation genetic effects. Given that $\mathbf{a} = \mathbf{Tu}_a$ and $\mathbf{d} = \mathbf{Vu}_d$, then under the assumption of no inbreeding (non-inbred population with Hardy–Weinberg equilibrium):

$$\mathrm{Cov}(\mathbf{a}) = \frac{\mathrm{TT}'}{2\sum p_j(1-p_j)}\sigma_a^2 = \mathrm{G}\sigma_\mathrm{a}^2 \quad \text{and} \quad \mathrm{Cov}(\mathbf{d}) = \frac{\mathrm{VV}'}{(2^2\sum_j p_j^2(1-p_j)^2)}\sigma_d^2 = \frac{\mathrm{VV}'}{4\sum_j p_j^2 q_j^2}\sigma_d^2 = \mathrm{D}\sigma_\mathrm{d}^2$$

Then, the MME equations to be solved for estimates of fixed, additive genetic and dominance deviation effects using these **G** and **D** matrices are:

$$\begin{bmatrix} X'X & X'Z & X'W \\ Z'X & Z'Z + \mathrm{G}^{-1}\alpha_1 & Z'W \\ W'X & W'Z & W'W + \mathrm{D}^{-1}\alpha_2 \end{bmatrix} \begin{bmatrix} \hat{b} \\ \hat{a} \\ \hat{d} \end{bmatrix} = \begin{bmatrix} X'y \\ Z'y \\ W'y \end{bmatrix} \tag{13.7}$$

where **Z** and **W** are defined in (13.2) and $\alpha_1 = \sigma_e^2 / \sigma_a^2$ and $\alpha_2 = \sigma_e^2 / \sigma_d^2$.

13.4.2 Estimating the biological genotypic additive values and dominance marker effects

Similar to Eqn 13.5, the model for biological genotypic additive values and dominance marker effects can be written as:

$$\mathbf{y} = \mathbf{Xb} + \mathbf{Tu}_{a*} + \mathbf{Vu}_{d*} + \mathbf{e} \tag{13.8}$$

where **y**, **b** and **e** are as defined in (13.5), but $\mathbf{u}_{a*}$ and $\mathbf{u}_{d*}$ are vectors for random genotypic SNP additive genetic and dominance effects, respectively; **T** is the same as in (13.5) with elements $t_{ij} = 2 - 2p$, $1 - 2p$, $-2p$ for the for SNP genotypes AA, AB, and BB, respectively, and p is the allele frequency of the minor allele. However, for the dominant component, **V** has elements $v_{ij} = -2pq$, $1 - 2pq$ and $-2pq$ for SNP genotypes AA, AB and BB, respectively.

Again, under the assumption of no inbreeding, there is no covariance between the genotypic additive and dominance values, which therefore are:

$$\mathrm{Cov}(a) = \frac{\mathrm{TT}'}{2\sum p_j(1-p_j)}\sigma_{a*}^2 = \mathrm{G}\sigma_{a*}^2 \quad \text{and} \quad \mathrm{Cov}(d) = \frac{\mathrm{VV}'}{(2\sum_j p_j(1-p_j))^2}\sigma_{d*}^2 = \frac{\mathrm{VV}'}{4\sum_j p_j^2 q_j^2}\sigma_{d*}^2 = \mathrm{D}\sigma_{\mathrm{d}*}^2$$

Note that σ^2_{a*} and σ^2_{d*} are the variances of genotypic additive and dominant values.

13.4.3 Relationship between genotypic additive and dominance variances and variance for additive genetic and dominance deviations

Variances from either model (Vitezica *et al.*, 2013) have the following components, where the summation is over *n* loci:

$$\sigma^2_{ua} \text{ (in model 13.5)} = \sum_n (2pq)\sigma^2_{ua} + \sum_n (2pq(q-p)^2)\sigma^2_{ud}$$

$$\sigma^2_{ua*} \text{ (in model 13.8)} = \sum_n (2pq)\sigma^2_{ua}$$

$$\sigma^2_{ud} \text{ (in model 13.5)} = \sum_n (2pq)^2\sigma^2_{ud} \text{ and}$$

$$\sigma^2_{d*} \text{ (in model 13.8)} = \sum_n (2pq(1-2)pq)\sigma^2_{ud}$$

From the above, it can be shown that $(\sigma^2_{ua*} + \sigma^2_{ud*}) = (\sigma^2_{ud} + \sigma^2_{ud})$ and if p = q = 0.5, then all variances are identical. Also, when d = 0, then $\sigma^2_{ua*} = \sigma^2_{ua}$; thus, conversion across the models is straightforward from the above identities.

However, in practice, prediction of additive genetic values with dominance is mostly in the context of estimating EBVs and dominance deviations; therefore subsequent discussion will be based on model 13.6 and the MME given in (13.7) and its application is hereby illustrated using the data in Example 13.3.

Example 13.3

The data on weaning weights in Example 13.1 have been modified, including more piglets and genotypic data for all piglets. Using the genotypic data, the aim is to predict the EBVs and dominance deviation effects for all piglets using the genomic and dominance relationship matrices computed from the SNP data. The same genetic parameters as in Example 13.1 are assumed.

Animal	Sire	Dam	Pen	WWT	SNP Genotype																			
1	0	0	–	–	2	2	0	0	1	1	1	0	1	1	0	0	1	0	1	2	0	1	0	1
2	0	0	–	–	1	1	1	1	2	0	1	0	2	2	0	1	0	1	1	1	0	0	0	0
3	0	0	–	–	1	2	1	0	2	0	1	1	2	2	0	1	2	0	2	2	0	0	1	0
4	0	0	–	–	1	1	0	0	2	0	2	2	2	2	1	1	0	1	0	2	1	0	1	0
5	1	2	1	17.0	1	1	1	1	1	0	2	0	1	2	0	0	0	1	1	2	0	0	0	1
6	3	4	1	20.0	2	1	0	0	2	0	1	1	2	2	0	1	1	1	1	2	0	0	2	0
7	6	5	1	18.0	2	1	0	0	1	0	2	1	1	2	0	1	0	1	1	2	0	0	1	1
8	0	5	1	13.5	2	2	0	0	1	0	2	1	1	2	0	0	1	0	2	1	0	0	0	1
9	3	8	2	20.0	2	2	0	0	1	0	2	2	2	2	0	1	2	0	2	1	0	0	0	0
10	3	8	2	15.0	1	2	0	0	2	0	2	2	2	2	0	1	2	0	2	1	0	0	0	0
11	6	8	2	25.0	2	1	0	0	1	0	1	0	1	2	0	0	1	0	2	1	0	0	1	0
12	6	8	2	19.5	2	1	0	0	1	0	2	1	1	2	0	1	0	1	1	2	0	0	1	1
13	7	10	1	22.5	2	1	0	0	2	0	2	2	2	2	0	1	1	1	1	1	0	0	1	0
14	9	12	1	16.0	2	1	0	0	1	0	2	2	2	2	0	1	1	0	2	1	0	0	0	1
15	7	11	2	24.5	2	1	0	0	0	0	2	0	1	2	0	0	0	0	2	1	0	0	0	1

The example data are analysed using Eqn13.6, therefore SNP allele frequencies are initially estimated so that matrices **G** and **D** can be computed.

Computing the SNP allele frequencies

Using the formula in Example 11.1, the allele frequencies for the 20 SNPs in Example 13.3 are: 0.833, 0.667, 0.100, 0.067, 0.667, 0.033, 0.833, 0.500, 0.767, 0.967, 0.033, 0.333, 0.400, 0.233, 0.700, 0.733, 0.033, 0.033, 0.267 and 0.233. The matrices **X**, **Z** and **W** were as computed in Example 13.1. The covariance matrix **G** for additive genetic effects (see (13.6)) is:

G =

1.177														
−0.106	1.038													
−0.059	0.045	0.957												
−0.464	0.160	−0.140	1.362											
0.230	0.507	−0.313	−0.025	1.015										
−0.221	0.056	0.276	0.391	−0.302	0.807									
−0.036	−0.106	−0.406	0.230	0.230	0.126	0.484								
0.241	−0.348	−0.129	−0.533	−0.013	−0.464	−0.106	0.518							
−0.198	−0.441	0.299	−0.279	−0.625	−0.209	−0.371	0.253	0.853						
−0.313	−0.209	0.530	−0.048	−0.568	−0.152	−0.487	0.137	0.738	0.969					
0.229	−0.013	0.033	−0.544	−0.025	0.045	−0.117	0.160	−0.106	−0.221	0.668				
−0.036	−0.106	−0.406	0.230	0.230	0.126	0.484	−0.106	−0.371	−0.487	−0.117	0.484			
−0.510	−0.059	−0.013	0.449	−0.417	0.345	0.010	−0.233	0.195	0.253	−0.244	0.010	0.576		
−0.256	−0.325	−0.106	−0.163	−0.337	−0.267	−0.082	0.195	0.449	0.333	−0.163	−0.082	0.137	0.565	
0.322	−0.094	−0.568	−0.625	0.414	−0.556	0.149	0.426	−0.186	−0.475	0.414	0.149	−0.498	0.102	1.027

The covariance matrix for dominance effects (D) as defined in (13.6) is:

D =

0.626														
−0.055	0.694													
−0.399	0.052	0.690												
−0.368	0.223	0.241	1.004											
0.297	0.302	−0.208	0.014	0.706										
−0.001	0.242	−0.176	−0.301	−0.018	1.161									
−0.027	0.106	0.197	−0.024	0.304	0.096	0.779								
0.195	−0.349	−0.196	−0.353	−0.077	0.036	0.062	0.581							
−0.063	0.244	0.157	0.240	0.109	−0.278	−0.009	−0.139	0.676						
−0.209	0.355	0.268	0.351	0.015	−0.218	−0.155	−0.285	0.530	0.641					
0.288	0.050	−0.333	0.126	0.170	−0.257	−0.022	0.161	0.003	−0.142	0.692				
−0.027	0.106	0.197	−0.024	0.304	0.096	0.779	0.062	−0.009	−0.155	−0.022	0.779			
0.056	0.430	−0.220	0.169	0.090	0.122	0.103	−0.187	−0.038	0.022	0.293	0.103	0.724		
0.212	0.180	−0.335	0.176	0.146	0.102	0.027	0.136	0.184	0.038	0.383	0.027	0.342	0.665	
−0.094	0.275	−0.092	0.270	0.136	−0.099	−0.188	−0.170	0.016	0.281	0.077	−0.188	0.141	0.052	0.865

The inverses of **G** and **D** used in (13.7) were obtained by adding 0.01 to the diagonal elements of both matrices to make them invertible. Solving (13.7) for the example data results in the following solutions:

Pen effects		
1	17.451	
2	20.812	
	Animal effects	Dominance effects
1	−0.095	0.197
2	0.572	0.925
3	−0.947	−1.007
4	0.228	0.504
5	0.028	0.087

Continued

Continued.

Pen effects		
6	1.556	0.369
7	0.744	−0.523
8	−0.932	−0.837
9	−1.315	−0.744
10	−2.093	−0.361
11	1.395	1.107
12	0.730	−0.536
13	0.378	1.798
14	−0.876	0.504
15	0.626	1.131

13.4.4 Inclusion of genomic inbreeding

Silió *et al.* (2013) suggested that inbreeding can be defined as the proportion of homozygous SNPs across all loci for each animal. When directional dominance causes inbreeding depression, Xiang *et al.* (2016) showed analytically, using GBLUP, that the inclusion of genomic inbreeding as a covariate is necessary to obtain correct estimates of dominance variance. They indicated that when there is directional dominance, the dominance effects of genes (which will be associated with markers in the genomic model), have, *a priori*, a positive value for traits that exhibit inbreeding depression or heterosis. Assuming $\mathbf{u}_{d*}$ as the vector of dominance marker effects, then it has the following prior distribution:

$$\mathbf{u}_{d*} \sim N(\mathbf{1}\mu_{ud}, \mathbf{I}\sigma^2_{ud*})$$

where μ_{ud} is the overall mean of dominance effects, which should be positive if there is heterosis due to dominance. Considering the model 13.8 for the genomic prediction of 'biological' genotypic additive and dominance effects, note that the elements of incidence matrices **T** and **V**, if they were unscaled, would be: 1, 0, −1 for SNP genotypes AA,AB and BB, respectively, for **T**, and 0, 1, 0 for SNP genotypes AA,AB and BB, respectively, for **V**.

It is usually assumed that $\mathbf{u}_{a*}$ and $\mathbf{u}_{d*}$ in (13.8) are vectors of random effects with zero means. However, when there is directional dominance, this is not true for $\mathbf{u}_{d*}$. If we define $\mathbf{u}_{d**} = \mathbf{u}_{d*} - E(\mathbf{u}_{d*})$, then the expectation of $\mathbf{u}_{d**}$ ($E(\mathbf{u}_{d**})$) is zero. Then $\mathbf{u}_{d*} = \mathbf{u}_{d**} + E(\mathbf{u}_{d*})$ and replacing $\mathbf{u}_{d*}$ with this term, Eqn 13.8 may be written as:

$$\begin{aligned} \mathbf{y} &= \mathbf{Xb} + \mathbf{Tu}_{a*} + \mathbf{V}(\mathbf{u}_{d**} + E(\mathbf{u}_{d*})) + \mathbf{e} \\ \mathbf{y} &= \mathbf{Xb} + \mathbf{Tu}_{a*} + \mathbf{Vu}_{d**} + \mathbf{V1}_{\mu ud*} + \mathbf{e} \end{aligned} \tag{13.9}$$

where $\mathbf{V1}_{\mu ud*}$ is the average of dominance effects for each individual. Note that the elements of **V1** are the row-sums of **V**, i.e. individual heterozygosities; given that if **V** is unscaled, it has a value of 1 at heterozygous loci for an individual, otherwise it is zero. Assume that, **h** represents the proportion of genotyped SNPs at which the individual is heterozygous. Let $\mathbf{h}_{\mu ud*} = \mathbf{V1}_{\mu ud*}$, then the vector inbreeding coefficients, **f** (the proportion of genotyped SNPs at which the individual is homozygous) can be computed as: $\mathbf{f} = \mathbf{1} - \mathbf{h}/N$, with N as the number of SNPs. Then $\mathbf{h} = (\mathbf{1} - \mathbf{f})N$ and $\mathbf{h}_{\mu ud*}$ can be rewritten as:

$$\mathbf{h}_{\mu ud*} = (\mathbf{1} - \mathbf{f})N_{\mu ud*} = \mathbf{1}N_{\mu ud*} + \mathbf{f}(-N_{\mu ud*})$$

where the term $\mathbf{1}N_{\mu ud*}$ is confounded with the overall of the model (μ) of model 13.8 while $\mathbf{f}(-N_{\mu ud*}) = \mathbf{f}b$, models the inbreeding depression, with b as the inbreeding depression parameter to be estimated. Therefore, Eqn 13.9 can be written as:

$$\mathbf{y} = \mathbf{Xb} + \mathbf{f}b + \mathbf{Tu}_{a*} + \mathbf{Vu}_{d**} + \mathbf{e} \tag{13.10}$$

with $\mathbf{f} = 1 - \mathbf{h}/N$. As stated earlier, **h** represents the proportion of genotyped SNPs at which the individual is heterozygous. However if **V** is scaled, **f** can be computed as:

$$f_j = 1 - \frac{\sum_{i=1}^{N} v_{ij}}{2\sum_{i}^{N} p_i q_i}$$

where v_{ij} is the element of **V** for the *i*th marker for the *j*th individual.

Note that inbreeding depression can also be modelled by fitting **h**, the proportion of heterozygous SNPs for an individual (Bolormaa *et al.*, 2015; Aliloo *et al.*, 2017) but both models are equivalent. With **V** scaled, **h** may be computed as:

$$h_j = \frac{\sum_{i=1}^{N} v_{ij}}{2\sum_{i}^{N} p_i q_i}$$

Using the same principles, inbreeding depression in model 13.6 for estimating breeding values and dominant deviations can be modelled (Vitezica *et al.*, 2018) as:

$$y = Xb + fb + Za + Wd + e \tag{13.11}$$

The importance of including inbreeding depression in the model for traits affected by dominance, such as fitness traits, was clearly demonstrated by Vitezica *et al.* (2018). They reported that inbreeding depression expressed as a change in phenotypic mean per 10% increase in inbreeding was equal to −1.23 piglets born in the pig population they studied. In addition, they reported that upward bias in the estimate of the dominance variance (0.38) when inbreeding was not included in the model compared with an estimate 0.18 in the model including inbreeding.

Example 13.4

Using the same data and genetic parameters in Example 13.3, animal genetic and dominance effects are to be estimated but accounting for inbreeding depression effects by fitting model 13.11. The MME to be solved are similar to (13.7), except that the fixed-effects component now includes a covariate term to model inbreeding depression.

Estimating proportion of heterozygous SNPs for each animal

Inbreeding depression will be modelled in this example using the proportion of heterozygous SNPs for each animal ($\mathbf{f} = 1 - \mathbf{h}/N$, see Eqn 13.10). Therefore, the proportion of heterozygous SNPs for the ith animal (h_i,) is initially computed from the genotypic data and subsequently vector **f**. The vector **f** for the animals with records (animals 5–15) is:

$$\mathbf{f}' = [0.550\ \ 0.650\ \ 0.550\ \ 0.700\ \ 0.850\ \ 0.850\ \ 0.650\ \ 0.550\ \ 0.650\ \ 0.700\ \ 0.800]$$

For instance, from the genotypic data, f_5 was computed as (1 − 9/20) = 0.550. The matrix X is the same as in Example 13.3 but is modified to include elements of **f** as a covariate term in the model and the remaining matrices are the same as Example 13.3. Solving the MME gives the following set of results:

Pen effects		
1	19.933	
2	23.518	
Inbreeding depression effect		
3	–3.767	
	Animal effects	Dominance effects
1	–0.125	0.131
2	0.460	0.946
3	–0.842	–0.996
4	0.104	0.540
5	–0.167	–0.020
6	1.442	0.381
7	0.577	–0.727
8	–0.816	–0.877
9	–1.043	–0.669
10	–1.830	–0.215
11	1.385	1.022
12	0.562	–0.741
13	0.407	1.747
14	–0.721	0.488
15	0.608	1.265

The estimates of breeding values and dominance effects were not very different from the model without inbreeding but the results indicate an inbreeding depression of –0.038kg for weaning weight per unit of inbreeding.

13.5 Epistasis

Epistasis refers to the interaction among additive and dominance genetic effects; for instance, additive by additive, additive by dominance, additive by additive by dominance, etc. The epistasis relationship matrix can be derived from **A** and **D** as:

A#A	for additive by additive
D#D	for dominance by dominance
AA#D	for additive by additive by dominance

where # represents the Hadamard product of the two matrices. The *ij* element of the Hadamard product of the two matrices is the product of the *ij* elements of the two matrices. Thus, if **M** = **A#B**, then $m_{ij} = (a_{ij})(b_{ij})$, where the matrices **A** and **B** should be of the same order.

The animal model in Eqn 13.2 can be expanded to include epistatic effects as:

$$\mathbf{y} = \mathbf{Xb} + \mathbf{Za} + \mathbf{Wd} + \mathbf{Saa} + \mathbf{e} \qquad (13.12)$$

where **aa** is the vector of interaction (epistatic) effects for additive by additive action. The matrices **Z**, **W** and **S** are identical. The evaluation can be carried out as described in Section 13.3 but the major limitation is obtaining the inverse of the epistatic relationship matrix for large data sets. However, VanRaden and Hoeschele (1991) presented a rapid method for obtaining the inverse of the epistatic relationship matrix when epistasis results from interactions between additive by additive ($\mathbf{A} \times \mathbf{A}$) genetic effects when the population is inbred or not. With the availability of genomic data, the method is likely to be of limited use and interested readers should see Chapter 12 of the third edition of this text, *Linear Models for the Prediction of Animal Breeding Values* (Mrode, 2014).

13.5.1 Epistatic genomic models

The availability of genotypic data has made the fitting of accounting for epistatic genetic effects more feasible for the various levels of interactions as the covariance matrices can be computed from the Hadamard products and traces of the genomic and dominance relationship matrices. For Eqn 13.6 on the GBLUP dominance genomic model, the computation of **G** and **D** matrices was illustrated. Given that **G** and **D** are the genomic relationships matrices for additive and dominance effects, then the genomic relationship matrices for epistatic additive by additive ($\mathbf{G}_{AA}$), epistatic dominance by dominance ($\mathbf{G}_{DD}$), and epistatic additive by dominance ($\mathbf{G}_{AD}$) (Vitezica *et al.*, 2018), respectively, are:

$$\mathrm{G}_{AA}\sigma^2_{aa} = \frac{\mathrm{G} \odot \mathrm{G}}{tr(\mathrm{G} \odot \mathrm{G})/N}\sigma^2_{aa}, \quad \mathrm{G}_{DD}\sigma^2_{dd} = \frac{\mathrm{D} \odot \mathrm{D}}{tr(\mathrm{D} \odot \mathrm{D})/N}\sigma^2_{dd} \quad \text{and} \quad \mathrm{G}_{AD}\sigma^2_{ad} = \frac{\mathrm{G} \odot \mathrm{D}}{tr(\mathrm{G} \odot \mathrm{D})/N}\sigma^2_{ad}$$

With the genomic covariance matrices defined, the genomic models that includes epistatic genetic effects are straightforward and are an extension of Eqns 13.5 and 13.6. Vitezica *et al.* (2018) reported that inbreeding effects were insignificant on estimates of epistatic genetic variances, even when they significantly affected dominance variance; so fitting inbreeding effects is not critical for epistatic effects.

GBLUP epistatic model

The GBLUP epistatic model is as defined in (13.12). The MME for the model that includes epistatic effect for additive by additive effects are:

$$\begin{pmatrix} \mathrm{X'X} & \mathrm{X'Z} & \mathrm{X'W} & \mathrm{X'S} \\ \mathrm{Z'X} & \mathrm{Z'Z}+\mathrm{G}^{-1}\alpha_1 & \mathrm{Z'W} & \mathrm{Z'S} \\ \mathrm{W'X} & \mathrm{W'Z} & \mathrm{W'W}+\mathrm{D}^{-1}\alpha_2 & \mathrm{W'S} \\ \mathrm{S'X} & \mathrm{S'Z} & \mathrm{S'W} & \mathrm{S'S}+\mathrm{G}_{aa}^{-1}\alpha_3 \end{pmatrix} \begin{pmatrix} \hat{\mathrm{b}} \\ \hat{\mathrm{a}} \\ \hat{\mathrm{d}} \\ \widehat{\mathrm{aa}} \end{pmatrix} = \begin{pmatrix} \mathrm{X'y} \\ \mathrm{Z'y} \\ \mathrm{W'y} \\ \mathrm{S'y} \end{pmatrix} \tag{13.13}$$

where **S**, incidence matrix that relates observations to epistatic effect, is the same as **Z** and **W**, which have been defined in (13.6) with $\alpha_1 = \sigma^2_e / \sigma^2_a$, $\alpha_2 = \sigma^2_e / \sigma^2_d$ and $\alpha_3 = \sigma^2_e / \sigma^2_{aa}$

Example 13.5
Using the data in Example 13.3, model 13.2 with epistatic effect for additive by additive effects (13.12) is fitted. It was assumed that the genetic variance for epistatic effect was 6 kg^2 and all other genetic parameters were as defined in Example 13.3. Thus $\alpha_1 = \sigma^2_e / \sigma^2_a = 1.333$, $\alpha_2 = \sigma^2_e / \sigma^2_d = 1.50$ and $\alpha_3 = \frac{\sigma^2_e}{\sigma^2_{aa}} = 20.00$.

The covariance matrix $\mathbf{G}_{AA}$ for additive by additive epistatic effects for the example data is:

G_{AA} =

1.812														
0.015	1.409													
0.005	0.003	1.198												
0.282	0.033	0.026	2.426											
0.069	0.336	0.128	0.001	1.347										
0.064	0.004	0.100	0.200	0.119	0.852									
0.002	0.015	0.216	0.069	0.069	0.021	0.306					Symmetric			
0.076	0.158	0.022	0.372	0.000	0.282	0.015	0.351							
0.051	0.254	0.117	0.102	0.511	0.057	0.180	0.084	0.951						
0.128	0.057	0.367	0.003	0.422	0.030	0.310	0.025	0.712	1.228					
0.069	0.000	0.001	0.387	0.001	0.003	0.018	0.033	0.015	0.064	0.584				
0.002	0.015	0.216	0.069	0.069	0.021	0.306	0.015	0.180	0.310	0.018	0.306			
0.340	0.005	0.000	0.264	0.227	0.156	0.000	0.071	0.050	0.084	0.078	0.000	0.434		
0.086	0.138	0.015	0.035	0.149	0.093	0.009	0.050	0.264	0.145	0.035	0.009	0.025	0.417	
0.136	0.012	0.422	0.511	0.224	0.404	0.029	0.237	0.045	0.295	0.224	0.029	0.324	0.014	1.379

Given that **S** = **Z** or **W**, then all the matrices in (13.13) have been defined from previous examples. Solving the MME gives the following solutions:

Pen effects			
1	17.453		
2	20.833		
Random effects			
	Animal	Dominance	Epistatic
1	−0.108	0.191	0.035
2	0.566	0.920	−0.026
3	−0.916	−0.975	−0.029
4	0.248	0.506	0.082
5	0.020	0.100	−0.029
6	1.545	0.351	0.044
7	0.732	−0.494	−0.056
8	−0.934	−0.836	0.002
9	−1.288	−0.710	−0.090
10	−2.045	−0.343	−0.153
11	1.352	1.085	0.065
12	0.718	−0.506	−0.057
13	0.387	1.768	0.078
14	−0.862	0.496	−0.029
15	0.585	1.094	0.126

Compared to the results from Example 13.2, the inclusion of epistatic effects has resulted in re-ranking of animals in terms of both the estimated breeding values and dominance deviation effects. Thus, it is essentially to fit epistatic effects for data where such an effect is expected to explain a significant portion of the total genetic variance.

14 Genetic and Genomic Models for Multibreed and Crossbred Analyses

14.1 Introduction

In many livestock production systems, crossbreeding is a common practice, either to improve the local breed by crossing with exotic superior breeds, such as in dairy cattle in many developing countries, or to improve specific traits such as fertility in the crossbred, or to produce commercial animals through the terminal crossing for use in the production environment, such as chicken and pigs. All these cases result in the production of animals with parents from different breeds, lines or populations. In dairy cattle breeding, computation of the genetic merit of crossbred cows is important for within-herd selection and in the creation of synthetic breeds by permitting the selection and use of superior crossbred bulls. However, in chicken and pig breeding, the breeding value of crossbred animals is not of interest *per se* but there is growing interest to optimize selection within the purebred animals in the nucleus for their performance in the crossbreds in a commercial setting. In this situation, an important parameter is the genetic correlation between purebred and crossbred performance (r_{pc}) as it influences the effect of selection based on purebred animals and the observed rate of genetic change on crossbred animals at the production level. Usually, this correlation is less than unity due to the different management systems at the commercial level. Therefore, the inclusion of crossbred information to evaluate purebred animals for the performance of crossbred at a commercial level is advantageous (Dekkers, 2007), even with a genetic correlation of 0.7 between purebred and crossbred performance.

When crossbred information is used in the evaluation of the genetic merits of animals, it is to be noted that SNP effects may be breed-specific because allele frequencies, linkage disequilibrium patterns between an SNP and a QTL, and allele substitution effects of a QTL may differ between breeds. When genomic information is available, it is possible to determine the effects of alleles from different breeds, assign breed origin to alleles (BOA) in crossbreds and, hence, be able to select purebred animals for crossbred performance more efficiently. An additive genomic prediction model that accounts for breed-specific SNP effects using crossbred information was presented by Ibánẽz-Escriche *et al.* (2009) and Christensen *et al.* (2014, 2015).

In a simulation study, Ibánẽz-Escriche *et al.* (2009) showed that with low SNP density, large training data size, and low breed relatedness, the BOA genomic model outperformed models in which SNP effects are assumed to be the same across breeds. In a three-way pig crossbreeding system, Sevillano *et al.* (2017) reported that the genomic models that accounted for BOA were especially relevant for traits with low r_{pc} with a higher prediction accuracy observed in average daily gain in one maternal breed, which had the lowest r_{pc} (0.30). Otherwise, the use of genomic relationship matrices based on breed-specific allele frequencies or allele frequencies averaged across breeds was equally effective according to prediction accuracies. Using dairy cattle data, Karaman *et al.* (2021) obtained optimum

© R.A. Mrode and I. Pocrnic 2023. *Linear Models for the Prediction of the Genetic Merit of Animals, 4th Edition* (R.A. Mrode and I. Pocrnic)
DOI: 10.1079/9781800620506.0014

results in terms of prediction accuracy when the admixed data was included with the purebred animals in the reference population and breed origin of alleles was included in the prediction method.

In this chapter, genetic models in the analysis of multibreeds and their crosses are initially examined based on the use of pedigree information focusing on two-breed crosses. Subsequently, using the principles of partial relationship matrices by Garcia-Cortes and Toro (2006) and the concept of partial genomic relationship matrices by Christensen *et al.* (2014), the genomic evaluation for purebred and crossbred performances is discussed, followed by the application of breed origin of alleles in the evaluation of crossbred animals.

14.2 Multibreed Analysis by Splitting the Breeding Values

The covariance matrix, **G**, for the additive values in a multibreed analysis, including both the purebreed contributions and the segregation deviations can be defined (Lo *et al.*, 1993) as:

$$g_{ii} = \sum_{p=1}^{np} f_p^i \sigma_p^2 + 2\sum_{p=1}^{np}\sum_{p'>p}^{np}(f_p^S f_{p'}^S + f_p^D f_{p'}^D)\sigma_{pp'}^2 + \frac{1}{2}g_{SD} \tag{14.1}$$

and

$$g_{ij} = \frac{1}{2}(g_{iS'} + g_{iD'}) \tag{14.2}$$

where np is the number of breeds in the founder generation, f_p^i refers to the proportion of genes from breed p for animal i, S and D are the sire and the dam of animal i, S' and D' are the sire and the dam of animal j, σ_p^2 is the additive variance component of breed p, and $\sigma_{pp}^2{}'$ is the segregation variance between breeds p and p'.

Given the definition of **G** in (14.1) and (14.2), the model for the analysis of multibreed populations is defined as usual:

$$\mathbf{y} = \mathbf{Xb} + \mathbf{Za} + \mathbf{e} \tag{14.3}$$

with $E(\mathbf{y}) = \mathbf{Xb}$ and $Var(\mathbf{y}) = \mathbf{ZGZ'} + \mathbf{I}\sigma_e^2$

However, because the purebreed and the segregation components are mixed in **G**, the additive covariance matrix **G** cannot be expressed as the numerator relationship matrix times the additive variance components as in the conventional animal model. Thus, **G** is difficult to estimate.

14.3 Equivalent Model

Garcia-Cortes and Toro (2006) presented an equivalent model. Equation 14.1 indicated that the g_{ii} can be partitioned into those related to the purebreed and others related to the segregation terms. Then the **G** matrix can be partitioned into several pieces by considering only one source of variability and assigning zero to the others. Thus, $\mathbf{G}^{(p)}$, the partial **G** associated with breed p, can be obtained by considering every genetic component as null apart from σ^2_p. Equations 14.1 and 14.2 become:

$$g_{ii}^{(p)} = f_p^i \sigma_p^2 + \frac{1}{2} g_{SD}^{(p)} \tag{14.4}$$

and

$$g_{ij}^{(p)} = \frac{1}{2}\left(g_{iS'}^{(p)} + g_{iD'}^{(p)}\right) \tag{14.5}$$

The same argument can be used to construct the partial **G** due to the segregation between breeds p and p', denoted $\mathbf{G}^{(pp')}$ when every genetic component is null apart from $\sigma^2_{pp'}$:

$$g_{ii}^{pp'} = 2(f_p^S f_{p'}^S + f_p^D f_{p'}^D)\sigma_{pp'}^2 + \frac{1}{2} g_{SD}^{(pp')} \tag{14.6}$$

and

$$g_{ij}^{pp'} = \frac{1}{2}(g_{iS'}^{(pp')} + g_{iD'}^{(pp')}) \tag{14.7}$$

From the above, it is obvious that $\mathbf{G}^{(p)}$ or $\mathbf{G}^{(pp')}$ each depend on a single variance component, therefore both can be expressed as $\mathbf{G}^{(p)} = \mathbf{A}_p\sigma^2_p$ and $\mathbf{G}^{(pp')} = \mathbf{A}_{pp'}\sigma^2_{pp'}$ where $\mathbf{A}_p$ and $\mathbf{A}_{pp}$ are partial numerator relationship matrices for purebreeds and segregation terms, respectively. Therefore, a model for both terms can be written as:

$$\mathbf{y} = \mathbf{Xb} + \sum_p \mathbf{Z}_p \mathbf{a}_p + \sum_p \sum_{p'>p} \mathbf{Z}_{pp'} \mathbf{a}_{pp'} + \mathbf{e} \tag{14.8}$$

where Zp is the incidence matrix relating records to the random additive genetic effects for the pth breed and $\mathbf{Z}_{pp'}$ relates records to the segregation effects between the breeds p and p′.

$$\mathbf{G} = \sum_p \mathbf{A}_p\sigma_p^2 + \sum_p \sum_{p'>p} \mathbf{A}_{pp'}\sigma_{pp'}^2 \quad \text{and} \quad \mathrm{Var}(\mathbf{y}) = \mathbf{Z}(\sum_p \mathbf{A}_p\sigma_p^2 + \sum_p \sum_{p'>p} \mathbf{A}_{pp'}\sigma_{pp'}^2)\mathbf{Z}' + \mathbf{R}$$

When a data set with two breeds is analysed, Eqn 14.8 becomes $\mathbf{y} = \mathbf{Xb} + \mathbf{Z}_1\mathbf{a}_1 + \mathbf{Z}_2\mathbf{a}_2 + \mathbf{Z}_{12}\mathbf{a}_{12} + \mathbf{e}$, where $\mathbf{a}_1$, $\mathbf{a}_2$ and $\mathbf{a}_{12}$ are the breeding values split by origin. The MME for (14.8) are:

$$\begin{pmatrix} \mathbf{X'X} & \mathbf{X'Z}_1 & \mathbf{X'Z}_2 & \mathbf{X'Z}_{12} \\ \mathbf{Z}_1'\mathbf{X} & \mathbf{Z}_1'\mathbf{Z}_1 + \mathbf{A}_1^-\alpha_1 & \mathbf{Z}_1'\mathbf{Z}_2 & \mathbf{Z}_1'\mathbf{Z}_{12} \\ \mathbf{Z}_2'\mathbf{X} & \mathbf{Z}_2'\mathbf{W}_1 & \mathbf{Z}_2'\mathbf{Z}_2 + \mathbf{A}_2^-\alpha_2 & \mathbf{Z}_2'\mathbf{Z}_{12} \\ \mathbf{Z}_{12}'\mathbf{X} & \mathbf{Z}_{12}'\mathbf{Z}_1 & \mathbf{Z}_{12}'\mathbf{Z}_2 & \mathbf{Z}_{12}'\mathbf{Z}_{12} + \mathbf{A}_{12}^-\alpha_{12} \end{pmatrix} \begin{pmatrix} \hat{\mathbf{b}} \\ \hat{\mathbf{a}}_1 \\ \hat{\mathbf{a}}_2 \\ \hat{\mathbf{a}}_{12} \end{pmatrix} = \begin{pmatrix} \mathbf{X'y} \\ \mathbf{Z}_1'\mathbf{y} \\ \mathbf{Z}_2'\mathbf{y} \\ \mathbf{Z}_{12}'\mathbf{y} \end{pmatrix} \tag{14.9}$$

where $\alpha_1 = \sigma^2_e/\sigma^2_{p1}$, $\alpha_2 = \sigma^2_e/\sigma^2_{p2}$ and $\alpha_{12} = \sigma^2_e/\sigma^2_{p12}$. Note that the sum of $\hat{\mathbf{a}}_1$, $\hat{\mathbf{a}}_2$ and $\hat{\mathbf{a}}_{12}$ is equal to $\hat{\mathbf{a}}$ in (14.3) and, hence, the equivalence of both models.

Note that generalized inverses of the partial **A** matrices are used in Eqn 14.9 to account for the fact that some animals will have zero contributions from some breeds or segregation components.

First, from Eqns 14.4 to 14.7, the partial **A** matrices can be computed recursively as: $\mathbf{A}_{ii}^{(p)} = f_p^i + \frac{1}{2}\mathbf{A}_{SD}^{(p)}$

$$\mathbf{A}_{ij}^{(p)} = \frac{1}{2}\left(\mathbf{A}_{iS'}^{(p)} + \mathbf{A}_{iD'}^{(p)}\right)$$

$$\mathbf{A}_{ii}^{pp'} = 2(f_p^S f_{p'}^S + f_p^D f_{p'}^D)\sigma_{pp'}^2 + \frac{1}{2}\mathbf{A}_{SD}^{(pp')} \qquad (14.10)$$

$$\mathbf{A}_{ij}^{pp'} = \frac{1}{2}(\mathbf{A}_{iS'}^{(pp')} + \mathbf{A}_{iD'}^{(pp')}$$

If c_i is defined as $c_i = f_p^i$ for the purebreed contributions and $c_i = 2(f_p^S f_{p'}^S + f_p^D f_{p'}^D)\sigma_{pp'}^2$ for segregation effects, the same algorithm based on the tabular method can be applied in both cases:

$$\mathbf{A}_{ii} = c_i + \frac{1}{2}\mathbf{A}_{SD}$$

and

$$\mathbf{A}_{ij} = \frac{1}{2}(\mathbf{A}_{iS'} + \mathbf{A}_{iD'})$$

The generalized inverses for the partial relationship matrices can be obtained as usual using Henderson (1976) rules outlined Section 3.5 in Chapter 3. Note that some rows of $\mathbf{A}_i$ will be null due to null contributions from some breeds or segregations. In such cases, the inverse of $\mathbf{A}_i$ will be based on the non-zero part of the relationship matrix but keeping the pattern of the null rows in the matrix.

Example 14.1

Table 14.1 has the example data set in Garcia-Cortes and Toro (2006) consisting of five crossbred animals reared in two herds and produced by crossing two breeds, each containing three purebred animals. It assumed that the residual variance was 4, while the genetic variances for breeds 1, 2 and the segregation effects were $\sigma^2_1 = 1$, $\sigma^2_2 = 2$ and $\sigma^2_{12} = 0.50$, respectively. The aim is to estimate the breed-specific breeding values for the two breeds and segregation effects using Eqn 14.9. Also, the data are analysed using (14.3) to illustrate the equivalence of models 14.3 and 14.8.

The model for the example data using Eqn 14.8 is:

$$\mathbf{y} = \mathbf{Xb} + \mathbf{Z}_1\mathbf{a}_1 + \mathbf{Z}_2\mathbf{a}_2 + \mathbf{Z}_{12}\mathbf{a}_{12} + \mathbf{e},$$

where the **b** vector has solutions for the fixed effects of herd and breed, and vectors $\mathbf{a}_1$, $\mathbf{a}_2$ and $\mathbf{a}_{12}$ are breeding values for breeds 1 and 2 and the segregation effects, respectively. The transpose of the design matrix **X** is: $\mathbf{X}' = \begin{bmatrix} 0 & 0 & 0 & 1 & 1 & 0 & 0 & 0 & 1 & 1 & 1 \\ 1 & 1 & 1 & 0 & 0 & 1 & 1 & 1 & 0 & 0 & 0 \\ 1 & 1 & 0 & 0 & 1 & 0.5 & 0 & 0.75 & 0.25 & 0.5 & 0.875 \\ 0 & 0 & 1 & 1 & 0 & 0.5 & 1 & 0.25 & 0.75 & 0.5 & 0.125 \end{bmatrix}$; noting that the elements of **X**′ in the last 2 rows represent breed effect and consist of the contribution of the genes from each breed to each individual. Given that every animal has an observation, the **Z** matrices are diagonal with the diagonal elements of $\mathbf{Z}_1$ or $\mathbf{Z}_2$ equal to 1 when there are genes contributed by breeds 1 or 2, respectively. Thus, the diagonal elements for $\mathbf{Z}_1$ and $\mathbf{Z}_2$ are [1 1 0 0 1 1 0 1 1 1 1] and [0 0 1 1 0 1 1 1 1 1 1], respectively. The diagonal element of $\mathbf{Z}_3$ equal to 1 when the segregation term (column h_{12} in Table 14.1) is greater than zero; therefore $\mathbf{Z}_3$ = [0 0 0 0 0 0 0 1 1 1 1].

Table 14.1. Pedigree and observations (Garcia-Cortes and Toro, 2006) for a group of animals with breed proportion values f_1 for breed 1, f_2 for breed 2 and h_{12} for segregation between both breeds.

Animal	Sire	Dam	Breed	Herd	Record	f_1	f_2	h_{12}
1	0	0	1	2	11.0	1	0	0
2	0	0	1	2	12.0	1	0	0
3	0	0	2	2	13.0	0	1	0
4	0	0	2	1	14.0	0	1	0
5	1	2		1	15.0	1	0	0
6	3	2		2	16.0	0.5	0.5	0
7	3	4		2	17.0	0	1	0
8	5	6		2	18.0	0.75	0.25	0.5
9	7	6		1	19.0	0.25	0.75	0.5
10	9	8		1	20.0	0.5	0.5	0.75
11	5	8		1	21.0	0.875	0.125	0.375

The partial relationship matrices $\mathbf{A}_i$ with zero rows when there is no information from the ith breed are:

$\mathbf{A}_1 =$

1.000										
0.000	1.000									
0.000	0.000	0.000								
0.000	0.000	0.000	0.000							
0.500	0.500	0.000	0.000	1.000			Symmetric			
0.000	0.500	0.000	0.000	0.250	0.500					
0.000	0.000	0.000	0.000	0.000	0.000	0.000				
0.250	0.500	0.000	0.000	0.625	0.375	0.000	0.875			
0.000	0.250	0.000	0.000	0.125	0.250	0.000	0.188	0.250		
0.125	0.375	0.000	0.000	0.375	0.312	0.000	0.531	0.219	0.594	
0.375	0.500	0.000	0.000	0.812	0.312	0.000	0.750	0.156	0.453	1.188

$\mathbf{A}_2 =$

0.000										
0.000	0.000									
0.000	0.000	1.000								
0.000	0.000	0.000	1.000							
0.000	0.000	0.000	0.000	0.000						
0.000	0.000	0.500	0.000	0.000	0.500					
0.000	0.000	0.500	0.500	0.000	0.250	1.000				Symmetric
0.000	0.000	0.250	0.000	0.000	0.250	0.125	0.250			
0.000	0.000	0.500	0.250	0.000	0.375	0.625	0.188	0.875		
0.000	0.000	0.375	0.125	0.000	0.312	0.375	0.219	0.531	0.594	
0.000	0.000	0.125	0.000	0.000	0.125	0.062	0.125	0.094	0.109	0.125

A_{12} =

0.000										
0.000	0.000									
0.000	0.000	0.000								
0.000	0.000	0.000	0.000							
0.000	0.000	0.000	0.000	0.000				Symmetric		
0.000	0.000	0.000	0.000	0.000	0.000					
0.000	0.000	0.000	0.000	0.000	0.000	0.000				
0.000	0.000	0.000	0.000	0.000	0.000	0.000	0.500			
0.000	0.000	0.000	0.000	0.000	0.000	0.000	0.000	0.500		
0.000	0.000	0.000	0.000	0.000	0.000	0.000	0.250	0.250	0.750	
0.000	0.000	0.000	0.000	0.000	0.000	0.000	0.250	0.000	0.125	0.375

The A^{-}_{i} matrices used in Eqn 14.9 are the inverses formed by the non-zero rows of each partial numerator relationship matrix. These A^{-}_{i} are multiplied by $\alpha_1 = 4/1 = 4$, $\alpha_2 = 4/2 = 2$ and $\alpha_{12} = 4/0.50 = 8$. All matrices needed for Eqn 14.9 have now been derived. Solving the MME with the solution for herd 1 constrained to zero gives the following answers in Table 14.2 under the sub-heading 'Breed-specific solutions'.

Using Eqn 14.3, the MME are:

$$\begin{pmatrix} X'X & X'Z \\ Z'X & Z'Z + G^{-1}\sigma_e^2 \end{pmatrix} \begin{pmatrix} \hat{b} \\ \hat{a} \end{pmatrix} = \begin{pmatrix} X'y \\ Z'y \end{pmatrix}$$

The design matrix **X** is the same as described above but **Z** is now a diagonal matrix with its diagonal elements equal to 1 for all animals with records, as in the usual case for an animal BLUP model. The **G** matrix is derived using Eqns 14.1 and 14.2 and is equal to $A_1\sigma^2_1 + A_2\sigma^2_2 + A_3\sigma^2_{12}$. The **G** matrix for the example data is:

G =

1.000										
0.000	1.000									
0.000	0.000	2.000								
0.000	0.000	0.000	2.000							
0.500	0.500	0.000	0.000	1.000					Symmetric	
0.000	0.500	1.000	0.000	0.250	1.500					
0.000	0.000	1.000	1.000	0.000	0.500	2.000				
0.250	0.500	0.500	0.000	0.625	0.875	0.250	1.625			
0.000	0.250	1.000	0.500	0.125	1.000	1.250	0.562	2.250		
0.125	0.375	0.750	0.250	0.375	0.938	0.750	1.094	1.406	2.156	
0.375	0.500	0.250	0.000	0.812	0.562	0.125	1.125	0.344	0.734	1.625

All matrices needed for the MME in Eqn 14.3 have been derived and solving the MME gives the solutions in Table 14.2 under the sub-heading 'Combined breed solutions'.

Table 14.2. The MME solutions for Example 14.1.

	Combined breed solutions		Breed-specific solutions	
Herd effect				
	0.000		0.000	
	–3.198		–3.198	
Breed effect				
1	16.613		16.613	
2	17.408		17.408	
Animal effect				
		Breed 1	Breed 2	Segregation
1	–0.293	–0.293	0.000	0.000
2	0.293	0.293	0.000	0.000
3	0.689	0.000	0.689	0.000
4	–0.689	0.000	–0.689	0.000
5	0.237	0.237	0.000	0.000
6	1.216	0.388	0.828	0.000
7	0.671	0.000	0.671	0.000
8	1.746	0.750	0.705	0.291
9	1.262	0.230	0.962	0.071
10	1.779	0.556	0.966	0.257
11	1.456	0.781	0.441	0.234

The results in Table 14.2 demonstrates the equivalence of Eqns 14.3 and 14.8 but the advantage of (14.8) is the ability to obtain breed-specific breeding values.

14.4 Random Regression Approach to Multibreed Analysis

Strandén and Mäntysaari (2013) presented a random regression model as an approximation for multibreed analysis, basing their derivation on the model of Garcia-Cortes and Toro (2006) in Section 14.2. Their approximation has the advantage that it permits the use of genomic data in the analysis of multibreed data.

The diagonal elements of the partial **A** matrices in the set of equations in (14.10) are proportional to $f^{\mathrm{i}}_{\mathrm{p}}$ and $h^{\mathrm{i}\prime}_{\mathrm{pp}}$ (with $h^{\mathrm{i}}_{\mathrm{pp}'} = 2(f^S_p f^S_{p'} + f^D_p f^D_{p'})$) when elements $a^{(p)}_{SD}$ and $a^{(pp')}_{SD}$ are zero. Assuming that this proportionality holds for all animals, Strandén and Mäntysaari (2013) approximated the diagonal elements $\mathbf{A}^{(p)*}_{ii}$ as $a^{(\mathrm{p})*}_{\mathrm{ii}} = f^{\mathrm{i}}_{\mathrm{p}} a_{\mathrm{ii}}$ and for $\mathbf{A}^{(pp')*}_{ii}$ as $a^{(\mathrm{pp}')*}_{\mathrm{ii}} = h^{\mathrm{i}}_{\mathrm{pp}}{}' a_{\mathrm{ii}}$. Extending the approximation to the off-diagonal elements, $\mathbf{A}^{(p)*}_{ij}$ could be expressed as $a^{(\mathrm{p})*}_{\mathrm{ij}} = \sqrt{(f^i_p f^i_p)} a_{ij}$ and $\mathbf{A}^{(pp)*}_{ij}$ as $a^{(\mathrm{pp}')*}_{\mathrm{ij}} = \sqrt{(h^i_{pp'} h^i_{pp'})} a_{ij}$.

The above approximation by Strandén and Mäntysaari (2013) is based on the assumption that the diagonal elements are functions of the breed proportions of the individual itself rather than that of the parents as in Garcia-Cortes and Toro (2006). If the approximated partial relationship matrices are expressed as $\mathbf{A}^*_p = \mathbf{F}_{\mathrm{p}}\mathbf{A}\mathbf{F}_{\mathrm{p}}$ and $\mathbf{A}^*_p = \mathbf{H}_{\mathrm{pp}'}\mathbf{A}\mathbf{H}_{\mathrm{pp}'}$, where the diagonal matrix $\mathbf{F}_{\mathrm{p}}$ has the square roots of the breed proportion of individuals and the diagonal matrix $\mathbf{H}_{\mathrm{pp}'}$ has the square roots of $h_{\mathrm{pp}'}$, then Eqn 14.8 can be written as:

$$\mathbf{y} = \mathbf{Xb} + \sum_p \mathbf{Za}^*_p + \sum_p \sum_{p'>p} \mathbf{Za}^*_{pp'} + \mathbf{e}$$

with the assumption now that $\mathbf{a}^*_p \sim N(\mathbf{0},\mathbf{A}^*_p\sigma^2_p)$ and $\mathbf{a}^*_{pp'} \sim N(\mathbf{0},\mathbf{A}^*_{pp'}\,\sigma^2_{pp'})$. However, an equivalent random regression model can be derived by factoring out the matrices $\mathbf{F}_p$ and $\mathbf{H}_{pp'}$ from the covariance structure of $\mathbf{a}^*_p$ and $\mathbf{a}^*_{pp'}$, such that they are part of the design matrices. Thus, if $\mathbf{Z}_p = \mathbf{ZF}_p$ and $\mathbf{Z}_p = \mathbf{ZH}_{pp'}$, Eqn 14.8 can be written as:

$$\mathbf{y} = \mathbf{Xb} + \sum_p \mathbf{Z}_p\mathbf{u}_p + \sum_p\sum_{p'>p} \mathbf{Z}_{pp'}\mathbf{u}_{pp'} + \mathbf{e} \qquad (14.11)$$

with the assumption that $\mathbf{u}_p \sim N(0,\, \mathbf{T}_p\mathbf{AT}_p\,\sigma^2_p)$ and $\mathbf{u}'_{pp} \sim N(0,\, \mathbf{T}_{pp'}\mathbf{AT}_{pp'}\,\sigma^2_{pp'})$, where $\mathbf{T}_p$ and $\mathbf{T}_{pp'}$ are diagonal incidence matrices with ones for animals with non-zero values in $\mathbf{F}_p$ and $\mathbf{H}'_{pp}$, respectively. Therefore, the $\mathbf{T}_p\mathbf{AT}_p$ matrix has the same elements as the ordinary $\mathbf{A}$ matrix but with zero rows and columns for the other purebred animals not from breed p. Note that total breeding value from random regression in (14.11) is computed as $\mathbf{a}^* = \sum_p \mathbf{F}_p\mathbf{u}_p + \sum_p\sum_{p'>p}\mathbf{H}_{pp'}\mathbf{u}_{pp'}$.

Example 14.2

Using Eqn 14.11, the example in Table 14.1 is analysed using the same genetic parameters to illustrate to approximate random regression approach of Strandén and Mäntysaari (2013).

The MME to be solved using Eqn 14.11 for the example data are:

$$\begin{pmatrix} \mathbf{X'X} & \mathbf{X'Z}_1 & \mathbf{X'Z}_2 & \mathbf{X'Z}_{12} \\ \mathbf{Z}_1'\mathbf{X} & \mathbf{Z}_1'\mathbf{Z}_1 + (\mathbf{T}_1\mathbf{AT}_1)^{-1}\alpha_1 & \mathbf{Z}_1'\mathbf{Z}_2 & \mathbf{Z}_1'\mathbf{Z}_{12} \\ \mathbf{Z}_2'\mathbf{X} & \mathbf{Z}_2'\mathbf{Z}_1 & \mathbf{Z}_2'\mathbf{Z}_2 + (\mathbf{T}_2\mathbf{AT}_2)^{-1}\alpha_2 & \mathbf{Z}_2'\mathbf{Z}_{12} \\ \mathbf{Z}_{12}'\mathbf{X} & \mathbf{Z}_{12}'\mathbf{Z}_1 & \mathbf{Z}_{12}'\mathbf{Z}_2 & \mathbf{Z}_{12}'\mathbf{Z}_{12} + (\mathbf{T}_{12}\mathbf{AT}_{12})^{-1}\alpha_{12} \end{pmatrix} \begin{pmatrix} \hat{\mathbf{b}} \\ \hat{\mathbf{u}}_1 \\ \hat{\mathbf{u}}_2 \\ \hat{\mathbf{u}}_{12} \end{pmatrix} = \begin{pmatrix} \mathbf{X'y} \\ \mathbf{Z}_1'\mathbf{y} \\ \mathbf{Z}_2'\mathbf{y} \\ \mathbf{Z}_{12}'\mathbf{y} \end{pmatrix}$$

The design matrix $\mathbf{X}$ and the variance ratios α_i are as defined in Example 14.1, but the diagonal elements of the $\mathbf{Z}$ matrices are the square roots of the proportion of genes contributed by i^{th} breed to each animal. The breed proportion for each animal is shown in columns titled f_1, f_2 and h_{12} in Table 14.2. Thus, the diagonal elements of the $\mathbf{Z}$ matrices are $\mathbf{Z}_1$ = [1 1 0 0 1 0.707 0 0.866 0.50 0.707 0.935], $\mathbf{Z}_2$ = [0 0 1 1 0 0.707 1 0.5 0.866 0.707 0.354] and $\mathbf{Z}_{12}$ = [0 0 0 0 0 0 0 0.707 0.707 0.866 0.612]. The $\mathbf{A}$ matrix is the usual numerator relationship. The diagonal elements of the $\mathbf{T}$ matrices are equal to 1 when the diagonal of the corresponding $\mathbf{Z}$ matrix is not zero, but otherwise zero. Thus, the diagonal elements of the $\mathbf{T}$ matrices are $\mathbf{T}_1$ = [1 1 0 0 1 1 0 1 1 1 1], $\mathbf{T}_2$ = [0 0 1 1 0 1 1 1 1 1 1] and $\mathbf{T}_{12}$ = [0 0 0 0 0 0 0 1 1 1 1].

For illustration purposes, the $\mathbf{A}$ matrix and $\mathbf{T}_1\mathbf{AT}_1$ are given below, noting the zero rows and columns for animals 3, 4 and 5 in $\mathbf{T}_1\mathbf{AT}_1$ that have no gene contribution from breed 1.

A =

1.000										
0.000	1.000									
0.000	0.000	1.000								
0.000	0.000	0.000	1.000							
0.500	0.500	0.000	0.000	1.000				symmetric		
0.000	0.500	0.500	0.000	0.250	1.000					
0.000	0.000	0.500	0.500	0.000	0.250	1.000				
0.250	0.500	0.250	0.000	0.625	0.625	0.125	1.125			
0.000	0.250	0.500	0.250	0.125	0.625	0.625	0.375	1.125		
0.125	0.375	0.375	0.125	0.375	0.625	0.375	0.750	0.750	1.188	
0.375	0.500	0.125	0.000	0.812	0.438	0.062	0.875	0.250	0.562	1.312

T_1AT_1 =

1.000										
0.000	1.000									
0.000	0.000	0.000								
0.000	0.000	0.000	0.000							
0.500	0.500	0.000	0.000	1.000				symmetric		
0.000	0.500	0.000	0.000	0.250	1.000					
0.000	0.000	0.000	0.000	0.000	0.000	0.000				
0.250	0.500	0.000	0.000	0.625	0.625	0.000	1.125			
0.000	0.250	0.000	0.000	0.125	0.625	0.000	0.375	1.125		
0.125	0.375	0.000	0.000	0.375	0.625	0.000	0.750	0.750	1.188	
0.375	0.500	0.000	0.000	0.812	0.438	0.000	0.875	0.250	0.562	1.312

Solving the MME with the solution of the first herd set to zero gives the following solutions in Table 14.3. Note that the breeding values for animals were obtained from the solutions from the MME as $\mathbf{a}^* = \Sigma_p \mathbf{F}_p \mathbf{u}_p + \Sigma_p \Sigma_{p'>p} \mathbf{H}_{pp'} \mathbf{u}_{pp'}$ where, as defined previously, the diagonal matrix $\mathbf{F}_p$ has the square roots of the breed proportion of individuals and the diagonal matrix $\mathbf{H}_{pp'}$ has the square roots of $h_{pp'}$.

Compared to the results in Table 14.2, some large differences in the solutions by the approximate approach are for animals 2 and 3. Strandén and Mäntysaari (2013) explained that these two animals had crossbred progeny which led to large differences in the genetic relationship matrices among animals 2 and 3 and other animals in the analysis. They explained that the approximation is likely to work well when the population is highly admixed, i.e. only a small proportion is purebred animals or F1 crossbred population. If the data set consists of lines of purebreed animals which are crossed to produce

Table 14.3. The MME solutions for Example 14.2.

	Combined breed solutions		Breed-specific solutions	
Herd effect				
	0.000		0.000	
	–3.051		–3.051	
Breed effect				
1	16.504		16.504	
2	17.462		17.462	
Animal effect				
		Breed 1	Breed 2	Segregation
1	–0.333	–0.333	0.000	0.000
2	0.147	0.147	0.000	0.00
3	0.298	0.000	0.298	0.000
4	–0.726	0.000	–0.726	0.000
5	0.104	0.104	0.000	0.000
6	1.219	0.437	0.782	0.000
7	0.427	0.000	0.427	0.000
8	1.888	0.721	0.792	0.375
9	1.286	0.257	0.838	0.190
10	1.944	0.535	1.019	0.390
11	1.597	0.723	0.542	0.332

F1 crossbreds, then the methodology is likely to perform poorest. Thus, if it is a crossbreeding system that is designed to achieve positive heterosis by maximizing breed composition between parents, then the approximation is not well suited to such a system. However, when genomic data are available, their method allows the application of genomic models (see Section 14.5). The application of their method in a highly admixed Nordic red dairy cattle (Makgahlela *et al.*, 2013) in a genomic model context (see Section 14.5) gave good results compared to when breed heterogeneity was not accounted for.

14.5 Multibreed Genomic Random Regression Model

When genotypes are available, relationships among animals can be computed using the genomic relationship matrix ($\mathbf{G}_M$) as described in Chapter 11. Thus, the model (14.11) can be applied for the analysis of genomic data using $\mathbf{G}_M$ instead of the usual $\mathbf{A}$ matrix. In a data set when all the genotyped animals are crossbreds and no purebred animals are genotyped, matrices $\mathbf{T}_p$ and $\mathbf{T}_{pp'}$ are identity matrices and, therefore, the MME for model Eqn 14.11 are just the usual GBLUP random regression model using a genomic relationship.

However, if the interest is to estimate SNP or marker effects ($\mathbf{v}$), Eqn 14.11 can be written as

$$\mathbf{y} = \mathbf{Xb} + \sum_{p} \mathbf{Z}_p \mathbf{Z}_M \mathbf{v}_p + \sum_{p} \sum_{p'>p} \mathbf{Z}_{pp'} \mathbf{Z}_M \mathbf{v}_{pp'} + \mathbf{e} \tag{14.12}$$

where the $\mathbf{Z}_M$ matrix has genotypes centralized as described in Chapter 11 and $\mathbf{v}_p \sim N(\mathbf{0}, \mathbf{I}\sigma^2_p)$ and $\mathbf{v}_{pp'} \sim N(\mathbf{0}, \mathbf{I}\sigma^2_{pp'})$. Equation 14.12 can be expressed as:

$$\mathbf{y} = \mathbf{Xb} + \sum_{p} \mathbf{W}_p \mathbf{v}_p + \sum_{p} \sum_{p'>p} \mathbf{W}_{pp'} \mathbf{v}_{pp'} + \mathbf{e} \tag{14.13}$$

where $\mathbf{W}_p = \mathbf{Z}_p\mathbf{Z}_M$ and $\mathbf{W}_{pp'} = \mathbf{Z}_{pp'}\mathbf{Z}_M$. Then, the MME for Eqn 14.13, considering two breeds and their cross, are:

$$\begin{pmatrix} \mathbf{X'X} & \mathbf{X'W}_1 & \mathbf{X'W}_2 & \mathbf{X'W}_{12} \\ \mathbf{W}_1'\mathbf{X} & \mathbf{W}_1'\mathbf{W}_1 + \alpha_1\mathbf{I} & \mathbf{W}_1'\mathbf{W}_2 & \mathbf{W}_1'\mathbf{W}_{12} \\ \mathbf{W}_2'\mathbf{X} & \mathbf{W}_2'\mathbf{W}_1 & \mathbf{W}_2'\mathbf{W}_2 + \alpha_2\mathbf{I} & \mathbf{W}_2'\mathbf{W}_{12} \\ \mathbf{W}_{12}'\mathbf{X} & \mathbf{W}_{12}'\mathbf{W}_1 & \mathbf{W}_{12}'\mathbf{W}_2 & \mathbf{W}_{12}'\mathbf{W}_{12} + \alpha_{12}\mathbf{I} \end{pmatrix} \begin{pmatrix} \hat{\mathbf{b}} \\ \widehat{\mathbf{v}_1} \\ \widehat{\mathbf{v}_2} \\ \widehat{\mathbf{v}_{12}} \end{pmatrix} = \begin{pmatrix} \mathbf{X'y} \\ \mathbf{W}_1'\mathbf{y} \\ \mathbf{W}_2'\mathbf{y} \\ \mathbf{W}_{12}'\mathbf{y} \end{pmatrix} \tag{14.14}$$

where $\alpha_1 = \sigma^2_e/\sigma^2_{p1}$, $\alpha_2 = \sigma^2_e/\sigma^2_{p2}$ and $\alpha_{12} = \sigma^2_e/\sigma^2_{p12}$. Note that σ^2_p and $\sigma^2_{pp'}$ are genetic variances for the marker effects and that the genomic breeding values $\widehat{\mathbf{u}_p}$ and $\widehat{\mathbf{u}_{pp'}}$ in (14.11) can be obtained as $\widehat{\mathbf{u}_p} = \mathbf{Z}_M\widehat{\mathbf{v}_p}$ and $\widehat{\mathbf{u}_{pp'}} = \mathbf{Z}_M\widehat{\mathbf{v}_{pp'}}$ respectively, from their marker solutions in (14.14).

14.6 Genomic Predictions for Crossbreds Using Genomic Breed Composition

14.6.1 Predicting genomic breed composition (GBC) of crossbreds

VanRaden *et al.* (2011, 2020) demonstrated that the genomic breed composition (GBC) of crossbreds can be estimated accurately using all SNP markers. The process involved using the usual genomic prediction models with the reference population consisting of the genotypes of purebred animals for the different breeds used to produce the crossbreds. For instance, assume that the aim is to estimate the GBC of crossbreds from three purebred populations – Holstein, Ayrshire and Jersey breeds – then the prediction process involves initially creating a dependent variable $\mathbf{y}_i$ (phenotype) for each of the three breeds, where *i* equals 1, 2 or 3 for Holstein, Ayrshire, and Jersey, respectively. In order to predict the fraction of Jersey genes, for example, it involves assigning 100% to $\mathbf{y}_3$ and 0% to $\mathbf{y}_1$ and $\mathbf{y}_2$ (VanRaden *et al.*, 2011).

Genomic evaluations are then run separately to predict the three breed fractions using the following model:

$$\mathbf{y}_i = \mathbf{1}'\mu_i + \mathbf{Z}\mathbf{g}_i + \mathbf{e}_i \tag{14.15}$$

where the **Z** matrix contains genotypes centred by base allele frequencies and it is of size *n* (the number of animals) times *k* (the number of SNP markers), μ_i is an intercept, the $\mathbf{g}_i$ vector is the random marker regression to predict the *i*th breed, and $\mathbf{e}_i$ is the random error. VanRaden (2008) used an approximate BayesA algorithm to estimate the marker effects.

The estimates of genomic breed composition (GBC) from the multiple regressions can result in breed composition exceeding 100% for a given breed or be negative for some breeds. Considering only purebred animals, the expected and average GBC equal 100% for the breed being analysed and 0% for other breeds, but some individual animals could fluctuate around those values because of genomic variation. Therefore, VanRaden *et al.* (2020) adjusted GBC for each animal to sum to 100%, with no estimates lower than 0% or higher than 100%. The adjusted GBC were referred to as breed base representation (BBR) and these were used for the computation of gEBVs for crossbred animals. They provided the following steps for adjusting GBC to obtain BBR: (i) sum GBC across breeds; (ii) adjust GBC mean by subtracting from each GBC value the sum of GBC divided by the number of breeds (N_b); (iii) obtain the range of adjusted GBC (maximum and minimum); (iv) compute standard deviation (SD) of adjusted GBC to be used for adjustment if any adjusted GBC are higher than 100 or lower than 0. Set the maximum: (largest adjusted GBC –(100/ N_b))/(100[1 – (1/N_b)]) and the minimum to: ((100/N_b) – smallest adjusted GBC)/(100/ N_b); (v) calculate BBR as (100/N_b) + [(adjusted GBC – (100/N_b)/SD]; and (vi) for animals with a BBR of 94% or higher, round BBR to 100% and set contributions from all other breeds to 0%. Thus, animals with contributions less than 6% from other breeds were assumed to be purebred.

14.6.2 Genomic prediction for crossbred using genomic breed compositions

The computation of gEBVs for crossbreds requires that the marker or SNP effects have been computed for all breeds used in the crossing breeding programme with any of the genomic prediction methods discussed in Chapter 11. Then, gEBVs are computed for each purebred animal from the breed marker effects but for crossbreed animals the marker effects are weighted by the BBR to obtain their gEBVs. This approach requires that the marker effects are on the same scale for all breeds, i.e. these have not been adjusted by any within-breed base adjustment.

In effect, crossbred predictions using this approach are essentially averages of genomic predictions computed using marker effects for each purebreed, which were weighted by the animal's BBR. VanRaden *et al.* (2020) reported that for crossbred animals, these predictions were more accurate than the average of parents' breeding values and slightly more accurate than predictions using only the predominant breed.

14.7 Genomic Prediction of Purebreed and Crossbreed Performances Using Breed Origin of Alleles

Using the concept of partial relationship matrices by Garcia-Cortes and Toro (2006) discussed in Section 14.2, Christensen *et al.* (2014) introduced the concept of partial genomic relationship matrices for the genomic evaluation of both purebred and crossbred performances involving two breeds and crosses between them. The use of partial genomic relationship matrices implies that the breed origin of alleles for crossbreds can be incorporated into the prediction process.

Christensen *et al.* (2014) derived their method by reformatting the two-breed crossbreeding model of Wei and van der Werf (1991) for the estimation of breeding values for both purebred and crossbred animals. Briefly, the model of Wei and van der Werf (1991) is reviewed, and the derivation of Christensen *et al.* (2014) is outlined, first using pedigree relationships, then followed by using genomic information.

The model of Wei and van der Werf (1991) is a trivariate model involving two breeds (1 and 2) and their crosses (3) and has the form:

$$\begin{aligned}\mathbf{y}_1 &= \mathbf{X}_1\boldsymbol{\beta}_1 + \mathbf{Z}_1\mathbf{a}_1 + \mathbf{e}_1\\ \mathbf{y}_2 &= \mathbf{X}_2\boldsymbol{\beta}_2 + \mathbf{Z}_2\mathbf{a}_2 + \mathbf{e}_2\\ \mathbf{y}_3 &= \mathbf{X}_3\boldsymbol{\beta}_3 + \mathbf{c}_3 + \mathbf{e}_3\end{aligned} \tag{14.16}$$

where $\mathbf{y}_1$, $\mathbf{y}_2$ and $\mathbf{y}_3$ are vectors of observations for breeds 1, 2 and crossbred (3) animals, respectively; $\mathbf{X}_1\boldsymbol{\beta}_1$, $\mathbf{X}_2\boldsymbol{\beta}_2$ and $\mathbf{X}_3\boldsymbol{\beta}_3$ are vectors of fixed effects for the three breed groups; $\mathbf{Z}_1$ and $\mathbf{Z}_2$ are incidence matrices assigning breeding values to records for breeds 1 and 2, respectively; $\mathbf{a}_1$ and $\mathbf{a}_2$ are breeding value vectors for purebred animal performance for breed 1 and 2 animals, respectively, and $\mathbf{c}_3$ is a vector of additive genetic effects for crossbred (3) animals and is related to the vectors of breeding values for purebred animals for crossbred performance (mating between breeds 1 and 2) as follows:

$$\mathbf{c}_3 = 0.5(\mathbf{Z}_{13}\mathbf{c}_1 + \mathbf{Z}_{23}\mathbf{c}_2) + \Phi_3 \tag{14.17}$$

where the $\mathbf{Z}_{13}$ and $\mathbf{Z}_{23}$ matrices assign purebred parents of breeds 1 and 2 to crossbred offspring, respectively, the $\mathbf{c}_1$ vector has breeding values for crossbred performance for breed 1 animals (crosses with breed 2 animals) and $\mathbf{c}_2$ is the corresponding vector for crossbred performance for breed 2 animals (crosses with breed 1 animals), and the $\boldsymbol{\Phi}_3$ vector has the Mendelian sampling effects.

The genetic covariance matrices for $\mathbf{a}_1$ and $\mathbf{c}_1$, $\mathbf{a}_2$ and $\mathbf{c}_2$ and $\boldsymbol{\Phi}_3$ are:

$$\text{Var}\begin{bmatrix}\mathbf{a}_1\\ \mathbf{c}_1\end{bmatrix} = \mathbf{S}^{(1)} \otimes \mathbf{A}_1; \text{Var}\begin{bmatrix}\mathbf{a}_2\\ \mathbf{c}_2\end{bmatrix} = \mathbf{S}^{(2)} \otimes \mathbf{A}_2 \text{ and } \text{Var}[\Phi_3] = \mathbf{D}_3$$

Note that these three vectors are independent with $\mathbf{A}_1$ and $\mathbf{A}_2$ being the additive relationship matrices for breeds 1 and 2, respectively, and $\otimes$ is the Kronecker product. Note that $\mathbf{S}^{(i)}$ is a 2 × 2 matrix containing the genetic variances for purebred ($s_{ii}^{(i)}$) and crossbred breeding values ($s_{33}^{(i)}$), and the covariance between the two ($s_{i3}^{(i)}$). The $\mathbf{S}^{(i)}$ matrix has the following structure, noting that the subscript for $\mathbf{S}$ relates to the coding for breeds:

$$\mathbf{S}^{(i)} = \begin{pmatrix} s_{ii}^{(i)} & s_{i3}^{(i)}\\ s_{3i}^{(i)} & s_{33}^{(i)}\end{pmatrix} \text{ and for breed 1, for example, } \mathbf{S}^{(1)} = \begin{pmatrix} s_{11}^{(1)} & s_{13}^{(1)}\\ s_{31}^{(1)} & s_{33}^{(1)}\end{pmatrix}.$$

The variance-covariance matrix of Mendelian sampling, $\mathbf{D}_3$, is a diagonal matrix with elements derived in the usual way as the variance of the breeding value for the crossbred animal deviated from the average breeding values of both parents which belong to different breeds. Thus:

$$
\begin{aligned}
(D_3)_{ii} &= \mathrm{Var}(\mathbf{c}_{3i}) - (\mathrm{Var}(\mathbf{c}_{1f(i)}) + \mathrm{Var}(\mathbf{c}_{2p(i)}))/4 \\
&= (s_{33}^{(1)} + s_{33}^{(2)})/2 - (s_{33}^{(1)}(A_1)_{f(i)f(i)} + s_{33}^{(2)}(A_2)_{p(i)p(i)})/4 \\
&= (s_{33}^{(1)}((1/2) - (A_1)_{f(i)f(i)}/4) + (s^{(2)}{}_{33}((1/2) - (A_2)_{p(i)p(i)}/4)
\end{aligned}
\qquad (14.18)
$$

where for the ith crossbred animal, f(i) and p(i) denote the breed 1 parent and breed 2 parent, respectively.

Formulating the trivariate equations in (14.16) using the concept of partial breed-specific partial relationship matrices of Garcia-Cortes and Toro (2006) (Section 14.2) begins by splitting the Mendelian sampling term into two Mendelian sampling terms, one for the breed 1 gametes and the other for the breed 2 gametes. Thus, the Mendelian sampling term in Eqn 14.17 for crossbred animals can be split into breed-of-origin effects as: $\Phi_3 = \Phi^{(1)}_3 + \Phi^{(2)}_3$, where $\Phi^{(1)}_3$ and $\Phi^{(2)}_3$ are independent.

Therefore, in matrix form, Eqn 14.18 can be formulated as $\mathbf{D}_3 = \mathrm{Var}(\Phi^{(1)}_3) + \mathrm{Var}(\Phi^{(2)}_3)$, where $\mathrm{Var}(\Phi^{(1)}_3) = s^{(1)}_{33}(0.5\mathbf{I}_{n3} - 0.25\mathrm{diag}(\mathbf{Q}_1))$ and $\mathrm{Var}(\Phi^{(2)}_3) = s^{(2)}_{33}(0.5\mathbf{I}_{n12} - 0.25\mathrm{diag}(\mathbf{Q}_2))$, where $\mathrm{diag}(\mathbf{Q}_1)$ and $\mathrm{diag}(\mathbf{Q}_2)$ are diagonal matrices with the diagonal elements of $(\mathbf{Z}_{13},\mathbf{A}_1\mathbf{Z}'_{13})$ and $(\mathbf{Z}_{23}\,\mathbf{A}_2\mathbf{Z}'_{23})$, respectively, and $\mathbf{I}_{n3}$ an identity matrix of size n_3, the number of crossbred animals.

Thus, the additive genetic effects for crossbred animals in (14.17) can then be expressed as:

$$
\mathbf{c}_3 = \mathbf{c}^{(1)}_3 + \mathbf{c}^{(2)}_3 \qquad (14.19)
$$

where $\mathbf{c}^{(1)}_3$ and $\mathbf{c}^{(2)}_3$ are independent indicating that the genetic effects for crossbred animals have been split into two breed-of-origin effects with $\mathbf{c}^{(1)}_3 = 0.5(\mathbf{Z}_{13}\mathbf{c}_1 + \Phi^{(1)}_3)$ and $\mathbf{c}^{(2)}_3 = 0.5(\mathbf{Z}_{23}\mathbf{c}_2 + \Phi^{(2)}_3)$.

Using (14.19), Eqn 14.16 can now be expressed as:

$$
\begin{aligned}
\mathbf{y}_1 &= \mathbf{X}_1\boldsymbol{\beta}_1 + \mathbf{Z}_1\mathbf{a}_1 + \mathbf{e}_1 \\
\mathbf{y}_2 &= \mathbf{X}_2\boldsymbol{\beta}_2 + \mathbf{Z}_2\mathbf{a}_2 + \mathbf{e}_2 \\
\mathbf{y}_3 &= \mathbf{X}_{12}\boldsymbol{\beta}_{12} + \mathbf{c}^{(1)}_3 + \mathbf{c}^{(2)}_3 + \mathbf{e}_{12}
\end{aligned}
\qquad (14.20)
$$

Illustrating with breed 1, the variance-covariance matrix of the genetic effects for $\mathbf{c}^{(1)}_3$, from the definition in (14.19) becomes:

$$
\begin{aligned}
\mathrm{Var}(\mathbf{c}^{(1)}_3) &= 0.25s^{(1)}_{33}\,\mathbf{Z}_{13}\mathbf{A}_1\mathbf{Z}'_{13} + s^{(1)}_{33}\,(0.5\mathbf{I}_{n3} - 0.25\mathrm{diag}(\mathbf{Z}_{13},\mathbf{A}_1\mathbf{Z}'_{13})) \\
&= s^{(1)}_{33}\,\mathbf{A}^{(1)}_3
\end{aligned}
$$

where $\mathbf{A}^{(1)}_3$ is a matrix with diagonal elements $(\mathbf{A}^{(1)}_3)_{i,i} = 0.5$ and off-diagonal element $(\mathbf{A}(^{1}_{3}))_{i,i'} = 0.25(\mathbf{Z}_{13},\mathbf{A}_1\,\mathbf{Z}'_{13})_{ii'}$, with $i \neq i'$ and the covariance matrix between $\mathbf{c}^{(1)}_3$ and $\mathbf{c}_1$ becomes:

$$
\mathrm{Cov}(\mathbf{c}^{(1)}_3, \mathbf{c}_1) = 0.5s^{(1)}_{13}\,\mathbf{Z}_{13}\mathbf{A}_1
$$

Putting all together, the variance-covariance matrix of breed 1-specific genetic effects for crossbred performance is:

$$
\mathrm{Var}\begin{pmatrix} \mathbf{c}_1 \\ \mathbf{c}^{(1)}_3 \end{pmatrix} = \mathbf{S}^{(1)}\mathbf{A}^{(1)}
$$

with $\mathbf{A}^{(1)}$ of dimension (n_1 (number of breed 1 animals) + n_3) denotes the breed 1-specific partial relationship matrix and is

$$\mathbf{A}^{(1)} = \begin{pmatrix} \mathbf{A}_1 & 0.5\mathbf{A}_1\mathbf{Z}'_{13} \\ 0.5\mathbf{Z}'_{13}\mathbf{A}_1 & \mathbf{A}_3^{(1)} \end{pmatrix}.$$

Similarly, for breed 2:

$\text{Var}\begin{pmatrix} \mathbf{c}_2 \\ \mathbf{c}_3^{(2)} \end{pmatrix} = \mathbf{S}^{(2)}\mathbf{A}^{(2)}$ and the breed 2-specific partial relationship matrix is:

$$\mathbf{A}^{(2)} = \begin{pmatrix} \mathbf{A}_2 & 0.5\mathbf{A}_2\mathbf{Z}'_{23} \\ 0.5\mathbf{Z}'_{23}\mathbf{A}_2 & \mathbf{A}_3^{(2)} \end{pmatrix}.$$

The recursive formulae to compute the breed 1 and breed 2-specific partial relationship matrices are similar to those presented by Garcia-Cortes and Toro (2006) in Eqn 14.10. For breed 1, these are:

$$\begin{aligned} \mathbf{A}_{ii}^{(1)} &= f_i^A + \frac{1}{2}\mathbf{A}_{f(i)p(i)}^{(1)} \\ \mathbf{A}_{ii'}^{(1)} &= \frac{1}{2}\left(\mathbf{A}_{f(i)i'}^{(1)} + \mathbf{A}_{p(i)i'}^{(1)}\right) \end{aligned} \qquad (14.21)$$

where f(i) and p(i) are the two parents of the ith animal, animal i' is not a descendant of i, and f_i^1 is the breed 1 proportion of individual i with the value of 1 for purebred 1 animals, 0 for purebred 2 animals and 0.5 for crossbred animals.

Note that the breed 1-specific partial relationship matrix $\mathbf{A}^{(1)}$ is of dimension ($n_1 + n_3$), whereas the vector of breeding values for breed 1, $\mathbf{a}_1$ is of dimension n_1. We need to introduce a conceptual random vector ($\mathbf{a}^{(1)}{}_3$) of dimension (n_3) which represents the effects of crossbred gametes for purebred animal performance, such that genetic variance-covariance matrices can be presented using Kronecker products. Thus, for breed 1, the genetic covariances can be presented as:

$$\text{Var}\begin{pmatrix} \mathbf{a}_1 \\ \mathbf{a}_3^{(1)} \\ \mathbf{c}_1 \\ \mathbf{c}_3^{(1)} \end{pmatrix} = \mathbf{S}^{(1)} \otimes \mathbf{A}^{(1)} = \begin{pmatrix} s_{11}^{(1)} & s_{13}^{(1)} \\ s_{31}^{(1)} & s_{33}^{(1)} \end{pmatrix} \otimes \mathbf{A}^{(1)} \qquad (14.22)$$

Similarly, for breed 2:

$$\text{Var}\begin{pmatrix} \mathbf{a}_2 \\ \mathbf{a}_3^{(2)} \\ \mathbf{c}_2 \\ \mathbf{c}_3^{(2)} \end{pmatrix} = \mathbf{S}^{(2)} \otimes \mathbf{A}^{(2)} \qquad (14.23)$$

14.7.1 Inverse of the partial relationship matrices

The specific partial relationship matrix for breed 1, for example, can be expressed as $\mathbf{A}^{(1)} = \mathbf{TDT}'$, where $\mathbf{T}$ is a lower triangular matrix with the same structure as described in Section 3.4 in Chapter 3, and the inverse $\mathbf{T}^{-1}$ is a lower triangular matrix with diagonal elements equal to 1 and in the lower diagonal, the only non-zero elements are –½ for progeny parent elements; $\mathbf{D}$ is a diagonal matrix with elements $\mathbf{D}_{ii} = 1-(\mathbf{A}^{(1)}_{f(i)f(i)} + \mathbf{A}^{(1)}_{p(i)p(i)})/4$ when animal i is breed 1, and $\mathbf{D}_{ii} = ½ - \mathbf{A}^{(1)}_{f(i)f(i)}/4$ when animal i is crossbred with breed 1 parent f(i). Therefore, the inverse of $\mathbf{A}^{(1)}$ can be expressed with the usual formula in Eqn (3.3) as $(\mathbf{A}^{(1)})^{-1} = (\mathbf{T}^{-1})'\mathbf{D}^{-1}\mathbf{T}^{-1}$.

The MME equations for (14.20) are:

$$\begin{pmatrix} \mathbf{X}_1'r^{11}\mathbf{X}_1 & & & & & & \\ 0 & \mathbf{X}_2'r^{22}\mathbf{X}_2 & & & & & \\ 0 & 0 & \mathbf{X}_3'r^{33}\mathbf{X}_3 & & & & \\ \mathbf{Z}_1'r^{11}\mathbf{X}_1 & 0 & 0 & \mathbf{Z}_1'r^{11}\mathbf{Z}_1 + s^{(1)11}\mathbf{A}^{(1)-1} & & & \\ 0 & 0 & \mathbf{Z}_{13}'r^{33}\mathbf{X}_3 & s^{(1)13}\mathbf{A}^{(1)-1} & \mathbf{Z}_{13}'r^{33}\mathbf{Z}_{13} + s^{(1)33}\mathbf{A}^{(1)-1} & & \\ 0 & \mathbf{Z}_2'r^{22}\mathbf{X}_2 & 0 & 0 & 0 & \mathbf{Z}_2'r^{22}\mathbf{Z}_2 + s^{(2)11}\mathbf{A}^{(2)-1} & \\ 0 & 0 & \mathbf{Z}_{23}'r^{33}\mathbf{X}_3 & 0 & \mathbf{Z}_{23}'r^{33}\mathbf{Z}_{13} & s^{(2)13}\mathbf{A}^{(2)-1} & \mathbf{Z}_{23}'r^{33}\mathbf{Z}_{23} + s^{(2)33}\mathbf{A}^{(2)-1} \end{pmatrix}$$

$$\begin{pmatrix} \beta_1 \\ \beta_2 \\ \beta_3 \\ \mathbf{g}_{11} \\ \mathbf{g}_{13} \\ \mathbf{g}_{22} \\ \mathbf{g}_{23} \end{pmatrix} = \begin{pmatrix} \mathbf{X}_1'r^{11}\mathbf{y}_1 \\ \mathbf{X}_2'r^{22}\mathbf{y}_2 \\ \mathbf{X}_3'r^{33}\mathbf{y}_3 \\ \mathbf{Z}_1'r^{11}\mathbf{y}_1 \\ \mathbf{Z}_{13}'r^{33}\mathbf{y}_3 \\ \mathbf{Z}_2'r^{22}\mathbf{y}_2 \\ \mathbf{Z}_{23}'r^{33}\mathbf{y}_3 \end{pmatrix} \tag{14.24}$$

where vector $\boldsymbol{\beta}_i$ has the fixed effects, vector $\mathbf{g}_{11} = \begin{pmatrix} \mathbf{a}1 \\ \mathbf{a}_3^{(1)} \end{pmatrix}$ has the breeding values for purebred performance for breed 1 ($\mathbf{a}_1$) and effects of crossbred gametes for purebred breed 1 performance ($\mathbf{a}_3^{(1)}$) as defined in (14.22); vector $\mathbf{g}_{22} = \begin{pmatrix} \mathbf{a}_2 \\ \mathbf{a}_3^{(2)} \end{pmatrix}$ has the corresponding values for breed 2; vector $\mathbf{g}_{13} = \begin{pmatrix} \mathbf{c}_1 \\ \mathbf{c}_3^{(1)} \end{pmatrix}$ has the breeding values for purebred animals for crossbred performance ($\mathbf{c}_1$, mating between breeds 1 and 2) and the genetic effects for crossbred animals $\mathbf{c}^{(1)}_3$ due to genes of breed 1 origin; vector $\mathbf{g}_{23} = \begin{pmatrix} \mathbf{c}_2 \\ \mathbf{c}_3^{(2)} \end{pmatrix}$ has the corresponding values for breed 2; the design matrices $\mathbf{X}_1, \mathbf{X}_2, \mathbf{X}_3, \mathbf{Z}_1, \mathbf{Z}_{13}, \mathbf{Z}_2$ and $\mathbf{Z}_{23}$ are as defined in (14.16) and (14.17), noting that the only non-zero elements in $\mathbf{Z}_{13}$ and $\mathbf{Z}_{23}$ are ½ linking parents and progeny; $s^{(1)ii}$ and $s^{(2)ii}$ are elements in the inverses of $\mathbf{S}^{(1)}$ and $\mathbf{S}^{(2)}$ as defined for breed 1 (Eqn 14.22) and breed 2 (Eqns 14.23), respectively, and $\mathbf{A}^{(1)-1}$ and $\mathbf{A}^{(2)-1}$ are the inverses of the breed-specific partial relationship matrices $\mathbf{A}^{(1)}$ and $\mathbf{A}^{(2)}$ for breeds 1 and 2, respectively, and r^{11}, r^{22} and r^{33} are the inverses of the residual variances for breeds 1, 2 and crossbreds animals, respectively. Note that for crossbred animals, the residual variance is composed of the half of the genetic variance (Mendelian sampling effects) plus an environmental error; that is $r_{33} = \sigma^2_{y3} - 0.25(\sigma^2_{f(1)} + \sigma^2_{p(2)})$, where $\sigma^2_{f(1)}$ and $\sigma^2_{p(2)}$ are the genetic variances of parents from breed 1 and breed 2, respectively, and $r^{33} = 1/\,r_{33}$. Note also that the additive genetic effects for crossbred animals equals $\mathbf{c}_3^{(1)} + \mathbf{c}_3^{(2)}$ as defined in Eqn 14.19.

Example 14.3

The data in Table 14.4 represents a data set from a two-breed terminal crossing scheme, with data coming from two herds. Assume the residual variance for breeds 1, 2 and crossbred animals is the same with a value of 4.0 and that the genetic covariance matrix between purebred animals of breed 1 and crossbred performance is $\mathbf{S}^{(1)} = \begin{pmatrix} 1.0 & 0.92 \\ 0.92 & 1.5 \end{pmatrix}$ and the corresponding genetic covariance matrix for purebred animals of breed 2 and crossbreds is $\mathbf{S}^{(2)} = \begin{pmatrix} 2.0 & 1.385 \\ 1.385 & 1.5 \end{pmatrix}$. Using Eqn (14.24), estimate the fixed-herd effects for the purebred animals (β_1 and $\boldsymbol{\beta}_2$ and crossbreds ($\boldsymbol{\beta}_3$), the breeding values for the purebred performance for breeds 1 ($\mathbf{g}_{11}$) and 2 ($\mathbf{g}_{22}$), the breeding values for crossbred performance for breed 1 animals mating with those of breed 2 ($\mathbf{g}_{13}$) and breed 2 animals mating with those of breed 1 ($\mathbf{g}_{23}$).

From the given parameters, $r^{11} = r^{22} = ¼ = 0.25$. The phenotypic variance for crossbreds = $(4 + 1.5)\text{kg}^2 = 5.5\text{kg}^2$. Therefore, $r_{33} = 5.5 - 0.25(1.0 + 2.0) = 4.75$ and $r^{33} = 1/4.75$.

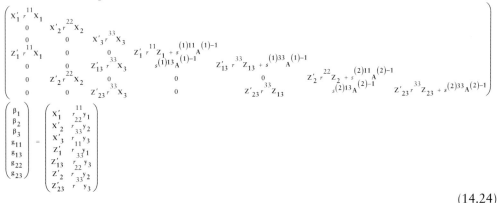

Table 14.4. Example data set from a two-breed terminal crossing scheme.

Animal	Sire	Dam	Breed*	Herd	Record
1	0	0	1	2	11.0
2	0	0	1	1	12.0
3	1	2	1	2	13.0
4	1	3	1	1	14.0
5	0	0	2	1	15.0
6	0	0	2	2	16.0
7	5	6	2	1	17.0
8	5	7	2	2	18.0
9	1	6	12	1	19.0
10	1	7	12	2	20.0
11	5	2	12	1	21.0
12	5	3	12	2	22.0

*Breed: 1 = breed1; 2 = breed 2; 12 = crossbred

The elements $s^{(1)ii}$ and $s^{(2)ii}$ in the MME (14.24) are obtained from the inverses of $\mathbf{S}^{(1)}$ and $\mathbf{S}^{(2)}$, respectively. The **X** matrices, which relate records to herd effects for purebreds 1 and 2 and the crossbreeds, are:

$$\mathbf{X}_1 = \begin{bmatrix} 0 & 1 \\ 1 & 0 \\ 0 & 1 \\ 1 & 0 \\ 0 & 0 \\ 0 & 0 \\ 0 & 0 \\ 0 & 0 \\ 0 & 0 \\ 0 & 0 \\ 0 & 0 \\ 0 & 0 \end{bmatrix}, \quad \mathbf{X}_2 = \begin{bmatrix} 0 & 0 \\ 0 & 0 \\ 0 & 0 \\ 0 & 0 \\ 1 & 0 \\ 0 & 1 \\ 1 & 0 \\ 0 & 1 \\ 0 & 0 \\ 0 & 0 \\ 0 & 0 \\ 0 & 0 \end{bmatrix} \quad \text{and} \quad \mathbf{X}_3 = \begin{bmatrix} 0 & 0 \\ 0 & 0 \\ 0 & 0 \\ 0 & 0 \\ 0 & 0 \\ 0 & 0 \\ 0 & 0 \\ 0 & 0 \\ 1 & 0 \\ 0 & 1 \\ 1 & 0 \\ 0 & 1 \end{bmatrix}$$

The matrices $\mathbf{Z}_1$ and $\mathbf{Z}_2$ have diagonal elements of 1 for breed 1 and breed 2 purebred animals with records, respectively, with all other elements being zero. The non-zero elements in the $\mathbf{Z}_{13}$ and $\mathbf{Z}_{23}$ matrices are those between parents and progeny from breeds 1 and 2, respectively. Thus:

$$\mathbf{Z}_{13} = \begin{bmatrix} 0 & 0 & 0 & 0 & 0 & 0 & 0 & 0 & 0 & 0 & 0 & 0 \\ 0 & 0 & 0 & 0 & 0 & 0 & 0 & 0 & 0 & 0 & 0 & 0 \\ 0 & 0 & 0 & 0 & 0 & 0 & 0 & 0 & 0 & 0 & 0 & 0 \\ 0 & 0 & 0 & 0 & 0 & 0 & 0 & 0 & 0 & 0 & 0 & 0 \\ 0 & 0 & 0 & 0 & 0 & 0 & 0 & 0 & 0 & 0 & 0 & 0 \\ 0 & 0 & 0 & 0 & 0 & 0 & 0 & 0 & 0 & 0 & 0 & 0 \\ 0 & 0 & 0 & 0 & 0 & 0 & 0 & 0 & 0 & 0 & 0 & 0 \\ 0 & 0 & 0 & 0 & 0 & 0 & 0 & 0 & 0 & 0 & 0 & 0 \\ 0.5 & 0 & 0 & 0 & 0 & 0 & 0 & 0 & 0 & 0 & 0 & 0 \\ 0.5 & 0 & 0 & 0 & 0 & 0 & 0 & 0 & 0 & 0 & 0 & 0 \\ 0 & 0.5 & 0 & 0 & 0 & 0 & 0 & 0 & 0 & 0 & 0 & 0 \\ 0 & 0 & 0.5 & 0 & 0 & 0 & 0 & 0 & 0 & 0 & 0 & 0 \end{bmatrix} \quad \text{and} \quad \mathbf{Z}_{23} = \begin{bmatrix} 0 & 0 & 0 & 0 & 0 & 0 & 0 & 0 & 0 & 0 & 0 & 0 \\ 0 & 0 & 0 & 0 & 0 & 0 & 0 & 0 & 0 & 0 & 0 & 0 \\ 0 & 0 & 0 & 0 & 0 & 0 & 0 & 0 & 0 & 0 & 0 & 0 \\ 0 & 0 & 0 & 0 & 0 & 0 & 0 & 0 & 0 & 0 & 0 & 0 \\ 0 & 0 & 0 & 0 & 0 & 0 & 0 & 0 & 0 & 0 & 0 & 0 \\ 0 & 0 & 0 & 0 & 0 & 0 & 0 & 0 & 0 & 0 & 0 & 0 \\ 0 & 0 & 0 & 0 & 0 & 0 & 0 & 0 & 0 & 0 & 0 & 0 \\ 0 & 0 & 0 & 0 & 0 & 0 & 0 & 0 & 0 & 0 & 0 & 0 \\ 0 & 0 & 0 & 0 & 0 & 0.5 & 0 & 0 & 0 & 0 & 0 & 0 \\ 0 & 0 & 0 & 0 & 0 & 0 & 0.5 & 0 & 0 & 0 & 0 & 0 \\ 0 & 0 & 0 & 0 & 0.5 & 0 & 0 & 0 & 0 & 0 & 0 & 0 \\ 0 & 0 & 0 & 0 & 0.5 & 0 & 0 & 0 & 0 & 0 & 0 & 0 \end{bmatrix}$$

The breed 1- and 2-specific partial relationship matrices $\mathbf{A}^{(1)}$ and $\mathbf{A}^{(2)}$ are generated using (14.21) and these are:

$A^{(1)}$ =

1.000											
0.000	1.000										
0.500	0.500	1.000									
0.750	0.250	0.750	1.250								
0.000	0.000	0.000	0.000	0.000							
0.000	0.000	0.000	0.000	0.000	0.000						
0.000	0.000	0.000	0.000	0.000	0.000	0.000			symmetric		
0.000	0.000	0.000	0.000	0.000	0.000	0.000	0.000				
0.500	0.000	0.250	0.375	0.000	0.000	0.000	0.000	0.500			
0.500	0.000	0.250	0.375	0.000	0.000	0.000	0.000	0.250	0.500		
0.000	0.500	0.250	0.125	0.000	0.000	0.000	0.000	0.000	0.000	0.500	
0.250	0.250	0.500	0.375	0.000	0.000	0.000	0.000	0.125	0.125	0.125	0.500

$A^{(2)}$ =

0.000											
0.000	0.000										
0.000	0.000	0.000									
0.000	0.000	0.000	0.000								
0.000	0.000	0.000	0.000	1.000							
0.000	0.000	0.000	0.000	0.000	1.000					symmetric	
0.000	0.000	0.000	0.000	0.500	0.500	1.000					
0.000	0.000	0.000	0.000	0.750	0.250	0.750	1.250				
0.000	0.000	0.000	0.000	0.000	0.500	0.250	0.125	0.500			
0.000	0.000	0.000	0.000	0.250	0.250	0.500	0.375	0.125	0.500		
0.000	0.000	0.000	0.000	0.500	0.000	0.250	0.375	0.000	0.125	0.500	
0.000	0.000	0.000	0.000	0.500	0.000	0.250	0.375	0.000	0.125	0.250	0.500

All matrices needed for the MME in (14.24) have now been formed. The $\mathbf{A}^{-}_{i}$ matrices used in Eqn 14.24 are the inverses formed by the non-zero rows of each breed-specific numerator relationship matrix. Solving the MME gives the solutions presented in Table 14.5.

14.7.2 Marker-based partial relationship matrix

The use of genomic information for the prediction of crossbred performance using a GBLUP approach involves the construction breed-specific partial genomic relationship matrices which are used instead of the partial relationship matrix described in the previous section. Also, the genomic breed-specific partial relationship matrices may be combined with the pedigree breed-specific partial relationship matrices for genomic prediction based on a single-step approach, which is discussed in Chapter 12.

The construction of the breed-specific partial genomic relationship matrix will be illustrated using the two-breed Example 14.3. Consider breed 1 and apply the VanRaden (2008) method but use only the breed-specific alleles. It is assumed that the marker genotypes for crossbred animals are phased such that it is known which allele originated from breed 1 or breed 2. Assume that the SNP genotypes for purebred 1 animals are stored in a matrix $\mathbf{M}^{(1)}$ and those for purebred 2 animals in matrix $\mathbf{M}^{(2)}$. Similarly, assume that breed 1 SNP genotypes for crossbred animals are stored in the matrix $\mathbf{Q}^{(1)}$ and corresponding SNP genotypes for breed 2 crossbred animals stored in $\mathbf{Q}^{(2)}$. Then the $\mathbf{Z}^{(1)}$ matrix of centralized SNP genotypes for purebred breed 1 animals is $\mathbf{Z}^{(1)} = \mathbf{M}^{(1)} - \mathbf{P}^{(1)}$, where the elements of the jth column of $\mathbf{P}^{(1)}$ matrix are twice the breed 1-specific allele frequency of the alternative allele of the jth SNP marker for purebred breed 1 animals. Thus, $z^{(1)}ij = (m^1_{ij} - p^{(1)}_{ij})$, which has value $0 - p^{(1)}_{ij}$, $2 - p^{(1)}_{ij}$ and $1 - p^{(1)}_{ij}$ for the two homozygotes (11 and 22) and the heterozygote (12), respectively. Note that the elements in $\mathbf{P}^{(1)}$, the breed 1-specific allele frequencies, are computed from SNP genotypes for purebred animals and breed 1-specific marker alleles for crossbred animals. The $\mathbf{Q}^{(1)}$ matrix of breed 1 SNP genotypes for crossbred animals has element $q^1_{ij} = 0$ or 1 if locus j of individual i has allele 1 or 2, respectively. The $\mathbf{W}^{(1)}$ matrix of centralized breed 1 SNP genotypes for crossbred animals ($\mathbf{W}^{(1)} = \mathbf{Q}^{(1)} - 0.5\mathbf{P}^{(1)}$) has element $w^{(1)}_{ij} = (m^{(1)}_{ij} - 0.5p^{(1)}_{ij})$, which has value $0 - 0.5p^{(1)}_{j}$ or $1 - 0.5p^{(1)}_{j}$ if loci j of individual i has breed 1 allele 1 or 2, respectively.

The marker-based breed-specific partial relationship matrix $\mathbf{G}^{(1)}$ for breed 1 is:

$$\mathbf{G}^{(1)} = \begin{pmatrix} \mathbf{G}^{(1)}_{1,1} & \mathbf{G}^{(1)}_{1,3} \\ \mathbf{G}^{(1)}_{3,1} & \mathbf{G}^{(1)}_{3,3} \end{pmatrix} \tag{14.25}$$

Thus, we have $\mathbf{G}^{(1)} = \frac{1}{k^{(1)}} \begin{pmatrix} \mathbf{Z}^{(1)} \\ \mathbf{W}^{(1)} \end{pmatrix} (\mathbf{Z}^{(1)'} \quad \mathbf{W}^{(1)'})$ with the submatrices are defined as:

$$\mathbf{G}^{(1)}_{1,1} = \frac{1}{k^{(1)}} \mathbf{Z}^{(1)} \mathbf{Z}^{(1)'}, \ \mathbf{G}^{(1)}_{1,3} = \frac{1}{k^{(1)}} \mathbf{Z}^{(1)} \mathbf{W}^{(1)'} \text{ and } \mathbf{G}^{(1)}_{3,3} = \frac{1}{k^{(1)}} \mathbf{W}^{(1)} \mathbf{W}^{(1)'}$$

where $k^{(1)} = 2\sum p^{(1)}_j (1 - p^{(1)}_j)$ and the subscript values 1 and 3 denote genotyped breed 1 and crossbred animals, respectively. The breed-specific partial genomic relationship matrix for breed 2 ($\mathbf{G}^{(2)}$) is constructed in a similar manner as illustrated for breed 1.

Note that the genomic matrix $\mathbf{G}^{(12)}$ does not exist. Therefore, for the genotyped crossbred animals, the genetic effect is the sum of two effects, with variance-covariance matrices proportional to $\mathbf{G}^{(1)}_{3,3}$ and $\mathbf{G}^{(2)}_{3,3}$, respectively. Thus, the genetic effect for crossbred animal i equals:

$$c_{3,i} = c^{(1)}_{3,i} + c^{(2)}_{3,i} \tag{14.26}$$

This is equivalent to the model described for crossbred animals by Ibánẽz-Escriche *et al.* (2009). Thus, Eqn (14.26) equals $\sum_{j=1}^{k} \left(w^{(1)}_{ij}\right) \propto^{(1)}_j + (w^{(2)}_{ij}) \propto^{(2)}_j$ with a SNP model where $\propto^{(1)}_j$ and $\propto^{(2)}_j$ are independent breed of origin-specific substitution effects for SNP $j = 1, \ldots, k$ within the two breeds 1 and 2 and $w^{(p)}_{ij}$ are elements of $\mathbf{W}^{(p)}$ matrix for the pth breed as defined in Eqn 14.25.

The application of the special partial genomic relationships in genomic prediction using GBLUP is straightforward using Eqn 14.24 by replacing the inverses of $\mathbf{A}^{(1)}$ and $\mathbf{A}^{(2)}$ with the inverses of $\mathbf{G}^{(1)}$ and $\mathbf{G}^{(2)}$, respectively.

Example 14.4

Using the example data in Table 14.4, assume that the genotype data in Table 14.6 is available on these animals.

From the breed column in Table 14.1, animals 1, 2 and 5 are purebreed animals belonging to breed 1, and animals 3, 4 and 7 are purebreed animals belonging to breed 2.

As explained in Section 14.7.1, the genotypes for the breed 1 purebreed animals are stored in matrix $\mathbf{M}^{(1)}$ and, correspondingly, those for breed 2 are in $\mathbf{M}^{(2)}$.

Assume that the genotypes of crossbred animals (6, 8, 9, 10, 11) are phased so that the allele inherited from breed 1 or 2 is known as given in Table 14.7.

The construction of $\mathbf{G}^{(1)}$ in Eqn 14.25 requires $\mathbf{Q}^{(1)}$, a matrix of marker alleles for the crossbred animal derived from the purebred 1 animals. As described in Section 14.7.1, matrix $\mathbf{Q}^{(1)}$ has elements $q^{(1)}_{ij}$ = 0 or 1 if loci j of crossbred individual i has breed 1 allele 1 or 2, respectively. Using the alleles inherited from the purebred animals in Table 14.7, the elements of $\mathbf{Q}^{(1)}$ and $\mathbf{Q}^{(2)}$ are shown in Table 14.8.

Table 14.5. The solutions to the MME for Example 14.3.

Herd effects				
Breed 1				
1	12.875			
2	11.924			
Breed 2				
1	15.767			
2	16.899			
Crossbreds				
1	20.002			
2	20.887			

Animal effect

	Breed 1	Crossbred 1	Breed 2	Crossbred 2
1	−0.060	−0.144	0.000	0.000
2	0.055	0.138	0.000	0.000
3	0.212	0.227	0.000	0.000
4	0.195	0.151	0.000	0.000
5	0.000	0.000	0.200	0.212
6	0.000	0.000	−0.207	−0.216
7	0.000	0.000	0.267	0.159
8	0.000	0.000	0.409	0.307
9	−0.028	−0.070	−0.101	−0.106
10	−0.028	−0.070	0.132	0.078
11	0.028	0.070	0.101	0.106
12	0.105	0.112	0.101	0.106

The animal effect solutions for breed 1, for example, represent the breeding values for the purebred performance of animals from 1 to 4, while the corresponding values for these animals under the column 'Crossbred 1' represent the breeding values for their crossbred performance. The breeding values for animals 9–10 under the column 'Breed 1' are seldom of interest as they present the genetic effects of crossbred animals under purebred management. However, the solutions for these animals under the column 'Crossbred 1' represent the genetic effects of breed 1-specific alleles for crossbred animals from 9 to 12. Note that, as defined in Eqn 14.19, the breeding values for crossbred animals can be calculated by the sum of genetic effects for animals from 9 to 12 under columns 'Crossbred 1' and 'Crossbred 2'. Thus, the breeding value for the crossbred animal 12 is 0.11196 + 0.10661 = 0.21857.

Table 14.6. Genotypes for animals in Table 14.4.

Animal	Genotype									
1	2	0	1	1	0	0	2	2	1	2
2	1	1	0	0	1	2	0	1	1	0
3	2	1	1	0	0	1	1	1	1	1
4	2	0	2	1	0	0	2	1	2	1
5	1	1	2	1	1	0	2	2	1	2
6	0	0	1	1	0	1	0	1	2	1
7	1	1	2	0	1	0	1	1	2	2
8	2	0	2	1	0	0	2	1	1	2
9	1	0	1	1	0	0	1	2	2	1
10	2	1	2	1	0	0	2	2	2	2
11	2	2	1	0	1	1	1	1	2	1
12	2	1	2	1	0	1	1	2	1	2

Table 14.7. Alleles inherited by crossbred animals from breeds 1 and 2.

Animal	Allele from breed 1										Allele from breed 2									
9	2	1	2	1	1	1	2	2	2	2	1	1	1	2	1	1	1	2	2	1
10	2	1	2	2	1	1	2	2	2	2	2	2	2	1	1	1	2	2	2	2
11	2	2	2	1	2	1	2	2	2	2	2	2	1	1	1	2	1	1	2	1
12	2	2	2	2	1	1	2	2	2	2	2	1	2	1	1	2	1	2	1	2

Table 14.8. The coded marker alleles for crossbred animals inherited from breeds 1 and 2.

Animal	$Q^{(1)}$										$Q^{(2)}$									
9	1	0	1	0	0	0	1	1	1	1	0	0	0	1	0	0	0	1	1	0
10	1	0	1	1	0	0	1	1	1	1	1	1	1	0	0	0	1	1	1	1
11	1	1	1	0	1	0	1	1	1	1	1	1	0	0	0	1	0	0	1	0
12	1	1	1	1	0	0	1	1	1	1	1	0	1	0	0	1	0	1	0	1

Therefore, the SNP frequencies for the alternative allele (allele 2) for breed 1 computed from SNP genotypes of purebreed animals 1, 2, 3 and 4 and the alleles of breed 1 inherited by crossbred animals are [0.917 0.333 0.667 0.333 0.167 0.250 0.750 0.750 0.750 0.667]. Corresponding SNP frequencies for breed 2 from SNP genotypes of purebreed animals 5, 6, 7 and 8 and the alleles of breed 1 inherited by crossbred animals are [0.583 0.333 0.750 0.333 0.167 0.250 0.500 0.667 0.750 0.750]. The allele frequency for the ith SNP marker for breed 1, for example, was computed as $\frac{\Sigma_j^n m_{ij}^{(1)} + \Sigma_j^{nc} q_{ij}^{(1)}}{2 * n + nc}$ where n = 4 is the number of breed 1 purebreed animals with genotypes, nc = 4 is the number of crossbred animals with genes from breed 1, and $m_{ij}^{(1)}$ and $q_{ij}^{(1)}$ are elements of $\mathbf{M}^{(1)}$ and $\mathbf{Q}^{(1)}$, respectively.

For the example data, the matrices $\mathbf{M}^{(1)}$ and $\mathbf{Q}^{(1)}$ have been computed in addition to $\mathbf{P}^{(1)}$, the breed 1-specific allele frequencies. Therefore, the centralized matrix $\mathbf{Z}^{(1)}$ of SNP genotypes for purebred 1 animals and $\mathbf{W}^{(1)}$ for the crossbred animals can be easily computed as in Section 14.7.1 as $\mathbf{Z}^{(1)} = \mathbf{M}^{(1)} - \mathbf{P}^{(1)}$ and $\mathbf{W}^{(1)} = \mathbf{Q}^{(1)} - 0.5\mathbf{P}^{(1)}$.

Table 14.9. The MME solutions for Example 14.4.

Herd effects				
Breed 1				
1			12.955	
2			12.054	
Breed 2				
1			15.801	
2				
			17.062	
Crossbreds				
1		20.070		
2		20.940		
Animal effect				
	Breed 1	Crossbred 1	Breed 2	Crossbred 2
1	−0.177	−0.282	0.000	0.000
2	−0.062	0.212	0.000	0.000
3	0.070	0.164	0.000	0.000
4	0.151	0.080	0.000	0.000
5	0.000	0.000	0.176	0.208
6	0.000	0.000	−0.485	−0.419
7	0.000	0.000	0.221	0.150
8	0.000	0.000	0.360	0.302
9	0.015	−0.037	−0.364	−0.281
10	−0.026	−0.104	0.213	0.178
11	0.014	0.005	−0.117	−0.135
12	0.015	−0.037	−0.005	−0.003

These solutions cannot be compared with those from Example 14.3 as genetic relationships were computed from pedigree information while the **G** matrices used in this example were based on assumed genotypes. The interpretation of the results is, however, the same as for Example 14.3.

For breed 1, the $\mathbf{Z}^{(1)}$ matrix (animals 1,2, 3 and 4) is:

$$\begin{bmatrix} 0.167 & -0.667 & -0.333 & 0.333 & -0.333 & -0.5 & 0.5 & 0.5 & -0.5 & 0.667 \\ -0.833 & 0.333 & -1.333 & -0.667 & 0.667 & 1.5 & -1.5 & -0.5 & -0.5 & -1.333 \\ 0.167 & 0.333 & -0.333 & -0.667 & -0.333 & 0.5 & -0.5 & -0.5 & -0.5 & -0.333 \\ 0.167 & -0.667 & 0.667 & 0.333 & -0.333 & -0.5 & 0.5 & -0.5 & 0.5 & -0.333 \end{bmatrix}$$

and $\mathbf{W}^{(1)}$ (animals 9, 10, 11, and 12) is:

$$\begin{bmatrix} 0.083 & -0.333 & 0.333 & -0.333 & -0.167 & -0.25 & 0.25 & 0.25 & 0.25 & 0.333 \\ 0.083 & -0.333 & 0.333 & 0.667 & -0.167 & -0.25 & 0.25 & 0.25 & 0.25 & 0.333 \\ 0.083 & 0.667 & 0.333 & -0.333 & 0.833 & -0.25 & 0.25 & 0.25 & 0.25 & 0.333 \\ 0.083 & 0.667 & 0.333 & 0.667 & -0.167 & -0.25 & 0.25 & 0.25 & 0.25 & 0.333 \end{bmatrix}$$

All matrices required to compute $\mathbf{G}^{(1)}$ in (14.24) have been illustrated. Therefore, for $\mathbf{G}^{(1)}$, the example data is:

$G^{(1)}$ =

1.815											
-0.613	1.849										
0.519	0.536	0.852									
1.271	-0.613	0.519	1.815								
0.000	0.000	0.000	0.000	0.000							
0.000	0.000	0.000	0.000	0.000	0.000						
0.000	0.000	0.000	0.000	0.000	0.000	0.000					
0.000	0.000	0.000	0.000	0.000	0.000	0.000	0.000				
0.888	-0.326	0.375	0.888	0.000	0.000	0.000	0.000	0.638			
1.057	-0.428	0.273	1.057	0.000	0.000	0.000	0.000	0.587	0.807		
0.735	0.064	0.494	0.735	0.000	0.000	0.000	0.000	0.561	0.510	1.028	
0.955	-0.259	0.443	0.955	0.000	0.000	0.000	0.000	0.536	0.756	0.731	.977

Similar principles are used to compute $\mathbf{G}^{(2)}$, which, for the example data, is:

$G^{(2)}$ =

0.000											
0.000	0.000										
0.000	0.000	0.000									
0.000	0.000	0.000	0.000								
0.000	0.000	0.000	0.000	3.882							
0.000	0.000	0.000	0.000	0.663	1.355						
0.000	0.000	0.000	0.000	2.734	0.819	2.890					
0.000	0.000	0.000	0.000	3.047	0.479	2.225	3.515				
0.000	0.000	0.000	0.000	0.246	0.593	-0.002	-0.008	0.661			
0.000	0.000	0.000	0.000	1.890	0.280	1.642	1.635	0.016	1.327		
0.000	0.000	0.000	0.000	-0.236	0.111	0.168	-0.165	-0.069	0.264	0.831	
0.000	0.000	0.000	0.000	0.885	0.253	0.637	0.956	-0.160	0.498	0.088	0.974

The $\mathbf{G}^{(i)-}$matrices used in the MME in this example are the inverses formed by the non-zero rows of each breed-specific genomic relationship matrix after adding 0.01 to the diagonal elements. Solving the MME gives the solutions presented in Table 14.9.

14.8 Analysing Crossbred Animals with Breed Origin of Alleles in Practice

In Section 14.7, the principles for the breed origin of alleles (BOA) in the genomic prediction of the genetic merits of animals have been discussed including the computation of the breed-specific partial genomic relationships. In practice, the initial step in the efficient use of the crossbred information-based BOA genomic model involves assigning breed origin to alleles of crossbred animals. When both parents are known, determining the breed origin of alleles in a two-way cross is relatively easy but more complicated in three-way or higher crosses such as in chicken and pigs. Bastiaansen *et al.* (2014) and Vandenplas *et al.* (2016) developed an approach to assign breed-of-origin to alleles in three-breed cross animals using a long-range phasing method. Haplotypes that were derived from the phasing were assigned to a breed if they were present in only one of the purebred populations, which subsequently allowed assigning the breed origin of alleles when that haplotype was observed in crossbred animals.

In summary, assigning breed origin to alleles of crossbred animals involves three steps (Vandenplas *et al.*, 2016): (i) the phasing of genotypes for purebred and crossbred animals; (ii) the assignment of breed origin to the phased haplotypes; and (iii) the assignment of breed origin to alleles of crossbred animals on the basis of the library of assigned haplotypes, the zygosity (homozygosity or heterozygosity) of their genotypes and their breed composition. The details of assigning breed origin to alleles are not covered in this text but details can be found in Vandenplas *et al.* (2016).

15 Analysis of Ordered Categorical Traits

15.1 Introduction

Some traits of economic importance in animal breeding, such as calving ease or litter size, are expressed and recorded in a categorical fashion. For instance, in the case of calving ease, births may be assigned to one of several distinct classes, such as difficult, assisted and easy calving, or litter size in pigs might be scored 1, 2, 3 or more piglets born per sow. Usually, these categories are ordered along a gradient. In the case of calving ease, for example, the responses are ordered along a continuum measuring the ease with which birth occurred. These traits are therefore termed ordered categorical traits. Such traits are not normally distributed, and animal breeders have usually attributed the phenotypic expression of categorical traits to an underlying continuous unobservable trait that is normally distributed, referred to as the liability (Falconer and Mackay, 1996). The observed categorical responses are therefore due to animals exceeding particular threshold levels (t) of the underlying trait. Thus, with m categories of responses, there are $m - 1$ thresholds such that $t_1 < t_2 < t_3, \ldots, t_{m-1}$. For traits such as survival to a particular age or stage, the variate to be analysed is coded 1 (survived) or 0 (not survived) and there is basically only one threshold.

Linear and non-linear models have been applied for the genetic analysis of categorical traits with the assumption of an underlying normally distributed liability. Usually, the non-linear (threshold) models are more complex and have higher computing requirements. The advantage of the linear model is the ease of implementation, as programs used for analysis of quantitative traits could be utilized without any modifications. However, Fernando *et al.* (1983) indicated that some of the properties of BLUP do not hold with categorical traits. Such properties include the invariance of BLUP to certain types of culling (selection) and the ability of BLUP to maximize the probability of correct pairwise ranking. Also, Gianola (1982) indicated that the variance of a categorical trait is a function of its expectation and the application of a linear model that has fixed effects in addition to an effect common to all observations results in heterogeneity of variance.

In a simulation study, Meijering and Gianola (1985) demonstrated that with no fixed effects and constant or variable number of offspring per sire, an analysis of a binary trait with either a linear or non-linear model gave similar sire rankings. This was independent of the heritability of the liability or incidence of the binary trait. However, with the inclusion of fixed effects and a variable number of progeny per sire, the non-linear model gave breeding values that were more similar to the true breeding values compared with the linear model. The advantage of the threshold model increased as the incidence of the binary trait and its heritability decreased. Thus, for traits with low heritability and low incidence, a threshold model might be the method of choice.

The principles required to apply a linear model for the analysis of categorical traits are the same as discussed in the previous chapters; therefore, the main focus of this chapter is on threshold models, assuming a normal distribution for the liability. Cameron

DOI: 10.1079/9781800620506.0015

(1997) illustrated the analysis of a binary trait with a threshold model using a logit function. In this chapter, sample data used for the illustration with the threshold model have also been analysed with a linear model for the purposes of comparison.

15.2 The Threshold Model

15.2.1 Defining some functions of the normal distribution

The use of the threshold model involves the use of some functions of the normal distribution and these are briefly defined. Assume the number of lambs born alive to ewes in the breeding season is scored using four categories. The distribution of liability for the number of lambs born alive with three thresholds (t_j) can be illustrated as in Fig. 15.1, where N_j is the number of ewes with the jth number of lambs and are those exceeding the threshold point t_{j-1}, when $j > 1$ and $j \leq m-1$.

With the assumption that the liability (l) is normally distributed ($l \sim N(0,1)$), the height of the normal curve at $t_j(\phi(t_j))$ is:

$$\phi(t_j) = \exp(-0.5t_j^{\,2})/\sqrt{2\pi} \tag{15.1}$$

For instance, given that $t_j = 0.779$, then $\phi(0.779) = 0.2945$.

The function $\Phi()$ is the standard cumulative distribution function of the normal distribution. Thus $\Phi(k)$ or Φ_k gives the areas under the normal curve up to and including the kth category. Given that there are m categories, then $\Phi_k = 1$ when the kth category equals m. For a variable x, for instance, drawn from a normal distribution, the value Φ_x can be computed, using a subroutine from the IMSL (1980) library. Thus, if x = 0.560, then $\Phi(0.560)$ = 0.7123.

$P(k)$ defines the probability of a response being observed in category k assuming a normal distribution. This is also the same probability that a response is between the

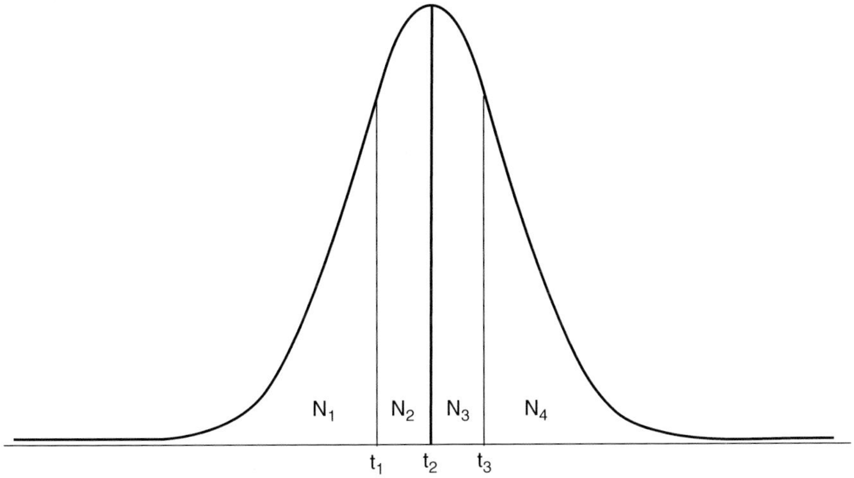

Fig. 15.1. The distribution of liability for number of lambs born alive with four categories and three thresholds.

thresholds defined by category k. Thus, $P(k)$ or P_k may be calculated as $P(k) = \Phi(k) - \Phi(k-1)$ with $\Phi(k-1) = 0$, when k = 1; or expressed in terms of thresholds defining the category k, $P_k = \Phi(t_k) - \Phi(t_{(k-1)})$. For instance, in Fig. 15.1, the probability of response in the k category (P_k) can be computed as:

$$P_1 = \Phi(t_1) \tag{15.2}$$

$$P_2 = \Phi(t_2) - \Phi(t_1) \tag{15.3}$$

$$P_3 = \Phi(t_3) - \Phi(t_2) \text{ and}$$

$$P_4 = 1 - \Phi(t_3)$$

15.2.2 Data organization and the threshold model

Usually, the data are organized into an s by m contingency table (Table 15.1), where the s rows represent individuals or herd–year subclasses of effects, such as herd, and the m columns indicate ordered categories of response. If the rows represent individuals, then all n_{jk} will be zero except one and the n_j = 1, for $j = 1,\ldots,s$.

The linear model for the analysis of the liability is:

$$\mathbf{y} = \mathbf{Xb} + \mathbf{Zu} + \mathbf{e}$$

where **y** is the vector of liability on a normal scale, **b** and **u** are vectors of fixed and random (sire or animal) effects, respectively, and **X** and **Z** are incidence matrices relating data to fixed effects and responses effects, respectively. Since **y** is not observed, it is not possible to solve for **u** using the usual MME.

Given that $\mathbf{H}' = [\mathbf{t}', \mathbf{b}', \mathbf{u}']$, where **t** is the vector for the threshold effects, Gianola and Foulley (1983) proceeded to find the estimator $\hat{\mathbf{H}}$ that maximizes the log of the posterior density L(**H**). The resulting set of equations involved in the differentiation were not linear with respect to **H**. They therefore provided the following non-linear iterative system of equations based on the first and second derivatives, assuming a normal distribution to obtain solutions for Δ**t**, Δ**b** and Δ**u**:

$$\begin{bmatrix} \mathbf{Q} & \mathbf{L'X} & \mathbf{L'Z} \\ \mathbf{X'L} & \mathbf{X'WX} & \mathbf{X'WZ} \\ \mathbf{Z'L} & \mathbf{Z'WX} & \mathbf{Z'WZ} + \mathbf{A}^{-1}\mathbf{G}^{-1} \end{bmatrix} \begin{bmatrix} \Delta\mathbf{t} \\ \Delta\mathbf{b} \\ \Delta\mathbf{u} \end{bmatrix} = \begin{bmatrix} \mathbf{p} \\ \mathbf{X'v} \\ \mathbf{Z'v} - \mathbf{A}^{-1}\mathbf{G}^{-1}\mathbf{u} \end{bmatrix} \tag{15.4}$$

Table 15.1. Ordered categorical data arranged as an s by m contingency table.

Categories[a]							
Subclasses	1	2	…	k	…	m	Totals[b]
1	n_{11}	n_{12}	…	n_{1k}	…	n_{1m}	$n_{1.}$
2	n_{21}	n_{22}	…	n_{2k}	…	n_{2m}	$n_{2.}$
⋮	⋮	⋮	⋮	⋮	⋮	⋮	⋮
j	n_{j1}	n_{j2}	…	n_{jk}	…	n_{jm}	$n_{j.}$
⋮	⋮	⋮	⋮	⋮	⋮	⋮	⋮
s	n_{s1}	n_{s2}	…	n_{sk}	…	n_{sm}	$n_{s.}$

[a]n_{jk} = number of counts in category k of response in row j.

[b]$n_{j.} = \sum_{k=1}^{m} n_{jk}$

with $\mathbf{G} = \mathbf{I}\sigma_s^2$ or $\mathbf{I}\sigma_u^2$ if a sire or an animal model is being fitted in a univariate situation and the vectors **v** and **p** are computed as in equations (15.7) and (15.12), respectively.

They presented equations for the calculation of the matrices in Eqn 15.4, which are outlined below. The calculation of most of these matrices involves P_{jk} (see Eqn 15.2) and it is initially described. P_{jk}, the response in the kth category under the conditions of the jth row, is:

$$P_{jk} = \Phi(t_k - a_j) - \Phi(t_{k-1} - a_j); k = 1, m-1; j = 1, \ldots, s \tag{15.5}$$

where $a_j = (\mathbf{x}_j\mathbf{b} + \mathbf{z}_j\mathbf{u})$, with x_j and z_j being the jth row of **X** and **Z**, respectively. This equation is no different from that in Section 15.2.1, but it shows that the distribution of response probabilities by category is a function of the distance between a_j and the threshold. Similarly, the height of the normal curve at t_k (Eqn 15.1) under the conditions of the jth row becomes:

$$\phi_{jk} = \phi(t_k - a_j) \tag{15.6}$$

The formulae for computing the various matrices and vectors in Eqn 15.4 are outlined below.

The jth element of vector **v** can be calculated as:

$$v_j = \sum_{k=1}^{m} n_{jk}\left(\frac{\phi_{j(k-1)} - \phi_{jk}}{p_{jk}}\right) \tag{15.7}$$

The elements of the matrix **W**, which is a weighting factor, are computed as:

$$w_{jj} = n_{j.}\sum_{k=1}^{m}\frac{\left(\phi_{j(k-1)} - \phi_{jk}\right)^2}{p_{jk}} \tag{15.8}$$

The matrix **Q** is an $(m - 1)$ by $(m - 1)$ banded matrix and the diagonal elements are calculated as:

$$q_{kk} = \sum_{j=1}^{s} n_{j.}\frac{P_{jk} + P_{j(k+1)}}{P_{jk}P_{j(k+1)}}\phi_{jk}^2, \quad \text{for } k = 1 \text{ to } (m-1) \tag{15.9}$$

and the off-diagonal elements are:

$$q_{(k+1)k} = -\sum_{j=1}^{s} n_{j.}\frac{\phi_{j(k+1)}\phi_{jk}}{P_{j(k+1)}}, \quad \text{for } k = 1 \text{ to } (m-2) \tag{15.10}$$

with the element $q_{k(k+1)} = q_{(k+1)k}$.

The matrix **L** is of order s by $(m - 1)$ and its jkth element is calculated as:

$$l_{jk} = -n_{j.}\phi_{jk}\left(\frac{\phi_{jk} - \phi_{j(k-1)}}{P_{jk}} - \frac{\phi_{j(k+1)} - \phi_{jk}}{P_{j(k+1)}}\right) \tag{15.11}$$

The vector **p** is accumulated over all subclasses and its elements are:

$$p_k = \left\{\sum_{j=1}^{s}\left[\frac{n_{jk}}{p_{jk}} - \frac{n_{j(k+1)}}{p_{j(k+1)}}\right]\phi_{jk}\right\}; \; k = 1, m-1 \tag{15.12}$$

The remaining matrices in Eqn 15.4 can be computed by matrix multiplication.

15.2.3 Numerical example

Example 15.1

The analysis of categorical traits is illustrated below, using the calving ease data described by Gianola and Foulley (1983) but with a relationship matrix included for the sires and the age of dam effect omitted from the model. The data consisted of calving ease scores from 28 male and female calves born in two herd–years from cows mated to four sires. Cows were scored for calving ease using three ordered categories: 1 = normal birth, 2 = slight difficulty and 3 = extreme difficulty. The data set is presented in Table 15.2.

The following pedigree was assumed for the four sires:

Animal	Sire	Dam
1	0	0
2	0	0
3	1	0
4	3	0

The sire variance used in the analysis was assumed to be $\frac{1}{19}$. In the underlying scale, residual variance equals one; therefore, $\sigma_e^2 / \sigma_s^2 = (4 - h^2) / h^2 = 19$. Thus the σ_s^2 assumed corresponded to a heritability of 0.20 on the underlying scale.

Table 15.2. Distribution of calving ease score by herd–year and sex of calf subclasses.

			Category of response[a]			
Herd	Sex of calf	Sire of calf	1	2	3	Total
1	Male	1	1	0	0	1
1	Female	1	1	0	0	1
1	Male	1	1	0	0	1
1	Female	2	0	1	0	1
1	Male	2	1	0	1	2
1	Female	2	3	0	0	3
1	Male	3	1	1	0	2
1	Female	3	0	1	0	1
1	Male	3	1	0	0	1
2	Female	1	2	0	0	2
2	Male	1	1	0	0	1
2	Male	1	0	0	1	1
2	Female	2	1	0	1	2
2	Male	2	1	0	0	1
2	Female	3	0	1	0	1
2	Male	3	0	0	1	1
2	Male	4	0	1	0	1
2	Female	4	1	0	0	1
2	Female	4	2	0	0	2
2	Male	4	2	0	0	2

[a]1, normal birth; 2, slight difficulty; 3, extreme difficulty.

The vectors of solutions in Eqn 15.4 for the example data are:

$\mathbf{t}' = (t_1 \ t_2)$, since there are two thresholds

$\mathbf{b}' = (h_1 \ h_2 \ \eta_1 \ \eta_2)$

$\mathbf{u}' = (u_1 \ u_2 \ u_3 \ u_4)$

where h_i and η_i represent solutions for level i of herd–year and the sex-of-calf effects, respectively; and $\mathbf{u}$ is the vector of solutions for sires.

The inverse of the relationship for the assumed pedigree is:

$$\mathbf{A}^{-1} = \begin{bmatrix} 1.3333 & 0.0000 & -0.6667 & 0.0000 \\ 0.0000 & 1.0000 & 0.0000 & 0.0000 \\ -0.6667 & 0.0000 & 1.6667 & -0.6667 \\ 0.0000 & 0.0000 & -0.6667 & 1.3333 \end{bmatrix}$$

For the example data, the transpose of matrix $\mathbf{X}$, which relates subclasses to herd–year and sex-of-calf effects, and that of matrix $\mathbf{Z}$, which relates subclasses to sires, are:

$$\mathbf{X}' = \begin{bmatrix} 1&1&1&1&1&1&1&1&1&0&0&0&0&0&0&0&0&0&0 \\ 0&0&0&0&0&0&0&0&0&1&1&1&1&1&1&1&1&1&1 \\ 1&0&1&0&1&0&1&0&1&0&1&1&0&1&0&1&1&0&0&1 \\ 0&1&0&1&0&1&0&1&0&1&0&0&1&0&1&0&0&1&1&0 \end{bmatrix}$$

and

$$\mathbf{Z}' = \begin{bmatrix} 1&1&1&0&0&0&0&0&0&1&1&1&0&0&0&0&0&0&0 \\ 0&0&0&1&1&1&0&0&0&0&0&0&1&1&0&0&0&0&0 \\ 0&0&0&0&0&0&1&1&1&0&0&0&0&0&1&1&0&0&0 \\ 0&0&0&0&0&0&0&0&0&0&0&0&0&0&0&1&1&1&1 \end{bmatrix}$$

Starting values for $\mathbf{t}$, $\mathbf{b}$ and $\mathbf{u}$ are needed to commence the iterative process. Let $\mathbf{b} = \mathbf{u} = 0$, but starting values for t_i can be computed from the proportion of records in all categories of response preceding t_i. In this example, there is only one category before t_1 and 0.679 of the records are in this category. The first two categories precede t_2 and 0.857 of the records are observed in both categories. Using these proportions, the values of t can be obtained from the usual table of standardized normal deviates of the normal distribution. From these proportions, $t_1 = 0.468$ and $t_2 = 1.080$ and these were used as starting values. However, using various starting values of t, Gianola and Foulley (1983) demonstrated that the system of equations converged rapidly. It seems, therefore, that the system of equations is not very sensitive to starting values for t. The calculations of the various matrices in the equations have been illustrated below using solutions obtained after the first iteration. The solutions obtained at the end of the first iteration and the updated estimates for the effects (which are now the starting values for the second iteration) are:

Solutions at the end of iteration one	Updated[a] estimates after iteration one
$\Delta t_1 = -0.026992$	$t_1 = 0.441008$
$\Delta t_2 = -0.035208$	$t_2 = 1.044792$
$\Delta \hat{h}_1 = 0.000000$	$\hat{h}_1 = 0.000000$
$\Delta \hat{h}_2 = 0.286869$	$\hat{h}_2 = 0.286869$
$\Delta \eta_1 = 0.000000$	$\eta_1 = 0.000000$
$\Delta \eta_2 = -0.358323$	$\eta_2 = -0.358323$
$\Delta u_1 = -0.041528$	$u_1 = -0.041528$
$\Delta u_2 = 0.057853$	$u_2 = 0.057853$
$\Delta u_3 = 0.039850$	$u_3 = 0.039850$
$\Delta u_4 = -0.065178$	$u_4 = -0.065178$

[a] The updated estimates were obtained as the sum of the starting values and the solutions at the end of the first iteration.

The following steps are involved in calculating P_{jk}, which is required to calculate subsequent matrices in Eqn 15.4 for the example data. In each round of iteration and for each subclass, i.e. for $j = 1,\ldots s$:

1. Initially calculate $(t_k - a_j)$ in Eqn 15.5 for $k = 1,\ldots\ m-1$. Therefore:

$$d_{jk} = (t_k - a_j) = t_k - x_j - z_j \text{ for } k = 1,\ldots\ m-1$$

where x_j and z_j are the jth rows of **X** and **Z**.

For the example data in the second iteration:

$$d_{11} = t_1 - \hat{h}_1 - \hat{\eta}_1 - \hat{u}_1$$
$$d_{11} = 0.441008 - 0 - 0 - (-0.041528) = 0.482536$$
$$d_{12} = t_2 - \hat{h}_1 - \hat{\eta}_1 - \hat{u}_1$$
$$d_{21} = 0.441008 - 0 - (-0.358323) - (-0.041528) = 0.840859$$
$$d_{22} = t_2 - \hat{h}_1 - \hat{\eta}_2 - \hat{u}_1$$
$$d_{22} = 1.044792 - 0 - (-0.358323) - (-0.041528) = 1.444643$$
$$\vdots$$
$$d_{201} = t_1 - \hat{h}_2 - \hat{\eta}_1 - \hat{u}_4$$
$$d_{201} = 0.441008 - 0.286869 - 0 - (-0.065178) = 0.219317$$
$$d_{202} = t_2 - \hat{h}_2 - \hat{\eta}_1 - \hat{u}_4$$
$$d_{202} = 1.044792 - 0.286869 - 0 - (-0.065178) = 0.823101$$

2. Using the values of d_{jk} computed above, calculate ϕ_{jk} (see Eqn 15.6) and Φ_{jk} for $k = 0,\ldots,m$. Note that in all cases, when $k = 0$, $\phi_{jk} = \Phi_{jk} = 0$ and when $k = m$, $\phi_{jk} = 0$ and $\Phi_{jk} = 1$.

In the second round of iteration for the example data:

$$\phi_{11} = \phi(0.482536) = 0.355099 \text{ and } \Phi_{11} = \Phi(0.482536) = 0.685288$$
$$\phi_{12} = \phi(1.086320) = 0.221135 \text{ and } \Phi_{12} = \Phi(1.086320) = 0.861331$$
$$\phi_{21} = \phi(0.840859) = 0.280142 \text{ and } \Phi_{21} = \Phi(0.840859) = 0.799787$$
$$\phi_{22} = \phi(1.444643) = 0.140516 \text{ and } \Phi_{22} = \Phi(1.444643) = 0.925721$$
$$\phi_{201} = \phi(0.219317) = 0.389462 \text{ and } \Phi_{201} = \Phi(0.219317) = 0.586799$$
$$\phi_{202} = \phi(0.823101) = 0.284311 \text{ and } \Phi_{202} = \Phi(0.823101) = 0.794775$$

3. Then calculate P_{jk} as $\Phi_{jk} - \Phi_{j(k-1)}$ for $k = 1,\ldots,m$

In the second round of iteration, for Example 15.1:

$$P_{11} = \Phi_{11} - \Phi_{10} = 0.685288 - 0 = 0.685288$$
$$P_{12} = \Phi_{12} - \Phi_{11} = 0.861331 - 0.685288 = 0.176044$$
$$P_{13} = \Phi_{13} - \Phi_{12} = 1.0 - 0.861331 = 0.138669$$
$$P_{21} = \Phi_{21} - \Phi_{20} = 0.799787 - 0 = 0.799787$$
$$P_{22} = \Phi_{22} - \Phi_{21} = 0.925721 - 0.799787 = 0.125934$$
$$P_{23} = \Phi_{23} - \Phi_{22} = 1.0 - 0.925721 = 0.074279$$
$$\vdots$$
$$P_{201} = \Phi_{201} - \Phi_{200} = 0.586799 - 0 = 0.586799$$
$$P_{202} = \Phi_{202} - \Phi_{201} = 0.794775 - 0.586799 = 0.207976$$
$$P_{203} = \Phi_{203} - \Phi_{202} = 1.0 - 0.794775 = 0.205225$$

The calculation of the remaining matrices in the MME can now be illustrated for the example data. The first elements of **W** using Eqn 15.8 for the example data are:

$$\mathrm{w}_{11} = 1\left[\frac{(0-0.355099)^2}{0.685288} + \frac{(0.355099-0.221135)^2}{0.176044} + \frac{(0.221135-0)^2}{0.138669}\right] = 0.638589$$

and

$$\begin{aligned}\mathbf{W} \;=\; & \mathrm{diag}[0.638589\;0.518748\;0.638589\;0.554385\;1.332860\;1.663156\\ & 1.323206\;0.548036\;0.661603\;1.233768\;0.710404\;0.710404\;1.293402\\ & 0.728641\;0.641496\;0.725614\;0.705526\;0.609417\;1.218834\;1.411052]\end{aligned}$$

For the vector **v**, the first element can be calculated from Eqn 15.7 as:

$$v_1 = \frac{1(0-0.355099)}{0.685288} + \frac{0(0.355099-0.221135)}{0.176044} + \frac{0(0.221135-0)}{0.138669} = -0.518175$$

and the transpose of **v** is:

$$\begin{aligned}\mathbf{v}' \;=\; & [-0.518175\;-0.350270\;-0.518175\;1.012257\;0.943660\;-1.179520\;0.120754\\ & 1.029729\;-0.561257\;-0.963633\;-0.677635\;1.366976\;1.039337\;-0.737615\\ & 0.751341\;1.304294\;0.505592\;-0.470090\;-0.940181\;-1.327414]\end{aligned}$$

The matrix **L** is order 20 × 2 for the example data. The elements in the first row of **L** from Eqn 15.11 can be calculated as:

$$l_{11} = (-1)(0.355099)\left[\frac{(0.355099-0)}{0.685288} - \frac{(0.221135-0.355099)}{0.176044}\right] = -0.454223$$

$$l_{12} = (-1)(0.221135)\left[\frac{(0.221135-0.355099)}{0.176044}\frac{(0-0.221135)}{0.138669}\right] = -0.184365$$

The matrix **L** has not been shown because it is too large but the elements of the last row, l_{201} and l_{202}, are −0.910795 and −0.500257, respectively.

The elements of **Q** calculated using Eqns 15.9 and 15.10 are:

$$q_{11} = \frac{1(0.355099)^2(0.685287+0.176044)}{(0.685286*0.176044)} + \frac{1(0.280142)^2(0.799787+0.125934)}{(0.799787*0.125934)} + \ldots$$
$$+\frac{2(0.389462)^2(0.586799+0.207976)}{(0.586799*0.207976)} = 25.072830$$

$$q_{12} = \frac{-[1(0.355099)(0.221135)]}{0.176044} + \frac{1(0.280142)(0.140516)}{0.125934} + \ldots$$
$$+\frac{2(0.389462)(0.284311)}{0.207976} = -12.566598$$

$$q_{22} = \frac{1(0.221135)^2(0.176044+0.138669)}{(0.176044*0.138669)} + \frac{1(0.140516)^2(0.125934+0.074279)}{(0.125934*0.074279)} + \ldots$$
$$+\frac{2(0.284311)^2*(0.207976+0.205225)}{(0.207976*0.205225)} = 17.928093$$

Since **Q** is symmetric, $q_{21} = q_{12}$.

Lastly, the elements of **p** can be calculated using Eqn 15.12 as:

$$p_1 = 0.355099\left(\frac{1}{0.685287} - \frac{0}{0.176044}\right) + 0.280142\left(\frac{1}{0.799787} - \frac{0}{0.125934}\right) + \ldots$$
$$+\ 0.3789462\left(\frac{2}{0.586799} - \frac{0}{0.207976}\right) = -0.288960$$

and

$$p_2 = 0.221135\left(\frac{0}{0.176044} - \frac{0}{0.138669}\right) + 0.140516\left(\frac{0}{0.125934} - \frac{0}{0.074279}\right) + \ldots$$
$$+\ 0.284311\left(\frac{0}{0.207976} - \frac{0}{0.205225}\right) = 0.48984$$

The matrices in Eqn 15.4 can now be obtained by matrix multiplication and $\mathbf{A}^{-1}$ is added to **Z′WZ**. The matrix $\mathbf{Z'WZ} + \mathbf{A}^{-1}\mathbf{G}^{-1}$ is illustrated below:

$$\mathbf{Z'WZ} + \mathbf{A}^{-1}\mathbf{G}^{-1} = \begin{bmatrix} 29.783773 & 0.000000 & -12.666731 & 0.000000 \\ 0.000000 & 24.572445 & 0.000000 & 0.000000 \\ -12.666731 & 0.000000 & 35.566685 & -12.666731 \\ 0.000000 & 0.000000 & -12.666731 & 29.278162 \end{bmatrix}$$

Then, Eqn 15.4 is:

$$\begin{bmatrix} 25.073 & -12.567 & -5.733 & -6.773 & -6.366 & -6.140 & -3.123 & -3.977 & -2.699 & -2.707 \\ -12.567 & 17.928 & -2.146 & -3.215 & -3.220 & -2.141 & -1.327 & -1.595 & -1.201 & -1.238 \\ -5.733 & -2.146 & 7.879 & 0.000 & 4.595 & 3.284 & 1.796 & 3.550 & 2.533 & 0.000 \\ -6.773 & -3.215 & 0.000 & 9.989 & 4.992 & 4.997 & 2.655 & 2.022 & 1.367 & 3.945 \\ -6.366 & -3.220 & 4.595 & 4.992 & 9.586 & 0.000 & 2.698 & 2.062 & 2.710 & 2.117 \\ -6.140 & -2.141 & 3.284 & 4.997 & 0.000 & 8.281 & 1.753 & 3.511 & 1.190 & 1.828 \\ -3.123 & -1.327 & 1.769 & 2.655 & 2.698 & 1.753 & 29.784 & 0.000 & -12.667 & 0.000 \\ -3.977 & -1.595 & 3.550 & 2.022 & 2.062 & 3.511 & 0.000 & 24.572 & 0.000 & 0.000 \\ -2.699 & -1.201 & 2.533 & 1.367 & 2.710 & 1.190 & -12.667 & 0.000 & 35.567 & -12.667 \\ -2.707 & -1.238 & 0.000 & 3.945 & 2.117 & 1.828 & 0.000 & 0.000 & -12.667 & 29.278 \end{bmatrix}$$

$$\begin{bmatrix} \Delta\hat{t}_1 \\ \Delta\hat{t}_2 \\ \Delta\hat{h}_1 \\ \Delta\hat{h}_2 \\ \Delta\hat{\eta}_1 \\ \Delta\hat{\eta}_2 \\ \Delta\hat{u}_1 \\ \Delta\hat{u}_2 \\ \Delta\hat{u}_3 \\ \Delta\hat{u}_4 \end{bmatrix} = \begin{bmatrix} -0.289 \\ 0.459 \\ -0.021 \\ -0.149 \\ -0.099 \\ -0.071 \\ -0.104 \\ -0.021 \\ 0.031 \\ -0.076 \end{bmatrix}$$

The equations were solved with the solutions for $\Delta\hat{h}_1$ and $\Delta\eta_1$ set to zero. The equations converged rapidly, and solutions at various different iteration numbers and the final solutions are given below. Solution from an analysis using a linear model with an α value of 19 are also shown:

	Iteration number				
Effects	1	2	3	7	Solutions from linear models
Threshold					
1	0.4410	0.4375	0.4378	0.4378 ± 0.44[a]	–
2	1.0448	1.0661	1.0675	1.0675 ± 0.47	–
Herd–year					
1	0.0000	0.0000	0.0000	0.0000 ± 0.00	0.0
2	0.2869	0.2763	0.2774	0.2774 ± 0.49	–0.0819
Sex of calf					
Male	0.0000	0.0000	0.0000	0.0000 ± 0.00	1.3457
Female	–0.3583	–0.3577	–0.3589	–0.3590 ± 0.48	1.6073
Sires					
1	–0.0415	–0.0431	–0.0434	–0.0434 ± 0.22	–0.0657
2	0.0579	0.0586	0.0592	0.0592 ± 0.21	0.0688
3	0.0399	0.0410	0.0412	0.0412 ± 0.22	–0.0557
4	–0.0652	–0.0653	–0.0660	–0.0660 ± 0.22	–0.0097

[a]Standard errors

The standard errors associated with the results from the last iteration were computed from the square root of the diagonals of the generalized inverse. Sire rankings from the linear model were similar to those from the threshold model but absolute values are different.

Usually of interest is calculating the probability of response in a given category under specific conditions. For instance, the proportion of calving in the jth category of response, considering only female calves in HYS subclass 1 for sire 1 can be estimated as:

$$P_{11} = \Phi(t_1 - \hat{h}_1 - \hat{\eta}_2 - \hat{u}_1) = \Phi(0.4378 - 0 - (-0.390) - (-0.0434))$$
$$= \Phi(0.8402) = 0.800$$

$$P_{12} = \Phi(t_2 - \hat{h}_1 - \eta_2 - \hat{u}_1) - \Phi(t_1 - \hat{h}_1 - \eta_2 - \hat{u}_1) = \Phi(1.0675 - 0 - (0.3590) - (-0.0434)) -$$
$$= \Phi(0.8402) = \Phi(1.4699) - \Phi(0.8402) = 0.129$$

$$P_{13} = 1 - \Phi(t_2 - \hat{h}_1 - \hat{\eta}_2 - \hat{u}_1) = 1 - \Phi(1.4699) = 0.071$$

Calculating this probability distribution by category of response for all sires gives the following:

	Probability in category of response		
	1	2	3
Sire 1	0.800	0.129	0.071
Sire 2	0.770	0.145	0.086
Sire 3	0.775	0.142	0.083
Sire 4	0.803	0.129	0.068

The results indicate that the majority of heifers calving in HYS subclass 1 for all four sires were normal, with a very low proportion of extreme difficulties.

Since sires are used across herds, the interest might be the probability distribution of heifer calvings for each sire across all herds and sexes. Such a probability for each sire in category 1 of response per herd–year–sex subclass (Z_{1kji}) can be calculated as follows:

$$Z_{1kji} = \Phi(t_1 - (\hat{h}_k + \hat{\eta}_j + \hat{u}_i)); \quad k = 1,2; \quad j = 1,2, \quad i = 1,\ldots,4$$

Since there are four herd–year–sex subclasses, the probability for sire i in category 1 (S_{1i}) can be obtained by weighting Z_{1kji} by factors that sum up to one. Thus:

$$S_{1i} = \sum_{i=1}^{4}\sum_{k=1}^{2}\sum_{m=1}^{2} a_{km} Z_{1ikm}$$

where $a_{km} = a_{11} + a_{12} + a_{21} + a_{22} = 1$. In the example data, $a_{11} = a_{12} = a_{21} = a_{22} = 0.25$.

Similarly, the probability for each sire in category 2 of response per herd–year–sex subclass (Z_{2kji^*}) can be calculated as:

$$Z_{2kji} = Z_{2kji^*} - Z_{1kji}$$

where:

$$Z_{1kji^*} = \Phi(t_2 - (\hat{h}_k + \hat{\eta}_j + \hat{u}_i)); \; k = 1,2; \; j = 1,2; \; i = 1,\ldots,4$$

Finally, the probability for each sire in category 3 of response per herd–year–sex subclass (Z_{3kji}) can be calculated as:

$$Z_{3kji} = 1 - Z_{2kji^*}$$

For Example 15.1, the probability distribution of heifer calvings for each sire across all herds and sexes in all categories are as follows:

	Probability in category of response		
	1	2	3
Sire 1	0.695	0.175	0.131
Sire 2	0.659	0.188	0.153
Sire 3	0.665	0.186	0.149
Sire 4	0.702	0.172	0.126

15.3 Joint Analysis of Quantitative and Binary Traits

Genetic improvement may be based on selecting animals on an index that combines both quantitative and categorical traits. Optimally, a joint analysis of the quantitative and categorical traits is required in the prediction of breeding values in such a selection scheme to adequately account for selection. A linear multivariate model might be used for such analysis. However, such an analysis suffers from the limitations associated with the use of a linear model for the analysis of discrete traits mentioned in Section 15.2. In addition, such a multivariate linear model will not properly account for the correlated effects of the quantitative traits on the discrete trait.

Foulley *et al.* (1983) presented a method of analysis to handle the joint analysis of quantitative and binary traits using a Bayesian approach. It involves fitting a linear model for the quantitative traits and a non-linear model for the binary trait. This section presents this methodology and illustrates its application to an example data set.

15.3.1 Data and model definition

Assume that a quantitative trait, such as birth weight, and a binary trait, such as calving difficulty (easy versus difficult calving), is being analysed. As in Section 15.2.2, the data for calving difficulty could be represented in an $s \times 2$ contingency table:

	Response category	
Row	Easy calving	Difficult calving
1	n_{11}	$n_{1.} - n_{11}$
2	n_{21}	$n_{2.} - n_{21}$
⋮	⋮	⋮
j	n_{j1}	$n_{j.} - n_{j1}$
⋮	⋮	⋮
s	n_{s1}	$n_{s.} - n_{s1}$

where the s rows refer to conditions affecting an individual record or grouped records. Note that n_{i1} or $n_{i.} - n_{i1}$ in the above table can be null, as responses in the two categories are mutually exclusive, but $n_i \neq 0$.

Assume that a normal function has been used to describe the probability of response for calving ease. Let $\mathbf{y}_1$ be the vector for observations for the quantitative trait, such as birth weight, and $\mathbf{y}_2$ be the vector of the underlying variable for calving difficulty. The model for trait 1 would be:

$$\mathbf{y}_1 = \mathbf{X}_1\boldsymbol{\beta}_1 + \mathbf{Z}_1\mathbf{u}_1 + \mathbf{e}_1 \tag{15.13}$$

and for the underlying variable for trait 2:

$$\mathbf{y}_2 = \mathbf{X}_2\boldsymbol{\beta}_2 + \mathbf{Z}_2\mathbf{u}_2 + \mathbf{e}_2 \tag{15.14}$$

where $\boldsymbol{\beta}_1$ and $\mathbf{u}_1$ are vectors of fixed effect and sire solutions for trait 1, and $\mathbf{X}_1$ and $\mathbf{Z}_1$ are the usual incidence matrices. The matrices $\mathbf{X}_2$ and $\mathbf{Z}_2$ are incidence matrices for the liability. The matrix $\mathbf{Z}_2 = \mathbf{Z}_1$ and $\mathbf{X}_2 = \mathbf{X}_1\mathbf{H}$, where $\mathbf{H}$ is an identity matrix if all factors affecting the quantitative traits also affect the liability. However, if certain fixed effects affecting the quantitative trait have no effect on the liability, $\mathbf{H}$ is obtained by deleting the columns of an identity matrix of appropriate order corresponding to such effects. It is assumed that:

$$\text{var}\begin{pmatrix}\mathbf{e}_1\\ \mathbf{e}_2\end{pmatrix} = \begin{pmatrix}\mathbf{R}_{11} & \mathbf{R}_{12}\\ \mathbf{R}_{21} & \mathbf{R}_{22}\end{pmatrix}$$

$$\text{var}\begin{pmatrix}\mathbf{u}_1\\ \mathbf{u}_2\end{pmatrix} = \mathbf{A}\otimes\mathbf{G} \tag{15.15}$$

where $\mathbf{G}$ is the genetic covariance matrix for both traits and $\mathbf{A}$ is the numerator relationship matrix.

Let $\boldsymbol{\theta}' = [\boldsymbol{\beta}_1, \boldsymbol{\tau}, \mathbf{u}_1, \boldsymbol{\nu}]$, the vector of location parameters in Eqns 15.13 and 15.14 to be estimated, where $\boldsymbol{\tau} = \boldsymbol{\beta}_2 - b\mathbf{H}\boldsymbol{\beta}_1$ and $\boldsymbol{\nu} = \mathbf{u}_2 - b\mathbf{u}_1$, where b is the residual regression coefficient of the underlying variate on the quantitative trait. The calculation of b is illustrated in the next section. Since the residual variance of liability is unity, the use of b is necessary to properly adjust the underlying variate for the effect of the residual covariance between both traits. The use of b can be thought of as correcting calving difficulty for other 'risk' factors affecting calving and, in this example, the birth weight of the calf. Thus Eqn 15.15 may be written as:

$$\text{var}\begin{pmatrix}\mathbf{u}_1\\ \mathbf{u}_2 - b\mathbf{u}_1\end{pmatrix} = \begin{pmatrix}\mathbf{u}_1\\ \boldsymbol{\nu}\end{pmatrix} = \mathbf{A}\otimes\mathbf{G}_c$$

where:

$$\mathbf{G}c = \begin{pmatrix}\mathrm{I} & 0\\ -b\mathrm{I} & \mathrm{I}\end{pmatrix}\begin{pmatrix}g_{11} & gg_{12}\\ g_{21} & gg_{22}\end{pmatrix}\begin{pmatrix}\mathrm{I} & -b\mathrm{I}\\ \mathrm{I} & \mathrm{I}\end{pmatrix}$$

with g_{ij} being the elements of $\mathbf{G}$.

Using a Bayesian approach, Foulley *et al.* (1983) calculated the mode of the posterior density of $\boldsymbol{\theta}$ by equating the derivatives of the log-posterior density of $\boldsymbol{\theta}$ to zero. The resulting system of equations were not linear in $\boldsymbol{\theta}$. They set up the following iterative system of equations for $\boldsymbol{\theta}$ to be estimated:

$$\begin{pmatrix} \mathbf{X}_1' \mathbf{R}_1^{-1}\mathbf{X}_1 & \mathbf{X}_1' \mathbf{R}_1^{-1}\mathbf{Z}_1 & 0 & 0 \\ \mathbf{Z}_1' \mathbf{R}_1^{-1}\mathbf{X}_1 & \mathbf{Z}_1' \mathbf{R}_1^{-1}\mathbf{Z}_1 + \mathbf{A}^{-1}\mathbf{g}_c^{11} & 0 & \mathbf{A}^{-1}\mathbf{g}_c^{12} \\ 0 & 0 & \mathbf{X}_2'\mathbf{W}^{[i-1]}\mathbf{X}_2 & \mathbf{X}_2'\mathbf{W}^{[i-1]}\mathbf{Z}_2 \\ 0 & \mathbf{A}^{-1}\mathbf{g}_c^{21} & \mathbf{Z}_2'\mathbf{W}^{[i-1]}\mathbf{X}_2 & \mathbf{Z}_2'\mathbf{W}^{[i-1]}\mathbf{Z}_2 + \mathbf{A}^{-1}\mathbf{g}_c^{21} \end{pmatrix}$$

$$\begin{pmatrix} \hat{\boldsymbol{\beta}}^{[i]} \\ \hat{\mathbf{u}}^{[i]} \\ \Delta\boldsymbol{\tau}^{[i]} \\ \Delta\mathbf{v}^{[i]} \end{pmatrix} = \begin{pmatrix} \mathbf{X}_1' \mathbf{R}_1^{-1}\mathbf{y}_1 \\ \mathbf{Z}_1' \mathbf{R}_1^{-1}\mathbf{y}_1 \\ \mathbf{X}_2' \mathbf{q}^{[i-1]} \\ \mathbf{Z}_2' \mathbf{q}^{[i-1]} \end{pmatrix} - \begin{pmatrix} 0 \\ \mathbf{A}^{-1}\mathbf{g}_c^{12}\mathbf{v}^{[i-1]} \\ 0 \\ \mathbf{A}^{-1}\mathbf{g}_c^{22}\mathbf{v}^{[i-1]} \end{pmatrix} \tag{15.17}$$

The matrices and vectors in Eqn 15.17 have been defined earlier, apart from **q** and **W**. Initially, P_{jk}, the probability of response in category k, given the conditions in the jth row, is defined for the category trait. With only two categories of response for calving difficulty, then from Eqn 15.5:

$$P_{j1} = \Phi(t - a_j) \quad \text{and} \quad P_{j2} = 1 - P_{j1}$$

with a_j regarded as the mean of the liability in the jth row or as defined in Eqn 15.5.

However, with only one threshold, the value of t by itself is of no interest; the probability of response in the first category for the jth row can then be written as:

$$P_{j1} = \Phi(t - a_j) = \Phi(\mu_j)$$

where μ_j can be defined as the expectation of $\mathbf{y}_{2j}$ given $\boldsymbol{\beta}$, $\mathbf{u}$ and $\mathbf{y}_{1j}$, and this is worked out in the next section.

The vector **q** is of order $s \times 1$ with elements:

$$\mathbf{q}_j = -\{n_{j1}d_{j1} + (n_{j.} - n_{j1})d_{j2}\}, \quad j = 1,\ldots,s \tag{15.18}$$

where $d_{j1} = -\phi(\mu_j)/P_{j1}$ and $d_{j2} = \phi\mu_j/(1 - P_{j1})$, with P_{j1} calculated as $\Phi(\mu_j)$.

W is an $s \times s$ diagonal matrix with the following elements:

$$w_{jj} = \mu_j q_j + n_{j1}d^2_{j1} + (n_{j.} - n_{j1})d^2_{j2}, \quad j = 1,\ldots,s \tag{15.19}$$

Calculating μ and the residual regression coefficient

From Eqn 15.14, the model for the jth row of the contingency table may be written as:

$$\mathbf{y}_{2j} = \mathbf{x}_{2j}'\boldsymbol{\beta}_2 + \mathbf{z}_{2j}'\mathbf{u}_2 + \mathbf{e}_{2j}$$

where $\mathbf{x}_{2j}'$ and $\mathbf{z}_{2j}'$ are vectors j of the $\mathbf{X}_2$ and $\mathbf{Z}_2$, respectively. Similarly, observations for trait 1, corresponding to the jth row of the contingency table, may be modelled as:

$$\mathbf{y}_{1j} = \mathbf{x}_{1j}'\boldsymbol{\beta}_1 + \mathbf{z}_{1j}'\mathbf{u}_1 + \mathbf{e}_{1j}$$

Let μ_j be the expectation of $\mathbf{y}_{2j}$ given $\boldsymbol{\beta}$, $\mathbf{u}$ and $\mathbf{y}_{1j^*}$. Thus:

$$\mu_j = \mathrm{E}(\mathbf{y}_{2j} \mid \boldsymbol{\beta}_1, \boldsymbol{\beta}_2, \mathbf{u}_1, \mathbf{u}_2, \mathbf{y}_{1j}) = \mathbf{x}_{2j}'\,\boldsymbol{\beta}_2 + \mathbf{z}_{2j}'\,\mathbf{u}_2 + \mathrm{E}(e_{2j} \mid e_{1j}) \tag{15.20}$$

given that e_{2j} is only correlated with e_{1j^*}. Assuming e_{2j} and e_{1j} are bivariately normally distributed:

$$\begin{aligned} E(e_{2j} \mid e_{1j}) &= \frac{\sigma_{e(2,1)}}{\sigma_{e1}^2}(e_{1j}) \\ &= r_{12}\left(\frac{\sigma_{e2}}{\sigma_{e1}}\right)e_{1j} \end{aligned} \tag{15.21}$$

where σ_{ei}^2 is the residual variance of trait i, $\sigma_{ei,k}$ and r_{ik} are the residual covariance and correlation between traits i and k, and σ_{ei} is the residual standard deviation of the ith trait. Similarly:

$$\text{var}(\mathbf{y}_{2j} \mid \boldsymbol{\beta}_1, \boldsymbol{\beta}_2, \mathbf{u}_1, \mathbf{u}_2, \mathbf{y}_{1j}) = \text{var}(e_{2j} \mid e_{1j}) = \sigma_{e2}^2(1 - r_{12}^2)$$

Since the unit of the conditional distribution of the underlying trait, given $\boldsymbol{\beta}_1$, $\boldsymbol{\beta}_2$, $\mathbf{u}_1$ $\mathbf{u}_2$, $\mathbf{y}_{1j}$ and is the standard deviation, then from the above equation:

$$\sigma_{e2} = \frac{1}{\sqrt{(1 - r_{12}^2)}}$$

Therefore, Eqn 15.21 can be written as:

$$E(e_{2j} \mid e_{1j}) = r_{12}\left(\frac{1}{\sigma_{e1}}\right)\frac{1}{\sqrt{1 - r_{12}^2}}e_{1j} = be_{1j} \tag{15.22}$$

In general, Eqn 15.20 can be expressed as:

$$\begin{aligned} \boldsymbol{\mu} &= \mathbf{X}_1\boldsymbol{\beta}_2 + \mathbf{Z}_2\mathbf{u}_2 + b\mathbf{e}_1 \\ &= \mathbf{X}_2\boldsymbol{\beta}_2 + \mathbf{Z}_2\mathbf{u}_2 + b(\mathbf{y}_1 - \mathbf{X}_1\boldsymbol{\beta}_1 - \mathbf{Z}_1\mathbf{u}_1) \end{aligned} \tag{15.23}$$

The above equation may be written as:

$$\begin{aligned} \boldsymbol{\mu} &= \mathbf{X}_2(\boldsymbol{\beta}_2 - b\mathbf{H}\boldsymbol{\beta}_1) + \mathbf{Z}_2(\mathbf{u}_2 - b\mathbf{u}_1) + b\mathbf{y}_1^* \\ \boldsymbol{\mu} &= \mathbf{X}_2\boldsymbol{\tau} + \mathbf{Z}_2\mathbf{v} + b\mathbf{y}_1^* \end{aligned} \tag{15.24}$$

with the solutions of factors affecting calving difficulty corrected for the residual relationship between the two traits and $\mathbf{y}_1^* = (\mathbf{y}_1 - \mathbf{X}_1\boldsymbol{\beta}_1 - \mathbf{Z}_1\mathbf{u}_1)$ or $\mathbf{y}_1^*$ may be calculated as: $\mathbf{y}_1^* = (\mathbf{y}_1 - \bar{y}_1)$, where $\bar{y}_1$ is the mean of $\mathbf{y}_1$.

15.3.2 Numerical application

Example 15.2

The bivariate analysis of a quantitative trait and a binary trait is illustrated using the data presented by Foulley *et al.* (1983) but with a sire–maternal grandsire relationship matrix included for the sires and pelvic opening omitted from the analysis. The data consisted of birth weight (BW) and calving difficulty (CD) on 47 Blonde d'Aquitaine heifers, with information on region of origin, sire of the heifer, calving season and sex of the calf included. Calving difficulty was summarized into two categories: easy or difficult calving. The data set is presented below:

Heifer origin	Sire	Season	Sex of calf	BW	CD[a]	Heifer origin	Sire	Season	Sex of calf	BWT	CD[a]
1	1	1	M	41.0	E	1	4	2	M	47.0	D
1	1	1	M	37.5	E	1	4	2	F	51.0	D
1	1	1	F	41.5	E	1	4	2	F	39.0	E
1	1	2	F	40.0	E	2	4	1	M	44.5	E
1	1	2	F	43.0	E	1	5	1	M	40.5	E
1	1	2	F	42.0	E	1	5	1	F	43.5	E
1	1	2	F	35.0	E	1	5	2	M	42.5	E
2	1	1	F	46.0	E	1	5	2	M	48.8	D
2	1	1	F	40.5	E	1	5	2	M	38.5	E
2	1	2	F	39.0	E	1	5	2	M	52.0	E
1	2	1	M	41.4	E	1	5	2	F	48.0	E
1	2	1	M	43.0	D	2	5	1	F	41.0	E
1	2	2	F	34.0	E	2	5	1	M	50.5	D
1	2	2	M	47.0	D	2	5	2	M	43.7	D
1	2	2	M	42.0	E	2	5	2	M	51.0	D
2	2	2	M	44.5	E	1	6	1	F	51.6	D
2	2	2	M	49.0	E	1	6	1	M	45.3	D
1	3	1	M	41.6	E	1	6	1	F	36.5	E
2	3	1	M	36.0	E	1	6	2	M	50.5	E
2	3	1	F	42.7	E	1	6	2	M	46.0	D
2	3	2	F	32.5	E	1	6	2	M	45.0	E
2	3	2	F	44.4	E	1	6	2	F	36.0	E
2	3	2	M	46.0	E	2	6	1	F	43.5	E
						2	6	1	F	36.5	E

[a]CD, calving difficulty; D, difficulty, E, easy

A summary of the data, in terms of marginal means of calving variables by level of factors considered, is shown in the following table:

Factor		Number	Birth weight (kg)	Frequency CD[a]
Heifer origin	1	30	43.02	0.267
	2	17	43.02	0.176
Calving season	1	20	42.23	0.200
	2	27	43.61	0.259
Sex of calf	M	25	–	0.360
	F	22	–	0.091
Sire of heifer	1	10	40.55	0.000
	2	7	42.99	0.286
	3	6	40.53	0.000
	4	4	45.38	0.500
	5	11	45.46	0.364
	6	9	43.43	0.333

[a]Frequency of calving difficulty

The following sire–maternal grandsire relationship matrix was assumed among the sires:

Bull	Sire	Maternal grandsire
1	0	0
2	0	0
3	1	0
4	2	1
5	3	2
6	2	3

The inverse of the sire–maternal grandsire relationship matrix obtained for the above pedigree using the rules in Section 2.5 is:

$$\mathbf{A}^{-1} = \begin{bmatrix} 1.424 & 0.182 & -0.667 & -0.364 & 0.000 & 0.000 \\ 0.182 & 1.818 & 0.364 & -0.727 & -0.364 & -0.727 \\ -0.667 & 0.364 & 1.788 & 0.000 & -0.727 & -0.364 \\ -0.364 & -0.727 & 0.000 & 1.455 & 0.000 & 0.000 \\ 0.000 & -0.364 & -0.727 & 0.000 & 1.455 & 0.000 \\ 0.000 & -0.727 & -0.364 & 0.000 & 0.000 & 1.455 \end{bmatrix}$$

The residual variance (σ_{e1}^2) for BW was assumed to be 20 kg^2 and the residual correlation (r_{12}) between BW and CD was assumed to be 0.459. Therefore, from Eqn 15.20, b equals 0.1155. The matrix G assumed was:

$$\mathbf{G} = \begin{pmatrix} 0.7178 & 0.1131 \\ 0.1131 & 0.0466 \end{pmatrix}$$

Therefore, from Eqn 15.16:

$$\mathbf{G}_c = \begin{pmatrix} 1 & 0 \\ -0.1155 & 1 \end{pmatrix} \begin{pmatrix} 0.7178 & 0.1131 \\ 0.1131 & 0.0466 \end{pmatrix} \begin{pmatrix} 1 & -0.1155 \\ 0 & 1 \end{pmatrix} = \begin{pmatrix} 0.7178 & 0.0302 \\ 0.0302 & 0.0300 \end{pmatrix}$$

Thus, the heritabilities for BW and CD are 0.14 and 0.18, respectively, with a genetic correlation of 0.62 between the two traits.

The model in Eqn 15.13 was used for the analysis of BW, thus $\boldsymbol{\beta}_1$ is the vector of solutions for origin of heifer, calving season and sex of calf and $\mathbf{u}_1$ is the vector of solutions for sire effects. The same effects were fitted for CD, with $\boldsymbol{\tau}$ being the vector of solutions for the fixed effects and $\mathbf{v}$ for the sire effects. Let $\boldsymbol{\theta}$ be as follows:

$$\begin{aligned} \boldsymbol{\beta}_1' &= (d_1, d_2, s_1, s_2, f_1, f_2) \\ \mathbf{u}_1' &= (\hat{u}_{11}, \hat{u}_{12}, \hat{u}_{13}, \hat{u}_{14}, \hat{u}_{15}, \hat{u}_{16}) \\ \boldsymbol{\tau}' &= (d_1', d_2', s_1', s_2', f_1', f_2') \\ \mathbf{v}' &= (\mathrm{v}_1, \mathrm{v}_2, \mathrm{v}_3, \mathrm{v}_4, \mathrm{v}_5, \mathrm{v}_6) \end{aligned}$$

where $d_i(d_i')$, $s_i(s_i')$ and $f_i(f_i')$ are level i of the effects of heifer origin, calving season and sex of calf, respectively; for BW (CD), $\hat{u}_{1j}$ and v_j are the solutions for the sire j for BW and CD, respectively.

The matrix $\mathbf{X}_1$, which relates records for BW to the effects of heifer origin, calving season and sex of calf, can be set by principles already outlined in previous chapters. For the example data, all fixed effects affecting BW also affect CD; therefore, **H** is an identity matrix and $\mathbf{X}_2 = \mathbf{X}_1$. Similarly, the matrix $\mathbf{Z}_1 = \mathbf{Z}_2$. The remaining matrix in Eqn 15.17 can be obtained through matrix multiplication and addition.

Equation 15.17 needs starting values for $\boldsymbol{\tau}$ and $\mathbf{v}$ to commence the iterative process. The starting values used were solutions ($\boldsymbol{\tau}^{(0)}$ and $\mathbf{v}^{(0)}$) from Eqn 15.17 with $\mathbf{W}^{[i-1]} = \mathbf{I}$, $\mathbf{q}^{[i-1]} =$ a vector of (0,1) variables (1, difficulty; 0, otherwise) and $\mathbf{V}^{[i-1]} = 0$. The solutions to Eqn 15.17 using these starting values are shown in Table 15.3, with equations for the second levels of calving season and sex of calf effects set to zero because of dependency in the systems of equations. Using these solutions, the calculation of $\mathbf{q}^{(0)}$ and $\mathbf{W}^{(0)}$ in the next round of iteration are illustrated for the first and last two animals in the example data.

Table 15.3. Solutions to Example 15.2 using Eqn 15.17.

		Iteration number					
Trait[a]	Factor	0	1	4	8	13	Linear model
BW	*Heifer origin*						
	1	41.6177	41.6183	41.6192	41.6192	41.6192	41.6175
	2	42.2069	42.2083	42.2109	42.2109	42.2109	42.2022
	Calving season						
	1	−1.2359	−1.2358	−1.2344	−1.2344	−1.2344	−1.2387
	2	0.0000	0.0000	0.0000	0.0000	0.0000	0.0000
	Sex of calf						
	Male	3.1728	3.1737	3.1690	3.1690	3.1690	3.1845
	Female	0.0000	0.0000	0.0000	0.0000	0.0000	0.0000
	Sire						
	1	−0.3497	−0.3488	−0.3592	−0.3592	−0.3592	−0.3268
	2	0.1201	0.1259	0.1303	0.1303	0.1303	0.1171
	3	−0.2852	−0.2870	−0.2948	−0.2948	−0.2948	−0.2641
	4	0.2022	0.2042	0.2126	0.2126	0.2126	0.1886
	5	0.2994	0.2899	0.2969	0.2969	0.2969	0.2688
	6	0.1794	0.1787	0.1815	0.1815	0.1815	0.1690
CD	*Heifer origin*						
	1	0.1343	−0.9466	−1.3936	−1.3936	−1.3936	0.1349
	2	0.0851	−1.1538	−1.7457	−1.7457	−1.7457	0.0876
	Calving season						
	1	−0.0317	0.1157	0.1404	0.1404	0.1404	−0.0311
	2	0.0000	0.0000	0.0000	0.0000	0.0000	0.0000
	Sex of calf						
	Male	0.2437	0.3852	0.8401	0.8401	0.8401	0.2410
	Female	0.0000	0.0000	0.0000	0.0000	0.0000	0.0000
	Sire						
	1	−0.0450	-0.0441	−0.0573	−0.0573	−0.0573	−0.0527
	2	0.0237	0.0308	0.0372	0.0372	0.0372	0.0285
	3	−0.0363	-0.0392	−0.0485	−0.0485	−0.0485	−0.0427
	4	0.0289	0.0318	0.0411	0.0411	0.0411	0.0350
	5	0.0188	0.0071	0.0152	0.0152	0.0152	0.0195
	6	0.0272	0.0270	0.0309	0.0309	0.0309	0.0323

[a]BW, birth weight; CD, calving difficulty

First, $\boldsymbol{\mu}$ in Eqn 15.18 is calculated for these animals using Eqn 15.24. For animals 1 and 2:

$$\mathbf{X}_2\boldsymbol{\tau} + \mathbf{Z}_2\mathbf{v} = (d_1' + \hat{s}_1' + f_1' + \hat{v}_1) = 0.1873 + -0.0874 + 0.2756 + (-0.1180) = 0.2575$$

Therefore, from Eqn 15.24, using the mean of birth weight, μ_1 is:

$$\mu_1 = 0.2575 + 0.1155(41 - 43.02) = 0.0242$$

and

$$\mu_2 = 0.2575 + 0.1155(37.5 - 43.02) = -0.3800$$

For animals 46 and 47:

$$\mathbf{X}_2\boldsymbol{\tau} + \mathbf{Z}_2\mathbf{v} = (d_2' + \hat{s}_1' + f_2' + v_6') = 0.1484 + -0.0874 + 0.0 + 0.0079 = 0.0690$$

Therefore, from Eqn 15.22:

$$\mu_{46} = 0.0690 + 0.1155(43.5 - 43.02) = 0.1244$$

and

$$\mu_{47} = 0.0690 + 0.1155(36.5 - 43.02) = -0.6841$$

Using Eqn 15.18, the elements of $\mathbf{q}$ for animals 1, 2, 46 and 47 are:

$$\begin{aligned}
\mathbf{q}(1) &= -\{0(-1)\phi(0.0242)/\Phi(0.0242) + (1-0)\phi(0.0242)/(1-\Phi(0.0242))\} \\
&= -\{0(-1)0.3988/0.5097 + 1(0.3988/0.4903)\} = -0.8134 \\
\mathbf{q}(2) &= -\{0(-1)\phi(0.3800)/\Phi(-0.3800) + (1-0)\phi(-0.3800)/(1-\Phi(-0.3800))\} \\
&= -\{0(-1)0.3712/0.3520 + 1(0.3712/0.6480)\} = -0.5727 \\
\mathbf{q}(46) &= -\{0(-1)\phi(0.1244)/\Phi(0.1244) + (1-0)\phi(0.1244)/(1-\Phi(0.1244))\} \\
&= -\{0(-1)0.3959/0.5495 + 1(0.3959/0.4505)\} = -0.8787 \\
\mathbf{q}(47) &= -\{0(-1)\phi(-0.6841)/\Phi(-0.6841) + (1-0)\phi(-0.6841)/(1-\Phi(-0.6841))\} \\
&= -\{0(-1)0.3157/0.2470 + 1(0.3157/0.7530)\} = -0.4193
\end{aligned}$$

The diagonal elements of $\mathbf{W}$ for each of the four animals above can be calculated using Eqn 15.19 as:

$$\begin{aligned}
w(1,1) &= 0.0242(-0.8134)\{0(-1)[\phi(0.0242)/\Phi(0.0242)]^2 + (1-0)[\phi(0.0242)/ \\
&\quad (1-\Phi(0.0242))]^2\} = 0.6419 \\
w(2,2) &= -0.3800(-0.5727)\{0(-1)[\phi(-0.3800)/\Phi(-0.3800)]^2 + (1-0) \\
&\quad [\phi(-0.3800)/(1-\Phi(-0.3800))]^2\} = 0.5458 \\
w(46,46) &= 0.1244(-0.8787)\{0(-1)[\phi(0.1244)/\Phi(0.1244)]^2 + (1-0)[\phi(0.1244)/ \\
&\quad (1-\Phi(0.1244))]^2\} = 0.6629 \\
w(47,47) &= -0.6841(-0.4193)\{0(-1)[\phi(-0.6841)/\Phi(-0.6841)]^2 + (1-0) \\
&\quad [\phi(-0.6841)/(1-\Phi(-0.6841))]^2\} = 0.4626
\end{aligned}$$

The equations were solved iteratively and were said to have converged at the 15th round of iteration when $\Delta'\Delta/20 \le 10^{-1}$, where $\Delta = \boldsymbol{\theta}^{(i)} - \boldsymbol{\theta}^{(i-1)}$. Solutions at convergence at the 13th round of iteration and at some intermediate rounds are shown in Table 15.3. Results from an analysis using a linear model fitting the same effects with the G matrix and residual variances of 20 kg^2 for BW, 1.036 for CD and residual covariance of 2.089 between the two traits are also presented.

The results indicate that the probability of a difficult calving is higher for a male calf than for a female calf. Similarly, there is a slightly higher probability for calving difficulty for calving in the first season.

In general, sire rankings from the threshold and linear models were similar, except for sires 2 and 6 slightly changing rankings in the two models. The ranking of sires for calving difficulty based on the results from the threshold model could be based on $\hat{\mathbf{u}}_2 = \mathbf{v} + b_1\hat{\mathbf{u}}_1$ using the information provided by BW. However, the interest might be on ranking sires in terms of probability of calving difficulty, under a given set of conditions. For instance, what is the probability that a heifer sired by the *j*th bull born in region 2, calving a male calf in season 1, will experience a calving difficulty? This probability (V_{211j}) can be calculated as:

$$V_{211j} = \Phi[\hat{d}_2' + \hat{s}_1' + \hat{f}_1' + \hat{v}_j + b_1(\hat{d}_2 + \hat{s}_1 + \hat{f}_1 - 43.02)] \tag{15.25}$$

Using the above equation, this probability for sire 1 is:

$$\begin{aligned} V_{211j} &= \Phi[-1.7452 + 0.1401 + 0.8411 + (-0.0577) + 0.1155(42.2112 \\ &\quad + (-1.2344) + 3.1690 - 43.02)] = 0.245 \end{aligned}$$

Similar calculations gave probabilities of 0.275, 0.247, 0.276, 0.268 and 0.273 for sires 2, 3, 4, 5 and 6, respectively. In general, there might be interest in the probability of difficult calving associated with using the *j*th sire across all regions of origin by season of calving and sex of calf subclasses. Such a probability can be calculated as:

$$V_{...j} = \Sigma_{ikl}\, \lambda_{ikl} V_{iklj} \tag{15.26}$$

with V_{iklj} estimated as Eqn 15.25 and λ_{ikl} is an arbitrary weight such that $\Sigma_{ikl}\, \lambda_{ikl} = 1$. For the example data, λ can be set to be equal to $\frac{1}{8}$, as there are eight region–season–sex of calf subclasses. The probabilities obtained using Eqn 15.26 with $\lambda = \frac{1}{8}$ were 0.167, 0.188, 0.169, 0.189, 0.183 and 0.187 for sires 1, 2, 3, 4, 5 and 6, respectively.

The analysis of a binary trait with a quantitative trait has been discussed and illustrated in this section. However, if the category trait has several thresholds, then the method discussed in Section 15.2 would be used for the analysis of the categorical trait.

16 Survival Analysis

16.1 Introduction

Survival is one of the most important functional economic traits in livestock production, affecting profitability through the rate of replacement and farm production levels. In dairy cattle, the average herd life or survival of dairy cows has an economic value approximately half that of protein yield on a genetic standard deviation basis (Visscher *et al.*, 1999). Consequently, most of the earlier research work on survival in terms of genetic evaluation and inclusion in breeding programmes has been in dairy cattle.

Various traits have been defined as the basis of evaluating survival in the dairy cow. These usually include some measure of survival for a period or length of life such as stayability until certain months of life defined as a binary trait (Everett *et al.*, 1976), or in terms of the length of life or length of productive life (VanRaden and Klaaskate, 1993), or number of lactations (Brotherstone *et al.*, 1997) or survival per lactation as a binary trait. Linear models are generally used – either a repeatability model (Madgwick and Goddard, 1989) or a multivariate model (Jairath *et al.*, 1998). Similar definitions of survival have been applied to other livestock species. The length of productive life between first farrowing and culling has been analysed in pigs (Tarrés *et al.*, 2006; Mészáros *et al.*, 2010). In rabbits, survival has been defined as the length of productive life, referring to the days between date of the first positive pregnancy diagnosis and date of culling or death (Piles *et al.*, 2006).

16.2 Functional Survival

Another important element of evaluating survival is the concept of functional survival or longevity. Functional longevity refers to survival that is independent of production such as milk yield for dairy cattle or litter size in pigs. The reasoning is that voluntary culling is based mostly on production, thus adjusting for production (usually at the phenotypic level) in the analysis of survival produces EBVs for animals that defines their ability to avoid involuntary culling.

16.3 Censoring

The traits used in survival analysis involve measuring the length of time between two events, usually a start and end point (also called 'failure'). However, at the time of analysis, some animals might still be alive, not having had the opportunity to reach the end point. Their measure of survival is based on their current status and does not therefore reflect their true measure of survival. This phenomenon is referred to as censoring and such records are regarded as censored. There are several types of censoring. When records

DOI: 10.1079/9781800620506.0016

are based on current values that are less than the unknown end point, this is called right censoring. Left censoring can occur when, for instance, an animal has been alive for a certain time before entering the study or the start of data collection. Interval censoring can occur when there is a break in data collection and the cow fails somewhere in that interval. However, the most common is right censoring and this is the only type of censoring considered in this chapter.

16.4 Models for Analysis of Survival

16.4.1 Linear models

The linear models described in Chapters 4 or 5 have been used by various researchers for the analysis of survival traits, including those defined as a binary trait (Everett *et al.*, 1976; Madgwick and Goddard, 1989; Jairath *et al.*, 1998).

One of the major limitations with analysis of survival traits using a linear model is the inability or the difficulty of accounting for censoring. Various authors have attempted to address this problem. Brotherstone *et al.* (1997) introduced the concept of lifespan, which is the number of lactations a cow has survived or is expected to survive. Thus if p_n is the probability of survival to lactation $n + 1$ of an animal that has survived to complete lactation n, the expected lifespan (LS) of a cow that has completed n lactations but has not had time to complete $n + 1$ is:

$$\text{LS} = n + p_n + p_n^* \; p_{n+1} + p_n^* \; p_{n+1}^* \; p_{n+2} +$$

Thus, if all p values above are constant and cows have completed their first lactation and have had no time restriction in the opportunity to express LS, then:

$$\text{Prob}(\text{Ls} = x) = (1 - p)P^{x-1} \quad \text{with} \quad x = (1, 2, 3 \ldots)$$

indicating that LS has a geometric distribution with mean = $1 + p/(1 - p)$ and variance = $p/(1 - p)^2$.

Similarly, VanRaden and Klaaskate (1993) evaluated survival using length of productive life, and censored records were predicted using phenotypic multiple regression. Madgwick and Goddard (1989) proposed a multi-trait model for the analysis of survival in each lactation, with observations in individual lactations treated as a different trait. Information on the current lactation of living cows can then be included as observed while their later (future) lactations are treated as missing records, hence accounting for all information.

While some of these linear models have included methods to predict expected survival for censored animals, these models are generally inadequate to handle time-dependent effects. Thus HYS effects, for instance, might be based on information from first calving, even for cows that have survived several lactations.

16.4.2 Random regression models for survival

Veerkamp *et al.* (1999) introduced the concept of fitting a random regression model (RRM) for the analysis of survival defined in terms of survival to the fourth lactation as another approach to handle censored records in a linear model. In addition, time-dependent variables could be fitted with an RRM. The records in lactations 1–4 were coded as 1 if next lactation was present or 0 otherwise. For censored animals, current lactations were coded as described but later (future) lactations were regarded as missing. Thus, for uncensored animals, there would be four observations and censored animals would have a number of observations equal to the current lactation at which they were censored. In addition to the fixed effects of HYS of calving, quadratic regressions for milk yield and age within herd, and a linear regression for Holstein percentage, they modelled the survival records of cows fitting a fixed cubic polynomial for lactation number and orthogonal polynomial of order 3 for additive animal genetic effects. It is not clear why a permanent environmental effect was not included in their model. They concluded that RRM could be considered as an alternative to a proportional hazard model in terms of handling time-dependent variables, but that the RRM was not very efficient at handling culling towards the end of lactation 4. This was attributed to lack of adequate data in the last lactation in the study. The same approach could be used to model survival defined in terms of days or months of productive life. The details of the methodology of fitting an RRM have been covered in Chapter 10; therefore only an outline is presented here.

Considering the data in Table 16.1 and assuming 60 months as the maximum length of productive life, the data can be analysed using an RRM considering herd and year–season–parity (YSP) as the only fixed (FIX) effects with the following model:

$$y_{tijk} = FIX_i + \sum_{k=0}^{nf} \phi_{jtk}\,\boldsymbol{\beta}_k + \sum_{k=0}^{nf} \phi_{jtk}\,\mathbf{u}_{jk} + \sum_{k=0}^{nf} \phi_{jtk}\,\mathbf{p}_{jk} + e_{tijk} \tag{16.1}$$

where y_{tijk} is the record for cow j, which is either 1 (alive) or 0 (dead) at time t (tth month of productive life) associated with the ith level of fixed effects (FIX_i); $\boldsymbol{\beta}_k$ are fixed regression coefficients; $\mathbf{u}_{jk}$ and $\mathbf{p}_{jk}$ are vectors of the kth random regression for animal and permanent environmental (pe) effects, respectively, for animal j; φ_{jtk} is the vector of the kth Legendre polynomial for the cow j at time t; nf is the order of polynomials fitted as fixed regressions; nr is the order of polynomials for animal and pe effects; and e_{tijk} is the random residual. Note that for cows 8, 10, 16 and 18, which are censored, their records consist of the number of observations equal to their last month alive when they were censored. Thus cows 8 and 10 have 40 and 22 observations consisting of ones, respectively.

Table 16.1. Length of productive life (LPL) in months for some cows reared in two herds.

Cow	Sire	Dam	Herd	Parity	YSP	Code	LPL
8	1	2	1	2	3	0	40
9	1	3	1	2	4	1	47
10	4	2	1	1	1	0	22
11	4	9	1	1	2	1	28
12	5	3	1	2	3	1	50
13	5	8	1	1	1	1	33
14	1	6	2	2	4	1	49
15	1	7	2	1	1	1	29
16	5	14	2	1	2	0	23
17	5	6	2	2	3	1	37
18	4	7	2	2	4	0	35
19	4	3	2	1	2	1	30

YSP, year–season–parity. Code: 1, uncensored; 0, censored

However, for uncensored cows, each has 60 observations consisting of ones and zeros. Thus, cow 9 has a record consisting of 47 ones and 13 zeros. The model in Eqn 16.1 can be fitted as described in Section 10.3.

16.4.3 Proportional hazard models

In view of the peculiarities associated with survival traits in terms of censoring of records and the presence of time-dependent covariates (i.e. whose values change with time), the proportional hazard model has been considered a more appropriate method of handling survival data. Its wide usage in the analysis of animal breeding data has been facilitated by the 'Survival kit' software by Ducrocq and Solkner (1998). A new version of the 'Survival kit', written in Fortran 90 with an R interface to make it user-friendly, has recently been released (Mészáros *et al.*, 2013). The new version offers the opportunity to account for the correlated nature of two random effects, either by specifying a known correlation coefficient or estimating it from the data. In addition to the computational complexities of the proportional hazard model, the other disadvantage of the method is the difficulty of applying it in a multi-trait situation with more than two traits. This is important as most direct measures of survival traits are obtained late in life; therefore, various traits, mostly linear or composite-type traits such as fore-udder attachment, udder depth, mammary composite and legs and feet composite, have been used as indirect predictors of survival.

Subsequently, in this section, it is assumed that censoring is random, such that the end time or censoring is independent for all individuals.

Defining some distributions

The basic idea is that survival time follows a distribution (for example, Fig. 16.1) and the goal is to use data to estimate the parameters of this distribution. Let T be the random continuous variable denoting the failure time (death) of an animal, then the survival function $S(t)$, which is the probability that the animal survives at least until time t, is:

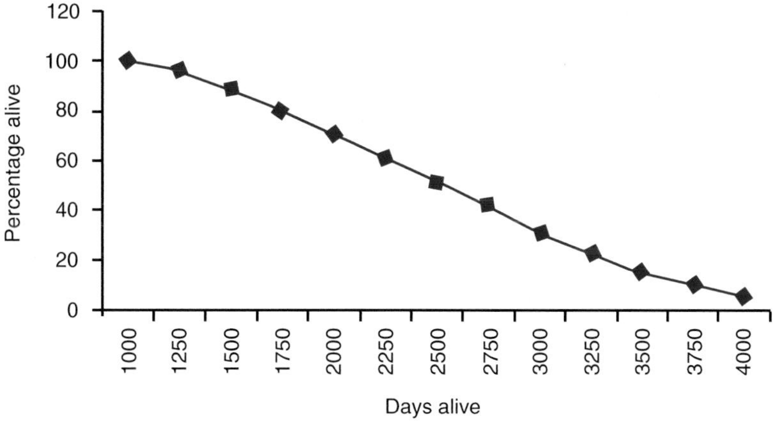

Fig. 16.1. Distribution of length of productive life for a group of Holstein dairy cows in the UK.

$$S(t) = \Pr(T \geq t) = 1 - \Pr(T < t) = 1 - F(t)$$

where $F(t)$ is the cumulative distribution of T and $S(t)$ can be regarded as the proportion of animals still alive at time t.

One of the approaches for modelling the survival function is through the hazard function $h(t)$, which measures the risk of failure of an individual at time t. It specifies the instantaneous rate of failure at time t, given that the individual has survived up to time t. The usefulness of the $h(t)$ stems from the fact it can provide the failure rate over time even when the exact nature of the survival curve is not known. It can be denoted as:

$$h(t) = \lim_{\Delta t \to 0} \frac{\Pr(t \leq T < t + \Delta t \mid T \geq t)}{\Delta t} = \frac{f(t)}{S(t)}$$

where $f(t)$ is the density function that equals $h(t)S(t)$. Another way of looking at the $h(t)$ is that for short periods of time (Δt), the probability that an animal fails is approximately equal to ($h(t)\Delta t$) (Kachman, 1999).

Exponential distribution

Several distributions can be used to define $h(t)$. If $h(t)$ is assumed to be constant over time, then this is an exponential distribution. This implies that the chance of an animal surviving, for instance, an additional two years, is the same, independent of how old the animal is. Assuming the exponential distribution, then $h(t) = \lambda$ and $S(t) = \exp(-\lambda t)$, where λ is the parameter of the exponential distribution.

Weibull distribution

The Weibull distribution, which is a two-parameter generalization of the exponential distribution, has also been used to model the hazard function to account for increasing or decreasing hazard function. With the Weibull distribution, $h(t)$ and $S(t)$ are:

$$h(t) = \rho\lambda(\lambda t)^{\rho-1} \quad \text{and} \quad S(t) = \exp(-(\lambda t)\rho)$$

with $\rho > 0$ and $\lambda > 0$. When $\rho = 1$, the Weibull distribution reduces to the exponential distribution. The Weibull distribution has a decreasing hazard function when $\rho < 1$ and an increasing hazard function when $\rho > 1$ (Fig. 16.2). Kachman (1999) showed that at a given λ, survival functions based on a Weibull model will all intersect at $t = 1/\lambda$, and that at $t = 1/\lambda$, the percentage survival is equal to $\exp(-1) \approx 37\%$. The role of the λ is to adjust the intercept.

Other possible distributions to model the hazard function include the gamma distribution, log-logistics and the log-normal distribution (Ducrocq, 1997). A summary of the commonly used distributions and parameters are given in Table 16.2.

16.4.4 Non-parametric estimation of the survival function

The survival function, $S(t)$, can be estimated from the parametric functions mentioned above. A non-parametric estimation of the survival function can be obtained using the Kaplan–Meier estimator (Kaplan and Meier, 1958). Let T_i represent failure times ordered from the

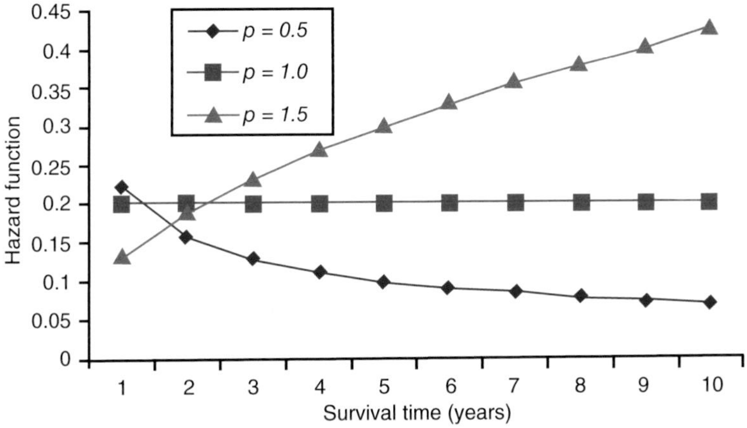

Fig. 16.2 The Weibull hazard function with a $\lambda = 0.20$ and with various ρ values.

Table 16.2. Some commonly used survival distributions and their parameters.

Distributions	$h(t)$	$S(t)$	$f(t)$
Exponential	λ	$\exp{-(\lambda t)}$	$\lambda\exp{-(\lambda t)\rho}$
Weibull	$\rho\lambda(\lambda t)^{\rho-1}$	$\exp{-(\lambda t)\rho}$	$\rho\lambda(\lambda t)^{\rho-1}\exp{-(\lambda t)\rho}$
Log-logistic	$\dfrac{\lambda\rho t^{\rho-1}}{1+\lambda t^{\rho}}$	$\dfrac{1}{1+\lambda t^{\rho}}$	$\dfrac{\lambda\rho t^{\rho-1}}{(1+\lambda t^{\rho})^2}$

first occurrence to the last. At T_i, let the number of animals that could have died (at risk) be denoted by n_i and the number that actually died as d_i. The Kaplan–Meier estimator then is:

$$\hat{S}(t) = \prod_{i|T_{i<t}} \left(\frac{n_i - d_i}{n_i} \right)$$

The usefulness of the Kaplan–Meier estimate of the survival function is that it could be used to check if the survival trait follows a particular parametric distribution. For instance, the appropriateness of a Weibull model can be evaluated by plotting $\log(-\log(\hat{S}(t)))$ versus $\log(t)$, where $\hat{S}(t)$ is the Kaplan–Meier estimate. This should result in a straight line with intercept $\rho\log(t)$ and slope ρ, given that:

$$S(t) = \exp(-(\lambda t)\rho) \rightarrow -\log(S(t)) = \lambda t^{\rho} \rightarrow \log(-\log(S(t))) = \log(\lambda) + \rho\log(t)$$

Similarly for the exponential distribution:

$$S(t) = \exp(-\lambda t) \rightarrow -\log(S(t)) = \lambda t$$

Therefore, the test for an exponential model will involve the plot of $-\log(\hat{S}(t))$ versus t, which should give straight line passing through the origin with slope λ.

16.4.5 Regression survival models

Initially, a fixed-effects survival model is considered to introduce the concept. Assume that **x** is a vector of risk fixed-effect factors or variables that influence failure time and **b** is the vector of corresponding solutions. One of the most popular procedures used to associate the hazard function $h(t)$ and $\mathbf{x}$ is the proportional hazard model (Cox, 1972; Ducrocq, 1997). The hazard function with vector of risk factors can be written as:

$$h(t;\mathbf{x}) = h_o(t)\exp(\mathbf{x}'\mathbf{b}) \tag{16.2}$$

where $h_o(t)$ is the baseline hazard function, representing the ageing process of the whole population. Thus, the hazard function has been factored into two parts. First, the baseline hazard function ($h_o(t)$), which is independent of the risk factors, and hence the ratio of the hazard functions of two animals, is equal to a constant at any time, i.e. their hazard functions are proportional (Ducrocq, 1997). Second, the remaining part of the equation, exp($\mathbf{x}'\mathbf{b}$), can be regarded as the scalar that does not depend on time and denotes the specific risk associated with animals with the factors $\mathbf{x}$ and acts multiplicatively on the baseline hazard function.

When $h_o(t) = \lambda$ = a constant, then the baseline hazard is exponential. When the baseline hazard function is left completely arbitrary, then the proportion model is termed a Cox model (Cox, 1972).

With the Weibull model, the baseline hazard function can be derived as:

$$\begin{aligned} h(t;\mathbf{x}) &= \rho\lambda(\lambda t)^{\rho-1}\exp(\mathbf{x}'\mathbf{b}) \\ &= \rho t^{\rho-1}\exp(\rho\log(\lambda)+\mathbf{x}'\mathbf{b}) \\ &= h_o(t)\exp(\mathbf{x}'\mathbf{b}) \end{aligned} \tag{16.3}$$

where $h_o(t) = \rho t^{\rho-1}$ models the baseline hazard function and exp($\mathbf{x}'\mathbf{b}$), the scalar, models the relative risk above or below the baseline risk. Note that the $\mathbf{x}'\mathbf{b}$ in Eqn 16.3 includes the intercept term such that $\mathbf{x} = (\mathbf{1}, \mathbf{x}')$ and $\mathbf{b} = (\rho\log(\lambda), \mathbf{b})$.

The corresponding survival function (Kachman, 1999) is:

$$S(t;\mathbf{x}) = \exp\{-t^{\rho}\exp(\mathbf{x}'\mathbf{b})\}$$

Stratified proportional hazard model

At times, the assumption of a single baseline hazard function for the whole population in proportional hazard models may be inappropriate. Therefore, data may be divided into subclasses on the basis of factors such as year or season of birth, treatment or region. Then for individuals in a subclass c, a baseline hazard function can be fitted as:

$$h(t;\ \mathbf{x},\ c) = h_{o,c}(t)\exp(\mathbf{x}'\mathbf{b})$$

Therefore, the hazards of two animals A1 and A2 in the same subclass with covariates $\mathbf{x}_{A1}$ and $\mathbf{x}_{A2}$, respectively, are proportional:

$$\frac{h(t;\ \mathbf{x}_{A1},\ c)}{h(t;\ \mathbf{x}_{A2},\ c)} = \exp[(\mathbf{x}'_{A1} - \mathbf{x}'_{A2})\mathbf{b}] = \text{constant}$$

and the baseline can have a known parametric form or be left arbitrary.

Accelerated failure time model

The accelerated failure time is another procedure to associate the hazard functions and the risk factors. Here it is assumed that the risk factors not only act multiplicatively but also accelerate or decelerate the failure time. Let $S_c(t)$ and $S_b(t)$ denote the survival functions of cows housed on concrete floors and floors with straw bedding, respectively. If it is assumed that the survival of cows housed on concrete floors is lower by a factor g than those housed on floors with bedding, then the accelerated failure time model assumes that $S_b(t) = S_c(\gamma t)$ and $\gamma > 0$ is the so-called accelerating factor.

Thus, Eqn 16.2 can be written as:

$$\begin{aligned} h(t;\ \mathbf{x}) &= h_o(\exp[\mathbf{x}'\mathbf{b}]t)\exp(\mathbf{x}'\mathbf{b}) \\ &= h_o(t^*)\exp(\mathbf{x}'\mathbf{b}) \end{aligned}$$

where this change in timescale from t to t^* denotes an acceleration or a deceleration depending on whether $\exp(\mathbf{x}'\mathbf{b})$ is smaller or greater than unity (Ducrocq, 1997).

Time-dependent risk factors

In the analysis of survival data that span a good length of time, it is possible that some of the risk factors may change with time. In livestock situations, the effect of such factors such as year–season of calving or herd management effects are likely to change over time. Such factors are termed time-dependent variables or risk factors. The proportional hazard model can be extended to incorporate time-dependent variables and Eqn 16.2 can then be written as:

$$h(t;\ \mathbf{x}(t)) = h_o(t)\exp(\mathbf{x}(t)'\mathbf{b})$$

where as usual $\mathbf{x}(t)$ represents a vector of risk factors, but some of them will be time-dependent variables. Ducrocq (2000) showed that it is possible to define time-dependent variables such as HYS as a sequence of indicator variables $\mathbf{x}(t) = (0 \ldots 1 \ldots 0)$ with $x_i(t) = 1$ if the observation is affected by the ith HYS at time t, or $x_i(t) = 0$ otherwise.

16.4.6 Mixed survival models

Mixed survival models, usually called frailty models, refer to the extension of the proportional hazard function to include random effects such as genetic effects. The random or frailty term $\mathbf{u}_m$ is defined as an unobserved random quantity that acts multiplicatively on the hazard of individuals or a group of animals (Ducrocq, 1997). The random vector $\mathbf{u}_m$ can be defined for individual animals or daughters of a sire m as in a sire model. With the simple transformation $\mathbf{a}_m = \log(\mathbf{u}_m)$, the frailty term can be included in the exponential part of the proportional hazards. Thus the mixed survival model can be written as:

$$\begin{aligned} h(t;\ \mathbf{x},\mathbf{z}) &= h_o(t)\exp(\mathbf{x}'\mathbf{b} + \mathbf{z}'\mathbf{a}) \text{ and} \\ S(t;\ \mathbf{x},\mathbf{z}) &= \exp\{-t^{\rho}\exp(\mathbf{x}'\mathbf{b} + \mathbf{z}'\mathbf{a})\} \end{aligned} \tag{16.4}$$

where $\mathbf{z}$ is an incidence matrix for random effects and the baseline hazard function can assume a parametric or arbitrary form. Ducrocq *et al.* (1988a, 1988b) and Ducrocq (1997) discussed the various distributions (gamma or log-gammas or inverse Gaussian) that have been assumed for the frailty term and various estimation procedures for the parameters of the frailty model. In the following section, the parametric model presented by Kachman (1999) is used to illustrate the prediction of $\mathbf{a}$ in the frailty model.

The parameters of interest in a survival model with or without the frailty term can be estimated using non-parametric, semi-parametric or parametric approaches. In this section, a brief outline of the parametric approach is presented. The basic parametric approach involves obtaining the joint likelihood of the survival time and the random effects, getting the marginal likelihood of survival time by integrating over the random effects or taking a second-order Taylor's series expansion of the joint log-likelihood. The joint log-likelihood for the Weibull function can be written (Kachman, 1999) as:

$$\begin{aligned} L(\mathbf{b},\mathbf{u},\rho) &= \sum_i \{\log h_o(t_i) + (\mathbf{x}_i\mathbf{b} + z_i\mathbf{a}) - H_o(t_i)\exp(\mathbf{x}_i\mathbf{b} + z_i\mathbf{a})\} \\ &\quad -1/2\log|\mathbf{G}| - 1/2\mathbf{a}'\mathbf{G}\mathbf{a} \end{aligned} \qquad (16.5)$$

The posterior mode estimates of the fixed and random effects can then be obtained by taking the first and second partial derivatives of Eqn 16.5. The resulting equations for the estimation are:

$$\begin{bmatrix} \mathbf{X'RX} & \mathbf{X'RZ} \\ \mathbf{Z'RX} & \mathbf{Z'RZ} + \mathbf{G}^{-1} \end{bmatrix} \begin{bmatrix} \hat{\mathbf{b}} \\ \hat{\mathbf{a}} \end{bmatrix} = \begin{bmatrix} \mathbf{X'y}^* \\ \mathbf{Z'y}^* \end{bmatrix} \qquad (16.6)$$

where $\mathbf{R}$ is a diagonal matrix with elements $r_{ii} = w_i \times \exp(\mathbf{x}_i\mathbf{b} + \mathbf{z}_i\mathbf{a})$, with $w_i = \exp(\rho^*\log(t_i))$ and t_i is the survival record for animal i, and $y_i^* = q_i - r_{ii}\{1 - (\mathrm{x}_i\mathbf{b} + \mathbf{z}_i\mathbf{a})\}$ with $q_i = 1$ for uncensored records or 0 if records are censored.

The use of Eqn 16.6 involves an iterative procedure with $d_i = (\mathbf{x}_i\mathbf{b} + \mathbf{z}_i\mathbf{a})$ being initially computed for record or individual i, then r_{ii} and y_i^* are calculated assuming that the estimate ρ is known for the data. Then Eqn 16.6 can be set up. Once all records have been processed, estimates of $\hat{\mathbf{b}}$ and $\hat{\mathbf{a}}$ are obtained by solving Eqn 16.6. The new estimates of $\hat{\mathbf{b}}$ and $\hat{\mathbf{a}}$ are then fed into the iterative procedure again until convergence is achieved.

Example 16.1

Presented in Table 16.1 is the length of productive life in months for a group of cows in two herds. The aim is to undertake a survival analysis using Eqn 16.6, fitting herd and year–season–parity as fixed risk factors and random animal effects. It is assumed that ρ is 1 and the genetic variance is 20. The full pedigree is incorporated into the analysis.

Considering the fixed effects, the design matrix X is:

$$\mathbf{X}' = \begin{pmatrix} 1 & 1 & 1 & 1 & 1 & 1 & 0 & 0 & 0 & 0 & 0 & 0 \\ 0 & 0 & 0 & 0 & 0 & 0 & 1 & 1 & 1 & 1 & 1 & 1 \\ 0 & 0 & 1 & 0 & 0 & 1 & 0 & 1 & 0 & 0 & 0 & 0 \\ 0 & 0 & 0 & 1 & 0 & 0 & 0 & 0 & 1 & 0 & 0 & 1 \\ 1 & 0 & 0 & 0 & 1 & 0 & 0 & 0 & 0 & 1 & 0 & 0 \\ 0 & 1 & 0 & 0 & 0 & 0 & 1 & 0 & 0 & 0 & 1 & 0 \end{pmatrix}$$

Thus given $\mathbf{M} = [\mathbf{X}, \mathbf{Z}]$, where $\mathbf{Z}$ is a diagonal matrix considering the cows with records and:

$$\mathbf{u} = \begin{pmatrix} \hat{\mathbf{b}} \\ \hat{\mathbf{a}} \end{pmatrix}$$

then vector $\mathbf{d} = \mathbf{Mu}$. For instance, if starting values for $\mathbf{u}$ in the first iteration were set to 0.1, then for the first animal, $d_1 = \mathbf{m}_1\mathbf{u} = 0.30$ with $\mathbf{m}_1 = (1\ 0\ 0\ 0\ 1\ 0\ 1\ 0\ 0\ 0\ 0\ 0\ 0\ 0\ 0\ 0\ 0\ 0)$.

Then for animal i, compute r_{ii} and y_i^* and set up the row of equations for ith animal. For animal 1, $r_{11} = \exp(1 * \log(40)) * \exp(0.3) = 53.994$ and $y_1^* = 0 - 53.994(1 - 0.3) = -37.760$.

Equation 16.6 is built up and solved after all animals are processed and the iteration continued until convergence. Due to dependencies in the system of equations, the first levels of herd and year–season–parity effects have been constrained to zero. The solutions obtained at convergence and the risk ratios (RRS) are:

Herd	Solution	RRS			
1	0.000	1.000			
2	−1.631	0.196			
Year–season–parity					
	Solution	RRS			
1	0.000	1.000			
2	−2.346	0.094			
3	−3.149	0.043			
4	−2.982	0.051			
Animal	Solution	RRS	Animal	Solution	RRS
1	−0.779	0.459	11	−0.706	0.494
2	−1.233	0.291	12	−0.476	0.621
3	−0.062	0.940	13	−1.902	0.149
4	−0.750	0.472	14	−0.178	0.837
5	−0.758	0.469	15	−0.842	0.431
6	0.238	1.269	16	−0.519	0.595
7	−0.328	0.720	17	−0.115	0.891
8	−1.477	0.228	18	−0.578	0.561
9	−0.533	0.587	19	−0.290	0.748
10	−1.753	0.173			

The estimates b_i can be expressed in relative risk (hazard) ratio by the transformation $\text{RRS}(b_i) = \exp(b_i)$. This expression gives the RRS of culling due to that effect and it follows from the assumption of the proportional hazard model:

$$h_o(t;\ \mathbf{x}_a) / h_o(t;\ \mathbf{x}_c) = \exp((\mathbf{x}'_a - \mathbf{x}'_c)b)$$

implying that the relative hazard for two animals with covariates described by $\mathbf{x}_a$ and $\mathbf{x}_c$, respectively, is independent of time and of other covariates. Thus, the RRS denotes the relative risk of a cow being culled in a certain fixed-effect class compared to a cow in a reference class with risk set to unity. The estimates of RRS in Table 16.1 indicate, for instance, that cows in herd 2 are 20% more likely to be culled compared to herd 1. Also, cows in YSP subclasses 2 and 3 are 10% and 4%, respectively, more likely to be culled compared to cows in YSP subclass 1. For the random animal effect, the RRS estimates indicate the relative risk of daughters of these animals being culled. Usually these estimates are transformed to relative breeding values; say, with mean 100 and standard deviation of 12, so they are comparable with breeding values of other traits.

The results can also be presented in several other forms. The interest may be to predict the percentage of live daughters for sires at 40 months of productive life; for instance, (i) in herd 1 and in the 4th YSP; or (ii) across all herds and YSPs. If (i), then for sire 1:

$$d_1 = \hat{b}_1 + \hat{b}_6 + \hat{a}_1 = 0.00 + -2.982 + -0.779 = -3.761$$

Then using Eqn 16.4, $S(40, \mathbf{x},\mathbf{z}) = \exp\{-40^{\rho}\exp(d_1)\} = 0.394$. For (ii), a weighted mean for fixed-effect solutions might be computed based on the number of daughters the sire has in each fixed-effect subclass. Thus for sire 1:

$$\begin{aligned} d_1 &= (2\hat{b}_1 + 2\hat{b}_2)/4 + (1\hat{b}_3 + 0\hat{b}_4 + 1\hat{b}_5 + 2\hat{b}_6)/4 + \hat{a}_1 \\ &= (2*0.0 + 2*0.196)/4 + (0.0 + 0 + -3.149 + 2*-2.982)/4 + -0.779 = -2.959 \end{aligned}$$

and $S(40,\ \mathbf{x}, z) = \exp\{-40^{\rho}\exp(d_1)\} = 0.126$

Equation 16.6 and its application in Example 16.1 was mainly to illustrate the basic principles of survival analysis using proportional hazard models with a frailty term. The parameter r has been assumed known and in practice this has to be estimated simultaneously, and usually more terms including time-dependent variables are included in the models. The 'Survival kit' (Ducrocq and Solkner, 1998; Mészáros *et al.*, 2013) is currently used for the genetic evaluation of survival traits at the national level by a number of countries. A summary of methods utilized for the evaluation of survival at the national level for the Holstein breed on the Interbull website (http://www-interbull.slu.se/national_ges_info2/framesida-ges.htm) indicates that eight countries (France, Germany, Italy, The Netherlands, Hungary, Slovenia, Spain and Switzerland) use proportional hazard models in their genetic evaluation systems. Similarly, seven countries (Canada, Denmark, Finland, Japan, New Zealand, Sweden and the UK) currently use a multi-trait animal model, while the USA, Israel and Australia employ a single-trait animal model. The only country that uses a random regression animal model is Belgium (Walloon region).

16.4.7 Group data survival model

When survival is defined as a discrete trait such as number of lactations completed or number of years completed, the Cox and Weibull models may not be suitable for the analysis of such traits. This is because these models assume continuity of the baseline hazard distribution and/or absence of ties between ordered failure times. Thus, with discrete survival traits, the grouped data version of the proportional hazards model introduced by Prentice and Gloeckler (1978) can be used. The group data proportional hazard model involves grouping failure time into intervals $Q_i = (q_i{-}1, q_i)$, $i = 1, \ldots, r$ with $q_0 = 0$, q_r = +infinite and failure times in Q_i are recorded as t_i. Thus, the regression vector is assumed to be time-dependent but fixed within each time interval. Grouped data models have been used in beef cattle (Phocas and Ducrocq, 2006) and rabbits (Piles *et al.*, 2006). Mészáros *et al.* (2010) demonstrated this grouped data model was more appropriate in length of productive life in pigs.

17 Estimation of Genetic Parameters

ROBIN THOMPSON

Rothamsted Research, Harpenden, UK

17.1 Introduction

In order to carry out prediction of breeding values, estimates of variance components are usually needed. In this chapter, the estimation of variance parameters is considered using univariate sire and animal models.

17.2 Univariate Sire Model

To motivate this work, the mixed-effect sire model introduced in Chapter 3 is used. This model (Eqn 4.15) has:

$$\mathbf{y} = \mathbf{Xb} + \mathbf{Zs} + \mathbf{e}$$

and

$$\begin{aligned} \text{var}(\mathbf{s}) &= \mathbf{A}\sigma_s^2 \\ \text{var}(\mathbf{y}) &= \mathbf{ZAZ}'\sigma_s^2 + \mathbf{R} \end{aligned}$$

where **A** is the numerator relationship matrix for sires, $\sigma_s^2 = 0.25\ \sigma_a^2$ and $\mathbf{R} = \mathbf{I}\sigma_e^2$. The aim is to estimate σ_s^2 and σ_e^2. The simplest case with this sire model is when **X** is a $n \times 1$ matrix with elements 1, **b** having one element representing an overall effect and the q sires being unrelated, so that **A** = **I**.

An analysis of variance can be constructed by fitting: (i) a model with the overall effect **b**; and (ii) a model with sire effects, these models giving residual sums of squares that can be put into an analysis of variance of the form:

Source	Degrees of freedom	Sums of squares
Overall	Rank (**X**) = 1	$\mathbf{y}'\mathbf{X}(\mathbf{X}'\mathbf{X})^{-1}\mathbf{X}'\mathbf{y} = F$
Sires	Rank (**Z**) – rank (**X**) = $q - 1$	$\mathbf{y}'\mathbf{Z}(\mathbf{Z}'\mathbf{Z})^{-1}\mathbf{Z}'\mathbf{y} - \mathbf{y}'\mathbf{X}(\mathbf{X}'\mathbf{X})^{-1}\mathbf{X}'\mathbf{y} = S$
Residual	n – rank (**Z**) = $n - q$	$\mathbf{y}'\mathbf{y} - \mathbf{y}'\mathbf{Z}(\mathbf{Z}'\mathbf{Z})^{-1}\mathbf{Z}'\mathbf{y} = R$

Essentially, the effects **b** and **s** are thought of as fixed effects to construct an unweighted analysis. If estimates of σ_s^2 and $\sigma_e^2 s$ are required, then the sums of squares S and R can be equated to their expectation $E(R) = (n - q)\sigma_e^2$ and $E(s) = (q-1)\sigma_e^2 + \text{trace}(\mathbf{Z}'\mathbf{SZ})\sigma_s^2$ where $\mathbf{S} = \mathbf{I} - \mathbf{X}(\mathbf{X}'\mathbf{X})^{-1}\mathbf{X}'$.

DOI: 10.1079/9781800620506.0017

17.3 Numerical Example of Sire Model

Consider the data in Table 17.1 for the pre-weaning gain (WWG) of beef calves. The objective is to illustrate the estimation of variance components on a very small example so that the calculations can be expressed concisely.

The model to describe the observations is:

$$y_{ij} = o + s_j + e_i$$

where y_{ij} = the WWG of the ith calf, o = the overall effect, s_j = random effect of the jth sire (j = 1, 2, 3) and e_i = random error effect (i = 1, 2, 3, 4).

In matrix notation, the model is the same as described in Eqn 4.1, with n = 4, p = 1 and q = 3.

The matrix **X** in the MME relates records to the overall effects. For the example data set, its transpose is:

$$X' = [1111]$$

The matrix **Z**, then, relates records to sires. In this case it is:

$$Z = \begin{bmatrix} 0 & 1 & 0 \\ 1 & 0 & 0 \\ 0 & 0 & 1 \\ 0 & 1 & 0 \end{bmatrix}$$

An analysis of variance can be constructed as:

Source	Degrees of freedom	Sums of squares (kg^2)
Overall	1	F = 48.3025
Sire	2	S = 0.4275
Residual	1	R = 0.1800

with:

$$y'X(X'X) - 1X'y = F = (2.9 + 4 + 3.5 + 3.5)2/4 = 48.3025$$

and

$$y'Z(Z'Z) - 1Z'y - y'X(X'X) - 1X'y = S = (4)2/1 + (2.9 + 3.5)2/2 + (3.5)2/1 - 48.3025$$
$$= 0.4275 \text{ and } y'y - y'Z(Z'Z) - 1Z'y = R = (2.9)2 + (4)2 + (3.5)2 + (3.5)2 - F - S = 0.18$$

Table 17.1. Pre-weaning gain (kg) for four beef calves.

Calf	Sire	WWG (kg)
4	2	2.9
5	1	4.0
6	3	3.5
7	2	3.5

In this case:

$$\mathbf{Z}=\begin{bmatrix}0&1&0\\1&0&0\\0&0&1\\0&1&0\end{bmatrix},\quad \mathbf{S}=\begin{bmatrix}0.75&-0.25&-0.25&-0.25\\-0.25&0.75&-0.25&-0.25\\-0.25&-0.25&0.75&-0.25\\-0.25&-0.25&-0.25&0.75\end{bmatrix}\quad \text{and}$$

$$\mathbf{Z'SZ}=\begin{bmatrix}0.75&-0.50&-0.25\\-0.50&1.00&-0.50\\-0.25&-0.50&0.75\end{bmatrix}$$

so that:

$$\mathrm{E(R)}=\sigma_e^2=0.18 \quad \text{and} \quad \mathrm{E(S)}=2\ \sigma_e^2+2.5\ \sigma_s^2=0.4275$$

Then, estimates of σ_e^2 and σ_s^2 are:

$$\sigma_e^2=0.18(\mathrm{kg}^2) \quad \text{and} \quad \sigma_s^2=0.027(\mathrm{kg}^2)$$

17.4 Extended Model

The model and analysis hold if the model is extended to allow **X** to represent an environmental effect with p levels. If sires are nested within levels of the environmental factor so that daughters of each sire are only associated with one level of the environmental factor, then the above analysis could be used. If, however, as usually happens, daughters of a sire are associated with more than one level, a slightly more complicated analysis is appropriate.

Source	Degrees of freedom	Sums of squares
Fixed effects	Rank (**X**) = p	$\mathbf{y'X(X'X)^{-1}X'y}$ = F
Sires corrected for fixed effects	Rank (**Z′SZ**) = df_S	$\mathbf{y'SZ(Z'SZ)^{-1}Z'Sy}$ = S
Residual	n – rank (**X**) – rank (**Z′SZ**) $= n - p - df_S = df_R$	**y′y** – F – S = R

Now, R and S have expectation:

$$\mathrm{E(R)}=df_R\sigma_e^2 \quad \text{and} \quad \mathrm{E(S)}=df_S\sigma_e^2+\mathrm{trace}(\mathbf{Z'SZ})\sigma_s^2$$

The term involved in the trace (**Z′SZ**) can sometimes have a simple interpretation. If **X** represents a fixed-effect matrix with p levels, then the ith diagonal element of **Z′SZ** is $n_{i.}-\Sigma n_{ij}^2/n_{.j}$ (summation is from j = 1 to p) where n_{ij} is the number of daughters of sire i in fixed-effect level j and $n_{.j}=\Sigma n_{ij}$ (summation is from j = 1 to p) and $n_{i.}=\Sigma n_{ij}$ (summation is from i = 1 to s). This number was called the effective number of daughters of sire i by Robertson and Rendel (1954) and measures the loss of information on a sire because his daughters are measured in different environmental classes. This method of analysis is called Henderson's method 3 (Henderson, 1953). These methods of analysis were very popular in that they related to sequential fitting of models and were relatively easy to compute. One problem is that the terms are generated under a fixed-effect model with $\mathbf{V}=\mathbf{I}\sigma_e^2$ and then sums of squares are equated to their expectation under a different variance model. Only in special balanced cases will estimation based on R and S lead to efficient estimates of σ_s^2 and σ_e^2. In general, B is based on **Z′Sy** with variance matrix

$\mathbf{Z'SZ}\sigma_e^2 + \mathbf{Z'SZAZ'SZ}\sigma_s^2$ and these can be transformed to df_S independent values $\mathbf{Q'Z'Sy}$ by using arguments similar to those used in Section 6.2 on the canonical transformation, where $\mathbf{Q}$ is a $df_S\, n$ matrix and $\mathbf{Q'Z'SZQ} = \mathbf{I}$ and $\mathbf{Q'Z'SZAZ'SZQ} = \mathbf{W}$, where $\mathbf{W}$ is a diagonal matrix of size df_S with ith diagonal element w_i. The variance matrix of $\mathbf{Q'Z'Sy}$ is then $\mathbf{I}\sigma_e^2 + \mathbf{W}\sigma_s^2$. Then, an analysis of variance can be constructed from squaring each of the df_S elements of $\mathbf{Q'Z'Sy}$ with ith sum of squares u_i with expectation $\sigma_e^2 + w_i\sigma_s^2$ and R is the residual sum of squares with expectation $\mathrm{E}(\mathrm{R}) = df_R\sigma_e^2$. The individual u_i are distributed as chi-squared variables with variance $\mathrm{E}(u_i)^2$. A natural scheme is to fit a linear model in σ_s^2 and σ_e^2 to u_i and R. One can also use an iterative scheme with the weight dependent on the estimated parameters.

17.5 Numerical Example

For the example with data in Table 17.1, it was shown that:

$$\mathbf{Z'SZ} = \begin{bmatrix} 0.75 & -0.50 & -0.25 \\ -0.50 & 1.00 & -0.50 \\ -0.25 & -0.50 & 0.75 \end{bmatrix}$$

so that the sires have 0.75, 1.0 and 0.75 effective daughters, respectively.

It can be found that with $\mathbf{A} = \mathbf{I}$:

$$\mathbf{Z'SZAZ'SZ} = \begin{bmatrix} 0.875 & -0.750 & -0.125 \\ -0.750 & 1.500 & -0.750 \\ -0.125 & -0.750 & 0.875 \end{bmatrix}$$

The algorithm in Appendix E, Section E.1, can be used to calculate the eigenvalues $\mathbf{Q}$ so that:

$$\mathbf{Q'Z'SZQ} = \mathbf{I} \quad \text{and} \quad \mathbf{Q'Z'SZAZ'SZQ} = \mathbf{W}$$

In this case:

$$\mathbf{Q} = \begin{bmatrix} -0.3333 & 0.6667 & -0.3333 \\ 0.7071 & 0.0000 & -0.7071 \end{bmatrix}$$

So $\mathbf{Q'Z'SZQ} = \mathbf{I}$ and $\mathbf{Q'Z'SZAZ'SZQ} = \mathbf{W}$ with:

$$\mathbf{W} = \begin{bmatrix} 1.5 & 0.0 \\ 0.0 & 1.0 \end{bmatrix}$$

The contrasts $\mathbf{Q'Z'Sy}$ are now:

$$\mathbf{Q'Z'S} = \begin{bmatrix} 0.5000 & -0.5000 & -0.5000 & 0.5000 \\ 0.0000 & 0.7071 & -0.7071 & 0.0000 \end{bmatrix}\mathbf{y}$$

So, the first contrast $(y_1 - y_2 - y_3 + y_4)/2 = (2.9 - 4.0 - 3.5 + 3.5)/2 = -1.1/2 = -0.55$ is a scaled contrast comparing sire 2 with sire 1 and sire 3, and the second contrast $(y_2 - y_3)/\sqrt{2} = (-4.0 = 3.5)/\sqrt{2} = (-0.5)/\sqrt{2}$ is a scaled contrast between sire 1 and sire 3.

An analysis of variance can be constructed:

Source	Degrees of freedom	Sums of squares (kg²)	Expected mean squares (kg²)
Overall	1	F = 48.3025	
Sire 2 compared with sires 1 and 3	1	$(-0.55)^2 = 0.3025$	$\sigma_e^2 + 1.5\sigma_s^2$
Sire 1 compared with sire 3	1	$(-0.5)^2/2 = 0.1250$	$\sigma_e^2 + \sigma_s^2$
Residual	1	R = 0.1800	σ_e^2

Fitting a linear model in σ_e^2 and σ_s^2 to the three sums of squares 0.3025, 0.1250 and 0.1800, gives estimates of $\sigma_e^2 = 0.143(\text{kg}^2)$ and $\sigma_s^2 = 0.079(\text{kg}^2)$. If a generalized linear model is fitted iteratively to the sum of squares with weights proportional to the variance of the sum of squares when the procedure converges, the estimate of σ_e^2 is 0.163 (kg^2) and of σ_s^2 is 0.047 (kg^2). The estimated variances of these estimates (from the inverse of the generalized least squares coefficient matrix) are 0.216 (kg^2) and 0.234 (kg^2).

17.6 Animal Model

It has been shown that estimates can be obtained from analysis of variance for some models. Now consider a more general model – the animal model introduced in Chapter 4. This linear model (Eqn 4.1) is:

$$\mathbf{y} = \mathbf{Xb} + \mathbf{Za} + \mathbf{e}$$

and the variance structure is defined, with:

$$\text{var}(\mathbf{e}) = \mathbf{I}\sigma_e^2 = \mathbf{R};\ \text{var}(\mathbf{a}) = \mathbf{A}\sigma_a^2 = \mathbf{G} \quad \text{and} \quad \text{cov}(\mathbf{a},\ \mathbf{e}) = \text{cov}(\mathbf{e},\ \mathbf{a}) = 0$$

where **A** is the numerator relationship matrix, and there is interest in estimating σ_a^2 and σ_e^2. A popular method of estimation is by restricted (or residual) maximum likelihood (REML) (Patterson and Thompson, 1971). This is based on a log-likelihood of the form:

$$\text{L}\alpha\left(\frac{1}{2}\right)\left\{-(\mathbf{y}-\mathbf{Xb})'\mathbf{V}^{-1}(\mathbf{y}-\mathbf{Xb}) - \text{logdet}(\mathbf{V}) - \text{logdet}(\mathbf{X}'\mathbf{V}^{-1}\mathbf{X})\right\}$$

where **b** is the generalized least squares (GLS) solution and satisfies:

$$\mathbf{X}'\mathbf{V}^{-1}\mathbf{Xb} = \mathbf{X}'\mathbf{V}^{-1}\mathbf{y}$$

There are three terms in L: the first is a weighted sum of squares of residuals; the second is a term that depends on the variance matrix; and a third depends on the variance matrix of the fixed effects and can be thought of as a penalty because fixed effects are estimated. MME (Chapter 3) play an important part in the analysis process.

For the particular model these can be written as (Eqn 4.4):

$$\begin{bmatrix} \mathbf{X}'\mathbf{X} & \mathbf{X}'\mathbf{Z} \\ \mathbf{Z}'\mathbf{X} & \mathbf{Z}'\mathbf{Z} + \mathbf{A}^{-1}\alpha \end{bmatrix} \begin{bmatrix} \hat{\mathbf{b}} \\ \hat{\mathbf{a}} \end{bmatrix} = \begin{bmatrix} \mathbf{X}'\mathbf{y} \\ \mathbf{Z}'\mathbf{y} \end{bmatrix}$$

with $\alpha = \sigma_e^2/\sigma_a^2$ or $(1 - h^2)/h^2$.

Extensive use is made of the prediction error matrix of **a**. In this case, the prediction error matrix is $\text{PEV} = \text{var}(\mathbf{a} - \hat{\mathbf{a}}) = \mathbf{C}^{22}\sigma_e^2$ (Eqn 4.14), where $\mathbf{C}^{22}$ is associated with the coefficient matrix of the MME.

Estimates of σ_a^2 and σ_e^2 are chosen to maximize L. It is useful to express relevant terms in this estimation process in terms of the projection matrix **P**:

$$\mathbf{P} = \mathbf{V}^{-1} - \mathbf{X}\left(\mathbf{X}'\mathbf{V}^{-1}\mathbf{X}\right)^{-1}\mathbf{X}'\mathbf{V}^{-1}$$

Then:

$$\text{L}\,\alpha\left(\frac{1}{2}\right)\{-\mathbf{y}'\mathbf{P}\mathbf{y} - \text{logdet}(\mathbf{V}) - \text{logdet}(\mathbf{X}'\mathbf{V}^{-1}\mathbf{X})\} \tag{17.1}$$

Estimation of a variance parameter $\theta_i\,(\theta_1 = \sigma_e^2, \theta_2 = \sigma_a^2)$ involves setting to zero the first derivatives:

$$\partial\text{L}/\partial\theta_i = \left(\frac{1}{2}\right)\{\mathbf{y}'\mathbf{P}(\partial\mathbf{V}/\partial\theta_i)\mathbf{P}\mathbf{y} - \text{trace}[\mathbf{P}(\partial\mathbf{V}/\partial\theta_i)]\}$$

These equations could be thought of as equating a function of data (the first term in the expression) to its expectation.

Normally, finding a maximum requires an iterative scheme. One suggested by Patterson and Thompson (1971) was based on using the expected value of the second differential matrix. In this case, these are:

$$\text{E}(\partial\text{L}^2/\partial\theta_i\,\partial\theta_j) = -\left(\frac{1}{2}\right)\text{trace}[\mathbf{P}(\partial\mathbf{V}/\partial\theta_i)\,\mathbf{P}(\partial\mathbf{V}/\partial\theta_j)]$$

Using the first and expected second differentials one can update θ using terms that depend on the solution of the MME and PEVs. For the particular animal model that is being considered, then:

$$\partial L/\partial\sigma_e^2 = \left(\frac{1}{2}\right)\{(\mathbf{y} - \mathbf{Xb} - \mathbf{Za})'(\mathbf{y} - \mathbf{Xb} - \mathbf{Za})/\sigma_e^4 - (n - p - q)/\sigma_e^4$$
$$- \text{trace}\left[\mathbf{C}^{22}\mathbf{A}^{-1}\right]/\sigma_a^4\} \tag{17.2}$$

$$\partial\text{L}/\partial\sigma_a^2 = \left(\frac{1}{2}\right)\{\mathbf{a}'\mathbf{A}^{-1}\mathbf{a}/\sigma_a^4 + q/\sigma_a^2 - \text{trace}[\mathbf{C}^{22}\mathbf{A}^{-1}]\sigma_e^2/\sigma_a^4\} \tag{17.3}$$

and

$$\text{E}\left(\partial\text{L}^2/\partial\sigma_e^4\right) = -\left(\frac{1}{2}\right)\left\{(n - p - q)/\sigma_e^4 + \text{trace}\left[\left(\mathbf{C}^{22}\mathbf{A}^{-1}\right)^2\right]/\sigma_a^4\right\}$$

$$\text{E}\left(\partial\text{L}^2/\partial\sigma_a^4\right) = -\left(\frac{1}{2}\right)\left\{+\text{trace}\left[\left\{\mathbf{I} - \mathbf{C}^{22}\mathbf{A}^{-1}\left(\sigma_e^4/\sigma_a^4\right)\right\}^2\right]/\sigma_a^4\right\}$$

$$\text{E}\left(\partial\text{L}^2/\partial\sigma_a^4\partial\sigma_e^2\right) = -\left(\frac{1}{2}\right)\left\{\text{trace}\left[\left\{\mathbf{I} - \mathbf{C}^{22}\mathbf{A}^{-1}\left(\sigma_e^2/\sigma_a^2\right)\right\}\left\{\mathbf{C}^{22}\mathbf{A}^{-1}\right\}\right]\sigma_a^4\right\}$$

Thinking of the variance parameters and the first differentials as vectors $\boldsymbol{\theta}$ and $\partial\mathbf{L}/\partial\boldsymbol{\theta}$ with ith (i = 1, 2) element $\boldsymbol{\theta}_i$ and $\partial\text{L}/\partial\boldsymbol{\theta}_i$, respectively, and **Einf**, the expected information matrix, a matrix with i,jth element $-\text{E}(\partial\text{L}^2/\partial\theta_i\partial\boldsymbol{\theta}_j)$, suggests an iterative scheme with the new estimate $\boldsymbol{\theta}_n$ satisfying:

$$\boldsymbol{\theta}_n = \boldsymbol{\theta} + \mathbf{Einf}^{-1}(\partial L / \partial \boldsymbol{\theta}) \quad (17.4)$$

There are two problems with this approach. First, the parameters might go negative and one would want estimates of variances to stay essentially positive. One popular way of avoiding this property is to note that at a maximum of the likelihood the first differentials are zero and to manipulate Eqns 17.1 and 17.2 in the form:

$$(n-p)\sigma_e^2 = (\mathbf{y} - \mathbf{Xb} - \mathbf{Za})'(\mathbf{y}) \quad (17.5)$$

$$q\sigma_a^2 = \mathbf{a}'\mathbf{A}^{-1}\mathbf{a} + \text{trace}\left[\mathbf{C}^{22}\mathbf{A}^{-1}\right]\sigma_e^2 \quad (17.6)$$

so that it can be seen that σ_e^2 is estimated from a sum of squares of residuals and σ_a^2 is estimated from a weighted sum of squares of predicted values and their PEV. This algorithm is an expectation maximization (EM) algorithm (Dempster *et al.*, 1977) and successive iterates are positive. The algorithm can be written in the form of the updating formula if **Einf** is replaced by a matrix that depends on the information derived, as if one could directly observe the residuals and breeding values rather than predicting them. This algorithm can be slow to converge in animal breeding applications.

A second problem is that the expected second differentials are difficult to calculate. Sometimes it is recommended to use observed second differentials. These are of the form:

$$(\partial L^2 / \partial\theta_i\partial\theta_j) = -\mathbf{y'P}(\partial\mathbf{V}/\partial\theta_i)\mathbf{P}(\partial\mathbf{V}/\partial\theta_j)\mathbf{Py} + \left(\frac{1}{2}\right)\text{trace}\left[\mathbf{P}(\partial\mathbf{V}/\partial\theta_i)\mathbf{P}(\partial\mathbf{V}/\partial\theta_j)\right]$$

But, again, these terms involve the complicated trace terms. One suggestion (Gilmour *et al.*, 1995) is to use the average of the expected and observed information terms. These are of the form:

$$A(\partial L^2 / \partial\theta_i\partial\theta_j) = -\left(\frac{1}{2}\right)\left\{\mathbf{y'P}(\partial\mathbf{V}/\partial\theta_i)\mathbf{P}(\partial\mathbf{V}/\partial\theta_j)\mathbf{Py}\right\}$$

These terms are similar to **y′Py** in that they could be thought of as a weighted sum of squares matrix with **y** replaced by two columns $(\partial\mathbf{V}/\partial\theta_i)\mathbf{Py}\,(i = 1, 2)$. In this particular case:

$$(\partial\mathbf{V}/\partial\sigma_e^2)\mathbf{Py} = (\mathbf{y} - \mathbf{Xb} - \mathbf{Za})/\sigma_e^2$$

and

$$(\partial\mathbf{V}/\partial\sigma_a^2)\mathbf{Py} = \mathbf{Za}/\sigma_a^2$$

As in the formation of **Einf,** we can construct and base an iterative scheme on Eqn 17.3 and on **Ainf**, a matrix with elements $-A(\partial L^2/\partial\theta_i\partial\theta_j)$. Once the iterative scheme has converged, then the asymptotic variance matrix of $\boldsymbol{\theta}$ can be estimated from $\mathbf{Ainf}^{-1}$ or $\mathbf{Einf}^{-1}$. The animal model and estimation procedure introduced can easily be extended to deal with other models, just as prediction procedures can be developed for a variety of models. Software for estimating variance parameters using this average information algorithm is described by Jensen and Madsen (1997) and Gilmour *et al.* (2003).

17.7 Numerical Example

Consider the data in Table 17.2 for the pre-weaning gain (WWG) of beef calves. This is very similar to the data of Table 4.1 with the data changed to give positive variance estimates.

Table 17.2. Pre-weaning gain (kg) for five beef calves.

Calf	Sex	Sire	Dam	WWG (kg)
4	Male	1	–	2.6
5	Female	3	2	0.1
6	Female	1	2	1.0
7	Male	4	5	3.0
8	Male	3	6	1.0

The model to describe the observations is:

$$y_{ijk} = p_i + a_j + e_{ijk}$$

where y_{ij} = the WWG of the jth calf of the ith sex, p_i = the effect of the ith sex, a_j = random effect of the jth calf and e_{ijk} = random error effect.

In matrix notation, the model is the same as described in Eqn 4.1.

Again, the objective is to illustrate the estimation of variance components σ_e^2 and σ_a^2 on a very small example so that the calculations can be expressed concisely.

In matrix notation, the model is the same as described in Eqn 4.1 with $n = 5$, $p = 2$ and $q = 8$, with the design matrices as given in Section 4.3. Now $\mathbf{y}' = [2.6, 0.1, 1.0, 3.0, 1.0]$ and, using initial estimates of $\sigma_e^2 = 0.4$ and $\sigma_a^2 = 0.2$, solutions to MME (Eqn 4.15) are:

Sex effects	
Male	2.144
Female	0.602
Animals	
1	0.117
2	−0.025
3	−0.222
4	−0.254
5	−0.135
6	0.032
7	0.219
8	−0.305

Then:

$$(\mathbf{y} - \mathbf{Xb} - \mathbf{Za})' = [0.2022 \; -0.3661 \; 0.3661 \; 0.6374 \; -0.8395]$$

$$\mathbf{C}^{22}\sigma_e^2 = \begin{bmatrix} 0.1884 & 0.0028 & 0.0131 & 0.0878 & 0.0180 & 0.0883 & 0.0554 & 0.0537 \\ 0.0028 & 0.1968 & -0.0041 & 0.0082 & 0.0949 & 0.0981 & 0.0479 & 0.0443 \\ 0.0131 & -0.0041 & 0.1826 & 0.0193 & 0.0805 & 0.0090 & 0.0504 & 0.0871 \\ 0.0878 & 0.0082 & 0.0193 & 0.1711 & 0.0188 & 0.0510 & 0.0971 & 0.0493 \\ 0.0180 & 0.0949 & 0.0805 & 0.0188 & 0.1712 & 0.0679 & 0.0879 & 0.0712 \\ 0.0883 & 0.0981 & 0.0090 & 0.0510 & 0.0679 & 0.1769 & 0.0609 & 0.0877 \\ 0.0554 & 0.0479 & 0.0504 & 0.0971 & 0.0879 & 0.0609 & 0.1767 & 0.0672 \\ 0.0537 & 0.0443 & 0.0871 & 0.0493 & 0.0712 & 0.0877 & 0.0672 & 0.1689 \end{bmatrix}$$

$\mathbf{y'Py}$ = 4.8193, logdet(V) = –2.6729 and logdet($\mathbf{X'V^{-1}X}$) = 2.6241 so L = –2.3852 from Eqn 17.1.

Then Eqns 17.2 and 17.3 give:

$$\partial L / \partial \sigma_e^2 = (0.5)\left\{(\mathbf{y}-\mathbf{Xb}-\mathbf{Za})'(\mathbf{y}-\mathbf{Xb}-\mathbf{Za})/\sigma_e^4 - (n-p-q)/\sigma_e^2 - \text{trace}\left[\mathbf{C}^{22}\mathbf{A}^{-1}\right]/\sigma_a^2\right\}$$

$$\partial L / \partial \sigma_e^2 = (0.5)\{8.8753-(-12.5000)-18.0733\} = 1.6510$$

$$\partial L / \partial \sigma_a^2 = (0.5)\left\{\mathbf{a'A^{-1}a}/\sigma_a^4 - q/\sigma_a^2 + \text{trace}\left[\mathbf{C}^{22}\mathbf{A}^{-1}\right]\sigma_e^2/\sigma_a^4\right\}$$

$$\partial L / \partial \sigma_a^2 = (0.5)\{6.3461-40.0000+36.1466\} = 1.2464$$

and

$$A\left(\partial L^2 / \partial \sigma_e^4\right) = -(0.5)\left\{(\mathbf{y}-\mathbf{Xb}-\mathbf{Za})'\mathbf{P}(\mathbf{y}-\mathbf{Xb}-\mathbf{Za})/\sigma_e^4\right\}$$

$$A\left(\partial L^2 / \partial \sigma_e^4\right) = -(0.5)16.5346 = -8.2673$$

$$A\left(\partial L^2 / \partial \sigma_a^4\right) = -(0.5)\{\mathbf{a'Z'PZa}\}/\sigma_a^4$$

$$A\left(\partial L^2 / \partial \sigma_a^4\right) = -(0.5)9.1163 = -4.5582$$

$$A\left(\partial L^2 / \partial \sigma_a^2 \partial \sigma_e^2\right) = -(0.5)\{\mathbf{a'Z'P}(\mathbf{y}-\mathbf{Xb}-\mathbf{Za})\}/\left(\sigma_e^2\sigma_a^4\right)$$

$$A\left(\partial L^2 / \partial \sigma_a^2 \partial \sigma_e^2\right) = -(0.5)11.3070 = -5.6535$$

and

$$\mathbf{Ainf} = \begin{bmatrix} 8.2673 & 5.6535 \\ 5.6535 & 4.5582 \end{bmatrix} \quad \text{so } \mathbf{Ainf}^{-1} = \begin{bmatrix} 0.7967 & -0.9882 \\ -0.9882 & 1.4450 \end{bmatrix}$$

Using Eqn 17.4 and replacing **Einf** by **Ainf**:

$$\boldsymbol{\theta}_n = \boldsymbol{\theta} + \mathbf{Ainf}^{-1}\left(\partial L / \partial \boldsymbol{\theta}\right) = \begin{bmatrix} 0.4 \\ 0.2 \end{bmatrix} + \begin{bmatrix} 0.7967 & -0.9882 \\ -0.9882 & 1.4450 \end{bmatrix}\begin{bmatrix} 1.6510 \\ 1.2464 \end{bmatrix} = \begin{bmatrix} 0.4 \\ 0.2 \end{bmatrix} + \begin{bmatrix} 0.0838 \\ 0.1695 \end{bmatrix}$$

so that new estimates of σ_e^2 and σ_a^2 are 0.4838 (kg^2) and 0.3695 (kg^2), respectively.

Table 17.3 gives six successive iterates and log-likelihood for this data.

In the last iteration:

$$\mathbf{Ainf}- = \begin{bmatrix} 2.4436 & -3.2532 \\ -3.2532 & 5.3481 \end{bmatrix}$$

Table 17.3. Estimates of σ_e^2 and σ_a^2 and **L**.

Iterate	σ_e^2 (kg^2)	σ_a^2 (kg^2)	L
1	0.4000	0.2000	–2.3852
2	0.4838	0.3695	–2.2021
3	0.4910	0.5126	–2.1821
4	0.4839	0.5500	–2.1817
5	0.4835	0.5514	–2.1817
6	0.4835	0.5514	–2.1817

so that the estimate of σ_e^2 is 0.4835 with standard error $\sqrt{2.4436} = 1.563$ and the estimate of σ_a^2 is 0.5514 with standard error $\sqrt{5.3481} = 2.313$.

By contrast, if estimates of $\sigma_e^2 = 0.4$ and $\sigma_a^2 = 0.2$ are used in conjunction with Eqns 17.5 and 17.6 then: $(n-p)\sigma_e^2 = (\mathbf{y} - \mathbf{Xb} - \mathbf{Za})'(\mathbf{y})$ so $3\sigma_e^2 = 1.9277$ so $\sigma_e^2 = 0.6426\left(\text{kg}^2\right)$ and $q\sigma_a^2 = \mathbf{a}'\mathbf{A}^{-1}\mathbf{a} + \text{trace}\left[\mathbf{C}^{22}\mathbf{A}^{-1}\right]\sigma_e^2$ so $8\sigma_a^2 = 0.2538 + 1.4458$ so $\sigma_a^2 = 0.2125\left(\text{kg}^2\right)$ with L = -2.3852. After 1000 iterations, the algorithm gives $\sigma_e^2 = 0.4842\left(\text{kg}^2\right)$ and $\sigma_a^2 = 0.5504\left(\text{kg}^2\right)$ with L = -2.1817, showing that this algorithm is slower to converge.

18 Use of Gibbs Sampling in Variance Component Estimation and Breeding Value Prediction

18.1 Introduction

Gibbs sampling is a numerical integration method and is one of several Markov chain Monte Carlo (MCMC) methods. They involve drawing samples from specified distributions; hence, they are called Monte Carlo and are referred to as Markov chain because each sample depends on the previous sample. Specifically, Gibbs sampling involves generating random drawings from marginal posterior distributions through iteratively sampling from the conditional posterior distributions. For instance, given that $\mathbf{Q}' = (Q_1, Q_2)$ and $P(Q_1, Q_2)$ is the joint distribution of Q_1 and Q_2, Gibbs sampling involves sampling from the full conditional posterior distributions of Q_1, $P(Q_1|Q_2)$ and Q_2, $P(Q_2|Q_1)$.

Thus, given that the joint posterior distribution is known to proportionality, the conditional distributions can be generated. However, defining the joint density involves the use of Bayes' thereom. In general, given that the probability of two events occurring together, $P(B, Y)$, is:

$$P(B,Y) = P(B)P(Y \mid B) = P(Y)P(B \mid Y)$$

then:

$$P(B \mid Y) = P(B)P(Y \mid B)/P(Y) \tag{18.1}$$

Equation 18.1 implies that inference about the variable B depends on the prior probability of its occurrence, $P(B)$. Given that observations on Y are available, this prior probability is then updated to obtain the posterior probability or density of B, $(P(B|Y)$. Equation 18.1 is commonly expressed as:

$$P(B \mid Y) \propto P(B)P(Y \mid B) \tag{18.2}$$

as the denominator is not a function of B. Therefore, the posterior density of B is proportional to the prior probability of B times the conditional distribution of Y given B. Assuming that B in Eqn 18.2 is replaced by $\mathbf{W}$, a vector of parameters, such that $\mathbf{W}' = (W_1, W_2, W_3)$, and that the joint posterior distribution is known to proportionality (Eqn 18.2), the full conditional probabilities needed for the Gibbs sampler can be generated for each parameter as $P(W_1|W_2, W_3, Y)$, $P(W_2|W_1, W_3, Y)$ and $P(W_3|W_1, W_2, Y)$. Assuming starting values $W_1^{[0]}$, $W_2^{[0]}$ and $W_3^{[0]}$, the implementation of the Gibbs sampler involves iterating the following loop:

1. Sample $W_1^{[i+1]}$ from $P\left(W_1 \mid W_2^{[i]}, W_3^{[i]}, Y\right)$
2. Sample $W_2^{[i+1]}$ from $P\left(W_2 \mid W_1^{[i+1]}, W_3^{[i]}, Y\right)$
3. Sample $W_3^{[i+1]}$ from $P\left(W_3 \mid W_2^{[i+1]}, W_3^{[i+1]}, Y\right)$

DOI: 10.1079/9781800620506.0018

Usually, the initial samples are discarded (the so-called burn-in period). In summary, the application of the Gibbs sampler involves defining the prior distributions and the joint posterior density and generating the full conditional posterior distributions and sampling from the latter.

The Gibbs sampler was first implemented by Geman and Geman (1984). In animal breeding, Wang *et al.* (1993, 1994) used Gibbs sampling for variance component estimation in sire and animal models. It has been implemented for the study of covariance components in models with maternal effects (Jensen *et al.*, 1994), in threshold models (Sorensen *et al.*, 1995) and in random regression models (Jamrozik and Schaeffer, 1997). It has recently been employed for the purposes of variance component estimation and breeding value prediction in linear threshold models (Heringstad *et al.*, 2002; Wang *et al.*, 2002). Detailed presentations of the Gibbs sampling within the general framework of Bayesian inference and its application for variance components estimation under several models have been published by Sorensen and Gianola (2002). In this chapter, the application of the Gibbs sampler for variance component estimation and prediction of breeding values with univariate and multivariate animal models are presented and illustrated.

18.2 Univariate Animal Model

Consider the following univariate linear model:

$$\mathrm{y} = \mathbf{Xb} + \mathbf{Zu} + \mathbf{e}$$

where terms are as defined in Eqn 4.1 but with $\mathbf{u} = \mathbf{a}$ in Eqn 4.1. The conditional distribution that generates the data, y, is:

$$\mathbf{y} \mid \mathbf{b}, \mathbf{u}, \sigma_e^2 \sim \mathrm{N}\left(\mathbf{Xb} + \mathbf{Zu} + \mathbf{R}\sigma_e^2\right) \tag{18.3}$$

18.2.1 Prior distributions

Prior distributions of $\mathbf{b}$, $\mathbf{u}$, σ_u^2 and σ_e^2 are needed to complete the Bayesian specification of the model (Wang *et al.*, 1993). Usually, a flat prior distribution is assigned to $\mathbf{b}$. Thus:

$$P(\mathbf{b}) \sim \text{constant} \tag{18.4}$$

This represents an improper or 'flat' prior distribution, denoting lack of prior knowledge about this vector. However, if there is information *a priori* about the value of $\mathbf{b}$ in terms of upper or lower limits, this can be incorporated in defining the posterior distribution of $\mathbf{b}$. Such a prior distribution will be called a proper prior distribution.

Assuming an infinitesimal model, the distribution of $\mathbf{u}$ is multivariate normal and is:

$$\mathbf{u} \mid \mathbf{A}, \sigma_u^2 \sim \mathrm{N}\left(\mathrm{O}, \mathbf{A}\sigma_u^2\right) \tag{18.5}$$

A scaled inverted chi-squared distribution (χ^2) is usually used as prior for the variance components (Wang *et al.*, 1993). Thus, for the residual variance:

$$P\left(\sigma_e^2 \mid \upsilon_e, s_e^2\right) \propto \left(\sigma_e^2\right)^{-\left(\frac{\upsilon_e}{2}+1\right)} \exp\left(-\frac{\upsilon_e s_e^2}{2\sigma_e^2}\right) \tag{18.6}$$

and the additive genetic variance:

$$P\left(\sigma_u^2 \mid v_u, s_u^2\right) \propto \left(\sigma_u^2\right)^{-\left(\frac{v_u}{2}+1\right)} \exp\left(-\frac{v_u s_u^2}{2\sigma_u^2}\right) \tag{18.7}$$

where v_e (v_u) is a 'degree of belief' parameter and $s_e^2\left(s_u^2\right)$ can be interpreted as a prior value of the appropriate variance component. Alternatively, prior uniform distribution could be assigned to the variance components such that:

$$P(\sigma_i^2) \propto \text{constant} \tag{18.8}$$

where $\sigma_i^2 = \sigma_e^2$ or σ_u^2 and an upper limit might be assigned for σ_i^2 based on prior knowledge. Setting v_e or v_u to –2 and s_e^2 or s_a^2 to 0 in Eqns 18.6 or 18.7 gives Eqn 18.8.

18.2.2 Joint and full conditional distributions

The joint posterior distribution of the parameters (**b**, **u**, σ_e^2 or σ_u^2) is proportional to the product of the likelihood function and the joint prior distribution. Using Eqns 18.3–18.7, the joint posterior distribution can be written as:

$$P\left(\mathbf{b},\mathbf{u},\sigma_u^2,\sigma_e^2 \mid \mathbf{y}\right) \propto \left(\sigma_e^2\right)^{-\left(\frac{n+v_e}{2}+1\right)} \exp\left[-\frac{(\mathbf{y}-\mathbf{Xb}-\mathbf{Zu})'(\mathbf{y}-\mathbf{Xb}-\mathbf{Zu})+v_e s_e^2}{2\sigma_e^2}\right] \tag{18.9}$$

$$\left(\sigma_u^2\right)^{-\left(\frac{m+v_u}{2}+1\right)} \exp\left[-\frac{\mathbf{u}'\mathbf{A}^{-1}\mathbf{u}+v_u s_u^2}{2\sigma_u^2}\right]$$

assuming n observations and m animals. Setting v_e or v_a and s_e^2 or s_a^2 to zero gives the joint posterior distributions for the uniform distribution in Eqn 18.8.

The full conditional posterior distribution of each parameter is obtained by regarding all other parameters in Eqn 18.9 as known. Thus, for **b**:

$$P\left(\mathbf{b} \mid \mathbf{u},\sigma_u^2,\sigma_e^2,\mathbf{y}\right) \propto \exp\left[-\frac{(\mathbf{y}-\mathbf{Xb}-\mathbf{Zu})'(\mathbf{y}-\mathbf{Xb}-\mathbf{Zu})}{2\sigma_e^2}\right] \tag{18.10}$$

A corresponding distribution to the above is:

$$\mathbf{Xb} \mid \mathbf{u},\sigma_u^2,\sigma_e^2,\mathbf{y} \sim \mathrm{N}(\mathbf{y}-\mathbf{Zu},\mathbf{I}\sigma_e^2)$$

or:

$$\mathbf{X'Xb} \mid \mathbf{u},\sigma_u^2,\sigma_e^2,\mathbf{y} \sim \mathrm{N}(\mathbf{X}'(\mathbf{y}-\mathbf{Zu}),\mathbf{X'X}\sigma_e^2)$$

Therefore:

$$\mathbf{b} \mid \mathbf{u},\sigma_u^2,\sigma_e^2,\mathbf{y} \sim \mathrm{N}(\hat{\mathbf{b}},(\mathbf{X'X})^{-1}\sigma_e^2)$$

where:

$$\hat{\mathbf{b}} = (\mathbf{X'X})^{-1}\mathbf{X}'(\mathbf{y}-\mathbf{Zu})$$

Thus, for the jth level of **b**:

$$\hat{b}_j \mid \mathbf{b}_{-j},\mathbf{u},\sigma_u^2,\sigma_e^2,\mathbf{y} \sim \mathbf{N}\left(\hat{b}_j,\left(\mathbf{x}_j'\mathbf{x}_j\right)^{-1}\sigma_e^2\right) \tag{18.11}$$

with $\hat{b}_j = \left(\mathbf{x}_j' \mathbf{x}_j\right)^{-1} \mathbf{x}_j'\left(\mathbf{y}_j - \mathbf{X}_{-j}\mathbf{b} - \mathbf{Zu}\right)$, which is equivalent to Eqn 4.5, $\mathbf{x}_j$ is the jth row of $\mathbf{X}$ and $\mathbf{b}_{-j}$ is the vector $\mathbf{b}$ with level j deleted.

Similarly, the distribution for the jth random effect is:

$$\mathbf{u}_j \mid \mathbf{b}, \mathbf{u}_{-j}, \sigma_u^2, \sigma_e^2, \mathbf{y} \sim \mathrm{N}\left(\hat{\mathbf{u}}_j, \left(\mathbf{z}_j'\mathbf{z}_j + \mathbf{A}_{j,j}^{-1}, \alpha\right)^{-1} \sigma_e^2\right) \tag{18.12}$$

with:

$$\hat{\mathbf{u}}_j = \left(\mathbf{z}_j'\mathbf{z}_j + \mathbf{A}_{j,j}^{-1}\alpha\right)^{-1} \mathbf{z}_j'\left(y - \mathbf{Xb} - \mathbf{A}_{j,-j}^{-1}, \alpha\mathbf{u}_{-j}\right)$$

which is equivalent to Eqn 4.8.

The full conditional distribution of the residual variance is derived from Eqn 18.9 by considering only terms that involve σ_e^2 and is in the scaled inverted χ^2 form (Wang *et al.*, 1993). Thus, for the residual variance:

$$P\left(\sigma_e^2 \mid \mathbf{b}, \mathbf{u}, \sigma_u^2, \mathbf{y}\right) \propto \left(\sigma_e^2\right)^{-\left(\frac{n+v_e}{2}+1\right)} \exp\left(-\frac{\overleftrightarrow{v}_e \overleftrightarrow{s}_e^2}{2\sigma_e^2}\right)$$

where $\overleftrightarrow{v}_e = n + v_e$ and $\overleftrightarrow{s}_e^2 = ((\mathbf{y} - \mathbf{Xb} - \mathbf{Zu})'(\mathbf{y} - \mathbf{Xb} - \mathbf{Z}\mathrm{u}) + v_e s_e^2) / \overleftrightarrow{v}_e$

Hence:

$$\sigma_e^2 \mid \mathbf{b}, \mathbf{u}, \sigma_u^2, \mathbf{y} \sim \overleftrightarrow{v}_e \overleftrightarrow{s}_e^2 \chi_{\overleftrightarrow{v}_e}^{-2} \tag{18.13}$$

which involve sampling from an inverted χ^2 distribution with scale parameter, $(\mathbf{y} - \mathbf{Xb} - \mathbf{Zu})'(\mathbf{y} - \mathbf{Xb} - \mathbf{Zu}) + v_e s_e^2$ and $\overleftrightarrow{\mathbf{v}}_e$ degrees of freedom.

Similarly, the full conditional distribution of σ_u^2 is also in the form of an inverted chi-square. Thus:

$$P\left(\sigma_u^2 \mid \mathbf{b}, \mathbf{u}, \sigma_e^2, \mathbf{y}\right) \propto \left(\sigma_u^2\right)^{-\left(\frac{m+v_u}{2}+1\right)} \exp\left(-\frac{\overleftrightarrow{v}_u \overleftrightarrow{s}_u^2}{2\sigma_u^2}\right)$$

where $\overleftrightarrow{v}_u = m + v_u$ and $\overleftrightarrow{s}_u^2 = \left(\left(\mathbf{u}'\mathbf{A}^{-1}\mathbf{u}\right) + v_u s_u^2\right) / \overleftrightarrow{v}_u$

Thus:

$$\sigma_u^2 \mid \mathbf{b}, \mathbf{u}, \sigma_u^2, \mathbf{y} \sim \overleftrightarrow{v}_u \overleftrightarrow{s}_u^2 \chi_{\overleftrightarrow{v}_u}^{-2} \tag{18.14}$$

which involves sampling from an inverted χ^2 distribution with scale parameter $\left(\mathbf{u}'\mathbf{A}^{-1}\mathbf{u}\right) + v_u s_u^2$ and $\overleftrightarrow{\mathbf{v}}_u$ degrees of freedom.

The Gibbs sampling then consists of setting initial values for **b**, **u**, σ_u^2 and σ_e^2 and iteratively sampling successively from Eqns 18.11 to 18.14, using updated values of the parameters from the i round in the i + 1 round. Assuming that k rounds of iteration were performed, then k is called the length of the chain. As mentioned earlier, the first j samples are usually discarded as the burn-in period. This is to ensure that samples saved are not influenced by the priors but are drawn from the posterior distribution. The size of j is determined rather arbitrarily, but a graphical illustration could help.

Several strategies can be implemented in using the Gibbs sampler and these have an effect on the degree of correlation between the values sampled. Details of various strategies are discussed in detail by Sorensen and Gianola (2002) and therefore are not presented here. One approach is to run a single long chain. A sample (**b**, **u**, σ_u^2, σ_e^2) is saved at every

dth iterate until a total t samples are saved and analysed. The larger d is, the lower the degree of autocorrelation between the samples. Another strategy, known as the multiple chain or short chain approach, involves carrying out several parallel t runs and saving the last nth sample from each run. Thus, this approach produces $m = nt$ samples. The different chains will produce different realizations, even if the same starting values are used. However, if the parameters in the model are highly correlated, it might be useful to utilize different starting values in the different chains.

Determining convergence with the Gibbs sampler is not very straightforward, but it is advisable, depending on the size of the problem, to run several chains and check convergence graphically.

18.2.3 Inferences from the Gibbs sampling output

The samples saved are usually analysed to estimate posterior means or variances of the posterior distribution. Detailed discussion of various estimation methods is given in Sorensen and Gianola (2002) and not presented here. Given that **w** is a vector of size k, containing the saved samples, then the posterior mean and variance can be computed, respectively, as:

$$\mu_f = \frac{\sum_{i=1}^{k} f(w_i)}{k} \tag{18.15}$$

and

$$\text{var}\left(\mu_f\right) = \frac{\sum_{i=1}^{k}\left(f(w_i) - \mu_f\right)^2}{k}$$

where $f(\mathbf{w})$ is a function of interest of the variables in **w**. For instance, in the linear animal model in Section 6.2, the function of interest would be the variance components (σ_u^2 and σ_e^2) and the vectors **b** and **u**.

The above estimates from the posterior distribution are associated with sampling variance (Monte Carlo variance). The larger the number of samples analysed, the smaller the sampling variance. It is usually useful to get an estimate of the sampling variance associated with the estimates from the posterior distributions. An empirical estimate could be obtained by running several independent runs and then computing the between-chain variance of the estimates obtained for each run. This is not computationally feasible in most practical situations and various methods are used to estimate this variance. A number of such estimators are fully discussed by Sorensen and Gianola (2002). A simple method that could be used involves calculating the batch effective chain size. Given a chain of size k, successive samples are grouped into b batches, each of size t. The average of the jth batch can be computed as:

$$\bar{u}_j = \frac{\sum_{i=1}^{t} f(w_i)}{t}$$

The batch estimator of the variance of μ in Eqn 18.14 is:

$$\text{var}_b(\mu) = \frac{\sum_{j=1}^{b}(\bar{u}_j - \mu)^2}{b(b-1)}$$

The batch effective chain size can be obtained as:

$$\psi_b = \frac{\sum_{i=1}^{k}[f(w_i) - \mu]^2}{(k-1)\text{var}_b(u)}$$

If samples are uncorrelated, then $\psi = k$. The difference between ψ and k gives an idea of the degree of the autocorrelation among the samples in the chain.

18.2.4 Numerical application

Example 18.1

Using the data in Example 4.1 and the variance components, the application of Gibbs sampling for estimation of variance components and the prediction of breeding values is illustrated. Uniform priors are assumed for the variance components such that $v_e = v_a = -2$ and $s_e^2 = s_u^2 = 0$. A flat prior is assumed for **b**, and **u** is assumed to be normally distributed.

First, sample $\mathbf{b}_1^{[1]}$, where the superscript in brackets denotes iteration number, using Eqn 18.11, with $\hat{\mathbf{b}}_1$ calculated using Eqn 4.5 and $(\mathbf{x}_j'\mathbf{x})^{-1}\sigma_e^2 = (3)^{-1}40 = 13.333$. From Eqn 4.5:

$$\hat{\mathbf{b}}_1 = [(4.5 + 3.5 + 5.0) - (0 + 0 + 0)]/3 = 4.333$$

Assuming the random number (RN) generated from a normal distribution, N(0,1), is 0.5855, then $\mathbf{b}_1$ from Eqn 18.11 is:

$$\mathbf{b}_1^{[1]} = 4.333 + 0.5855\sqrt{13.333} = 6.4714$$

Then, sample $\mathbf{b}_2$ using Eqn 18.11 with $(\mathbf{x}_j'\mathbf{x})^{-1}\sigma_e^2 = (2)^{-1}40 = 20$ and $\hat{\mathbf{b}}_2$ is:

$$\hat{\mathbf{b}}_2 = [(2.9 + 3.9) - (0 + 0)]/2 = 3.40$$

Assuming the RN from N(0,1) is 0.7095, then:

$$\mathbf{b}_2^{[1]} = 3.40 + 0.7095\sqrt{20} = 6.5728$$

The vector of solution $\mathbf{u}_j$ for animal j is sampled using Eqn 18.12, with $\hat{\mathbf{u}}_j$ calculated using Eqn 4.8. Thus, for animal 1:

$$\hat{\mathbf{u}}_1 = 0 \quad \text{and} \quad (\mathbf{z}_1'\mathbf{z}_1 + \mathbf{A}_{1,1}^{-1}\,\alpha)^{-1}\sigma_e^2 = (3.667)^{-1}40 = 10.908$$

The value of $\left(\mathbf{z}_1'\mathbf{z}_1 + \mathbf{A}_{1,1}^{-1}\,\alpha\right)^{-1}$ is taken from the diagonal element of the coefficient matrix of the MME for Example 4.1. Assuming the RN from N(0,1) is –0.1093:

$$\mathbf{u}_1^{[1]} = 0 + \left(-0.1093\sqrt{10.908}\right) = -0.3610$$

For animal 2, $\hat{\mathbf{u}}_2$ from Eqn 4.8 = 0.090, and $\left(\mathbf{z}_1'\mathbf{z}_1 + \mathbf{A}_{1,1}^{-1}\,\alpha\right)^{-1}\sigma_e^2 = (4)^{-1}40 = 10$. Then, from Eqn 18.12, assuming RN from N(0,1) is –0.4535:

$$\mathbf{u}_2^{[1]} = 0.090 + \left(-0.4535\sqrt{10}\right) = -1.3438$$

Similarly, given that $\hat{\mathbf{u}}_3$ from Eqn 4.8 = 0.336, $\left(\mathbf{z}_1'\mathbf{z}_1 + \mathbf{A}_{1,1}^{-1}\ \alpha\right)^{-1}\sigma_e^2 = (4)^{-1}40 = 10$ and RN = 0.6059, then:

$$\mathbf{u}_3^{[1]} = 0.336 + 0.6059\sqrt{10} = 2.2519$$

For animal 4, $\hat{\mathbf{u}}_4 = -0.526$ from Eqn 4.8, $(\mathbf{z}_1'\mathbf{z}_1 + \mathbf{A}_{1,1}^{-1}\ \alpha)^{-1}\sigma_e^2 = (4.667)^{-1}40 = 8.571$ and RN = −1.8180, then:

$$\mathbf{u}_4^{[1]} = -0.526 + \left(-1.8180\sqrt{8.571}\right) = -5.8480$$

Similar calculations using Eqn 18.12 with RNs equal to 0.6301, −0.2762, −0.2842 and −0.9193, gave estimates of $\mathbf{u}_5^{[1]}$, $\mathbf{u}_6^{[1]}$, $\mathbf{u}_7^{[1]}$ and $\mathbf{u}_8^{[1]}$ to be, 2.2921 −2.1022 −2.8204 and −2.8346, respectively.

The vector of residuals, $\hat{\mathbf{e}} = \mathbf{y} - \mathbf{Xb} - \mathbf{Zu}$, is:

$$\begin{pmatrix} \hat{e}_4 \\ \hat{e}_5 \\ \hat{e}_6 \\ \hat{e}_7 \\ \hat{e}_8 \end{pmatrix} = \begin{pmatrix} 4.5 \\ 2.9 \\ 3.9 \\ 3.5 \\ 5.0 \end{pmatrix} - \begin{pmatrix} 6.471 \\ 6.573 \\ 6.573 \\ 6.471 \\ 6.471 \end{pmatrix} - \begin{pmatrix} -5.848 \\ 2.292 \\ -2.102 \\ -2.820 \\ -2.835 \end{pmatrix} = \begin{pmatrix} 3.877 \\ -5.965 \\ -0.571 \\ -0.151 \\ 1.363 \end{pmatrix}$$

and $\hat{e}'\hat{e}$ = 52.816. Sampling from the inverted χ^2 distribution with three degrees of freedom (Eqn 18.13) gave an estimate of 29.768 for the residual variance.

Using Eqn 18.14, sampling for σ_u^2 is again from the inverted χ^2 distribution, with $\mathbf{u}'\mathbf{A}^{-1}\mathbf{u}$ = 78.819 and degrees of freedom being 6. An estimate of 11.464 was obtained for σ_u^2. Note that it is easier to compute $\mathbf{u}'\mathbf{A}^{-1}\mathbf{u}$ using Eqn 3.3. Thus, $\mathbf{u}'\mathbf{A}^{-1}\mathbf{u} = \mathbf{u}'(\mathbf{T}^{-1})'\mathbf{D}^{-1}\mathbf{T}^{-1}\mathbf{u}$ = $\mathbf{m}'\mathbf{Dm}$ where $\mathbf{m} = \mathbf{T}^{-1}\mathbf{u}$, with $\mathbf{m}$ being a vector of Mendelian sampling for animals calculated using Eqn 3.2.

The next round of iteration is then commenced using the updated values computed for the parameters.

18.3 Multivariate Animal Model

In this section, the Gibbs sampling algorithm developed by Jensen *et al.* (1994) for models with maternal genetic effects is generalized for a multivariate situation. Given that animals are ordered within traits, the multivariate model for two traits could be written as:

$$\begin{pmatrix} \mathbf{y}_1 \\ \mathbf{y}_2 \end{pmatrix} = \begin{pmatrix} \mathbf{X}_1 & 0 \\ 0 & \mathbf{X}_2 \end{pmatrix}\begin{pmatrix} \mathbf{b}_1 \\ \mathbf{b}_2 \end{pmatrix} + \begin{pmatrix} \mathbf{Z}_1 & 0 \\ 0 & \mathbf{Z}_2 \end{pmatrix}\begin{pmatrix} \mathbf{u}_1 \\ \mathbf{u}_2 \end{pmatrix} + \begin{pmatrix} \mathbf{e}_1 \\ \mathbf{e}_2 \end{pmatrix}$$

where terms are as defined in Eqn 5.1, with $\mathbf{u} = \mathbf{a}$. The conditional distribution of the complete data, given that animals are ordered within traits, is:

$$\left\langle \begin{matrix} \mathbf{y}_1 \\ \mathbf{y}_2 \end{matrix} \middle| \mathbf{b}_1, \mathbf{b}_2, \mathbf{u}_1, \mathbf{u}_2, \mathbf{R} \right\rangle \sim N\left[\begin{matrix} \mathbf{X}_1\mathbf{b}_1 + \mathbf{Z}_1\mathbf{u}_1 \\ \mathbf{X}_2\mathbf{b}_2 + \mathbf{Z}_2\mathbf{u}_2 \end{matrix}, \mathbf{R} \otimes \mathbf{I}\right] \tag{18.16}$$

It is assumed that:

$$\left\langle \begin{matrix} \mathbf{u}_1 \\ \mathbf{u}_2 \end{matrix} \middle| \mathbf{G}, \mathbf{A} \right\rangle \sim N\left[\begin{matrix} 0 \\ 0 \end{matrix}, \mathbf{G} \otimes \mathbf{A}\right] \tag{18.17}$$

where $\mathbf{G}$ is the genetic covariance matrix and $\mathbf{A}$ is the numerator relationship matrix.

18.3.1 Prior distributions

Assume that proper uniform distributions are defined for the fixed effects:

$$P(\mathbf{b}_1) \propto \text{constant}; \quad P(\mathbf{b}_2) \propto \text{constant}$$

with:

$$\mathbf{b}_i(\min) \le \mathbf{b}_i \le \mathbf{b}_i(\max)$$

An inverted Wishart distribution (Jensen *et al.*, 1994) is used as prior distribution for the genetic and residual covariances. Thus, the prior distribution for the residual covariance is:

$$P(\mathbf{R} \mid \mathbf{V}_e, v_e) \propto |\mathbf{R}|^{-\frac{1}{2}(v_e+p+1)} \exp\left[-\frac{1}{2}\text{tr}\left(\mathbf{R}^{-1}\mathbf{V}_e^{-1}\right)\right] \tag{18.18}$$

The above is a p-dimensional inverse Wishart distribution (IW_2), where p is the order of $\mathbf{R}$, $\mathbf{V}_e$ is a parameter of the prior distribution and v_e are the degrees of freedom. If $\mathbf{V}_e = 0$ and $v_e = -(p + 1)$, the above reduces to a uniform distribution. Similarly, for the genetic covariance, the following prior distribution is assumed:

$$P(\mathbf{G} \mid \mathbf{V}_u, v_u) \propto |\mathbf{G}|^{-\frac{1}{2}(v_u+p+1)} \exp\left[-\frac{1}{2}\text{tr}\left(\mathbf{G}^{-1}\mathbf{V}_u^{-1}\right)\right] \tag{18.19}$$

with terms $\mathbf{V}_u$ and v_u equivalent to $\mathbf{V}_e$ and v_e, respectively, in Eqn 18.18.

The joint posterior distribution assuming n traits and using Eqns 18.16–18.19, is:

$$\begin{aligned} &P(\mathbf{b}_1,\ldots,\mathbf{b}_n,\mathbf{u}_1,\ldots,\mathbf{u}_n,\mathbf{R},\mathbf{G}) \\ &\quad \propto p(\mathbf{y}_1,\ldots,\mathbf{y}_n \mid \mathbf{b}_1,\ldots,\mathbf{b}_n,\mathbf{u}_1,\ldots\mathbf{u}_n,\mathbf{R})\,p(\mathbf{u}_1,\ldots\mathbf{u}_n,\mid \mathbf{G})\,p(\mathbf{G})\,p(\mathbf{R}) \end{aligned} \tag{18.20}$$

18.3.2 Conditional probabilities

Using the same principles as those for obtaining Eqns 18.11 and 18.12, the conditional distribution for the level k of the ith trait is:

$$\mathbf{b}_{i,k} \mid \mathbf{b}_{i,-k},\mathbf{b}_j,\mathbf{u},\mathbf{R}_e,\mathbf{G},\mathbf{y} \sim N\left(\mathbf{b}_{i,k},\left(\mathbf{x}'_{i,k}\ \mathbf{r}^{ii}\mathbf{x}_{i,k}\right)^{-1}\right); \qquad j = 1,n \text{ and } j \ne i \tag{18.21}$$

with:

$$\begin{aligned} \hat{\mathbf{b}}_{i,k} &= \left(\mathbf{x}'_{i,k}\ \mathbf{r}^{ii}\mathbf{x}_{i,k}\right)^{-1} \mathbf{x}'_{i,k}\left(\mathbf{r}^{ii}\mathbf{y}_i + \mathbf{r}^{ij}\mathbf{y}_j\right) - \mathbf{r}^{ii}\left(\mathbf{x}'_{i,-k}\ \mathbf{b}_{i,-k} + \mathbf{z}_i\mathbf{u}_i\right) \\ &\quad - \mathbf{r}^{ij}\left(\mathbf{x}_j\ \mathbf{b}_j + \mathbf{z}_j\ \mathbf{u}_j\right); \quad j = 1,n \text{ and } j \ne i \end{aligned}$$

Similarly, for the random animal effect, the conditional distribution for animal k of the ith trait is:

$$\mathbf{u}_{i,k} \mid \mathbf{u}_{i,-k},\mathbf{u}_j,\mathbf{b},\mathbf{R}_e,\mathbf{G},\mathbf{y} \sim N\left(\hat{\mathbf{u}}_{i,k},\left(\mathbf{r}^{ii}\mathbf{z}'_{i,k}\mathbf{z}_{i,k} + \mathbf{g}^{ii}\mathbf{A}^{-1}_{k,k}\right)^{-1}\right), j = 1,n \text{ and } j \ne i \tag{18.22}$$

with:

$$\begin{aligned} \hat{\mathbf{u}}_{i,k} \;=\; &\left(\mathbf{z}'_{i,k}\ \mathbf{r}^{ii}\mathbf{z}_{i,k} + \mathbf{A}^{-1}_{k,k}\ \mathbf{g}^{ii}\right)^{-1}\left\{\mathbf{z}'_{i,k}\left(\mathbf{r}^{ii}\mathbf{y}_i + \mathbf{r}^{ij}\mathbf{y}_j - \mathbf{r}^{ii}x_i\mathbf{b}_i + \mathbf{r}^{ij}x_j\mathbf{b}_j\right)\right. \\ &\left. -\left(\mathbf{z}'_{i,k}\ \mathbf{r}^{ij}\mathbf{z}_{i,k} + \mathbf{A}^{-1}_{k,k}\mathbf{g}^{ij}\mathbf{u}_{j,k}\right) - \mathbf{A}^{-1}_{k,s}\left(\mathbf{g}^{ii}\ \mathbf{u}_{i,s} + \mathbf{g}^{ij}\ \mathbf{u}_{j,s}\right)\right\} \end{aligned}$$

where s represents the known parents of the kth animal.

However, instead of sampling for each level of fixed or random effects for one trait at a time, it is more efficient to implement block sampling for each level of fixed or random effect across all traits at once. The conditional distribution for level k of a fixed effect required for block sampling, assuming $n = 2$, is:

$$\left\langle \begin{matrix} \mathbf{b}_{1,k} \\ \mathbf{b}_{2,k} \end{matrix} \middle| \mathbf{b}_{-k}, \mathbf{u}, \mathbf{R}, \mathbf{G}, \mathbf{y} \right\rangle \sim N\left[\begin{matrix} \hat{\mathbf{b}}_{1,k} \\ \hat{\mathbf{b}}_{2,k} \end{matrix}, \left(\mathbf{X}'_k \ \mathbf{R}^{-1}\mathbf{X}_k\right)^{-1} \right] \tag{18.23}$$

where:

$$\begin{pmatrix} \hat{\mathbf{b}}_{1,k} \\ \hat{\mathbf{b}}_{2,k} \end{pmatrix} = (\mathbf{X}'_k \ \mathbf{R}^{-1}\mathbf{X}_k)^{-1}(\mathbf{X}'_k \ \mathbf{R}^{-1}(\mathbf{y}_k - \mathbf{X}_{-k} \ \ \mathbf{b}_{-k} - \mathbf{Z}\hat{\mathbf{u}}))$$

which is equivalent to Eqn 5.4.

For the random animal effect, block sampling for animal k, assuming $n = 2$, the conditional distribution is:

$$\left\langle \begin{matrix} \mathbf{u}_{1,k} \\ \mathbf{u}_{2,k} \end{matrix} \middle| \mathbf{b}, \mathbf{u}_{j,-k}, \mathbf{R}, \mathbf{G}, \mathbf{y} \right\rangle \sim N\left[\begin{matrix} \hat{\mathbf{u}}_{1,k} \\ \hat{\mathbf{u}}_{2,k} \end{matrix}, (\mathbf{Z}'_k \ \mathbf{R}^{-1}\mathbf{Z}_k + \mathbf{A}^{-1}_{k,k} \otimes \mathbf{G}^{-1})^{-1} \right] \tag{18.24}$$

where:

$$\begin{pmatrix} \hat{\mathbf{u}}_{1,k} \\ \hat{\mathbf{u}}_{2,k} \end{pmatrix} = (\mathbf{Z}'_k \ \mathbf{R}^{-1}\mathbf{Z}_k + \mathbf{A}^{-1} \otimes \mathbf{G}^{-1})^{-1}\{(\mathbf{Z}'_k \ \mathbf{R}^{-1}(\mathbf{y}_k - \mathbf{X}\mathbf{b}) - \mathbf{A}^{-1} \otimes \mathbf{G}^{-1}(\hat{\mathbf{u}}_s + \hat{\mathbf{u}}_d)\}$$

where s and d are the sire and dam of the kth animal.

From Eqn 18.20, the full conditional distribution of the residual variance is:

$$P(\mathbf{R} \mid \mathbf{b}, \mathbf{u}, \mathbf{y}) \propto P(\mathbf{R})P(\mathbf{y} \mid \mathbf{b}, \mathbf{u}, \mathbf{R})$$

Including the prior distribution, the above can be expressed (Jensen *et al.*, 1994) as:

$$P(\mathbf{R} \mid \mathbf{b}, \mathbf{u}, \mathbf{y}) \propto |\mathbf{R}|^{-\frac{1}{2}(v_e+p+1+m)} \exp\left[-\frac{1}{2}\mathrm{tr}\left\{\mathbf{R}^{-1}\left(\mathbf{S}_e^2 + \mathbf{V}_e^{-1}\right)\right\}\right]$$

where m is the number of records and $\mathbf{S}_e^2$ is:

$$\mathbf{S}_e^2 = \begin{pmatrix} \hat{\mathbf{e}}'_1 \ \hat{\mathbf{e}}_1 & \hat{\mathbf{e}}'_1 \ \hat{\mathbf{e}}_2 \\ \hat{\mathbf{e}}'_2 \ \hat{\mathbf{e}}_1 & \hat{\mathbf{e}}'_2 \ \hat{\mathbf{e}}_2 \end{pmatrix}$$

assuming that $n = 2$ and $\hat{\mathbf{e}}_i = \mathbf{y}_i - \mathbf{X}_i\mathbf{b}_i - \mathbf{Z}_i\mathbf{u}_i$, $i = 1, n$.

Thus:

$$\mathbf{R} \mid \mathbf{b}, \mathbf{u}, \mathbf{y} \sim \mathrm{IW}_2((\mathbf{S}_e^2 + \mathbf{V}_e^{-1})^{-1}, \mathbf{v}_e + m) \tag{18.25}$$

which is in the form of a p-dimensional inverted Wishart distribution with $\mathbf{v}_e + m$ degrees of freedom and scale parameter $\left(\mathbf{S}_e^2 + \mathbf{V}_e^{-1}\right)$.

Similarly, the conditional distribution for the additive genetic variance is:

$$P(\mathbf{G} \mid \mathbf{b}, \mathbf{u}, \mathbf{y}) \propto P(\mathbf{G})P(\mathbf{u} \mid \mathbf{G})$$

Including the prior distribution, the above can be expressed (Jensen *et al.*, 1994) as:

$$P(\mathbf{G} \mid \mathbf{b},\mathbf{u},\mathbf{y}) \propto |\mathbf{G}|^{-\frac{1}{2}(v_u+p+1+q)} \exp\left[-\frac{1}{2}\mathrm{tr}\{\mathbf{G}^{-1}(\mathbf{S}_u^2+\mathbf{V}_u^{-1})\}\right]$$

where q is the number of animals and, assuming $n = 2$, $\mathbf{S}_u^2$ is:

$$\mathbf{S}_u^2 = \begin{pmatrix} \mathbf{u}_1'\ \mathbf{A}^{-1}\ \mathbf{u}_1 & \mathbf{u}_1'\ \mathbf{A}^{-1}\ \mathbf{u}_2 \\ \mathbf{u}_2'\ \mathbf{A}^{-1}\ \mathbf{u}_1 & \mathbf{u}_2'\ \mathbf{A}^{-1}\ \mathbf{u}_2 \end{pmatrix}$$

Thus:

$$\mathbf{G} \mid \mathbf{b},\mathbf{u},\mathbf{y} \sim \mathrm{IW}_2\left(\left(\mathbf{S}_u^2+\mathbf{V}_u^{-1}\right)^{-1}, \mathbf{v}_u + q\right) \tag{18.26}$$

Which, again, is in the form of a p-dimensional inverted Wishart distribution with $\mathbf{v}_u + q$ degrees of freedom and scale parameter $\left(\mathbf{S}_u^2+\mathbf{V}_u^{-1}\right)$.

18.3.3 Numerical illustration

Example 18.2

Using the data in Example 5.1 and the variance components, the application of Gibbs sampling to estimating variance components and predicting breeding values is illustrated. Uniform priors are assumed for the variance components such that $\mathbf{v}_e = \mathbf{v}_u = -3$ and $\mathbf{V}_e = \mathbf{V}_u = 0$. A flat prior is assumed for **b**, and **u** is assumed to be normally distributed.

Processing data and accumulating right-hand side (rhs) and diagonals (Diag) for level j of sex of calf effects as:

$$\mathrm{rhs}_{1j} = \mathrm{rhs}_{1j} + \mathbf{R}^{11}(\mathbf{y}_1 - \mathbf{u}_{1i}) + \mathbf{R}^{12}(\mathbf{y}_2 - \mathbf{u}_{2i})$$
$$\mathrm{rhs}_{2j} = \mathrm{rhs}_{2j} + \mathbf{R}^{21}(\mathbf{y}_2 - \mathbf{u}_{1i}) + \mathbf{R}^{22}(\mathbf{y}_2 - \mathbf{u}_{2i})$$
$$\mathbf{Diag}_j = \mathbf{Diag}_j + \mathbf{R}$$

When all data have been read, calculate solutions for level j of sex effect as:

$$\begin{pmatrix} \hat{b}_{1j} \\ \hat{b}_{2j} \end{pmatrix} = \mathbf{Diag}_j^{-1} \begin{pmatrix} rhs_{1j} \\ rhs_{2j} \end{pmatrix}$$

Sample $\mathbf{b}_j$ in Eqn 18.23 as:

$$\mathbf{b}_j = \begin{pmatrix} \hat{b}_{1j} \\ \hat{b}_{2j} \end{pmatrix} + \left\{\mathrm{CHOL}\left(\mathbf{Diag}_j^{-1}\right)\right\}\mathbf{h}$$

where **h** is the vector of normal deviates from a population of mean zero and variance 1 and CHOL is the Cholesky decomposition of the inverse of the matrix **Diag.**

Next, process data and accumulate right-hand side (rhs) and diagonals (Diag) for animal i as:

$$\mathrm{rhs}_{1i} = \mathrm{rhs}_{1i} + \mathbf{R}^{11}(\mathbf{y}_1 - \mathbf{b}_{1j}) + \mathbf{R}^{12}(\mathbf{y}_2 - \mathbf{b}_{2j})$$
$$\mathrm{rhs}_{2i} = \mathrm{rhs}_{2i} + \mathbf{R}^{21}(\mathbf{y}_2 - \mathbf{b}_{1j}) + \mathbf{R}^{22}(\mathbf{y}_2 - \mathbf{b}_{2j})$$
$$\mathbf{Diag}_i = \mathbf{Diag}_i + \mathbf{R}$$

When all data have been read, calculate solutions for animal i as:

$$\begin{pmatrix} \hat{u}_{1i} \\ \hat{u}_{2i} \end{pmatrix} = \mathbf{Diag}_i^{-1} \begin{pmatrix} rhs_{1i} \\ rhs_{2i} \end{pmatrix}$$

Sample $\mathbf{u}_i$ in Eqn 18.24 as:

$$\mathbf{u}_i = \begin{pmatrix} \hat{u}_{1i} \\ \hat{u}_{2i} \end{pmatrix} + \{\mathrm{CHOL}(\mathbf{Diag}_i^{-1})\}\mathbf{h}$$

All data is then processed to obtain residual effects as:

$$\hat{\mathbf{e}} = \begin{pmatrix} \hat{\mathbf{e}}_1 \\ \hat{\mathbf{e}}_2 \end{pmatrix} = \begin{pmatrix} \mathbf{y}_1 - \mathbf{X}_1\hat{\mathbf{b}}_1 - \mathbf{Z}_1\hat{\mathbf{u}}_1 \\ \mathbf{y}_2 - \mathbf{X}_2\hat{\mathbf{b}}_2 - \mathbf{Z}_2\hat{\mathbf{u}}_2 \end{pmatrix}$$

and calculate residual sums of squares, $\mathbf{S}_e^2 = \hat{\mathbf{e}}\hat{\mathbf{e}}'$. Then compute $\mathbf{T} = (\mathbf{S}_e^2 + \mathbf{V}_e^{-1})^{-1}$. Cholesky decomposition of $\mathbf{T}$ is carried out to obtain $\mathbf{LL}'$, where $\mathbf{L}$ is a lower triangular matrix. Sampling from a Wishart distribution with $\mathbf{L}$ as the input matrix and $\mathbf{v}_e + m$ degrees of freedom (Eqn 18.25) generates a new sample value of $\mathbf{R}$.

Similarly, to compute a new sample value of $\mathbf{G}$ using Eqn 18.26, first compute $\mathbf{T}^{-1} = (\mathbf{S}_u^2 + \mathbf{V}_u^{-1})^{-1}$. Decompose $\mathbf{T}$ to obtain $\mathbf{LL}'$ and sample from a Wishart distribution with $\mathbf{L}$ as the input matrix and $\mathbf{v}_u + q$ degrees of freedom. Another cycle of sampling is then initiated until the desired length of chain is achieved. Post-processing of results can be carried out, as discussed in Section 18.2.3.

19 Solving Linear Equations

19.1 Introduction

Different methods can be used to solve the MME covered in the previous chapters. These various methods could broadly be divided into three main categories:

1. Direct inversion (Section 19.3)
2. Iteration on the MME (Section 19.4)
3. Iteration on the data (Section 19.5)

The manner in which the MME are set up depends on the method to be used in solving these equations. As shown in Section 19.5, the third method, for instance, does not involve setting up the MME directly.

19.2 Absorption

In some cases, the dimensionality of the MME might be reduced through a process of absorption before the application of either of the three methods above to solve the MME. Absorption is therefore a computational technique used to reduce computing resource needs when solving a large system of linear equations. In animal breeding, absorption is commonly used when a blocking factor with a large number of levels, such as herd–year–season effects, is a term in the model. This therefore reduces the size of the system of equations to be solved. It is also commonly used in the estimation of accuracies or reliabilities, to absorb a major fixed effect with a great number of levels into the animal-by-animal block of equations in the MME and inverting the animal block of equations to estimate reliability in an iterative approach (see Appendix D).

Thus, given the MME for Example 4.1, let the matrix $\mathbf{P} = (\mathbf{I} - \mathbf{X}(\mathbf{X}'\mathbf{X})^{-1}\mathbf{X}')$. The absorption of the sex-of-calf effect into the set of equations for animal effects involves multiplying the set of equations for animal effect by the matrix **P**, as follows.

$$\mathbf{Z'PZ} + \mathbf{A}^{-1}\alpha = \mathbf{Z'Py} \tag{19.1}$$

and then solving these equations for animal effect. The matrix **P** is sometimes called the projection matrix. Expanding 19.1 gives $\mathbf{Z'PZ} = \mathbf{Z'Z} - \mathbf{Z'X}(\mathbf{X'X})^{-1}\mathbf{X'Z}$ and $\mathbf{Z'Py} = \mathbf{Z'y} - \mathbf{Z'X}(\mathbf{X'X})^{-1}\mathbf{X'y}$. This implies the absorption process involves multiplying the inverse of the block diagonal for the sex-of-calf effects by the off-diagonal block matrix between sex-of-calf effects and animal effects and then by its transpose. The product is then deviated from the animal-by-animal block effect. In addition, the inverse of the block diagonal for the sex-of-calf effects is multiplied by the off-diagonal block matrix between sex-of-calf effects and then multiplied by the right-hand side for the sex-of-calf effects. This is then deviated from the right-hand side for the animal effects. Then, Eqn 19.1 can be solved by other methods described in subsequent sections to compute breeding values for animals.

DOI: 10.1079/9781800620506.0019

In practice, the data are usually sorted by the fixed with large number of levels to be absorbed and each subclass for the fixed effect is absorbed into the MME for the other factors in the analysis, after all data for that subclass have been processed. This implies that the size of MME to be formed is reduced as equations for major fixed effect are never formed. In the case of the data in Example 4.1, the data were sorted by the sex-of-calf effects. The matrix **P** for this Example 4.1 is:

$$\begin{bmatrix} 0.667 & 0.000 & 0.000 & -0.333 & -0.333 \\ 0.000 & 0.500 & -0.500 & 0.000 & 0.000 \\ 0.000 & -0.500 & 0.500 & 0.000 & 0.000 \\ -0.333 & 0.000 & 0.000 & 0.667 & -0.333 \\ -0.333 & 0.000 & 0.000 & -0.333 & 0.667 \end{bmatrix}$$

The left-hand side for Eqn 19.1 using the same variance components as in example 4.1 is:

$Z'PZ+ A^{-1}\alpha =$

$$\begin{bmatrix} 3.667 & & & & & & & \\ 1.000 & 4.000 & & & & & & \\ 0.000 & 1.000 & 4.000 & & & \text{symmetric} & & \\ -1.333 & 0.000 & 0.000 & 4.333 & & & & \\ 0.000 & -2.000 & -2.000 & 1.000 & 5.500 & & & \\ -2.000 & -2.000 & 1.000 & 0.000 & -0.500 & 5.500 & & \\ 0.000 & 0.000 & 0.000 & -2.333 & -2.000 & 0.000 & 4.667 & \\ 0.000 & 0.000 & -2.000 & -0.333 & 0.000 & -2.000 & -0.333 & 4.667 \end{bmatrix}$$

and the transpose right-hand side for Eqn 19.1, $(\mathbf{Z'Py})'$, is [0.000 0.000 0.000 0.167 –0.500 0.500 –0.833 0.667]. The main differences between $\mathbf{Z'PZ}+ \mathbf{A}^{-1}\alpha$ shown above and the left-hand side of the full MME in Example 4.1 are in the diagonal elements for calves with records and in the off-diagonal elements between calves in the same subclass for sex effects. Solving Eqn 19.1 by direct inversion gives the same solutions obtained for Example 4.1.

19.3 Direct Inversion

The solutions to the MME in the various examples given so far in this book have been based on this method. It involves setting up the MME and inverting the coefficient matrix. Solutions are obtained by multiplying the right-hand side (RHS) by the inverse of the coefficient matrix. Thus, **b**, the vector of solution, is calculated as:

$$\hat{\mathbf{b}} = \mathbf{C}^{-1}\mathbf{y}$$

where **C** is the coefficient matrix and **y** is the RHS. Since the coefficient matrix is symmetrical, only the upper triangular portion is usually set up and inverted. The major limitation of this approach is that it can only be applied to small data sets in view of the memory requirements and computational difficulties of inverting large matrices.

19.4 Iteration on the Mixed-model Equations

This involves setting up the MME and iterating on these equations until convergence is achieved at a predetermined criterion. The iterative procedures are based on the general theory for solving simultaneous equations. For instance, given two simultaneous equations with unknown parameters, b_1 and b_2, the first equation can be solved for b_1 in terms of b_2. This value of b_1 can then be substituted in the second equation to solve for b_2. The value of b_2 is then substituted in the first equation to calculate b_1. This is the principle on which the iterative procedures are based. In the iterative procedure, the above process is continued until the solutions for the *b*s are more or less the same in each round of iteration and the equations are said to have converged. There are various iterative procedures that can be used, and some are described below.

19.4.1 Jacobi iteration

One of the simplest methods is Jacobi iteration or total step iteration.

Consider the following set of simultaneous equations:

$$\begin{bmatrix} c_{11} & c_{12} & c_{13} \\ c_{21} & c_{22} & c_{23} \\ c_{31} & c_{32} & c_{33} \end{bmatrix} \begin{bmatrix} b_1 \\ b_2 \\ b_3 \end{bmatrix} = \begin{bmatrix} y_1 \\ y_2 \\ y_3 \end{bmatrix}$$

These equations can also be written as:

$$\begin{aligned} c_{11}b_1 + c_{12}b_2 + c_{13}b_3 &= y_1 \\ c_{21}b_1 + c_{22}b_2 + c_{23}b_3 &= y_2 \\ c_{31}b_1 + c_{32}b_2 + c_{33}b_3 &= y_3 \end{aligned}$$

or as:

$$\mathbf{Cb} = \mathbf{y} \tag{19.2}$$

The system of equations is rearranged so that the first is solved for b_1, the second for b_2 and the third for b_3. Thus:

$$\begin{aligned} b_1^{r+1} &= (1/c_{11})(y_1 - c_{12}b_2^r - c_{13}b_3^r) \\ b_2^{r+1} &= (1/c_{22})(y_2 - c_{21}b_1^r - c_{23}b_3^r) \\ b_3^{r+1} &= (1/c_{33})(y_3 - c_{31}b_1^r - c_{32}b_2^r) \end{aligned} \tag{19.3}$$

The superscript *r* refers to the number of the round of iteration. In the first round of iteration, *r* equals 1 and b_1 to b_3 could be set to zero or an assumed set of values that are used to solve the equations to obtain a new set of solutions (**b** terms). The process is continued until two successive sets of solutions are within previously defined allowable deviations and the equations are said to converge. One commonly used convergence criterion is the sum of squares of differences between the current and previous solutions divided by the sum of squares of the current solution. Once this is lower than a predetermined value, for instance 10^{-9}, the equations are considered to have converged.

From the set of equations above, the solution for b_i was obtained by dividing the adjusted RHS by the diagonal (a_{ii}). It is therefore mandatory that the diagonal element, often called the pivot element, is not zero. If a zero pivot element is encountered during the iterative process, the row containing the zero should be exchanged with a row below it in which the element in that column is not zero. To avoid the problem of encountering a zero pivot element and generally improving the efficiency of the iterative process, it is sometimes recommended that the system of equations should be ordered such that the coefficient of b_1 of the greatest magnitude occurs in the first equation, and the coefficient of b_2 of the greatest magnitude in the remaining equations occurs in the second equation, etc.

The iterative procedure described above is usually called Jacobi iteration as all new solutions in the current (r) round of iteration are obtained using solutions only from the previous ($r - 1$) round of iteration. The Jacobi iterative procedure is inefficient in handling systems of equations that are not constrained (i.e. with no restrictions placed on the solutions for the levels of an effect) and convergence is not guaranteed (Maron, 1987; Misztal and Gianola, 1988). When a random animal effect is involved in the system of equations with relationships included, it is usually necessary to use a relaxation factor of below 1.0, otherwise equations may not converge (Groeneveld, 1990). The relaxation factor refers to a constant estimated on the basis of the linear changes in the solutions during the iteration process and applied to speed up the solutions towards convergence. When iterating on the data (Section 19.5), the Jacobi iterative procedure involves reading only one data file, even with several effects in the model. With large data sets this has the advantage of reducing memory requirement and processing time compared with the Gauss–Seidel iterative procedure (see Section 19.4.2).

The Jacobi iterative procedure can be briefly summarized as follows.

Following Ducrocq (1992), Eqn 19.2 can be written as:

$$[\mathbf{M} + (\mathbf{C} - \mathbf{M})]\mathbf{b} = \mathbf{y}$$

if $\mathbf{M}$ is the diagonal matrix containing the diagonal elements of $\mathbf{C}$; then the algorithm for Jacobi iteration is:

$$\mathbf{b}^{(r+1)} = \mathbf{M}^{-1}(\mathbf{y} - \mathbf{C}\mathbf{b}^{(r)}) + \mathbf{b}^{(r)} \qquad (19.4)$$

When a relaxation factor (w) is applied, the above equation becomes:

$$\mathbf{b}^{(r+1)} = w[\mathbf{M}^{-1}(\mathbf{y} - \mathbf{C}\mathbf{b}^{(r)})] + \mathbf{b}^{(r)}$$

Another variation of the Jacobi iteration, called second-order Jacobi, is usually employed in the analysis of large data sets and it can increase the rate of convergence. The iterative procedure for second-order Jacobi is:

$$\mathbf{b}^{(r+1)} = \mathbf{M}^{-1}(\mathbf{y} - \mathbf{C}\mathbf{b}^{(r)} + \mathbf{b}^{(r)} + w(\mathbf{b}^{(r)} - \mathbf{b}^{(r-1)}))$$

Example 19.1

Using the coefficient matrix and the RHS for Example 4.1, Jacobi iteration (Eqn 19.3) is carried out using only the non-zero element of the coefficient matrix. Solutions for sex effect (**b** vector) and random animal effect (**u** vector) are shown below with the round of iteration. The convergence criterion (CONV) was the sum of squares of differences between the current and previous solutions divided by the sum of squares of the current solution.

	Rounds of iteration									
Effects	0[a]	1	2	3	4	16	17	18	19	20
b_1	4.333	4.333	4.381	4.370	4.368	4.358	4.358	4.358	4.358	4.358
b_2	3.400	3.400	3.433	3.365	3.414	3.404	3.404	3.404	3.404	3.404
$\hat{u}_1$	0.000	0.267	0.164	0.185	0.131	0.099	0.099	0.099	0.099	0.099
$\hat{u}_2$	0.000	0.000	−0.073	−0.003	−0.039	−0.018	−0.018	−0.018	−0.018	−0.018
$\hat{u}_3$	0.000	−0.033	−0.080	−0.049	−0.070	−0.041	−0.041	−0.041	−0.041	−0.041
$\hat{u}_4$	0.167	−0.138	−0.007	−0.035	0.000	−0.008	−0.008	−0.008	−0.008	−0.008
$\hat{u}_5$	−0.500	−0.411	−0.248	−0.265	−0.204	−0.185	−0.185	−0.185	−0.185	−0.185
$\hat{u}_6$	0.500	0.345	0.318	0.237	0.236	0.178	0.178	0.178	0.177	0.177
$\hat{u}_7$	−0.833	−0.406	−0.390	−0.301	−0.295	−0.249	−0.249	−0.249	−0.249	−0.249
$\hat{u}_8$	0.667	0.400	0.286	0.232	0.207	0.183	0.183	0.183	0.183	0.183
CONV	1.000	2.3^{-2}	3.9^{-3}	1.4^{-3}	5.9^{-4}	4.2^{-8}	1.6^{-8}	1.0^{-8}	4.1^{-9}	3.0^{-9}

[a]Starting values

The starting solutions for sex effect were the mean yield for each sex subclass and, for animals with records, starting solutions were the deviation of their yields from the mean yield of their respective sex subclass and zero for ancestors. The final solutions obtained after the 20th round of iteration were exactly the same as obtained in Section 4.2 by direct inversion of the coefficient matrix. The solutions for sex effect were obtained using Eqn 19.2. Thus, in the first round of iteration the solution for males was:

$$b_1 = \frac{1}{c_{11}}\left(\sum_{k=1}^{m} y_k - (1)\hat{u}_4 - (1)\hat{u}_7 - (1)\hat{u}_8\right)$$

where c_{ii} is the diagonal element of the coefficient matrix for level i of sex effect and m is the number of records for males.

$$b_1 = 1/3(13.0 - 0.167 - (-0.833) - 0.667) = 4.333$$

However, using Eqn 19.2 to obtain animal solutions caused the system of equations to diverge. A relaxation factor (w) of 0.8 was therefore employed and solutions for animal j were computed as:

$$\hat{u}_j^r = w\left[\left(\frac{1}{c_{ll}}\right)\left(y_j - c_{li}\hat{b}_i^{r-1} - \sum_k c_{lt}\hat{u}_k^{r-1}\right) - \hat{u}_j^{r-1}\right] + \hat{u}_j^{r-1}$$

where $l = j + n$, $t = k + n$, with $n = 2$; the total number of levels of fixed effect, c_{lt} and c_{li}, for instance, are the elements of the coefficient matrix between animals j and k, and animal j and level i of sex effect, respectively. Thus, in the first round of iteration, solutions for animals 1 and 8 are calculated as:

$$\begin{aligned}
\hat{u}_1^1 &= w[\{1/c_{33}(y_1 - (1)\hat{u}_2 - (-1.333)\hat{u}_4 - (-2)\hat{u}_6)\} - \hat{u}_1^0] + \hat{u}_1^0 \\
&= w[\{1/3.667(0 - 0 - (-0.223) - (-1))\} - 0] + 0 \\
&= 0.8(0.334 - 0) + 0 = 0.267
\end{aligned}$$

and

$$\begin{aligned}
\hat{u}_8^1 &= w[\{1/c_{1010}(y_8 - (1)b_1 - (-2)\hat{u}_3 - (-2)\hat{u}_6)\} - \hat{u}_8^0] + \hat{u}_8^0 \\
&= w[\{1/5(5 - 4.333 - 0 - (-1))\} - 0.667] + 0.667 \\
&= 0.8(0.333 - 0.667) + 0.667 = 0.400
\end{aligned}$$

19.4.2 Gauss–Seidel iteration

Another iterative procedure commonly used is Gauss–Seidel iteration. This is similar to Jacobi iteration except that most current solutions are calculated from the most recent available solution rather than the solution from the previous round of iteration. Using the same set of simultaneous equations as in Eqn 19.2, solutions for b_1, b_2 and b_3 in the first round of iteration become:

$$\begin{aligned} b_1^{r+1} &= (1/c_{11})(y_1 - c_{12}b_2^r - c_{13}b_3^r) \\ b_2^{r+1} &= (1/c_{22})(y_2 - c_{21}b_1^{r+1} - c_{23}b_3^r) \\ b_3^{r+1} &= (1/c_{33})(y_3 - c_{31}b_1^{r+1} - c_{32}b_2^{r+1}) \end{aligned} \qquad (19.5)$$

Thus the solution for b_2 in the $r + 1$ round of iteration is calculated using the most recent solution for $b_1\,(b_1^{r+1})$ instead of the previous solution (b_1^r), and the current solution for b_3 is calculated from the current solutions for $b_1\,(b_1^{r+1})$ and $b_2\,(b_2^{r+1})$. If, in Eqn 19.4, **L** is strictly the lower triangular of **C** and **D** the diagonal of **C**, then Eqn 19.4 becomes the Gauss–Seidel iteration when **M** = **L** + **D**. The convergence criteria could equally be defined as discussed in Section 19.4.1. Generally, equations are guaranteed to converge with the Gauss–Seidel iterative procedure. However, when iterating on the data, this iterative procedure involves reading one data file for each effect in the model. With large data sets, the setting up of data files for each effect could result in large memory requirement and the reading of several files in each round of iteration could increase processing time.

Example 19.2

Using the same coefficient matrix, RHS and starting values as in Example 19.1 above, the Gauss–Seidel iteration (Eqn 19.5) is carried out for the same number of iterations as in Jacobi's method and the results are shown below. The convergence criterion is as defined in Example 19.1.

	Rounds of iteration									
Effects	0	1	2	3	4	16	17	18	19	20
b_1	4.333	4.333	4.400	4.372	4.364	4.359	4.359	4.359	4.359	4.359
b_2	3.400	3.400	3.392	3.403	3.407	3.405	3.405	3.405	3.405	3.405
$\hat{u}_1$	0.000	0.333	0.194	0.149	0.115	0.098	0.098	0.098	0.098	0.098
$\hat{u}_2$	0.000	−0.083	−0.035	−0.006	−0.008	−0.019	−0.019	−0.019	−0.019	−0.019
$\hat{u}_3$	0.000	−0.021	−0.136	−0.109	−0.076	−0.041	−0.041	−0.041	−0.041	−0.041
$\hat{u}_4$	0.167	−0.119	0.001	0.004	−0.003	−0.009	−0.009	−0.009	−0.009	−0.009
$\hat{u}_5$	−0.500	−0.376	−0.261	−0.218	−0.199	−0.186	−0.186	−0.186	−0.186	−0.186
$\hat{u}_6$	0.500	0.392	0.254	0.204	0.185	0.177	0.177	0.177	0.177	0.177
$\hat{u}_7$	−0.833	−0.364	−0.284	−0.260	−0.253	−0.250	−0.250	−0.250	−0.250	−0.250
$\hat{u}_8$	0.667	0.282	0.167	0.164	0.171	0.182	0.183	0.183	0.183	0.183
CONV	1.000	1.9^{-2}	3.4^{-3}	3.1^{-4}	1.0^{-4}	7^{-10}	4^{-10}	2^{-10}	1^{-10}	8^{-11}

CONV, convergence criterion

The solutions obtained are the same as those obtained from Jacobi iteration and by direct inversion of the coefficient matrix in Example 4.1. In addition, the equations converged faster than when using Jacobi iteration and no relaxation factor was applied.

Iterating on the MME could be carried out as described above, once the equations have been set up, using only the stored non-zero elements of the coefficient matrix. In practice, it may be necessary to store the non-zero elements and their rows and columns on disk for large data sets because of the memory requirement, and these are read in each round of iteration.

19.5 Iterating on the Data

This is the most commonly used methodology in national genetic evaluations, which usually involve millions of records. Schaeffer and Kennedy first presented this method in 1986. It does not involve setting up the coefficient matrix directly, but it involves setting up equations for each level of effects in the model as the data and pedigree files are read and solved using either Gauss–Seidel or Jacobi iteration or a combination of both or a variation of any of the iterative procedures such as second-order Jacobi. Presented below are the basic equations for the solutions of various effects under several models and these form the basis of the iterative process for each of the models.

The equation for the solution of level i for a fixed effect in the model in a univariate animal situation is Eqn 4.5, which is derived from the MME and can be generalized as:

$$\hat{b}_i = \frac{\sum_{k=1}^{n_i} y_{ki} - \sum_{j=1}^{m} \hat{w}_{ij}}{n_i} \qquad (19.6)$$

where y_{ki} is the kth record in level i, m is the total number of levels of other effects within subclass i of the fixed effect, $\hat{w}_{ij}$ is the solution for the jth level, and n_i is the number of records in fixed-effect subclass i. However, when there are many fixed effects in the model, the above formula may be used to obtain solutions for the major fixed effect with many levels such as HYS, while the vector of solutions (**f**) for other minor fixed effects with few levels may be calculated as:

$$\mathbf{f} = (\mathbf{X'X})^{-1}\mathbf{X'}(\mathbf{y} - \hat{\mathbf{w}} - \hat{\mathbf{b}}) \qquad (19.7)$$

where $\mathbf{y}$ is the vector of observations, $(\mathbf{X'X})^{-1}$ is the inverse of the coefficient matrix for the minor fixed effects, and $\hat{\mathbf{w}}$ and $\hat{\mathbf{b}}$ are vectors of solutions for effects as defined in Eqn 19.5. The matrix $\mathbf{X'X}$ could be set up in the first round of iteration and stored in the memory for use in subsequent rounds of iterations.

The solution ($\hat{u}$) for the level j (animal j) of the random animal effect in the univariate animal model is calculated using Eqn 4.8, which can be rewritten (replacing n_3 by k) as:

$$\hat{u}_j = [n_1\alpha(\hat{u}_s + \hat{u}_d) + n_2 yd + \sum_o\{k_o\alpha(\hat{u}_o - 0.5(\hat{u}_{mo}))\}] / \text{diag}_j \qquad (19.8)$$

with:

$$\text{diag}_j = 2(n_1)\alpha + n_2 + \sum_o\{(k_o/2)\alpha\}$$

where $\hat{u}_s$, $\hat{u}_d$ and $\hat{u}_o$ are solutions or EBVs for the sire, dam and oth progeny of animal j, respectively; $\hat{u}_{mo}$ is the solution of the mate of animal j with respect to progeny o; yd is yield deviation, i.e. yield of animal j corrected for all other effects in the model; $n_1 = 1$ or

$^2/_3$ if both or one parent of animal j is known; n_2 is the number of records; k_o is 1 or $^2/_3$ if the other parent of progeny o (mate of animal j) is known or not known; and $\alpha = \sigma_e^2 / \sigma_a^2$.

In the multivariate animal model situation with equal design and random animal effect as the only random effect in addition to residual effects, the solutions for the levels of fixed effect and animal effects are obtained using Eqns 6.4 and 6.8, respectively, which are derived from the MME (Eqn 6.3).

For maternal animal model equations, the solutions for fixed effects could be calculated using Eqn 8.3. The equations for animal and genetic maternal effects are based on Eqn 8.4, given earlier. From Eqn 8.4, the solution ($\hat{u}$) for direct effect for animal i is:

$$\begin{aligned}\hat{u}_i &= [n_1\alpha_1(\hat{u}_s + \hat{u}_d) + n_1\alpha_2(\hat{m}_s + \hat{m}_d) - n_4\alpha_2(\hat{m}_i) - (k_o/2)\alpha_2(\hat{m}_i) \\ &\quad + n_2(y_i - b_j - \hat{m}_d - \hat{pe}_d) + \sum_o\{k_o\alpha_1(\hat{u}_o - 0.5(\hat{u}_{mo}))\} \\ &\quad + \sum_o\{k_o\alpha_2(\hat{m}_o - 0.5(\hat{m}_{mo}))\}] / \text{diag}_i\end{aligned} \tag{19.9}$$

with:

$$\text{diag}_i = 2(n_1)\alpha_1 + n_2 + \Sigma_o\{(k_o/2)\alpha_1\}$$

where $\hat{m}_i$, $\hat{m}_s$, $\hat{m}_d$, $\hat{m}_o$ and $\hat{m}_{mo}$ are solutions for genetic maternal effects for animal i, sire, dam, oth progeny of animal i and mate of animal i, respectively; y_i is the yield for animal i; b_j is the solution for fixed effect j; $\hat{pe}_d$ is the permanent environmental effect for the dam of animal i; n_1, n_2 and k_o are as defined above and $n_4 = 2(n_1)$; and α terms are as defined in Eqn 8.4.

The solution (m) for genetic maternal effect for animal i from Eqn 8.4 is:

$$\begin{aligned}m_i &= [n_1\alpha_2(\hat{u}_s + \hat{u}_d) + n_1\alpha_3(\hat{m}_s + \hat{m}_d) - n_4\alpha_2(\hat{u}_i) - (k_o/2)\alpha_2(\hat{u}_i) \\ &\quad + n_2(y_i - b_j - \hat{u}_i - \hat{m}_d - \hat{pe}_d) + \sum_o\{k_o\alpha_2(\hat{u}_o - 0.5(\hat{u}_{mo}))\} \\ &\quad + \sum_o\{k_o\alpha_3(\hat{m}_o - 0.5(\hat{m}_{mate}))\}] / \text{diag}_i\end{aligned} \tag{19.9}$$

with:

$$\text{diag}_i = 2(n_1)\alpha_3 + n_2 + \Sigma_o\{(k_o/2)\alpha_3\}$$

Solutions for permanent environmental effect are obtained using Eqn 8.5.

The computational procedure for a reduced animal model was presented by Schaeffer and Wilton (1987) using a bivariate analysis. The procedure is similar to the animal model described above except that records for non-parents are written twice, one record for each parent. Consequently, the residual variance of non-parental records (r_2) is multiplied by 2, that is:

$$r_2 = 2(\sigma_e^2 + d(\sigma_a^2)) = 2(1 + d\alpha^{-1})\sigma_e^2$$

where $d = \frac{1}{2}$ or $\frac{3}{4}$ if both or one parent is known and the contribution of non-parents' records to the diagonal of their parents is 0.5 instead of 0.25 (see Example 7.2).

The equations for solutions for levels of fixed and random effects are similar to those defined earlier. From Eqn 8.3, if the residual variance for parental records is defined as r_1, the contribution of parental records to the RHS for level i of a major fixed effect is:

$$\text{RHS}_i = \sum_{k=1}^{n_i}(r_1^{-1}(y_{ik} - \hat{w}_{kj})) \tag{19.11}$$

where n_i is the number of parental records in level i of fixed effect and $\hat{w}_{kj}$ is the solution for the jth level of other effects in the model affecting record k. The contribution of non-parental records to the RHS is included as:

$$\text{RHS}_i = \text{RHS}_i + \sum_{k=1}^{m_i}(r_{2k}^{-1}(y_{ki} - 0.5(\hat{u}_s + \hat{u}_d) - \hat{w}_{ki})) \tag{19.12}$$

where m_i is the number of non-parental records in level i of fixed effect, $\hat{u}_s$ and $\hat{u}_d$ are solutions for the sire and dam of the non-parent with record k, r_{2k}^{-1} is the inverse of the residual variance for the non-parental record k and $\hat{w}_{kj}$ is the solution for level j of other effects in the model apart from random animal effects affecting record k. Then:

$$b_i = \frac{\text{RHS}_i}{\sum_{k=1}^{n_i} r_1^{-1} + \sum_{j=1}^{m_i} r_2^{-1}}$$

The equation for the breeding value of the jth animal, which is a parent with its own yield record, a non-parental record from progeny i and information from another progeny (o), who is itself a parent, is:

$$\begin{aligned}\hat{u}_j \;=\; & [n_1\alpha(\hat{u}_s + \hat{u}_d) + n_2 r_1^{-1}(yd_j) + n_3 r_2^{-1}(yd_i - (0.5)\hat{u}_{mi}) \\ & + \sum_o \{k_o\alpha(\hat{u}_o - 0.5(\hat{u}_{mo}))\}] / \text{diag}_j\end{aligned} \tag{19.13}$$

with:

$$\text{diag}_j = 2(n_1)\alpha + n_2 r_1^{-1} + (0.5)n_3 r_2^{-1} + \sum_o\{(k_o/2)\alpha\}$$

where yd_j and yd_i are yield deviations for animal j and progeny i, which is a non-parent, $\hat{u}_{mi}$ is the breeding value for the mate of animal j with respect to the ith progeny (non-parent), n_2 is the number of observations (records) on animal j, n_3 is the number of non-parental records, r_1^{-1} and r_2^{-1} are as defined earlier and all other terms are as defined in Eqn 19.8. Note that contributions from the oth progeny in the above equation refer to those progeny of animal j who are themselves parents and that non-parental records are adjusted for half the breeding value of the mate of animal j. If animal j has no non-parental records from its progeny, Eqn 19.13 is the same as Eqn 19.8.

The principles of evaluation based on iterating on the data are illustrated below using a univariate animal model and a reduced animal model with maternal effects.

19.5.1 Animal model without groups

Example 19.3

Using the same data as in Example 4.1 (Table 4.1) on the weaning weight of beef calves, parameters and model, the principles of predicting breeding values and estimating solutions for fixed effects iterating on the data are illustrated using Gauss–Seidel iteration.

Data arrangement

Gauss–Seidel iteration requires the data files to be sorted by the effect to be solved for. The pedigree file is needed when solving for animal solutions. The pedigree file is created and ordered in such a manner that contributions to the diagonal and RHS of an animal from the pedigree due to the number of parents known (see type 1 record below) and from progeny accounting for whether mate is known (type 2 record), can be accumulated while processing the animal. Thus, initially, a pedigree file is created consisting of two types of records:

1. Type 1 record for all animals in the data comprising the animal identity, record type, and sire and dam identities;
2. Type 2 record for each parent in the data comprising the parent identity, record type, identities for progeny and other parent (mate) if known. The type 2 records are used to adjust the contribution of the progeny to each parent for the mate's breeding value when solving for animal solutions.

The pedigree file is sorted by animal and record type. The sorted pedigree file for the example data is given below.

Animal	Code	Sire or progeny	Dam or mate
1	1	0	0
1	2	4	0
1	2	6	2
2	1	0	0
2	2	5	3
2	2	6	1
3	1	0	0
3	2	5	2
3	2	8	6
4	1	1	0
4	2	7	5
5	1	3	2
5	2	7	4
6	1	1	2
6	2	8	3
7	1	4	5
8	1	3	6

Second, a data file is set up consisting of animal identity, fixed effects, covariates and traits. If there is a major fixed effect with many levels, two data files need to be set up, one sorted by the major fixed effects such as herd or HYS (file A), to be used when solving for the major fixed effect, and the other sorted by animal identity (file B), to be used to solve for animal solutions. Assuming sex effect to be the major fixed effect in the example data, the data sorted by sex are as follows:

Calf	Sex	Weaning weight gain (kg)
4	Male	4.5
7	Male	3.5
8	Male	5.0
5	Female	2.9
6	Female	3.9

Iteration stage

Let $\hat{\mathbf{b}}$ and $\hat{\mathbf{a}}$ be vectors of solutions for sex and animal effects. Starting values for sex and animal effect are assumed to be the same as in Example 19.1.

SOLVING FOR FIXED EFFECTS In each round of iteration, file A is read one level of sex effect at a time with adjusted right-hand sides (ARHS) and diagonals (DIAG) accumulated for the *i*th level as:

$$\begin{aligned} \text{ARHS}_i &= \text{ARHS}_i + y_{ik} - \hat{u}_k \\ \text{DIAG}_i &= \text{DIAG}_i + 1 \end{aligned}$$

At the end of the *i*th level, the solution for the level is computed as:

$$\hat{b}_i = \text{ARHS}_i \,/\, \text{DIAG}_i$$

The above step, essentially, involves adjusting the yields for animal effects using previous solutions and calculating solutions for each level of sex effect. For example, the solution for level 1 of sex effect in the first round of iteration is:

$$\hat{b}_1 = [(4.5 - 0.167) + (3.5 - (-0.833)) + (5.0 - 0.667)] / 3 = 4.333$$

After calculating solutions for fixed effect in the current round of iteration, file B and the pedigree file are processed to compute animal solutions.

SOLVING FOR ANIMAL SOLUTIONS DIAG and ARHS are accumulated as data for each animal and read from the pedigree file or from both the pedigree file and file B for animals with records. When processing type 1 records in the pedigree file for the *k*th animal, the contribution to the DIAG and ARHS according to the number of parents known is as follows:

Number of parents known		
None	One (sire (*s*))	Both
$\text{ARHS}_k = 0$	$\text{ARHS}_k = (\frac{2}{3})\,\alpha(\hat{u}_s)$	$\text{ARHS}_k = \alpha(\hat{u}_s + \hat{u}_d)$
$\text{DIAG}_k = \alpha$	$\text{DIAG}_k = (\frac{4}{3})\alpha$	$\text{DIAG}_k = 2\alpha$

where $\hat{u}_s$ and $\hat{u}_d$ are current solutions for the sire and dam, respectively.

When processing type 2 records in the pedigree file for the *k*th animal, the contribution to the DIAG and ARHS according to whether the mate of animal *k* is known or not is as follows:

Mate unknown	Mate known
$\text{ARHS}_k = \text{ARHS}_k + (\frac{2}{3})\ \alpha(\hat{u}_o)$	$\text{ARHS}_k = \text{ARHS}_k + \alpha(\hat{u}_s - 0.5\ \hat{u}_m)$
$\text{DIAG}_k = \text{DIAG}_k + (\frac{1}{3})\alpha$	$\text{DIAG}_k = \text{DIAG}_k + (\frac{1}{2})\alpha$

where $\hat{u}_o$ and $\hat{u}_m$ are current solutions for the progeny and mate, respectively, of the *k*th animal. If the *k*th animal has a yield record:

$$\begin{aligned} \text{ARHS}_k &= \text{ARHS}_k + y_{ik} - \hat{b}_i \\ \text{DIAG}_k &= \text{DIAG}_k + 1 \end{aligned}$$

where $\hat{b}_i$ are current solutions for level *i* of sex effect.

When all pedigree and yield records for the *k*th animal have been processed, the solution for the animal is computed as:

$$\hat{u}_k = \text{ARHS}_k / \text{DIAG}_k$$

For the example data, the solutions for animal 5 in the first round of iteration is computed as follows:

Contribution to diagonal from pedigree is:

$$\text{DIAG}_5 = (2 + 0.5)\alpha = 5.00$$

Accounting for yield record, diagonal becomes:

$$\text{DIAG}_5 = 5.00 + 1 = 6.00$$

Contribution to RHS from yield is:

$$\text{ARHS}_5 = 2.9 - 3.40 = -0.5$$

Contribution to RHS from parents and progeny (pedigree) is:

$$\begin{aligned} \text{ARHS}_5 &= \text{ARHS}_5 + \alpha(\hat{u}_2 + \hat{u}_3) + \alpha(\hat{u}_7 - 0.5(\hat{u}_4)) \\ &= -0.5 + 2(-0.083 + (-0.021)) + 2(-0.833 - 0.5(-0.119) \\ &= -2.255 \end{aligned}$$

and

$$\hat{u}_5 = -2.255 / 6.00 = -0.376$$

When all animals have been processed, the current round of iteration is completed. However, the iteration process is continued for sex and animal effects until convergence is achieved. The convergence criterion can be defined as in Section 19.4.1. In this example, solutions were said to have converged when the sum of squares of differences between the current and previous solutions divided by the sum of squares of the current solution was less than 10^{-7}. The solutions for all effects in the first round of iteration and at convergence at the 20th iteration are as follows:

	Solutions	
Effects	At round 1	At convergence
Sex		
Male	4.333	4.359
Female	3.400	3.404
Animal		
1	0.333	0.098
2	−0.083	−0.019
3	−0.021	−0.041
4	−0.119	−0.009
5	−0.376	−0.186
6	0.392	0.177
7	−0.364	−0.249
8	0.282	0.183

These solutions are the same as those obtained by direct inversion of the coefficient matrix in Section 4.2 or iterating on the coefficient matrix in Section 19.3. However, as stated earlier, the advantage of this method is that the MME are not set up and therefore memory requirement is minimal and can be applied to large data sets.

19.5.2 Animal model with groups

Example 19.4
With unknown parents assigned to phantom groups, the procedure is very similar to that described in Section 19.5.1, with no groups in the model except in the way the pedigree file is set up and animal solutions are computed. Using the same data, parameters and model as in Example 4.4, the methodology is illustrated below.

Data preparation

The pedigree file is set up as described in Section 19.5.1 with ancestors with unknown parentage assigned to groups. The assignment of unknown parents for the example pedigree has been described in Section 4.6. However, there is also an additional column for each animal indicating the number of unknown parents for each animal.

The pedigree with unknown parents assigned to groups and the additional column indicating the number of unknown parents is as follows:

Calf	Sire	Dam	Number of unknown parents
1	9	10	2
2	9	10	2
3	9	10	2
4	1	10	1
5	3	2	0
6	1	2	0
7	4	5	0
8	3	6	0

and the ordered pedigree for the analysis is:

Animal	Code	Sire or progeny	Dam or mate	Number of unknown parents
1	1	9	10	2
1	2	4	10	1
1	2	6	2	0
2	1	9	10	2
2	2	5	3	0
2	2	6	1	0
3	1	9	10	2
3	2	5	2	0

Continued

Continued.

Animal	Code	Sire or progeny	Dam or mate	Number of unknown parents
3	2	8	6	0
4	1	1	10	1
4	2	7	5	0
5	1	3	2	0
5	2	7	4	0
6	1	1	2	0
6	2	8	3	0
7	1	4	5	0
8	1	3	6	0
9	2	1	10	2
9	2	2	10	2
9	2	3	10	2
10	2	1	9	2
10	2	2	9	2
10	2	3	9	2
10	2	4	1	1

The arrangement of yield data is the same as in Section 19.5.1 in the animal model analysis without groups.

Iterative stage

SOLVING FOR FIXED EFFECTS This is exactly as described for the animal model without groups in Section 19.5.1, with yield records adjusted for other effects in the model and solutions for fixed effects computed.

SOLVING FOR ANIMAL SOLUTIONS Solutions for animals are computed one at a time as both pedigree and data file sorted by animals are read, as described for the animal model without groups. Therefore, only the differences in terms of the way diagonals and ARHSs are accumulated are outlined.

For the *k*th animal in the pedigree file, calculate:

$$w_k = \alpha(4/(2 + \text{no. of unknown parents}))$$

For the type 1 record in the pedigree file for the *k*th animal:

$$\text{ARHS}_k = \text{ARHS}_k + (\hat{u}_s + \hat{u}_d)0.5w_k$$

$$\text{DIAG}_k = \text{DIAG}_k + w_k$$

For the type 2 record in the pedigree file for the *k*th animal:

$$\text{ARHS}_k = \text{ARHS}_k + (\hat{u}_o - 0.5\hat{u}_m)0.5w_k$$

Accumulation of ARHSs from the data file is as specified in Section 19.5.1 in the model without groups.

The solution for the *k*th animal is computed as $\text{ARHS}_k/\text{DIAG}_k$ when all records for the animal in the pedigree and data file have been read. The solutions in the first round of iteration and at convergence without and with constraint on group solutions, as in Example 4.4, are as follows:

Effects	Solutions		
	At round 1	At convergence	At convergence[a]
Sex			
Male	4.333	4.509	5.474
Female	3.400	3.364	4.327
Animal			
1	0.333	0.182	−0.780
2	−0.083	0.026	−0.937
3	−0.021	−0.014	−0.977
4	−0.119	−0.319	−1.287
5	−0.376	−0.150	−1.113
6	0.392	0.221	−0.741
7	−0.364	−0.389	−1.355
8	0.282	0.181	−0.782
9	0.153	0.949	0.000
10	−0.176	−0.820	−1.795

[a]With solutions for groups constrained to those in Example 4.4

When the solutions for groups are constrained as those in Example 4.4, this method gives the same solutions. However, when there is no constraint on group solutions, the ranking of animals is the same and linear differences between levels of effects are more or less the same as when there is a constraint on group solutions.

19.5.3 Reduced animal model with maternal effects

Example 19.5

The principles of genetic evaluation iterating on the data with a reduced animal model with maternal effects are illustrated using the same data set, parameters and model as in Example 8.1. The genetic parameters were:

$$\text{var}\begin{bmatrix}\mathbf{a}\\ \mathbf{m}\\ \mathbf{p}\\ \mathbf{e}\end{bmatrix}=\begin{bmatrix}g_{11} & g_{12} & 0 & 0\\ g_{21} & g_{22} & 0 & 0\\ 0 & 0 & \sigma^2_{pe} & 0\\ 0 & 0 & 0 & \sigma^2_e\end{bmatrix}=\begin{bmatrix}150 & -40 & 0 & 0\\ -40 & 90 & 0 & 0\\ 0 & 0 & 40 & 0\\ 0 & 0 & 0 & 350\end{bmatrix}$$

and

$$\mathbf{G}^{-1}=\begin{bmatrix}g^{11} & g^{12}\\ g^{21} & g^{22}\end{bmatrix}=\begin{bmatrix}0.00756 & 0.00336\\ 0.00336 & 0.01261\end{bmatrix}$$

The inverse of the residual variance for parental records is $1/\sigma^2_e = r^{-1}_{pa} = 0.002857$ and for non-parental records is $1/(\sigma^2_e + dg_{11}) = r^{-1}_{np}$, where d = 3/4 or 1/2 when one or both parents are known and the inverse of the variance due to permanent environmental effect = $1/\sigma^2_{pe} = 0.025$.

Data arrangement

The pedigree file is set up as described in Section 19.5.1 but only for animals that are parents. The pedigree file for the example data is:

Animal	Code	Sire or progeny	Dam or mate
1	1	0	0
1	2	5	2
1	2	9	6
2	1	0	0
2	2	5	1
2	2	6	3
3	1	0	0
3	2	6	2
3	2	8	5
4	1	0	0
4	2	7	6
5	1	1	2
5	2	8	3
6	1	3	2
6	2	7	4
6	2	9	1
7	1	4	6
8	1	3	5
9	1	1	6

A data file is set up consisting of a code to identify parents and non-parents. For non-parents, one record is set up for each parent, comprising the parent, a code indicating it is a non-parent, the animal that has the yield record, the other parent (mate), the sire and dam of the animal with the yield record, fixed effects, covariates (if any) and traits. A single record is set up for parents, comprising the animal, a code indicating it is a parent, the animal again, a field set to zero corresponding to the column for the other parent in non-parents' records, the sire and dam of the animal, fixed effects, covariates (if any) and traits. The data file may be sorted in three sequences if there is a major fixed effect in the model: sorted by major fixed effect, such as HYS (file A); sorted by animal (file B); and sorted by dam code (file C). For the example, file A is:

Parent/ animal	Code[a]	Animal	Mate	Sire	Dam	Herd	Sex	Birth weight (kg)
5	0	5	0	1	2	1	Male	35.0
6	0	6	0	3	2	1	Female	20.0
7	0	7	0	4	6	1	Female	25.0
8	0	8	0	3	5	1	Male	40.0
9	0	9	0	1	6	2	Male	42.0
3	1	10	2	3	2	2	Female	22.0
2	1	10	3	3	2	2	Female	22.0
3	1	11	7	3	7	2	Female	35.0

Continued

Continued.

Parent/ animal	Code[a]	Animal	Mate	Sire	Dam	Herd	Sex	Birth weight (kg)
7	1	11	3	3	7	2	Female	35.0
8	1	12	7	8	7	3	Female	34.0
7	1	12	8	8	7	3	Female	34.0
9	1	13	2	9	2	3	Male	20.0
2	1	13	9	9	2	3	Male	20.0
3	1	14	6	3	6	3	Female	40.0
6	1	14	3	3	6	3	Female	40.0

[a]0, parental record; 1, non-parental record

Iteration stage

The solution vectors for herd $(\hat{\mathbf{hd}})$, sex $(\hat{\mathbf{b}})$, direct animal effect $(\hat{\mathbf{u}})$, genetic maternal effect $(\hat{\mathbf{m}})$ and permanent environmental effect $(\hat{\mathbf{pe}})$ are initially set to zero.

SOLVING FOR FIXED EFFECTS Data file A is read at each round of iteration, one herd at a time, with ARHS and DIAG accumulated for the ith herd as:

$$\text{ARHS}_i = \text{ARHS}_i + r_{pa}^{-1}(y_{ijklt} - \hat{b}_j - \hat{u}_k - \hat{m}_l - \hat{pe}_t)$$

for parental records (Eqn 19.11):

$$\text{ARHS}_i = \text{ARHS}_i + r_{np}^{-1}(y_{ijklt} - \hat{b}_j - 0.5(\hat{u}_s + \hat{u}_d) - \hat{m}_l - \hat{pe}_t)$$

for non-parent records (Eqn 19.12):

$$\text{DIAG}_i = \text{DIAG}_i + r_n^{-1}$$

where r_n^{-1} is the inverse of the residual variance of the nth record being read.

At the end of records for the ith herd, the solution is computed as:

$$\hat{hd}_i = \text{ARHS}_i / \text{DIAG}_i$$

In the first round of iteration, the solution for the first herd is:

$$\begin{aligned}
\hat{hd}_1 &= [r_{pa}^{-1}(y_1 - \hat{b}_1 - \hat{u}_5 - \hat{m}_2 - \hat{pe}_2) + (y_2 - \hat{b}_2 - \hat{u}_6 - \hat{m}_2 - \hat{pe}_2) \\
&\quad +(y_3 - \hat{b}_2 - \hat{u}_7 - \hat{m}_6 - \hat{pe}_6) + (y_4 - \hat{b}_1 - \hat{u}_8 - \hat{m}_5 - \hat{pe}_5)] / 4(r_{pa}^{-1}) \\
&= [r_{pa}^{-1}((35-0-0-0-0) + (20-0-0-0-0) + (25-0-0-0-0) \\
&\quad +(40-0-0-0-0)] / 4(r_{pa}^{-1}) \\
&= 0.3432 / 0.01144 = 30.00
\end{aligned}$$

While reading data file A, ARHSs consisting of yield adjusted for previous animal, maternal and permanent environmental solutions are accumulated for each level of sex effect. Thus, for the jth level of sex effect:

$$\text{ARHS}_j = \text{ARHS}_j + r_{pa}^{-1}(y_{ijklt} - \hat{u}_k - \hat{m}_l - \hat{pe}_t)$$

for parent records:

$$\text{ARHS}_j = \text{ARHS}_j + r_{np}^{-1}(y_{ijklt} - 0.5(\hat{u}_s + \hat{u}_d) - \hat{m}_l - \hat{pe}_t)$$

and for non-parent records:

$$\text{DIAG}_j = \text{DIAG}_j + r_n^{-1}$$

After reading file A, the solution for the *j* sex class is computed as:

$$\text{ARHS}_j = \text{ARHS}_j - nr_{ij}^{-1}\hat{hd}_i$$

$$\hat{b}_j = \text{ARHS}_j / \text{DIAG}_j$$

where $\hat{hd}_i$ is the current solution of herd *i* and nr_{ij}^{-1} is the sum of the inverse of the residual variance for records of the *j*th level of sex effect in herd *i*. The latter is accumulated while reading file A. For the example data, solutions for sex effect in the first round of iteration are:

$$\begin{aligned}\hat{b}_1 &= \text{ARHS}_1 - 2r_{pa}^{-1}(\hat{hd}_1) - r_{pa}^{-1}(\hat{hd}_2) - 2r_{np}^{-1}(\hat{hd}_3) / [3r_{pa}^{-1} + 2r_{np}^{-1}] \\ &= (0.38134 - 2r_{pa}^{-1}(30.0) - r_{pa}^{-1}(33.638) - 2r_{np}^{-1}(31.333) / 0.01092 \\ &= 3.679\end{aligned}$$

After obtaining solutions for fixed effects in the current round of iteration, the solutions for animals are solved for.

SOLVING FOR ANIMAL SOLUTIONS As described in Section 19.5.1, animal solutions are computed one at a time as the pedigree file and file B are read. Briefly, for a type 1 record in the pedigree file for the *k*th animal, contributions to DIAG and ARHS according to the number of parents known (Eqn 19.9) are:

Number of parents known		
None	One (sire (*s*))	Both
$\text{ARHS}_k = 0$	$\text{ARHS}_k = \frac{2}{3}g^{11}(\hat{u}_s)$	$\text{ARHS}_k = g^{11}(\hat{u}_s + \hat{u}_d)$
$\text{DIAG}_k = g^{11}$	$\text{DIAG}_k = \frac{4}{3}g^{11}$	$\text{DIAG}_k = 2g^{11}$

where $\hat{u}_s$ and $\hat{u}_d$ are current solutions for direct effects for the sire and dam of the animal *k*.

The ARHS is augmented by contributions from the maternal effect as a result of the genetic correlation between animal and maternal effects. These contributions are from the sire, dam and the *k*th animal (see Eqn 19.10) and these are:

Number of parents known		
None	One (sire (*s*))	Both
–	$\text{ARHS}_k = \text{ARHS}_k + (\hat{m}_s)\frac{2}{3}g^{12}$	$\text{ARHS}_k = \text{ARHS}_k + (\hat{m}_s + \hat{m}_d)g^{12}$
$\text{ARHS}_k = \text{ARHS}_k - (\hat{m}_k)g^{12}$	$\text{ARHS}_k = \text{ARHS}_k - (\hat{m}_k)\frac{4}{3}g^{12}$	$\text{ARHS}_k = \text{ARHS}_k - (\hat{m}_k)2g^{12}$

where $\hat{m}_s$, $\hat{m}_d$ and $\hat{m}_k$ are current maternal solutions for the sire and dam of animal *k*, respectively.

In processing a type 2 record in the pedigree file for the kth animal, contributions to DIAG and ARHS according to whether the mate of k is known are:

Mate is unknown	Mate is known
$\text{ARHS}_k = \text{ARHS}_k + \frac{2}{3} g^{11} (\hat{u}_o)$	$\text{ARHS}_k = \text{ARHS}_k + (\hat{u}_o - 0.5\ \hat{u}_{ma})g^{11}$
$\text{DIAG}_k = \text{DIAG}_k + \frac{1}{3} g^{11}$	$\text{DIAG}_k = \text{DIAG}_k + \frac{1}{2} g^{11}$

where $\hat{u}_o$ and $\hat{u}_{ma}$ are current solutions for direct effects for the progeny and mate of the animal k.

Accounting for contributions from the maternal effect to ARHS:

Mate is unknown	Mate is known
$\text{ARHS}_k = \text{ARHS}_k + \frac{2}{3} g^{12} (\hat{m}_o)$	$\text{ARHS}_k = \text{ARHS}_k + (\hat{m}_o) - 0.5\ \hat{m}_{ma})g^{12}$
$\text{ARHS}_k = \text{ARHS}_k - (\hat{m}_k) 1/3g^{12}$	$\text{ARHS}_k = \text{ARHS}_k - (\hat{m}_k) \text{S} g^{12}$

where $\hat{m}_o$ and $\hat{m}_{ma}$ are current maternal solutions for the progeny and mate of the animal k.

If the animal has a yield record:

$$\text{DIAG}_k = \text{DIAG}_k + r_n^{-1} \text{ if it is a parent}$$

or:

$$\text{DIAG}_k = \text{DIAG}_k + (r_n^{-1})0.5 \text{ if it is a non} - \text{parent}$$

The diagonals of non-parents are multiplied by 0.5 instead of 0.25 because records of non-parents have been written twice (see Section 19.5).

Contributions to the RHS are accumulated as:

$$\text{ARHS}_k = \text{ARHS}_k + r_{pa}^{-1}(y_{ijklt} - \hat{hd}_i - \hat{b}_j - \hat{m}_l - \hat{pe}_t) \text{ for parent records}$$

and

$$\text{ARHS}_k = \text{ARHS}_k + r_{np}^{-1}(y_{ijklt} - \hat{hd}_i - \hat{b}_j - 0.5(\hat{u}_{ma}) - \hat{m}_l - \hat{pe}_t \text{ for non-parent records.}$$

In the equations above, $\hat{hd}_i$, $\hat{b}_j$, $\hat{m}_l$, $\hat{pe}_t$ and $\hat{u}_{ma}$ are current solutions for herd i, jth level for sex effect, lth maternal effect level, tth level of permanent environment effect and animal solution for the other parent (mate), respectively. The solution for animal k is computed as usual when all records for the animal in the pedigree and data file have been read as:

$$\hat{u}_k = \text{ARHS}_k / \text{DIAG}_k$$

The solution for animal 2 in the example data in the first round of iteration is as follows:

The contribution to the diagonal from pedigree is:

$$\text{DIAG}_2 = (1 + \frac{1}{2} + \frac{1}{2})0.00756 = 0.01512$$

The contribution to the diagonal from yield is:

$$\text{DIAG}_2 = \text{DIAG}_2 + 2(0.00059) = 0.01512 + 0.00118 = 0.0163$$

The contribution to the ARHS from the pedigree is zero since both parents are unknown and solutions for progeny are zero in the first round of iteration. The contribution to ARHS from yield record is:

$$\text{ARHS}_2 = r_{np}^{-1}(y_{10} - \hat{h}d_2 - \hat{b}_2 - \hat{u}_3 - \hat{m}_3 - \hat{p}e_3)$$
$$+r_{np}^{-1}(y_{13} - \hat{h}d_3 - \hat{b}_1 - \hat{u}_9 - \hat{m}_2 - \hat{p}e_2)$$
$$\text{ARHS}_2 = r_{np}^{-1}(22 - (-2.567) - 33.600 - 0 - 0 - 0)$$
$$+r_{np}^{-1}(20 - 3.679 - 31.333) = -0.02818$$

Therefore:

$$\hat{u}_2 = -0.02818 / 0.0163 = -1.729$$

After processing all animals in the pedigree and data file in the current round of iteration, equations for maternal effects are set and solved as described below.

SOLUTIONS FOR MATERNAL EFFECT Solutions for maternal effects are computed using both the pedigree file and the data file sorted by dam. Records for the *l*th animal are read in from the pedigree file and from file C if it is a dam that has progeny with a yield record, while accumulating DIAG and ARHS. For the type 1 record in the pedigree file for animal *l*, contributions to ARHS and DIAG according to the number of parents known are as follows:

Number of parents known		
None	One (dam(*d*))	Both
$\text{ARHS}_l = 0$	$\text{ARHS}_l = \frac{2}{3} g^{22}(\hat{m}_d)$	$\text{ARHS}_l = g^{22}(\hat{m}_s + \hat{m}_d)$
$\text{DIAG}_l = g^{22}$	$\text{DIAG}_l = \frac{4}{3} g^{22}$	$\text{DIAG}_l = 2g^{22}$

Taking into account contributions from animal effects to the ARHS due to genetic correlation gives:

Number of parents known		
None	One (dam(*d*))	Both
–	$\text{ARHS}_l = \text{ARHS}_l + (\hat{u}_d)\frac{2}{3} g^{12}$	$\text{ARHS}_l = \text{ARHS}_l + (\hat{u}_s + \hat{u}_d)g^{12}$
$\text{ARHS}_l = \text{ARHS}_l - (\hat{u}_l)g^{12}$	$\text{ARHS}_l = \text{ARHS}_l - (\hat{u}_l)\frac{4}{3} g^{12}$	$\text{ARHS}_l = \text{ARHS}_l - (\hat{u}_l)2g^{12}$

For the type 2 record in the pedigree file for animal *l*, contributions to the ARHS and DIAG according to whether the mate of animal *l* is known or not are:

Mate is unknown	Mate is known
$\text{ARHS}_l = \text{ARHS}_l + (\frac{2}{3})g^{22}(\hat{m}_o)$	$\text{ARHS}_l = \text{ARHS}_l + g^{22}(\hat{m}_o - 0.5\hat{m}_{ma})$
$\text{DIAG}_l = \text{DIAG}_l + (\frac{1}{3})g^{22}$	$\text{DIAG}_l = \text{DIAG}_l + (1/2)g^{22}$

Taking into account contributions from animal effect (see Eqn 19.7) gives:

Mate is unknown	Mate is known
$\text{ARHS}_l = \text{ARHS}_l + (\frac{2}{3})g^{12}(\hat{u}_o)$	$\text{ARHS}_l = \text{ARHS}_l + (\hat{u}_o - 0.5\hat{u}_{ma})g^{12}$
$\text{ARHS}_l = \text{ARHS}_l - (\hat{u}_l)(\frac{1}{3})g^{12}$	$\text{ARHS}_l = \text{ARHS}_l - (\hat{u}_l)(\frac{1}{2})g^{12}$

For the animal l, which is a dam with progeny having yield records, DIAG and ARHS from the pedigree is augmented with information from yield as:

$$\text{DIAG} = \text{DIAG} + r_n^{-1}$$

and

$$\text{ARHS}_l = \text{ARHS}_l + r_{pa}^{-1}(y_{ijklt} - \hat{h}d_i - \hat{b}_j - \hat{u}_k - \hat{p}e_t) \text{ for parent records}$$

and

$$\text{ARHS}_l = \text{ARHS}_l + r_{np}^{-1}(y_{ijklt} - \hat{h}d_i - \hat{b}_j - 0.5(\hat{u}_s + \hat{u}_d) - \hat{p}e_t) \text{ for non-parent records}$$

After processing all records from pedigree and yield records for the lth animal, the solution for the maternal effect is computed as:

$$\hat{m}_l = \text{ARHS}_l / \text{DIAG}_l$$

The calculation of the solution for animal 5 in the first round of iteration is as follows. The contribution from a type 1 record in the pedigree is:

$$\begin{aligned}\text{ARHS}_5 &= (\hat{m}_1 + \hat{m}_2)g^{22} + (\hat{u}_1 + \hat{u}_2)g^{12} - (\hat{u}_5 2g^{12})\\ &= (0.0217 + -1.7027)0.01261 + (0 + (-1.7294))0.00336 - ((-0.5831)(2)0.0336)\\ &= -0.02309\\ \text{DIAG}_5 &= (2)0.01261 = 0.02522\end{aligned}$$

The contribution from a type 2 record in the pedigree is:

$$\begin{aligned}\text{ARHS}_5 &= \text{ARHS}_5 + (\hat{m}_8 - \frac{1}{2}\hat{m}_3)g^{22} + (\hat{u}_8 - \frac{1}{2}\hat{u}_3)g^{12} - (\hat{u}_5 \frac{1}{2} g^{12})\\ &= -0.02309 + (0 - \frac{1}{2}(0.4587))0.01261\\ &\quad +(1.4382 - \frac{1}{2}(0.8960))0.00336 - ((-0.5831)(\frac{1}{2})0.00336)\\ &= -0.021675\\ \text{DIAG}_5 &= \text{DIAG}_5 + \frac{1}{2}g^{22} = 0.02522 + 0.0063 = 0.03153\end{aligned}$$

The contribution from yield of progeny (animal 8) for dam 5 is:

$$\begin{aligned}\text{ARHS}_5 &= \text{ARHS}_5 + r_{pa}^{-1}(\hat{y}_8 - \hat{h}d_1 - \hat{b}_1 - \hat{u}_8 - \hat{p}e_5)\\ &= -0.021675 + r_{pa}^{-1}(40 - 30.00 - 3.679 - (1.4382) - 0)\\ &= -0.007724\\ \text{DIAG}_5 &= \text{DIAG}_5 + r_{pa}^{-1} = 0.03153 + 0.002857 = 0.034387\end{aligned}$$

and the solution is:

$$\hat{m}_5 = -0.007724 / 0.034387 = -0.225$$

Solutions for permanent environmental effects are solved for after processing all animals for maternal effects in the current round of iteration.

SOLVING FOR PERMANENT ENVIRONMENTAL (pe) EFFECTS Only the data file sorted by dams is required to obtain solutions for pe effects.

The records for the t th dam are read from file C while ARHS and DIAG are accumulated as:

$$\text{ARHS}_t = \text{ARHS}_t + r_{pa}^{-1}(y_{ijklt} - \hat{hd}_i - \hat{b}_j - \hat{u}_k - \hat{m}_l) \text{ for parent records}$$

and:

$$\text{ARHS}_t = \text{ARHS}_t + r_{np}^{-1}(y_{ijklt} - \hat{hd}_i - \hat{b}_j - 0.5(\hat{u}_s + \hat{u}_d) - \hat{m}_l) \text{for non-parent records.}$$
$$\text{DIAG}_t = \text{DIAG}_t + r_n^{-1}$$

At the end of records for the tth dam, solutions are computed as:

$$\hat{pe}_t = \text{ARHS}_t / (\text{DIAG}_t + 1/\sigma_p^2)$$

The solution for permanent environmental effect for animal 5 in the first round of iteration is:

$$\begin{aligned}\text{ARHS}_5 &= r_{pa}^{-1}(y_8 - \hat{hd}_1 - \hat{b}_1 - \hat{u}_8 - \hat{m}_5)\\ &= r_{pa}^{-1}(40 - 3.679 - 30.0 - 1.4822 - (-0.2246))\\ &= 0.01459\\ \text{DIAG}_5 &= r_{pa}^{-1} + 0.025 = 0.02786\end{aligned}$$

and:

$$\hat{pe}_5 = 0.01459 / 0.02786 = 0.524$$

Further iterations are carried out until convergence is achieved. The convergence criteria defined in Section 19.4.1 could also be used. The solutions for the first round of iteration and at convergence are shown below.

	Solutions	
Effects	At round 1	At convergence
Herd		
1	30.000	30.563
2	33.600	33.950
3	31.333	31.997
Sex of calf		
Male	3.679	3.977
Female	−2.657	−2.872
Animal		
1	0.000	0.564
2	−1.729	−1.246
3	0.896	1.166
4	0.000	−0.484

Continued

Continued.

Effects	Solutions	
	At round 1	At convergence
5	−0.583	0.630
6	−0.554	−0.859
7	−0.020	−1.156
8	1.438	1.918
9	−0.396	−0.553
Maternal		
1	0.022	0.261
2	−1.703	−1.582
3	0.459	0.735
4	0.046	0.586
5	−0.225	−0.507
6	0.425	0.841
7	0.788	1.299
8	−0.224	−0.158
9	0.255	0.659
Permanent environment		
2	−1.386	−1.701
5	0.524	0.415
6	0.931	0.825
7	0.527	0.461

These solutions are exactly the same obtained as those obtained in Section 8.3 by directly inverting the coefficient matrix.

BACK-SOLVING FOR NON-PARENTS The solutions for direct animal and maternal effects for non-parents are calculated after convergence has been achieved, as described in Section 8.3. The solutions for non-parents for this example have been calculated in Section 8.3.

19.6 Preconditioned Conjugate Gradient Algorithm

Berger *et al.* (1989) investigated the use of the plain or Jacobi conjugate gradient iterative scheme for solving MME for the prediction of sire breeding values. They indicated that plain conjugate gradient was superior to a number of other iterative schemes, including Gauss–Seidel. Strandén and Lidauer (1999) implemented the use of the preconditioned conjugate gradient (PCG) in genetic evaluation models for the routine evaluation of dairy cattle with very large data. In the PCG method, the linear systems of equations (Eqn 19.2, for instance) is made simpler by solving an equivalent system of equations:

$$\mathbf{M}^{-1}\mathbf{C}\mathbf{b} = \mathbf{M}^{-1}\mathbf{r}$$

where **M** is a symmetric, positive definite, preconditioner matrix that approximates **C** and **r** is the right-hand side. In the plain conjugate gradient method, the preconditioner **M** is an identity matrix.

The implementation of the PCG method requires storing four vectors of size equal to the number of unknowns in the MME: a vector of residuals (**e**), a search direction vector (**d**), a solution vector (**b**) and a work vector (**v**). The PCG method can be implemented with less memory by storing the solution vector on disk and reading it in during the iteration. The pseudo-code for the PCG method (Lidauer *et al.*, 1999) is outlined below, assuming that starting values are:

$$\mathbf{b}^{(0)} = 0, \quad \mathbf{e}^{(0)} = \mathbf{r} - \mathbf{Cb} = \mathbf{r}, \quad \mathbf{d}^{(0)} = \mathbf{M}^{-1}\mathbf{e}^{(0)} = \mathbf{M}^{-1}\mathbf{r}^{(0)}$$

For $k = 1,2, \ldots, n$

$$\mathbf{v} = \mathbf{C}\mathbf{d}(k-1)$$

$$\omega = \mathbf{e}'^{(k-1)}\mathbf{M}^{-1}\mathbf{e}^{(k-1)} / (\mathbf{d}'^{(k-1)}\mathbf{v})$$

$$\mathbf{b}^{(k)} = \mathbf{b}^{(k-1)} + \omega\mathbf{d}^{(k-1)}$$

$$\mathbf{e}^{(k)} = \mathbf{e}^{(k-1)} - \omega\mathbf{v}$$

$$\mathbf{v} = \mathbf{M}^{-1}\mathbf{e}^{(k)}$$

$$\beta = \mathbf{e}'^{(k)}\mathbf{v} / (\mathbf{e}'^{(k-1)}\mathbf{M}^{-1}\mathbf{e}^{(k-1)})$$

$$\mathbf{d}^{(k)} = \mathbf{v} + \beta\mathbf{d}^{(k-1)}$$

If not converged, continue iteration until converged, and $\boldsymbol{\omega}$ and $\boldsymbol{\beta}$ are step sizes in the PCG method.

19.6.1 Computation strategy

The major task in the PCG algorithm above is calculating **Cd**, where **C** is the coefficient matrix of the MME. The vector **d** is the search direction vector and every iteration of the PCG minimizes the distance between the current and the true solutions in the search direction. Strandén and Lidauer (1999) presented an efficient computation strategy for computing **Cd** for a multivariate model. Assuming, for instance, that data are ordered by animals, the MME for the multivariate model (Eqn 5.2) can be written as:

$$\begin{pmatrix} \sum_{i=1}^{N}\mathbf{x}_i\mathbf{R}_i^{-1}\mathbf{x}_i' & \sum_{i=1}^{N}\mathbf{x}_i\mathbf{R}_i^{-1}\mathbf{z}_i' \\ \sum_{i=1}^{N}\mathbf{z}_i\mathbf{R}_i^{-1}\mathbf{x}_i' & \sum_{i=1}^{N}\mathbf{z}_i\mathbf{R}_i^{-1}\mathbf{z}' + \mathbf{A}^{-1}\otimes\mathbf{G}^{-1} \end{pmatrix}\begin{pmatrix}\hat{\mathbf{b}} \\ \hat{\mathbf{a}}\end{pmatrix} = \begin{pmatrix} \sum_{i=1}^{N}\mathbf{x}_i\mathbf{R}_i^{-1}\mathbf{y}_i \\ \sum_{i=1}^{N}\mathbf{z}_i\mathbf{R}_i^{-1}\mathbf{y}_i \end{pmatrix}$$

where N is the number of animals with records, $\mathbf{x}_i'$ and $\mathbf{z}_i'$ are matrices having rows with l_i equal to the number of traits observed on animal i. Denote $\mathbf{w}' = [\mathbf{x}_i' \ \mathbf{z}_i']$ and **V** as:

$$\mathbf{V} = \begin{pmatrix} 0 & 0 \\ 0 & \mathbf{A}\otimes\mathbf{G} \end{pmatrix}$$

Computing **Cd** then implies calculating:

$$\sum_{i=1}^{N}\mathbf{w}_i\ \mathbf{R}_i^{-1}\mathbf{w}_i'\ \mathbf{d} + \mathbf{V}^{-1}\mathbf{d} = \sum_{i=1}^{N}\mathbf{v}_i + \mathbf{v}_d \qquad (19.14)$$

If solving the MME with iteration on the data for a univariate model without any regression effects, this calculation can be achieved by accumulating for each individual i, the product $\mathbf{v}_i = \mathbf{T}_i\mathbf{d}$, where the coefficients in $\mathbf{T}_i = \mathbf{w}_i\mathbf{R}_i^{-1}\mathbf{w}_i'$ can be deduced without performing any of the products, as **w***i* contains zeros and ones only and $\mathbf{R}_i^{-1}$ is a scalar or $\mathbf{R}_i^{-1}$ is factored out (Eqn 4.4). For a multivariate model, the principles for computing $\mathbf{T}_i$ are, essentially, the same but with scalar contributions replaced by matrix $\mathbf{R}_i$. Strandén and Lidauer (1999) suggested the following three-step method for calculating the product $\mathbf{w}_i\mathbf{R}_i^{-1}\mathbf{d}$:

$$\mathbf{s}_i \leftarrow \mathbf{w}_i'\mathbf{d}; \quad \mathbf{s}_i^* \leftarrow \mathbf{R}_i^{-1}\mathbf{s}_i; \quad \mathbf{v}_i \leftarrow \mathbf{w}_i\mathbf{s}_i^*$$

where vectors $\mathbf{s}_i$ and $\mathbf{s}_i^*$ are of size equal to the number of traits observed on individual i (l_i). They demonstrated that this three-step approach reduced substantially the number of floating-point operations (multiplications) compared with a multivariate accumulation technique as used by Groeneveld and Kovac (1990). For instance, given that q_i is the number of effects over traits observed for individual i, the number of floating-point operations were 720 with $l_i = 3$ and $q_i = 15$ using the multivariate accumulation technique compared with 78 with the three-point approach. They also suggested that $\mathbf{v}_d = \mathbf{V}^{-1}\mathbf{d}$ in Eqn 19.14 can be evaluated in a two-step approach:

$$\mathbf{x} \leftarrow (\mathbf{I} \otimes \mathbf{A}^{-1})\mathbf{d}; \quad \mathbf{v}_d \leftarrow (\mathbf{G}^{-1} \otimes \mathbf{I})\mathbf{x}$$

19.6.2 Numerical application

Example 19.6

The application of PCG to solve MME is illustrated using data for Example 4.1 for a univariate model and iterating on the data.

Computing starting values

Initially, the pedigree is read and diagonal elements of $\mathbf{A}^{-1}$ multiplied by a are accumulated for animals, where the variance ratio α is 2, as in Example 4.1. This is straightforward and has not been illustrated, but elements for animals 1–8 stored in a vector $\mathbf{h}$ are:

$$\mathbf{h}' = [3.667 \quad 4.0 \quad 4.0 \quad 3.667 \quad 5.0 \quad 5.0 \quad 4.0 \quad 4.0]$$

Second, read through the data as shown in Table 4.1 and accumulate right-hand side ($\mathbf{r}$) for all effects, diagonals for the levels of sex of calf effect and add contribution of information from data to diagonals from $\mathbf{A}^{-1}\alpha$ for animals. Assuming that diagonals for all effects are stored as diagonal elements of $\mathbf{M}$, such that the first two elements are for the two levels of sex of calf effect and the remaining elements for animals 1–8, then $\mathbf{r}$ and $\mathbf{M}$ are:

$$\mathbf{r}' = [13.0 \quad 6.8 \quad 0.0 \quad 0.0 \quad 0.0 \quad 4.5 \quad 2.9 \quad 3.9 \quad 3.5 \quad 5.0]$$

and

$$\mathbf{M} = \text{diag}[3.0 \quad 2.0 \quad 3.667 \quad 4.0 \quad 4.0 \quad 4.667 \quad 6.0 \quad 6.0 \quad 5.0 \quad 5.0]$$

The starting values for PCG can now be calculated. Thus:

$$\mathbf{b}^{(0)} = 0, \quad \mathbf{e}^{(0)} = \mathbf{r} - \mathbf{C}\mathbf{b}^{(0)} = \mathbf{y} \quad \text{and} \quad \mathbf{d}^{(0)} = \mathbf{M}^{-1}\mathbf{r}$$

Thus:

$$\mathbf{d}^{(0)'} = [4.333 \quad 3.4 \quad 0.0 \quad 0.0 \quad 0.0 \quad 0.964 \quad 0.483 \quad 0.650 \quad 0.70 \quad 1.0]$$

Iterative stage

Reading through the data and performing the following calculations in each round of iteration, start the PCG iterative process. Calculations are shown for the first round of iteration.

The vector $\mathbf{v} = \mathbf{Cd}$ is accumulated as data are read. For the ith level of fixed effect:

$$\mathbf{v}(i) = \mathbf{v}(i) + 1(\mathbf{d}(i)) + 1(\mathbf{d}(\text{anim}_k))$$

where anim_k refers to the animal k associated with the record. Thus, for the level 1 of sex of calf effect:

$$\mathbf{v}(1) = 3(4.333) + \mathbf{d}(\text{anim}_4) + \mathbf{d}(\text{anim}_7) + \mathbf{d}(\text{anim}_8) = 15.663$$

As each record is read, calculate:

$$z = 4\,/\,(2 + \text{ number of unknown parents for animal with record})$$

$$xx = -0.5(z)\alpha \text{ if either parent is known, otherwise } xx = 0$$

$$xm = 0.25(z)\alpha \text{ if both parents are known, otherwise } xm = 0$$

If only one parent, p, of animal k is known, then accumulate:

$$\mathbf{v}(\text{anim}_k) = \mathbf{v}(\text{anim}_k) + 1(\mathbf{d}(i)) + \mathbf{M}_{k,k}(\mathbf{d}(\text{anim}_k)) + xx(\mathbf{d}(\text{anim}_p)) \tag{19.15}$$

where $\mathbf{d}(i)$ refers to the ith level of the fixed effect and $\mathbf{M}_{k,k}$ the diagonal element of $\mathbf{M}$ for animal k.

Accumulate the contribution to the known parent, p, of k at the same time:

$$\mathbf{v}(\text{anim}_p) = \mathbf{v}(\text{anim}_p) + xx(\mathbf{d}(\text{anim}_k))$$

If both parents p and j of animal k are known, then accumulate for animal k as:

$$\mathbf{v}(\text{anim}_k) = \mathbf{v}(\text{anim}_k) + 1(\mathbf{d}(i)) + \mathbf{M}_{k,k}(\mathbf{d}(\text{anim}_k)) + xx(\mathbf{d}(\text{anim}_p) + \mathbf{d}(\text{anim}_j)) \tag{19.15}$$

Accumulate for both parents as:

$$\mathbf{v}(\text{anim}_p) = \mathbf{v}(\text{anim}_p) + xx(\mathbf{d}(\text{anim}_k))$$
$$\mathbf{v}(\text{anim}_p) = \mathbf{v}(\text{anim}_p) + xm(\mathbf{d}(\text{anim}_j))$$
$$\mathbf{v}(\text{anim}_j) = \mathbf{v}(\text{anim}_j) + xx(\mathbf{d}(\text{anim}_k))$$
$$\mathbf{v}(\text{anim}_j) = \mathbf{v}(\text{anim}_j) + xm(\mathbf{d}(\text{anim}_p))$$

After processing all animals with records, the contribution for animals in the pedigree without records is accumulated. The equations for accumulating contributions for these animals is the same as shown above except that the coefficient for $\mathbf{d}(i)$ in Eqns 19.14 and 19.15 is zero instead of one, indicating no contribution from records.

For example, for animal 4 with only the sire known:

$$\mathbf{v}(4) = \mathbf{v}(4) + \mathbf{d}(1) + \mathbf{M}_{4,4}(\mathbf{d}(\text{anim}_4)) + (-2\,/\,3)\alpha(\mathbf{d}(\text{anim}_1)) = 8.833$$

Add contribution from progeny when processing the record for animal 7:

$$\mathbf{v}(4) = 8.833 + -1.0\alpha(\mathbf{d}(\text{anim}_7)) + 0.25\alpha(\mathbf{d}(\text{anim}_5)) = 7.917$$

The vector **v** for all effects is:

$$\mathbf{v}' = [15.664 \quad 7.933 \quad -2.586 \quad -2.267 \quad -2.317 \quad 7.917 \quad 5.864 \quad 5.300 \quad 4.938 \quad 8.033]$$

Next ω is computed using matrix multiplication and scalar division as:

$$\omega = 95.1793 / 120.255 = 0.7915$$

The solution vector is then computed as $\mathbf{b}^{(1)} = \mathbf{b}^{(0)} + \omega\mathbf{d}^{(0)}$. The vector $\mathbf{b}^{(1)}$ is:

$$\mathbf{b}'^{(1)} = [3.430 \quad 2.691 \quad 0.0 \quad 0.0 \quad 0.0 \quad 0.763 \quad 0.383 \quad 0.514 \quad 0.554 \quad 0.791]$$

The updated vector of residuals $\mathbf{e}^{(1)}$ is computed as $\mathbf{e}^{(0)} - \omega\mathbf{v}$. For the example data $\mathbf{e}^{(1)}$ is:

$$\mathbf{e}'^{(1)} = [0.602 \; 0.521 \; 2.047 \; 1.794 \; 1.834 \; -1.766 \; -1.741 \; -0.295 \; -0.408 \; -1.358]$$

The vector **v** is then computed as $\mathbf{M}^{-1}\mathbf{e}(1)$. For the example data, **v** is:

$$\mathbf{v}' = [0.201 \quad 0.260 \quad 0.558 \; 0.449 \; 0.458 \; -0.378 \; -0.290 \; -0.049 \; -0.082 \; -0.272]$$

Next, compute the scalar β. The denominator of β is equal to the numerator of ω and this has already been computed. Using the example data:

$$\beta = 4.634 / 95.179 = 0.0487$$

Finally, $\mathbf{d}^{(1)}$, the search-direction vector for the next iteration is computed as $\mathbf{v} + \beta\mathbf{d}^{(0)}$. This vector for the example data is:

$$\mathbf{d}'^{(1)} = [0.412 \; 0.426 \; 0.558 \; 0.449 \; 0.458 - 0.331 - 0.267 - 0.017 - 0.048 - 0.223]$$

The next cycle of iteration is continued until the system of equations converges. Convergence can either be monitored using the criteria defined in Example 19.1 or the relative difference between the right-hand and left-hand sides:

$$c_d^{(r)} = \frac{\| \mathbf{y} - \mathbf{C}\mathbf{b}^{(r+1)} \|}{\| \mathbf{y} \|}$$

where:

$$\| \mathbf{x} \| = \left(\sum_i x_i^2 \right)^{1/2}$$

Using the convergence criteria used in Example 19.1, the iteration was stopped at the tenth iteration when equations converged to 8.3^{-07}. Some intermediary and final solutions are shown in the following table.

	Iteration number				
Effects	1	3	5	7	10
Sex of calf					
Male	3.430	3.835	4.280	4.367	4.359
Female	2.691	3.122	3.154	3.377	3.404

Continued

Continued.

Effects	Iteration number				
	1	3	5	7	10
Animals					
1	0.000	0.475	0.170	0.092	0.098
2	0.000	0.224	0.116	0.012	–0.019
3	0.000	0.272	0.058	–0.056	–0.041
4	0.763	0.390	0.032	–0.029	–0.009
5	0.383	0.249	–0.072	–0.155	–0.186
6	0.514	0.547	0.435	0.194	0.177
7	0.554	0.193	–0.178	–0.231	–0.249
8	0.791	0.537	0.334	0.171	0.183

The equations converged at the tenth round of iteration compared with 20 iterations on the data in Example 19.3.

Appendix A: Introduction to Matrix Algebra

The basic elements of matrix algebra necessary for the understanding of the principles for the prediction of breeding values are briefly covered in this appendix. Little or no previous knowledge of matrix algebra is assumed. For a detail study of the matrix algebra, the reader should see Searle (1982).

A.1 Matrix: A Definition

A matrix is a rectangular array of numbers set in rows and columns. These elements are called the elements of a matrix. The matrix B, for instance, consisting of two rows and three columns may be represented as:

$$\mathbf{B} = \begin{bmatrix} b_{11} & b_{12} & b_{13} \\ b_{21} & b_{22} & b_{23} \end{bmatrix}$$

or:

$$\mathbf{B} = \begin{bmatrix} 2 & 4 & 5 \\ 6 & 8 & 9 \end{bmatrix}$$

The element b_{ij} is called the ij element of the matrix, the first subscript referring to the row the element is in and the second to the column. The order of a matrix is the number of rows and columns. Thus, a matrix of r rows and c columns has order r × c (read as r by c). The matrix **B**, above, is of the order 2 × 3 and can be written as $\mathbf{B}_{2\times3}$.

A matrix consisting of a single row of elements is called a row vector. A row vector consisting of three elements may be represented as:

$$\mathbf{c} = \begin{bmatrix} 2 & 6 & -4 \end{bmatrix}$$

Only one subscript is needed to specify the position of an element in a row vector. Thus, the ith element in the row vector **c**, above, refers the element in the ith column. For instance, the $\mathbf{c}_3 = -4$.

Similarly, a matrix consisting of a single column is called a column vector. Again, only one subscript is needed to specify the position of an element, which refers to the row the element is in, since there is only one column. A column vector **d** with four elements can be shown as:

$$\mathbf{d} = \begin{bmatrix} -20 \\ 60 \\ 8 \\ 2 \end{bmatrix}$$

A scalar is matrix with one row and one column.

DOI: 10.1079/9781800620506.appx

A.2 Special Matrices

A.2.1 Square matrix

A matrix with an equal number of rows and columns is referred to as a square matrix. Shown below is a square matrix, **G**, of order 3 × 3.

$$G = \begin{bmatrix} 2 & 1 & 6 \\ 4 & 2 & 7 \\ 0 & 4 & 8 \end{bmatrix}$$

The ij elements in a square matrix with i equal to j are called the diagonal elements. Other elements of the square matrix are called off-diagonal or non-diagonal elements. Thus, the diagonal elements in the G matrix, above, are 2, 2 and 8.

A.2.2 Diagonal matrix

A square matrix having zero for all of its off-diagonal elements is referred to as a diagonal matrix. For example, a diagonal matrix, B, can be shown as:

$$B = \begin{bmatrix} 3 & 0 & 0 \\ 0 & 4 & 0 \\ 0 & 0 & 18 \end{bmatrix}$$

When all the diagonal elements of a diagonal matrix are one, it is referred to as an identity matrix. Given below is an identity matrix, **I**.

$$I = \begin{bmatrix} 1 & 0 & 0 & 0 \\ 0 & 1 & 0 & 0 \\ 0 & 0 & 1 & 0 \\ 0 & 0 & 0 & 1 \end{bmatrix}$$

A.2.3 Triangular matrix

A square matrix with all elements above the diagonal being zero is called a lower triangular matrix. When all the elements below the diagonal are zeros, it is referred to as an upper triangular matrix. For instance, the matrices **D**, a lower triangular matrix, and **E**, an upper triangular matrix, can be illustrated as:

$$D = \begin{bmatrix} 4 & 0 & 0 \\ 1 & 3 & 0 \\ -2 & 7 & 9 \end{bmatrix};\ E = \begin{bmatrix} 3 & 9 & 1 \\ 0 & 4 & 8 \\ 0 & 0 & 6 \end{bmatrix}$$

The transpose (see A.3.1) of an upper triangular matrix is a lower triangular matrix and *vice versa*.

A.2.4 Symmetric matrix

A symmetric matrix is a square matrix with the elements above the diagonal equal to the corresponding elements below the diagonal, i.e. element ij is equal to element ji. The matrix **A**, below, is an example of a symmetric matrix.

$$\mathbf{A} = \begin{bmatrix} 2 & -4 & 0 \\ -4 & 6 & 3 \\ 0 & 3 & 7 \end{bmatrix}$$

A.3 Basic Matrix Operations

A.3.1 Transpose of a matrix

The transpose of a matrix A, usually written as $\mathbf{A}'$ or $\mathbf{A}^T$, is the matrix whose ji elements are the ij elements of the original matrix, that is $a'_{ji} = a_{ij}$. In other words, the columns of $\mathbf{A}'$ are the rows of **A**, and rows of $\mathbf{A}'$ are the columns of **A**. For instance, the matrix **A** and its transpose **A** are illustrated below.

$$\mathbf{A} = \begin{bmatrix} 3 & 2 \\ 1 & 1 \\ 4 & 0 \end{bmatrix}; \ \mathbf{A}' = \begin{bmatrix} 3 & 1 & 4 \\ 2 & 1 & 0 \end{bmatrix}$$

Note that **A** is not equal to $\mathbf{A}'$ but the transpose of a symmetric matrix is equal to the symmetric matrix. Also, $(\mathbf{AB})' = \mathbf{B}'\mathbf{A}'$, where **AB** refers to the product (see A.3.3) of **A** and **B**.

A.3.2 Matrix addition and subtraction

Two matrices can be added together *only* if they have the same number of rows and columns, i.e. they are of the same order and they are said to be conformable for addition. Given that **W** is the sum of the matrices **X** and **Y**, then $w_{ij} = x_{ij} + y_{ij}$. For example, if **X** and **Y**, both of order 2 × 2 are as illustrated below,

$$\mathbf{X} = \begin{bmatrix} 40 & 10 \\ 39 & -25 \end{bmatrix}; \ \mathbf{Y} = \begin{bmatrix} -2 & 20 \\ 4 & 40 \end{bmatrix}$$

then the matrix **W**, the sum of **X** and **Y**, is:

$$\mathbf{W} = \begin{bmatrix} 40+(-2) & 10+20 \\ 39+4 & -25+40 \end{bmatrix} = \begin{bmatrix} 38 & 30 \\ 43 & 15 \end{bmatrix}$$

Matrix subtraction follows the same principles used for matrix addition. If **B** = **X** – **Y**, then $b_{ij} = x_{ij} - y_{ij}$. Thus, the matrix B obtained by subtracting Y from X above is:

$$\mathrm{B} = \mathrm{X} - \mathrm{Y} = \begin{bmatrix} 40-(-2) & 10-20 \\ 39-4 & -25-40 \end{bmatrix} = \begin{bmatrix} 42 & -10 \\ 35 & -65 \end{bmatrix}$$

A.3.3 Matrix multiplication

Two matrices can be multiplied only if the number of columns in the first matrix equals the number of rows in the second. The order of the product matrix is equal to the number of rows of the first matrix multiplied by the number of columns in the second. Given that C = **AB**, then

$$\mathrm{C} = c_{ij} = \sum_{j=1}^{m}\sum_{i=1}^{n}\sum_{k=1}^{z} a_{ik} b_{kj}$$

where m = number of columns in **B**, n = number of rows in **A**, and z = number of rows in B. Let:

$$\mathbf{A} = \begin{bmatrix} 1 & 4 & -1 \\ 2 & 5 & 0 \\ 3 & 6 & 1 \end{bmatrix};\ \mathbf{B} = \begin{bmatrix} 2 & 5 \\ 4 & 3 \\ 6 & 1 \end{bmatrix}$$

then **C** can be obtained as:

c_{11} = 1(2) + 4(4) + −1(6) = 12 (row 1 of **A** multiplied by column 1 of **B**)
c_{21} = 2(2) + 5(4) + 0(6) = 24 (row 2 of **A** multiplied by column 1 of **B**)
c_{31} = 3(2) + 6(4) + 1(6) = 36 (row 3 of **A** multiplied by column 1 of **B**)
c_{12} = 1(5) + 4(3) + −1(1) = 16 (row 1 of **A** multiplied by column 2 of **B**)
c_{22} = 2(5) + 5(3) + 0(1) = 25 (row 2 of **A** multiplied by column 2 of **B**)
c_{23} = 3(5) + 6(3) + 1(1) = 34 (row 3 of **A** multiplied by column 2 of **B**)

$$\mathrm{C} = \begin{bmatrix} 12 & 16 \\ 24 & 25 \\ 36 & 34 \end{bmatrix}$$

Note that C has order 3 × 2, where 3 equals the number of rows of A and 2 the number of columns in B. Also note that **AB** is not equal to **BA** but **IA** = **AI** = **A**, where **I** is an identity matrix.

If **M** is the product of a scalar g and a matrix **B**, then **M** = $b_{ij}g$; that is, each element of **M** equals the corresponding element in **B** multiplied by g.

A.3.4 Direct product of matrices

Given a matrix **G** of order n × m and A of order t × s, the direct product is:

$$\mathbf{G} \otimes \mathbf{A} = \begin{bmatrix} g_{11}\mathbf{A} & g_{12}\mathbf{A} \\ g_{21}\mathbf{A} & g_{22}\mathbf{A} \end{bmatrix}$$

The direct product is also known as the Kronecker product and is of the order nt × ms. For instance, assuming that:

$$\mathbf{G} = \begin{bmatrix} 10 & 5 \\ 5 & 20 \end{bmatrix} \text{ and } \mathbf{A} = \begin{bmatrix} 1 & 0 & 2 \\ 0 & 1 & 4 \\ 2 & 4 & 1 \end{bmatrix}$$

the Kronecker product is:

$$\mathbf{G} \otimes \mathbf{A} = \begin{bmatrix} 10 & 0 & 20 & 5 & 0 & 10 \\ 0 & 10 & 40 & 0 & 5 & 20 \\ 20 & 40 & 10 & 10 & 20 & 5 \\ 5 & 0 & 10 & 20 & 0 & 40 \\ 0 & 5 & 20 & 0 & 20 & 80 \\ 10 & 20 & 5 & 40 & 80 & 20 \end{bmatrix}$$

The Kronecker product is useful in multiple-trait evaluations.

A.3.5 Matrix inversion

An inverse matrix is one which, when multiplied by the original matrix, gives an identity matrix as the product. The inverse of a matrix A is usually denoted as A^{-1}, and from the above definition, $A^{-1}A = I$, where I is an identity matrix. Only square matrices can be inverted, and for a diagonal matrix the inverse is calculated simply as the reciprocal of the diagonal elements. For instance, the diagonal matrix B and its inverse are:

$$\mathbf{B} = \begin{bmatrix} 3 & 0 & 0 \\ 0 & 4 & 0 \\ 0 & 0 & 18 \end{bmatrix} \quad \text{and} \quad \mathbf{B}^{-1} = \begin{bmatrix} 1/3 & 0 & 0 \\ 0 & 1/4 & 0 \\ 0 & 0 & 1/18 \end{bmatrix}$$

For a 2 × 2 matrix, the inverse is easy to calculate and is illustrated below. Let:

$$\mathbf{A} = \begin{bmatrix} a_{11} & a_{12} \\ a_{21} & a_{22} \end{bmatrix}$$

First, calculate the determinant, which is the difference between the product of the two diagonal elements and the two off-diagonal elements $(a_{11}a_{22} - a_{12}a_{21})$. Second, the inverse is obtained by reversing the diagonal elements, multiplying the off-diagonal elements by –1 and dividing all elements by the determinant. Thus:

$$\mathbf{A}^{-1} = \frac{1}{a_{11}a_{22} - a_{12}a_{21}} \begin{bmatrix} a_{22} & -a_{12} \\ -a_{21} & a_{11} \end{bmatrix}$$

For instance, given that:

$$\mathbf{A} = \begin{bmatrix} 8 & 4 \\ 6 & 4 \end{bmatrix} \quad \text{then} \quad \mathbf{A}^{-1} = \frac{1}{(8)(4) - (6)(4)} \begin{bmatrix} 4 & -4 \\ -6 & 8 \end{bmatrix} = \begin{bmatrix} 0.50 & -0.50 \\ -0.75 & 1.00 \end{bmatrix}$$

Note that $\mathbf{A}^{-1}\mathbf{A} = \mathbf{I} = \mathbf{A}\mathbf{A}^{-1}$, as stated earlier. Calculating the inverse of a matrix becomes more cumbersome as the order increases and they are usually obtained using computer programs. The methodology has not been covered in this text. It is obvious from the above that an inverse of a non-diagonal matrix cannot be calculated if the determinant is equal to zero. A square matrix with a determinant equal to zero is said to be singular and does not have an inverse. A matrix with a non-zero determinant is said to be non-singular.

Note that $(\mathbf{AB})^{-1} = \mathbf{B}^{-1}\mathbf{A}^{-1}$. The inverses of matrices may be required when solving linear equations. Thus, given the following linear equation:

$$\mathbf{Ab} = \mathbf{y}$$

pre-multiplying both sides by $\mathbf{A}^{-1}$ gives the vector of solutions **b** as:

$$\mathbf{b} = \mathbf{A}^{-1}\mathbf{y}$$

A.3.6 Rank of a matrix

The rank of a matrix is the number of linearly independent rows or columns. A square matrix with the rank equal to the number of rows or columns is said to be of full rank. In some matrices, some of the rows or columns are a linear combination of other rows or columns; therefore the rank is less than the number of rows or columns. Such a matrix is not of full rank. Consider the following set of equations:

$$\begin{aligned} 3x_1 + 2x_2 + 1x_3 &= y_1 \\ 4x_1 + 3x_2 + 0x_3 &= y_2 \\ 7x_1 + 5x_2 + 1x_3 &= y_3 \end{aligned}$$

The third equation is the sum of the first and second equations; therefore the vector of solutions, $\mathbf{x}$ ($\mathbf{x}' = [x_1\ x_2\ x_3]$), cannot be estimated due to the lack of information. In other words, if the system of equations were expressed in matrix notation as:

$$\begin{bmatrix} 3 & 2 & 1 \\ 4 & 3 & 0 \\ 7 & 5 & 1 \end{bmatrix} \begin{bmatrix} x_1 \\ x_2 \\ x_3 \end{bmatrix} = \begin{bmatrix} y_1 \\ y_2 \\ y_3 \end{bmatrix}$$

i.e. as:

$$\mathbf{Dx} = \mathbf{y}$$

a unique inverse does not exist for **D** because of the dependency in the rows. Only two rows are linearly independent in **D** and it is said of to be of rank 2, usually written as $r(\mathbf{D})=2$. When a square matrix is not of full rank, the determinant is zero and hence a unique inverse does not exist.

A.3.7 Generalized inverses

While an inverse does not exist for a singular matrix, a generalized inverse can, however, be calculated. A generalized inverse for a matrix **D** is usually denoted as $\mathbf{D}^-$ and satisfies the expression:

$$\mathbf{DD}^-\mathbf{D} = \mathbf{D}$$

Generalized inverses are not unique and may be obtained in several ways. One of the simple ways to calculate a generalized inverse of a matrix, say **D** in Section A.3.6, is to initially obtain a matrix **B** of full rank as a subset of **D**. Set all elements of **D** to zero. Calculate the inverse of **B** and replace the elements of **D** with corresponding elements of **B** and the result is $\mathbf{D}^-$. For instance, for the matrix **D**, above, the matrix **B**, a full rank subset of **D**, is:

$$\mathrm{B} = \begin{bmatrix} 3 & 2 \\ 4 & 3 \end{bmatrix} \quad \textit{and} \quad \mathrm{B}^{-1} = \begin{bmatrix} 3 & -2 \\ -4 & 3 \end{bmatrix}$$

Replacing elements of **D** with the corresponding elements of **B** after all elements of **D** have been set to zero gives $\mathbf{D}^-$ as:

$$\mathrm{D}^- = \begin{bmatrix} 3 & -2 & 0 \\ -4 & 3 & 0 \\ 0 & 0 & 0 \end{bmatrix}$$

A.3.8 Eigenvalues and eigenvectors

Eigenvalues are also referred to as characteristic or latent roots and are useful in simplifying multivariate evaluations when transforming data. The sum of the eigenvalues of a square matrix equals its trace (sum of the diagonal elements of a square matrix) and their product equals its determinant (Searle, 1982). For symmetric matrices, the rank equals the number of non-zero eigenvalues.

For a square matrix, **B**, the eigenvalues are obtained by solving:

$$|\mathrm{B} - d\mathrm{I}| = 0$$

where the vertical lines denote finding the determinant.

With the condition specified in the above equation, B can be represented as:

$$\mathrm{BL} = \mathrm{LD}$$

$$\mathrm{B} = \mathrm{LDL}^{-1} \qquad \text{(a.1)}$$

where **D** is a diagonal matrix containing the eigenvalues of **B** and **L** is a matrix of corresponding eigenvectors. The eigenvector (k) is found by solving:

$$(\mathrm{B} - d_k\mathrm{I})l_k = 0$$

where d_k is the corresponding eigenvalue.

For symmetric matrices, **L** is orthogonal (i.e. $\mathbf{L}^{-1} = \mathbf{L}'$; $\mathbf{LL}' = \mathbf{I} = \mathbf{L}'\mathbf{L}$), therefore, given that **B** is symmetric, (a.1) can be expressed as:

$$\mathrm{B} = \mathrm{LDL}'$$

Usually, eigenvalues and eigenvectors are calculated by means of computer programs.

Appendix B: Fast Algorithms for Calculating Inbreeding Based on the L Matrix

In this appendix, two algorithms based on the **L** matrix for calculating inbreeding are discussed.

B.1 Meuwissen and Luo Algorithm

The algorithm given by Quaas (1976) involves the calculation of one column of **L** at a time. The algorithm requires $n(n + 1)/2$ operations and computational time is proportional to n^2, where n is the size of the data set. It suffers from the disadvantage of not being readily adapted for updating when a new batch of animals is available without restoring a previously stored **L**. Meuwissen and Luo (1992) presented a faster algorithm, which involves computing the elements of **L** row by row.

The fact that each row of **L** is calculated independently of other rows makes it suitable for updating. The row i of **L** for animal i gives the fraction of genes the animal derives from its ancestors. If s_i and d_i are the sire and dam of animal i, then $l_{is_i} = l_{id_i} = 0.5$. The ith row of **L** can be calculated by proceeding through a list of i's ancestors from the youngest to the oldest and updating continually as $l_{is_j} = l_{is_j} + 0.5\, l_{ij}$ and $l_{id_j} = l_{id_j} + 0.5 l_{ij}$, where j is an ancestor of i. The fraction of genes derived from an ancestor is:

$$l_{ij} = \sum_{k \varepsilon \mathbf{P}_j} 0.5 l_{ik}$$

where $\mathbf{P}_j$ is a set of identities of the progeny of j. However, $l_{ij} = 0$ only when k is not an ancestor of i or k is not equal to i. Thus, if **AN** is the set of identities of the number of ancestors of i, then:

$$l_{ij} = \sum_{k \varepsilon \mathbf{AN} \cap P_j} 0.5 l_{ik}$$

i.e. the summation of $0.5l_{ik}$ is over those k animals that are both ancestors of i and progeny of j. This forms the basis of the algorithm given below for the calculation of the row i of **L**, one row at a time. As each row of **L** is calculated, its contribution to the diagonal elements of the relationship matrix (a_{ii}) is accumulated. Initially, set row i of **L** and a_{ii} to zero. The list of ancestors whose contributions to a_{ii} are yet to be included are added to the vector **AN** (if not already there) as each row of **L** is being processed.

The algorithm is:

$$F_0 = -1$$

For $i = 1, N$ (all rows of **L**):

$\mathbf{AN}_i = i$

$l_{ii} = 1$

$\mathbf{D}_{ii} = [0.5 - 0.25(F_{s_i} + F_{d_i})]$ if both parents are known; otherwise use the appropriate formula (see Chapter 3)

© R.A. Mrode and I. Pocrnic 2023. *Linear Models for the Prediction of the Genetic Merit of Animals, 4th Edition* (R.A. Mrode and I. Pocrnic)
DOI: 10.1079/9781800620506.appx

Do while $\mathbf{AN}_i$ is not empty.

$j = \max(\mathbf{AN}_i), (j =$ youngest animal in $\mathbf{AN}_i)$

If s_j is known, add s_j to $\mathbf{AN}_i$:

$$l_{is_j} = l_{is_j} + 0.5l_{ij}$$

If d_j is known, add d_j to $\mathbf{AN}_i$:

$$l_{id_j} = l_{id_j} + 0.5l_{ij}$$

$$a_{ii} = a_{ii} + l_{ij}^2 \mathbf{D}_{jj}$$

Delete j from $\mathbf{AN}_i$.

End while:

$$F_i = a_{ii} - 1$$

B.1.1 Illustration of the algorithm

Using the pedigree in Table 3.1, the algorithm is illustrated for animals 1 and 5.

For animal 1:

$$a_{11} = 0$$

$$\mathbf{AN}_1 = 1, l_{11} = 1$$

Since both parents are unknown:

$$\mathbf{D}_{11} = 1$$

Processing animals in $\mathbf{AN}_1$:

$$j = \max(\mathbf{AN}_1) = 1$$

Both parents of j are unknown:

$$a_{11} = a_{11} + l_{11}^2 \mathbf{D}_{11} = (1^2)1 = 1$$

Delete animal 1 from $\mathbf{AN}_1$; $\mathbf{AN}_1$ is now empty.

$$F_1 = 1 - 1 = 0$$

For animal 5:

$$a_{55} = 0$$

$$\mathbf{AN}_5 = 5,\ l_{55} = 1$$

$\mathbf{D}_{55} = 0.5$, since neither parent is inbred.

Processing animals in $\mathbf{AN}_5$:

$$j = \max(\mathbf{AN}_5) = 5$$

Add sire and dam of 5 (animals 4 and 3) to AN_5:

$$l_{54} = l_{54} + 0.5l_{55} = 0.5$$
$$l_{53} = l_{53} + 0.5l_{55} = 0.5$$
$$a_{55} = a_{55} + l_{55}^2\mathbf{D}_{55} = 1^2(0.5) = 0.5$$

Delete animal 5 from $\mathbf{AN}_5$; animals 4 and 3 left in $\mathbf{AN}_5$.
Next animal in $\mathbf{AN}_5$:

$$j = \max(\mathbf{AN}_5) = 4$$

Add sire of 4 (animal 1) to $\mathbf{AN}_5$:

$$l_{51} = l_{51} + 0.5l_{54} = 0.25$$
$$a_{55} = a_{55} + l_{55}^2\mathbf{D}_{44} = 0.5 + (0.5)^2(0.75) = 0.6875$$

Delete animal 4 from $\mathbf{AN}_5$; animals 3 and 1 left in $\mathbf{AN}_5$.
Next animal in $\mathbf{AN}_5$:

$$j = \max(\mathbf{AN}_5) = 3$$

Since animal 1, the sire of j, is already in $\mathbf{AN}_5$, add only the dam of 3 (animal 2) to $\mathbf{AN}_5$:

$$l_{51} = l_{51} + 0.5l_{53} = 0.25 + (0.5)0.5 = 0.5$$
$$l_{52} = l_{52} + 0.5l_{53} = 0 + (0.5)0.5 = 0.25$$
$$a_{55} = a_{55} + l_{53}^2\mathbf{D}_{33} = 0.6875 + (0.5)^2 0.5 = 0.8125$$

Delete animal 3 from $\mathbf{AN}_5$; animals 1 and 2 left in $\mathbf{AN}_5$.
Next animal in $\mathbf{AN}_5$:

$$j = \max(\mathbf{AN}_5) = 2$$

Both parents are unknown:

$$a_{55} = a_{55} + l_{52}^2\mathbf{D}_{22} = 0.8125 + (0.25)^2 1 = 0.875$$

Delete animal 2 from $\mathbf{AN}_5$; animal 1 left in $\mathbf{AN}_5$.
Next animal in $\mathbf{AN}_5$:

$$j = \max(\mathbf{AN}_5) = 1$$

Both parents are unknown:

$$a_{55} = a_{55} + l_{51}^2\mathbf{D}_{11} = 0.875 + (0.5)^2 1 = 1.125$$

Delete 1 from $\mathbf{AN}_5$; $\mathbf{AN}_5$ is empty.

$$F_5 = 1.125 - 1 = 0.125$$

which is the same inbreeding coefficient as that obtained for animal 5 in Section 2.2.

B.2 Modified Meuwissen and Luo Algorithm

The approach of Meuwissen and Luo, given above, was modified by Quaas (1995) to improve its efficiency. The disadvantage of the above method is that, while calculating a row of **L** at a time (Henderson, 1976), it is accumulating diagonal elements of **A**, as in Quaas (1976), and this necessitates tracing the entire pedigree for *i*, but what is really needed is only the common ancestors. Thus, a more efficient approach is to accumulate $a_{s_id_i}$ as $\sum_k l_{s_ik} l_{d_ik} \mathbf{D}_{kk}$ (Henderson, 1976) and calculate F_i as $0.5a_{s_id_i} = \sum_k l_{s_ik} l_{d_ik}(0.5\mathbf{D}_{kk})$. Instead of computing the *i*th row of **L**, only the non-zero elements in the rows for the sire and dam of *i* are calculated. Quaas (1995) suggested setting up a separate ancestor list ($\mathbf{AS}_{s_i}$) for s_i and another ($\mathbf{AD}_{d_i}$) for d_i; then $\mathrm{F}_i = 0.5a_{s_id_i} = \sum_k \varepsilon_{s_i} U_{d_i} l_{s_ik} l_{d_ik}(0.5\mathbf{D}_{kk})$.

Similar to the approach of Meuwissen and Luo (1992), the two lists can be set up simultaneously while processing the *i*th animal by continually adding the parents of the next youngest animal in either list to the appropriate list. If the next youngest in each list is the same animal, say *k*, then it is a common ancestor and F_i is updated as $\mathrm{F}_i = \mathrm{F}_i + l_{s_ik} l_{d_ik}(0.5D_{kk})$. When ancestors of one of the parents have been processed, the procedure can be stopped, and it is not necessary to search both lists completely. The algorithm for this methodology is:

$\mathrm{F}_0 = -1$

For $i = 1, N$:

$\mathrm{F}_i = 0$

If s_i is known, add s_i to $\mathbf{AS}_{s_i}$, $l_{s_is_i} = 1$
If d_i is known, add d_i to $\mathbf{AD}_{di}$, $l_{d_id_i} = 1$
Do while $\mathbf{AS}_{s_i}$ not empty and $\mathbf{AD}_{d_i}$ not empty.

$j = \max(\mathbf{AS}_{s_i}), k = \max(\mathbf{AD}_{d_i})$.

If $j > k$, then (next youngest *j* is in $\mathbf{AS}_{si}$).
If s_j is known, add s_j to $\mathbf{AS}_{s_i}$; $l_{s_is_i} = l_{s_is_i} + 0.5l_{s_ij}$.
If d_j is known, add d_j to $\mathbf{AS}_{s_i}$; $l_{s_id_i} = l_{s_id_i} + 0.5l_{s_ij}$.
Delete *j* from $\mathbf{AS}_{s_i}$.
Else if $k > j$ then (next youngest *k* is in $\mathbf{AD}_{d_i}$).
If s_k is known, add s_k to $\mathbf{AD}_{d_i}$: $l_{d_is_k} = l_{d_is_k} + 0.5l_{d_ik}$.
If d_k is known, add d_k to $\mathbf{AD}_{d_i}$; $l_{d_id_k} = l_{d_id_k} + 0.5l_{d_ik}$.
Delete *k* from $\mathbf{AD}_{di}$.
Else (next youngest ancestor $j = k$ is a common ancestor).
If s_j is known, add s_j to $\mathbf{AS}_{s_i}$: $l_{s_is_i} = l_{s_is_j} + 0.5l_{s_ij}$.
add s_j to $\mathbf{AD}_{d_i}$: $l_{d_id_k} = l_{d_id_k} + 0.5l_{d_ik}$.
If d_j is known, add d_j to $\mathbf{AS}_{s_i}$; $l_{s_id_i} = l_{s_id_j} + 0.5l_{s_ij}$.
add d_j to $\mathbf{AD}_{d_i}$; $l_{d_id_j} = l_{d_id_j} + 0.5l_{d_ij}$.

$F_i = F_i + l_{s_ij} l_{d_ij} 0.5(\mathbf{D}_{jj})$.

Delete *j* from $\mathbf{AN}_{\hat{s}_i}$ and $\mathbf{AD}_{d_i}$.

End if
End while
End do

B.2.1 Illustration of the algorithm

Using the pedigree in Table 2.1, the algorithm is illustrated for animal 5, which is inbred. For animal 5:

Both parents known, $s = 4$ and $d = 3$.

Add4 to $\mathbf{AD}_4$; $l_{44} = 0.5$

Add3 to $\mathbf{AD}_3$; $l_{33} = 0.5$

Processing animals in $\mathbf{AS}_4$ and $\mathbf{AD}_3$:

$j = 4;\ k = 3$

$j > k$ therefore.

Add sire of 4, animal 1 to $\mathbf{AS}_4$; $l_{41} = l_{41} + 0.5l_{44} = 0.5$.
Delete animal 4 from $\mathbf{AS}_4$.

Next, animals in $\mathbf{AS}_4$ and $\mathbf{AD}_3$:

$j = 1,\ k = 3$

$k > j$ therefore.

Add sire of 3, animal 1 to $\mathbf{AD}_3$; $l_{31} = l_{31} + 0.5l_{33} = 0.5$.
Add dam of 3, animal 2 to $\mathbf{AD}_3$; $l_{32} = l_{32} + 0.5l_{33} = 0.5$.
Delete 3 from $\mathbf{AD}_3$.

Next animals in $\mathbf{AS}_4$ and $\mathbf{AD}_3$:

$j = 1, k = 2$

$k > j$

Both parents of 2 are unknown.
Delete 2 from $\mathbf{AD}_3$.

Next animals in $\mathbf{AS}_4$ and $\mathbf{AD}_3$:

$j = 1, k = 1$

$j = k$

Both parents are unknown.

$$F_5 = F_5 + l_{41}l_{31}0.5(\mathbf{D}_{11}) = 0.5(0.5)(0.5)(1) = 0.125$$

which is the same inbreeding coefficient as that obtained from the algorithm in Section B.1.

Appendix C

C.1 Outline of the Derivation of the Best Linear Unbiased Prediction (BLUP)

Consider the following linear model:

$$\mathbf{y} = \mathbf{Xb} + \mathbf{Za} + \mathbf{e} \qquad \text{(c.1)}$$

where the expectations are:

$$\mathrm{E}(\mathbf{y}) = \mathbf{Xb};\ \mathrm{E}(\mathbf{a}) = \mathrm{E}(\mathbf{e}) = 0$$

and

$$\mathrm{var}(\mathbf{a}) = \mathbf{A}\sigma_a^2 = \mathbf{G},\ \mathrm{var}(\mathbf{e}) = \mathbf{R} \text{ and } \mathrm{cov}(\mathbf{a},\ \mathbf{e}) = \mathrm{cov}(\mathbf{e},\ \mathbf{a}) = 0$$

Then, as shown in Section 4.2:

$$\mathrm{var}(\mathbf{y}) = \mathbf{V} = \mathbf{ZGZ}' + \mathbf{R},\ \mathrm{cov}(\mathbf{y},\ \mathbf{a}) = \mathbf{ZG} \text{ and } \mathrm{cov}(\mathbf{y},\ \mathbf{e}) = \mathbf{R}$$

The prediction problem involves both **b** and **a**. Suppose we want to predict a linear function of **b** and **a**, say $\mathbf{k'b} + \mathbf{a}$, using a linear function of **y**, say $\mathbf{L'y}$, and $\mathbf{k'b}$ is estimable. The predictor $\mathbf{L'y}$ is chosen such that:

$$\mathrm{E}(\mathbf{L'y}) = \mathrm{E}(\mathbf{k'b} + \mathbf{a})$$

i.e. it is unbiased and the prediction error variance (PEV) is minimized (Henderson, 1973). Now, PEV (Henderson, 1984) is:

$$\begin{aligned} \mathrm{PEV} &= \mathrm{var}(\mathbf{L'y} - \mathbf{k'b} + \mathbf{a}) \\ &= \mathrm{var}(\mathbf{Ly} - \mathbf{a}) \\ &= \mathbf{L}'\,\mathrm{var}(\mathbf{y})\mathbf{L} + \mathrm{var}(\mathbf{a}) - \mathbf{L}'\,\mathrm{cov}(\mathbf{y},\ \mathbf{a}) - \mathrm{cov}(\mathbf{a},\ \mathbf{y})\mathbf{L} \\ &= \mathbf{L'VL} + \mathbf{G} - \mathbf{L'ZG} - \mathbf{ZG'L} \end{aligned} \qquad \text{(c.2)}$$

Minimizing PEV subject to $\mathrm{E}(\mathbf{L'y}) = \mathrm{E}(\mathbf{k'b} + \mathrm{a})$ and solving (see Henderson, 1973, 1984 for details of derivation) gives:

$$\mathbf{L'y} = \mathbf{k}'(\mathbf{X'V}^{-1}\mathbf{X})^{-1}\mathbf{X'V}^{-1}\mathbf{y} - \mathbf{GZ'V}^{-1}(\mathbf{y} - \mathbf{X}(\mathbf{X'V}^{-1}\mathbf{X})^{-1}\mathbf{X'V}^{-1}\mathbf{y})$$

Let $\hat{\mathbf{b}} = (\mathbf{X'V}^{-1}\mathbf{X})\mathbf{XV}^{-1}\mathbf{y}$, the generalized least square solution for **b**, then the predictor can be written as:

$$\mathbf{L'y} = \mathbf{k'}\hat{\mathbf{b}} + \mathbf{GZ'V}^{-1}(\mathbf{y} - \mathbf{X}\hat{\mathbf{b}}) \qquad \text{(c.3)}$$

which is the BLUP of $\mathbf{k'b}$ + **a**.

Note that if $\mathbf{k'b}$ = 0, then:

$$\mathbf{L'y} = \mathrm{BLUP}(\mathbf{a}) = \mathbf{GZ'V}^{-1}(\mathbf{y} - \mathbf{X}\hat{\mathbf{b}}) \qquad \text{(c.4)}$$

which is equivalent to the selection index. Thus, BLUP is the selection index with the GLS solution of b substituted for **b**.

DOI: 10.1079/9781800620506.appx

C.2 Proof that $\hat{b}$ and $\hat{a}$ from MME Are the GLS of b and BLUP of a, Respectively

In computation terms, the use of Eqn c.3 to obtain the BLUP of **k′b** + **a** is not feasible because the inverse of **V** is required. Henderson (1950) formulated the MME that are suitable for calculating solutions for **b** and **a**, and showed later that $\mathbf{k}'\hat{\mathbf{b}}$ and $\hat{\mathbf{a}}$, where $\hat{\mathbf{b}}$ and $\hat{\mathbf{a}}$ are solutions from the MME, are the best linear unbiased estimator (BLUE) of **k′b** and BLUP of **a**, respectively.

The usual MME for Eqn c.1 are:

$$\begin{bmatrix} \mathbf{X}'\mathbf{R}^{-1}\mathbf{X} & \mathbf{X}'\mathbf{R}^{-1}\mathbf{Z}' \\ \mathbf{Z}'\mathbf{R}^{-1}\mathbf{X} & \mathbf{Z}'\mathbf{R}^{-1}\mathbf{Z}+\mathbf{G}^{-1} \end{bmatrix} \begin{bmatrix} \hat{\mathbf{b}} \\ \hat{\mathbf{a}} \end{bmatrix} = \begin{bmatrix} \mathbf{X}'\mathbf{R}^{-1}\mathbf{y} \\ \mathbf{Z}'\mathbf{R}^{-1}\mathbf{y} \end{bmatrix} \quad \text{(c.5)}$$

The proof that $\hat{\mathbf{b}}$ from the MME is the GLS of b and therefore $\mathbf{k}'\hat{\mathbf{b}}$ is the BLUE of **k′b** was given by Henderson *et al.* (1959). From the second row of Eqn c.5:

$$(\mathbf{Z}'\mathbf{R}^{-1}\mathbf{Z}+\mathbf{G}^{-1})\hat{\mathbf{a}} = \mathbf{Z}'\mathbf{R}^{-1}(\mathbf{y}-\mathbf{X}\hat{\mathbf{b}})$$

$$\hat{\mathbf{a}} = (\mathbf{Z}'\mathbf{R}^{-1}\mathbf{Z}+\mathbf{G}^{-1})^{-1}\mathbf{Z}'\mathbf{R}^{-1}(\mathbf{y}-\mathbf{X}\hat{\mathbf{b}}) \quad \text{(c.6)}$$

From the first row of Eqn c.5:

$$\mathbf{X}'\mathbf{R}^{-1}\mathbf{X}\hat{\mathbf{b}} + \mathbf{X}'\mathbf{R}^{-1}\mathbf{Z}\hat{\mathbf{a}} = \mathbf{X}'\mathbf{R}^{-1}\mathbf{y}$$

Substituting the solution for $\hat{\mathbf{a}}$ into the above equation gives:

$$\mathbf{X}'\mathbf{R}^{-1}\mathbf{X}\mathbf{b} + \mathbf{X}'\mathbf{R}^{-1}\mathbf{Z}(\mathbf{W}\mathbf{Z}'\mathbf{R}^{-1})(\mathbf{y}-\mathbf{X}\mathbf{b}) = \mathbf{X}\mathbf{R}^{-1}\mathbf{y}$$

where $\mathbf{W} = (\mathbf{Z}\mathbf{R}^{-1}\mathbf{Z}+\mathbf{G}^{-1})^{-1}$:

$$\mathbf{X}\mathbf{R}^{-1}\mathbf{X}\mathbf{b} - (\mathbf{X}\mathbf{R}^{-1}\mathbf{Z})(\mathbf{W}\mathbf{Z}'\mathbf{R}^{-1})\mathbf{X}\mathbf{b} = \mathbf{X}'\mathbf{R}^{-1}\mathbf{y} - \mathbf{X}'\mathbf{R}^{-1}\mathbf{Z}\mathbf{W}\mathbf{Z}'\mathbf{R}^{-1}\mathbf{y}$$

$$\mathbf{X}'(\mathbf{R}^{-1} - \mathbf{R}^{-1}\mathbf{Z}\mathbf{W}\mathbf{Z}'\mathbf{R}^{-1})\mathbf{X}\mathbf{b} = \mathbf{X}'(\mathbf{R}^{-1} - \mathbf{R}^{-1}\mathbf{Z}\mathbf{W}\mathbf{Z}'\mathbf{R}^{-1})\mathbf{y}$$

$$\mathbf{X}'\mathbf{V}^{-1}\mathbf{X}\mathbf{b} = \mathbf{X}'\mathbf{V}^{-1}\mathbf{y}$$

with $\mathbf{V}^{-1} = \mathbf{R}^{-1} - \mathbf{R}^{-1}\mathbf{Z}\mathbf{W}\mathbf{Z}'\mathbf{R}^{-1}$:

$$\hat{\mathbf{b}} = (\mathbf{X}'\mathbf{V}^{-1}\mathbf{X})^{-1}\mathbf{X}'\mathbf{V}^{-1}\mathbf{y} \quad \text{(c.7)}$$

It can be shown that:

$$\mathbf{V}^{-1} = \mathbf{R}^{-1} - \mathbf{R}^{-1}\mathbf{Z}\mathbf{W}\mathbf{Z}'\mathbf{R}^{-1}$$

by pre-multiplying the right-hand side by **V** and obtaining an identity matrix (Henderson *et al.*, 1959):

$$\begin{aligned}
\mathbf{V}[\mathbf{R}^{-1} - \mathbf{R}^{-1}\mathbf{Z}\mathbf{W}\mathbf{Z}'\mathbf{R}^{-1}] &= (\mathbf{R}+\mathbf{Z}\mathbf{G}\mathbf{Z}')(\mathbf{R}^{-1} - \mathbf{R}^{-1}\mathbf{Z}\mathbf{W}\mathbf{Z}'\mathbf{R}^{-1}) \\
&= \mathbf{I} + \mathbf{Z}\mathbf{G}\mathbf{Z}'\mathbf{R}^{-1} - \mathbf{Z}\mathbf{W}\mathbf{Z}'\mathbf{R}^{-1} - \mathbf{Z}\mathbf{G}\mathbf{Z}'\mathbf{R}^{-1}\mathbf{Z}\mathbf{W}\mathbf{Z}'\mathbf{R}^{-1} \\
&= \mathbf{I} + \mathbf{Z}\mathbf{G}\mathbf{Z}'\mathbf{R}^{-1} - \mathbf{Z}(\mathbf{I} + \mathbf{G}\mathbf{Z}'\mathbf{R}\mathbf{Z})\mathbf{W}\mathbf{Z}'\mathbf{R}^{-1} \\
&= \mathbf{I} + \mathbf{Z}\mathbf{G}\mathbf{Z}'\mathbf{R}^{-1} - \mathbf{Z}\mathbf{G}(\mathbf{G}^{-1} + \mathbf{Z}'\mathbf{R}\mathbf{Z})\mathbf{W}\mathbf{Z}'\mathbf{R}^{-1} \\
&= \mathbf{I} + \mathbf{Z}\mathbf{G}\mathbf{Z}'\mathbf{R}^{-1} - \mathbf{Z}\mathbf{G}(\mathbf{W}^{-1})\mathbf{W}\mathbf{Z}'\mathbf{R}^{-1} \\
&= \mathbf{I} + \mathbf{Z}\mathbf{G}\mathbf{Z}'\mathbf{R}^{-1} - \mathbf{Z}\mathbf{G}\mathbf{Z}'\mathbf{R}^{-1} \\
&= \mathbf{I}
\end{aligned}$$

Thus, the solution for **b** from the MME is equal to the GLS solution for **b** in Eqn c.3.

The proof that $\hat{\mathbf{a}}$ from the MME is equal to $\mathbf{GZ'V^{-1}(y - X\hat{b})}$ in Eqn c.3 was given by Henderson (1963). Replace $\mathbf{V}^{-1}$ in $\mathbf{GZ'V^{-1}(y - X\hat{b})}$ by $\mathbf{R^{-1} - R^{-1}ZWX'R^{-1}}$, thus:

$$\begin{aligned}
\mathbf{GZ'V^{-1}(y - X\hat{b})} &= \mathbf{GZ'(R^{-1} - R^{-1}ZWZ'R^{-1})(y - X\hat{b})} \\
&= \mathbf{G(Z'R^{-1} - Z'R^{-1}ZWZ'R^{-1})(y - X\hat{b})} \\
&= \mathbf{G(I - Z'R^{-1}ZW)Z'R^{-1}(y - X\hat{b})} \\
&= \mathbf{G(W^{-1} - Z'R^{-1}Z)WZ'R^{-1}(y - X\hat{b})} \qquad \text{(See Eqn c.6)} \\
&= \mathbf{G((Z'RZ + G^{-1}) - Z'R^{-1}Z)WZ'R^{-1}(y - X\hat{b})} \\
&= \mathbf{GZ'R^{-1}Z + I - GZ'R^{-1}Z)WZ'R^{-1}(y - X\hat{b})} \\
&= \mathbf{(I)WZ'R^{-1}(y - X\hat{b})} \\
&= \mathbf{WZ'R^{-1}(y - X\hat{b}) = \hat{a}}
\end{aligned}$$

Thus, the BLUP of $\mathbf{k'b + a = k'\hat{b} + \hat{a}}$, where $\hat{\mathbf{b}}$ and $\hat{\mathbf{a}}$ are solutions to the MME.

C.3 Deriving the Equation for Progeny Contribution (PC)

Considering an individual *i* that has one record with both sire (*s*) and dam (*d*) known, the MME for the three animals can be written (assuming the sire and dam are ancestors with unknown parents) as:

$$\begin{bmatrix} u_{ss}\ \boldsymbol{\alpha} & u_{sd}\ \boldsymbol{\alpha} & u_{si}\ \boldsymbol{\alpha} \\ u_{ds}\ \boldsymbol{\alpha} & u_{dd}\ \boldsymbol{\alpha} & u_{di}\ \boldsymbol{\alpha} \\ u_{is}\ \boldsymbol{\alpha} & u_{id}\ \boldsymbol{\alpha} & 1 + u_{ii}\ \boldsymbol{\alpha} \end{bmatrix} \begin{bmatrix} \hat{a}_s \\ \hat{a}_d \\ \hat{a}_i \end{bmatrix} = \begin{bmatrix} 0 \\ 0 \\ 1'y \end{bmatrix} \qquad \text{(c.8)}$$

where the *u* terms are elements of $\mathbf{A}^{-1}$.

From Eqn c.8, the equation for solution of the sire is:

$$u_{ss}\boldsymbol{\alpha}\hat{a}_s = 0 - u_{sd}\boldsymbol{\alpha}\hat{a}_d - u_{si}\boldsymbol{\alpha}\hat{a}_i$$

$$u_{ss}\boldsymbol{\alpha}\hat{a}_s = PC$$

with:

$$PC = 0 - u_{sd}\ \boldsymbol{\alpha}\hat{a}_d - u_{si}\ \boldsymbol{\alpha}\hat{a}_i$$

When the mate is known:

$$PC = 0 - \frac{1}{2}\boldsymbol{\alpha}\hat{a}_d + (1)\boldsymbol{\alpha}\hat{a}_i$$

$$PC = \boldsymbol{\alpha}(\hat{a}_i - \frac{1}{2}\hat{a}_d) = 0.5\boldsymbol{\alpha}(2\hat{a}_i - \hat{a}_d)$$

In general, assuming sire s has *k* progeny:

$$PC_s = 0.5\boldsymbol{\alpha}\sum_k u_{prog}(2\hat{a}_i - \hat{a}_m) / \sum_k u_{prog}$$

where u_{prog} is 1 when the mate of *s* is known or $\frac{2}{3}$ when the mate is not known.

Appendix D: Methods for Obtaining Approximate Reliability for Genetic Evaluations

D.1 Computing Approximate Reliabilities for an Animal Model

Presented below is a method published by Meyer (1989) for obtaining approximate values of repeatability or reliability for genetic evaluations from an animal model and has been used to estimate reliabilities in the national dairy evaluation system in Canada. The reliability for each animal is derived from the corresponding diagonal element in the MME, adjusting for selected off-diagonal coefficients. For instance, the section of the coefficient matrix (**C**) pertaining to an animal i with parents f and m and with a record in a subclass h of a major fixed effect as HYS could be represented as:

$$\begin{bmatrix} c_{ii} & -\alpha & -\alpha & 1 \\ -\alpha & c_{ff} & 0.5\alpha & 0 \\ -\alpha & 0.5\alpha & c_{mm} & 0 \\ 1 & 0 & 0 & n_h \end{bmatrix}$$

where n_h is the number of records in subclass h of the major fixed effect $\alpha = \sigma_e^2 / \sigma_a^2$ and if this were the complete coefficient matrix for this animal, $\mathbf{C}^{-1}$ and, hence, true reliability could be obtained using partition matrix results. Thus, the coefficient c^{ii} can be calculated as the reciprocal of the ith diagonal element of **C** after absorbing all other rows and columns. For animal i:

$$c^{ii} = (c_{ii} - 1/n_h - \alpha^2(c_{ff} + c_{mm} - \alpha)/(c_{ff}c_{mm} - \frac{1}{4}\alpha^2))^{-1}$$

and for parent f:

$$c^{ff} = (c_{ff} - Q - (\frac{1}{2}\alpha - Q)^2 / (c_{mm} - Q))^{-1}$$

with:

$$Q = \alpha^2(c_{ii} - 1/n_h)^{-1}$$

Exchange m for f for parent m. However, if there are other off-diagonals for animal i, the above equations will yield approximations of the diagonal elements of **C** and hence reliability. Based on the above principle of forming and inverting the submatrix of the MME for each animal, Meyer outlined three steps for calculating approximate r^2, which were similar to the true r^2 from her simulation study. These steps are:

1. Diagonal elements (D) of animals with records are adjusted for the effect of the major fixed effects such as HYS. Thus:

$$D_{1i} = D_{0i} - 1/n_h$$

DOI: 10.1079/9781800620506.appx

and for animals without records:

$$D_{1i} = D_{0i}$$

where D_{0i} is the diagonal element for animal i in the MME and, in general, its composition, depending on the amount of information available on the animal, is:

$$D_{0i} = x_i + n_i\alpha + n_{1i}\alpha / 3 + n_{2i}\alpha / 2$$

where x_i = 1 if the animal has a record and otherwise it is zero, n_i equals 1 or $\frac{4}{3}$ or 2 if none or one or both parents are known, n_{1i} and n_{2i} are number of progeny with one or both parents known, respectively.

2. Diagonal elements for parents (f and m) are adjusted for the fact that the information on their progeny is limited. For each progeny i with only one parent known, adjust the diagonal element of the parent as:

$$D_{2f} = D_{1f} - \alpha^2(\frac{4}{9}D_{1i}^{-1})$$

and if both parents are known, adjust the diagonal of parent f as:

$$D_{2f} = D_{1f} - \alpha^2 D_{1i}^{-1}$$

Replace subscript f with m for the other parent. For animals that are not parents:

$$D_{2i} = D_{1i}$$

3. Adjustment of progeny diagonals for information on parents. This involves initially un-adjusting the diagonals of the parents for the contribution of the ith progeny in question by reversing step two before adjusting progeny diagonals for parental information. If only one parent f is known, the diagonal is un-adjusted, initially, as:

$$D_{2f}^{*} = D_{2f} + \alpha^2(\frac{4}{9}D_{2i}^{-1})$$

and if both parents are known as:

$$D_{2f}^{*} = D_{2f} - \alpha^2 D_{2i}^{-1}$$

for parent f. Exchange m for f in the above equation to calculate for parent m. Adjustment of progeny i diagonal, then, is:

$$D_{3i} = D_{2i} - \alpha^2 \frac{4}{9} D_{2f}^{*-1}$$

if only parent f is known and:

$$D_{3i} = D_{2i} - \alpha^2((D_{2f}^{*} + D_{2m}^{*} - \alpha) / (D_{2f}^{*}D_{2m}^{*} - \frac{1}{4}\alpha^2))$$

when both parents f and m are known. For animals with unknown parents:

$$D_{3i} = D_{2i}$$

Reliability for progeny i is calculated as:

$$r^2 = \text{const.}(1 - \alpha D_{3i}^{-1})$$

where const. is a constant of between 0.90 and 0.95 from Meyer simulation studies which gave the best estimate of r^2.

D.2 Computing Approximate Reliabilities for Random Regression Models

Meyer and Tier (2003) extended the method in Appendix D.1 to estimate reliabilities for multivariate and random regression models. They outlined several steps.

D.2.1 Determining value of observation for an animal

Compute the diagonal block (D*i*) for animal *i* in the MME, based on the information from the data, as:

$$\mathbf{D}_i = \mathbf{Z}_i'\mathbf{R}_i^{-1}\mathbf{Z}_i$$

However, to account for the limited subclass sizes of contemporary group effect, such as HTD in dairy cattle, $\mathbf{D}_i$ can better be calculated as:

$$\mathbf{D}_i = \mathbf{Z}_i'(\mathbf{R}_i^{-1} - \mathbf{R}_i^{-1}(\mathbf{S}_i^{-1})\mathbf{R}_i^{-1})\mathbf{Z}_i$$

where $\mathbf{Z}_i$ and $\mathbf{R}_i^{-1}$ are submatrices of $\mathbf{Z}$ and $\mathbf{R}^{-1}$ for animal *i*, and $\mathbf{S}_i$ is the block of coefficient matrix pertaining to the contemporary groups of which animal *i* is a member. Then the permanent environmental (pe) effects are also absorbed into the block corresponding to animal genetic effects:

$$\mathbf{D}_i^* = \mathbf{D}_i - \mathbf{Z}_i'\mathbf{R}_i^{-1}\mathbf{Q}_i(\mathbf{Q}_i'\mathbf{R}_i^{-1}\mathbf{Q}_i + \mathbf{P}^{-1})\mathbf{Q}_1'\mathbf{R}_i^{-1}\mathbf{Z}_i$$

where $\mathbf{Q}_i$ is a submatrix of the matrix $\mathbf{Q}$ defined in Section 10.3. Limited subclass effects of pe can be accounted for by using weights $w_m = (n_m - k)/n_m \le 1$, for the *m*th record, with n_m the size of the subclass to which the record belongs and *k* the number of 'repeated' records it has in that subclass. Then $\mathbf{R}_i$ in the above equation is replaced with $\mathbf{R}_i^* = \mathrm{Diag}(w_m\sigma_e^2)$

D.2.2 Value of records on descendants

In this second step, the contributions from progeny and other descendants are accumulated for each animal, processing the pedigree from youngest to the oldest. Let $\mathbf{E}_i$ be the block of contributions for animal *i* that has n_i progeny. Then:

$$\mathbf{E}_i = \frac{1}{3}\mathbf{G}^{-1} - \frac{4}{9}\mathbf{G}^{-1}\left(\mathbf{D}_i^* + \sum_{k=1}^{n_i}\mathbf{E}_k + \frac{4}{3}\mathbf{G}^{-1}\right)^{-1}\mathbf{G}^{-1}$$

This block is accumulated for both sire and dam of the *i*th animal. This equation can be derived by assuming each progeny has only one parent known and that the parent has no other information; then, the MME are set up for the animal and the parent and the equations for the animal are absorbed into those of the parent. The above equation will give an overestimate of the individual's contribution to its parents if it were in a contemporary group with many of its half-sibs. This can be discounted by weighting contributions with a factor dependent on the proportion of sibs in a subclass. Let $\mathbf{H}_i$ be a diagonal matrix of weights $w_m < 1$, with $w_m = \sqrt{(n_m - s_m)/n_m}$,where n_m is the total number of records in the subclass for trait m and s_m the total number of sibs of animal i in the subclass. Calculate $\mathbf{D}_i^{**} = \mathbf{H}_i\mathbf{D}_i^*\mathbf{H}_i$ and then replace $\mathbf{D}_i^*$ with $\mathbf{D}_i^{**}$.

D.2.3 Value of records on ancestors

In the third step, contributions from parents, ancestors and collateral relatives are accumulated for each animal, processing the pedigree from oldest to youngest. However, in step two, contributions from descendants were accumulated for all animals. Hence, $\mathbf{E}_j$ for parent j of animal i includes the contribution from animal i. The contributions of animal i have to be removed from $\mathbf{E}_j$ to avoid double-counting. The corrected block is:

$$\mathbf{E}_j^* = \frac{1}{3}\mathbf{G}^{-1} - \frac{4}{9}\mathbf{G}^{-1}(-\mathbf{E}_i + \mathbf{F}_i + \frac{4}{3}\mathbf{G}^{-1})^{-1}\mathbf{G}^{-1}$$

where $\mathbf{F}_j$ is the sum of contributions from all sources of information for parent j. As parents are processed in the pedigree before progeny, $\mathbf{F}_j$ is always computed before the contribution of parent j to animal i is required. For animal i, $\mathbf{F}_j$ is:

$$\mathbf{F}_i = \sum_{j=1}^{t_i}\mathbf{E}_j^* + \mathbf{D}_i^* + \sum_{k=1}^{n_i}\mathbf{E}_k$$

with t_i = 0, 1 or 2 denoting the number of parents of animal i that are known.

The matrix $\mathbf{T}_i$ of the approximate PEV and PEC for the genetic effects for animal i is:

$$\mathbf{T}_i = (\mathbf{F}_i + \mathbf{G})^{-1}$$

The approximate reliability for a linear function of EBVs for animal i, then, is:

$$r_i^2 = 1 - \mathbf{k}'\mathbf{T}_i\mathbf{k} / \mathbf{k}'\mathbf{G}\mathbf{k}$$

with $\mathbf{k}$ calculated as described in Section 10.3.4.

Appendix E

E.1 Canonical Transformation: Procedure to Calculate the Transformation Matrix and Its Inverse

The simplification of a multivariate analysis into n single-trait analyses using canonical transformation involves transforming the observations of several correlated traits into new uncorrected traits (Section 7.2). The transformation matrix **Q** can be calculated by the following procedure, which has been illustrated by the **G** and **R** matrices for Example 7.1 in Section 7.2.2.

The **G** and **R** matrices are, respectively:

$$\begin{matrix} \text{WWG} \\ \text{PWG} \end{matrix} \begin{bmatrix} 20 & 18 \\ 18 & 40 \end{bmatrix} \quad \text{and} \quad \begin{matrix} \text{WWG} \\ \text{PWG} \end{matrix} \begin{bmatrix} 40 & 11 \\ 11 & 30 \end{bmatrix}$$

where WWG is the pre-weanng gain and PWG is the post-weaning gain.

1. Calculate the eigenvalues (**B**) and eigenvectors (**U**) of **R**:

$$\mathbf{R} = \mathbf{UBU'}$$

For the above **R**:

$$\mathbf{B} = \text{diag}(47.083, 22.917)$$

and

$$\mathbf{U} = \begin{bmatrix} 0.841 & -0.541 \\ 0.541 & 0.841 \end{bmatrix}$$

2. Calculate **P** and **PGP′**:

$$\mathbf{P} = \mathbf{U}\sqrt{\mathbf{B}^{-1}}\ \mathbf{U'}$$

$$\mathbf{P} = \begin{bmatrix} 0.1642 & -0.0288 \\ -0.0288 & 0.1904 \end{bmatrix} \quad \text{and} \quad \mathbf{PGP'} = \begin{bmatrix} 0.403 & 0.264 \\ 0.264 & 1.269 \end{bmatrix}$$

3. Calculate the eigenvalues (**W**) and eigenvectors (**L**) of **PGP′**:

$$\mathbf{PGP'} = \mathbf{LWL'}$$

$$\mathbf{W} = \text{diag}(0.3283, 1.3436)$$

and

$$\mathbf{L} = \begin{bmatrix} 0.963 & 0.271 \\ -0.271 & 0.963 \end{bmatrix}$$

DOI: 10.1079/9781800620506.appx

4. The transformation matrix Q can be obtained as:

$$\mathbf{Q} = \mathbf{L'P}$$

$$\mathbf{Q} = \begin{bmatrix} 0.1659 & -0.0792 \\ 0.0168 & 0.1755 \end{bmatrix} \quad \text{and} \quad \mathbf{Q}^{-1} = \begin{bmatrix} 5.7651 & 2.6006 \\ -0.5503 & 5.4495 \end{bmatrix}$$

E.2 Canonical Transformation with Missing Records and Same-incidence Matrices

Ducrocq and Besbes (1993) presented a methodology for applying canonical transformation when all effects in the model affect all traits but there are missing traits for some animals. The principles of the methodology are briefly discussed and illustrated by an example.

Let **y**, the vector of observations, be partitioned as $\mathbf{y}' = [\mathbf{y}_v, \mathbf{y}_m]$ and $\mathbf{u} = [\mathbf{b}', \mathbf{a}']$, where $\mathbf{y}_v$ and $\mathbf{y}_m$ are vectors of observed and missing records, respectively; **b** is the vector of fixed effects and **a** is the vector of random effects. Assuming that the distribution of **y** given **u** is multivariate normal, Ducrocq and Besbes (1993) showed that the following expectation maximization (EM) algorithm gives the same solutions for **a** and **b** as when the usual multivariate MME are solved:

E step: at iteration *k*, calculate $\hat{\mathbf{y}}^{[k]} = E[\mathbf{y} \mid \mathbf{y}_v, \hat{\mathbf{u}}^{[k]}]$
M step: calculate $\hat{\mathbf{u}}^{[k+1]} =$ BLUE and BLUP solutions of **b** and **a**, respectively, given $\hat{\mathbf{y}}^{[k]}$

The E step implies doing nothing to observed records but replacing the missing observations by their expectation given the current solutions for **b** and **a**, and the observed records. The equation for the missing records for animal *i* is:

$$\hat{\mathbf{y}}_{im}^{[k]} = \mathbf{x}'_{im}\mathbf{b}^{[k]} + \hat{\mathbf{a}}_{im}^{[k]} + \hat{\mathbf{e}}_{im}^{[k]} \tag{e.1}$$

If **X** is the matrix that relates fixed effects to animals, $\mathbf{x}'_{im}$ denotes the row of **X** corresponding to missing records for animal *i* and $\hat{\mathbf{e}}_{im}^{[k]}$ is the regression of the residuals of missing records on the current estimates of the residuals for observed traits. Thus:

$$\hat{\mathbf{e}}_{im}^{[k]} = E\left[\mathbf{e}_{im} \mid \mathbf{y}_{iv}, \ \mathbf{u} = \hat{\mathbf{u}}^{[k]}\right] = \mathbf{R}_{mv}\mathbf{R}_{vv}^{-1}\left[\mathbf{y}_{iv} - \mathbf{x}'_{iv}\ \mathbf{b}^{[k]} - \hat{\mathbf{a}}_{iv}^{[k]}\right]$$

where $\mathbf{R}_{mv}$ and $\mathbf{R}_{vv}$ are submatrices obtained through partitioning of **R**, the residual covariance matrix. $\mathbf{R}_{vv}$ represents the residual variance of observed traits and $\mathbf{R}_{mv}$ is the covariance between missing traits and observed traits. If three traits are considered, for example, and trait 2 is missing for animal *i*, then $\mathbf{R}_{vv}$ is the submatrix obtained by selecting in **R** the elements at intersection of rows 1 and 3 and columns 1 and 3. The submatrix $\mathbf{R}_{mv}$ is the element at the intersection of row 2 and columns 1 and 3. Once the missing observations have been estimated, records are now available on all animals and the analysis can be carried out as usual, applying canonical transformation as when all records are observed.

The application of the method in genetic evaluation involves the following steps at each iteration *k*, assuming **Q** is the transformation matrix to canonical scale and $\mathbf{Q}^{-1}$ the back-transforming matrix:

1. For each animal *i* with missing observations:
(1a) calculate $\hat{\mathbf{y}}_{im}^{[k]}$, given $\hat{\mathbf{b}}^{[k]}$ and $\hat{\mathbf{a}}^{[k]}$ using Eqn e.1;
(1b) transform $\hat{\mathbf{y}}_i$ to the canonical scale: $\hat{\mathbf{y}}_i^* = \mathbf{Q}\hat{\mathbf{y}}_i$.

2. Solve the MME to obtain solutions in the canonical scale: $\hat{\mathbf{b}}^{*[k+1]}$ and $\hat{\mathbf{a}}^{*[k+1]}$.
3. Back-transform using Q–1 to obtain $\hat{\mathbf{b}}^{[k+1]}$ and $\hat{\mathbf{a}}^{[k+1]}$.
4. If convergence is not achieved, go to 1.

Ducrocq and Besbes (1993) showed that it is possible to update **y** (step 1) without back-transforming to the original scale (step 3) in each round of iteration. Suppose that the vector of observations for animal *i* with missing records, $\mathbf{y}_i$, is ordered such that observed records precede missing values: $\mathbf{y}_i' = [\mathbf{y}_{iv}', \mathbf{y}_{im}']$, and rows and columns of **R**, **Q** and $\mathbf{Q}^{-1}$ are ordered accordingly. Partition **Q** as $(\mathbf{Q}_v \mid \mathbf{Q}_m)$ and $\mathbf{Q}^{-1}$ as:

$$\mathrm{Q}^{-1} = \begin{bmatrix} \mathrm{Q}^{\mathrm{v}} \\ \mathrm{Q}^{\mathrm{m}} \end{bmatrix}$$

then from Eqn e.1, the equation for $\mathbf{Q}\hat{\mathbf{y}}_i$ or $\hat{\mathbf{y}}_i^*$ (see 1b) is:

$$\hat{\mathbf{y}}_i^* = \mathbf{Q}_v \mathbf{y}_{iv} + \mathbf{Q}_m \left[\mathbf{x}_{im}' \hat{\mathbf{b}}^{[k]} + \hat{\mathbf{a}}_{im}^{[k]} + \mathbf{R}_{mv} \mathbf{R}_{vv}^{-1} \left(\mathbf{y}_{iv} - \mathbf{x}_{iv}' \hat{\mathbf{b}}^{[k]} - \hat{\mathbf{a}}_{iv}^{[k]} \right) \right] \qquad \text{(e.2)}$$

However:

$$\begin{bmatrix} \hat{\mathrm{b}}_{\mathrm{iv}} \\ \hat{\mathrm{b}}_{\mathrm{im}} \end{bmatrix} = \mathrm{Q}^{-1}\hat{\mathrm{b}}^{*} = \begin{bmatrix} \mathrm{Q}^{\mathrm{v}}\hat{\mathrm{b}}^{*} \\ \mathrm{Q}^{\mathrm{m}}\hat{\mathrm{b}}^{*} \end{bmatrix}$$

and a similar expression exists for $\hat{\mathbf{a}}$. Substituting these values for $\hat{\mathbf{b}}$ and $\hat{\mathbf{a}}$ in Eqn e.2:

$$\begin{aligned} \hat{\mathbf{y}}_i^* &= \left(\mathbf{Q}_v + \mathbf{Q}_m \mathbf{R}_{mv} \mathbf{R}_{vv}^{-1} \right) \mathbf{y}_{iv} + \left(\mathbf{Q}_m \mathbf{Q}_m - \mathbf{Q}_m \mathbf{R}_{mv} \mathbf{R}_{vv}^{-1} \mathbf{Q}^v \right) \left(\mathbf{x}_i' \hat{\mathbf{b}}^{*[k]} + \hat{\mathbf{a}}^{*[k]} \right) \\ &= \mathbf{Q}_1 \mathbf{y}_{iv} + \mathbf{Q}_2 \left(\mathbf{x}_i' \hat{\mathbf{b}}^{*[k]} + \hat{\mathbf{a}}^{*[k]} \right) \end{aligned} \qquad \text{(e.3)}$$

with $\mathbf{Q}_1 = \mathbf{Q}_v + \mathbf{Q}_m \mathbf{R}_{mv} \mathbf{R}_{vv}^{-1}$ and $\mathbf{Q}_2 = \left(\mathbf{Q}_m \mathbf{Q}^m - \mathbf{Q}_m \mathbf{R}_{mv} \mathbf{R}_{vv}^{-1} \mathbf{Q}^v \right)$

Thus, for an animal with missing records, $\hat{\mathbf{y}}_i^*$ in Eqn e.3 is the updated vector of observation transformed to canonical scale (steps 1a and 1b above) and this is calculated directly without back-transformation to the original scale (step 3). The matrices $\mathbf{Q}_1$ and $\mathbf{Q}_2$ in Eqn e.3 depend on the missing pattern and if there are *n* missing patterns, *n* such matrices of each type must be set up initially and stored for use at each iteration.

E.2.1 Illustration

Using the same genetic parameters and data as for Example 5.3, the above methodology is employed to estimate sex effects and predict breeding values for pre-weaning weight and post-weaning gain iterating on the data (see Section 17.5). From Section E.1, **Q** is:

$$\mathrm{Q} = \begin{bmatrix} 0.1659 & -0.0792 \\ 0.0168 & 0.1755 \end{bmatrix} \quad \text{and} \quad \mathrm{Q}^{-1} = \begin{bmatrix} 5.7651 & 2.6006 \\ -0.5503 & 5.4495 \end{bmatrix}$$

Partitioning **Q** and $\mathbf{Q}^{-1}$ as specified above gives the following matrices:

$$\mathrm{Q}_v = \begin{bmatrix} 0.1659 \\ 0.0168 \end{bmatrix}, \quad \mathrm{Q}_m = \begin{bmatrix} -0.0792 \\ 0.1755 \end{bmatrix},$$

$$\mathrm{Q}^v = \begin{bmatrix} 5.7651 & 2.6006 \end{bmatrix} \quad \text{and} \quad \mathrm{Q}^m = \begin{bmatrix} -0.5503 & 5.4495 \end{bmatrix}$$

From the residual covariance matrix in Section E.1:

$$\mathbf{R}_{mv}\mathbf{R}_{vv}^{-1} = 11/40 = 0.275$$

The matrices $\mathbf{Q}_1$ and $\mathbf{Q}_2$, respectively, are:

$$\mathbf{Q}_1 = \begin{bmatrix} 0.1659 \\ 0.0168 \end{bmatrix} + \begin{bmatrix} -0.0792 \\ 0.1755 \end{bmatrix} 0.275 = \begin{bmatrix} 0.1441 \\ 0.0654 \end{bmatrix}$$

and

$$\mathbf{Q}_2 = \left[\begin{bmatrix} -0.0792 \\ 0.1755 \end{bmatrix} \begin{bmatrix} -0.5503 & 5.4495 \end{bmatrix} - \begin{bmatrix} -0.0792 \\ 0.1755 \end{bmatrix} 0.275 \begin{bmatrix} 5.7651 & 2.6006 \end{bmatrix} \right]$$

$$= \begin{bmatrix} 0.1691 & -0.3750 \\ -0.3748 & 0.8309 \end{bmatrix}$$

Employing steps 1–4, given earlier, to the data in Example 5.2, using the various transformation matrices given above and solving for sex and animal solutions by iterating on the data (see Section 17.5), gave the following solutions on the canonical scale at convergence. The solutions on the original scale are also presented.

	Canonical scale		Original scale	
Effects	VAR1	VAR2	WWG	PWG
Sex				
Male	0.180	1.265	4.326	6.794
Female	0.124	1.108	3.598	5.968
Animal				
1	0.003	0.053	0.154	0.288
2	−0.006	−0.010	−0.059	−0.054
3	0.003	−0.030	−0.062	−0.163
4	0.002	0.007	0.027	0.037
5	−0.010	−0.097	−0.307	−0.521
6	0.001	0.088	0.235	0.477
7	−0.011	−0.084	−0.280	−0.452
8	0.013	0.076	0.272	0.407
9	0.009	0.010	0.077	0.051

VAR1, $\mathbf{Q}y_1$, VAR2, $\mathbf{Q}y_2$ with WWG = $y1$ and PWG = y_2

These are similar to the solutions obtained from the multivariate analysis in Section 6.3 or the application of the Cholesky transformation in Section 7.3. The advantage of this methodology is that the usual univariate programmes can easily be modified to incorporate missing records.

The prediction of the missing record (PWG) for animal 4 using solutions on canonical and original scales at convergence is illustrated below.

Using Eqn e.1:

$$\begin{aligned} y_{42} &= \hat{b}_{12} + \hat{a}_{42} + \mathbf{R}_{mv}\mathbf{R}_{vv}^{-1}(y_{41} - \hat{b}_{11} - \hat{a}_{41}) \\ &= 6.836 + 0.016 + 0.275(4.5 - 4.366 - 0.007) \\ &= 6.9 \end{aligned}$$

where y_{ij} and $\hat{a}_{ij}$ are the record and EBV, respectively, for animal i and trait j, and $\hat{b}_{kj}$ is the fixed-effect solution for level k for trait j.

Using Eqn e.2:

$$\begin{bmatrix} \hat{y}^*_{41} \\ \hat{y}^*_{42} \end{bmatrix} = Q_1\, y_{41} + Q_2\left(x'\hat{b}^* + \hat{a}^*_4\right)$$

$$\begin{bmatrix} \hat{y}^*_{41} \\ \hat{y}^*_{42} \end{bmatrix} = \begin{bmatrix} 0.648 \\ 0.294 \end{bmatrix} + Q_2 \begin{bmatrix} 0.183 \\ 1.273 \end{bmatrix} + Q_2 \begin{bmatrix} 0.000 \\ 0.003 \end{bmatrix}$$

$$= \begin{bmatrix} 0.648 \\ 0.294 \end{bmatrix} + \begin{bmatrix} -0.446 \\ 0.989 \end{bmatrix} = \begin{bmatrix} 0.202 \\ 1.283 \end{bmatrix}$$

These predicted records for animal 4 are on the canonical scale and they are used for the next round of iteration if convergence has not been achieved. These predicted records can be transformed to the original scale as:

$$\begin{bmatrix} \hat{y}_{41} \\ \hat{y}_{42} \end{bmatrix} = Q^{-1} \begin{bmatrix} 0.202 \\ 1.283 \end{bmatrix} + \begin{bmatrix} 4.5 \\ 6.9 \end{bmatrix}$$

The record for WWG is as observed and predicted missing record for PWG is the same as when using Eqn e.1.

E.3 Cholesky Decomposition

Any positive semi-definite symmetric matrix **R** can be expressed in the form **TT′**, where **T** is a lower triangular matrix. The matrix **T** can be calculated using the following formulae. The ith diagonal element of **T** is calculated as:

$$t_{ii} = r_{ii} - \sqrt{\sum_{j=1}^{i-1} t_{ij}^2}$$

and the lower off-diagonal element of the ith row and the kth column of **T** as:

$$t_{ik} = \frac{1}{t_{kk}}\left(r_{ik} - \sum_{j=1}^{k-1} t_{ij} t_{kj}\right)$$

Appendix F: Procedure for Computing De-regressed Breeding Values

The de-regressed breeding values (DRB) of bulls used in multi-trait across-country evaluations (MACE) are obtained by solving Eqn 6.15 for **y** considering data from only one country at a time. Jairath *et al.* (1998) presented an algorithm for calculating DRP. For instance, Eqn 5.15 for country *i* can be written as:

$$\begin{pmatrix} \mathbf{1}'\mathbf{R}_i^{-1}\mathbf{1} & \mathbf{1}'\mathbf{R}_i^{-1} & 0 & 0 \\ \mathbf{R}_i^{-1}\mathbf{1} & \mathbf{R}_i^{-1}+\mathbf{A}_{nn}^{-1}\boldsymbol{\alpha}_i & \mathbf{A}_{np}^{-1}\boldsymbol{\alpha}_i & \mathbf{A}_{ng}^{-1}\boldsymbol{\alpha}_i \\ 0 & \mathbf{A}_{pn}^{-1}\boldsymbol{\alpha}_i & \mathbf{A}_{pp}^{-1}\boldsymbol{\alpha}_i & \mathbf{A}_{pg}^{-1}\boldsymbol{\alpha}_i \\ 0 & \mathbf{A}_{gn}^{-1}\boldsymbol{\alpha}_i & \mathbf{A}_{gp}^{-1}\boldsymbol{\alpha}_i & \mathbf{A}_{gg}^{-1}\boldsymbol{\alpha}_i \end{pmatrix} \begin{pmatrix} \hat{\boldsymbol{\mu}}_i \\ \mathbf{Qg}_i+\hat{\mathbf{s}}_i \\ \hat{\mathbf{p}}_i \\ \hat{\mathbf{g}}_i \end{pmatrix} = \begin{pmatrix} \mathbf{1}'\mathbf{R}_i^{-1}\mathbf{y}_i \\ \mathbf{R}_i^{-1}\mathbf{y}_i \\ 0 \\ 0 \end{pmatrix} \qquad \text{(f.1)}$$

where $\mathbf{p}_i$ is the vector of identified parents without EBV and $\mathbf{A}_{jj}^{-1}$ are blocks of the inverse of the relationship (see Chapter 4, Section 4.6) with $j = n$, p and g for animals with records, ancestors and genetic groups, respectively, and $\boldsymbol{\alpha}_i = (4-h_i^2)/h_i^2$, the ratio of residual variance to sire variance for the *i*th country. The de-regression of EBV involves solving Eqn f.1 for $\mathbf{y}_i$. The constant $\boldsymbol{\mu}_i$ and vectors $\mathbf{s}_i$, $\mathbf{p}_i$, $\mathbf{g}_i$ and $\mathbf{y}_i$ are unknown but $\mathbf{a}_i$, the vector of genetic evaluations for sires, is known, as well as matrices Q, $\mathbf{R}^{-1}{}_i$ and $\mathbf{A}_{jj}^{-1}$. Let $\mathbf{a}_i = \mathbf{1}\boldsymbol{\mu}_i + \mathbf{Qg}_i + \mathbf{s}_i$. The following iterative procedure can be used to compute the vector of DRB, $\mathbf{y}_i$:

1. Set $\mathbf{1}\boldsymbol{\mu}_i$, $\mathbf{p}_i$, $\mathbf{s}_i$, and $\mathbf{g}_i$ to 0.
2. Calculate $\mathbf{Qg}_i + \mathbf{s}_i = \mathbf{a}_i - \mathbf{1}\boldsymbol{\mu}_i$.
3. Compute:

$$\begin{pmatrix} \hat{\mathbf{p}}_i \\ \hat{\mathbf{g}}_i \end{pmatrix} = -\begin{pmatrix} \mathbf{A}_{pp}^{-1} & \mathbf{A}_{pg}^{-1} \\ \mathbf{A}_{gp}^{-1} & \mathbf{A}_{gg}^{-1} \end{pmatrix}^{-1} \begin{pmatrix} \mathbf{A}_{pn}^{-1} \\ \mathbf{A}_{gn}^{-1} \end{pmatrix} (\mathbf{Q}\hat{\mathbf{g}}_i + \hat{\mathbf{s}}_i).$$

4. Generate:

$$\mathbf{R}_i^{-1}\mathbf{y}_i = \mathbf{R}_i^{-1}\mathbf{1}\boldsymbol{\mu}_i + (\mathbf{R}_i^{-1} + \mathbf{A}_{nn}^{-1})(\mathbf{Q}\hat{\mathbf{g}}_i + \mathbf{s}_i) + \mathbf{A}_{pn}^{-1}\mathbf{p}_i\boldsymbol{\alpha}_i + \mathbf{A}_{gn}^{-1}\hat{\mathbf{g}}_i\boldsymbol{\alpha}_i$$

and $\mathbf{1}'\mathbf{R}_i^{-1}\mathbf{y}_i$.

5. Now calculate:

$$\boldsymbol{\mu}_i(\mathbf{1}'\mathbf{R}_i^{-1}\mathbf{1})^{-1}\mathbf{1}'\mathbf{R}_i^{-1}\mathbf{y}_i$$

6. Continue at step two until convergence is achieved.
7. Then, compute DRB as $\mathbf{y}_i = \mathbf{R}_i(\mathbf{R}_i^{-1}\mathbf{y}_i)$.

Using the data for country 1 in Example 6.5, the de-regression steps above are illustrated in the first iteration. For country 1, $\boldsymbol{\alpha}_1 = 206.50/20.5 = 10.0732$ and, considering only the bulls with evaluations, $\mathbf{R}_1 = \text{diag}(0.0172, 0.0067, 0.0500, 0.0400)$.

10.1079/9781800620506.appx

The pedigree structure (see Example 5.5) used for the de-regression of breeding values in country 1 is:

Bull	Sire	MGS	MGD
1	5	G2	G3
2	6	7	G4
3	5	2	G4
4	1	G2	G4
5	G1	G2	G3
6	G1	G2	G3
7	G1	G2	G3

The matrix $\mathbf{A}_1^{-1}$ was calculated according to the rules in Section 5.5.2.
In the first round of iteration, the transpose of the vector $\mathbf{Qg}_1 + \mathbf{s}_1$ in step two above is:

$$(\mathbf{Qg}_1 + \mathbf{s}_1)' = (9.0\ 10.1\ 15.8\ -4.7)$$

The vector of solutions for $\mathbf{p}_1$ and $\mathbf{g}_1$ in step three is computed as:

$$\begin{bmatrix} \hat{\mathbf{p}}_1 \\ \hat{\mathbf{g}}_1 \end{bmatrix} = \begin{bmatrix} 17.094 & 0.000 & 0.000 & -5.037 & -0.839 & -0.839 & 1.831 \\ 0.000 & 13.736 & 1.831 & -5.037 & -2.518 & -2.518 & 1.831 \\ 0.000 & 1.831 & 10.989 & -5.037 & -2.518 & -2.518 & 0.916 \\ -5.037 & -5.037 & -5.037 & 8.555 & 3.777 & 3.777 & 0.000 \\ -0.839 & -2.518 & -2.518 & 3.777 & 4.568 & 2.728 & 0.839 \\ 0.839 & -2.518 & -2.518 & 3.777 & 2.728 & 3.728 & 0.000 \\ 1.831 & 1.831 & 0.916 & 0.000 & 0.839 & 0.000 & 3.671 \end{bmatrix}^{-1} \begin{bmatrix} 6.716 & -1.831 & 7.326 & 0.000 \\ 0.000 & 7.326 & 0.000 & 0.000 \\ 0.000 & 3.663 & 0.000 & 0.000 \\ 0.000 & 0.000 & 0.000 & 0.000 \\ 1.679 & 0.000 & 0.000 & 3.357 \\ 3.357 & 0.000 & 0.000 & 0.000 \\ -1.679 & 2.747 & 3.663 & 3.357 \end{bmatrix} \begin{bmatrix} 9.0 \\ 10.1 \\ 15.8 \\ -4.7 \end{bmatrix} = \begin{bmatrix} 16.330 \\ 12.861 \\ 12.622 \\ 23.481 \\ -9.801 \\ 12.375 \\ -0.564 \end{bmatrix}$$

The transpose of the vector $(\mathbf{R}_1^{-1}\mathbf{y}_1)$ in step four is: (30.2 9.0 10.1 15.8) and:

$$\mathbf{1}'\mathbf{R}_1^{-1}\mathbf{y}_1 = 2235.50$$

Therefore, in the first round of iteration (step four):

$$\boldsymbol{\mu}_1 = 2235.50 / 253 = 8.835$$

Convergence was achieved after about six iterations. The transpose of the vector $(\mathbf{R}_1^{-1}\mathbf{y}_1)$ after convergence is:

$$(\mathbf{R}_1^{-1}\mathbf{y}_1)' = (563.928\ 1495.751\ 385.302\ -241.278)$$

with $\mathbf{R}_1^{-1} = \text{diag}(0.0172, 0.0067, 0.050, 0.04)$, the transpose of the vector of DRB calculated in step seven is:

$$\mathbf{y}_1' = (9.7229\ 9.9717\ 9.2651\ -8.5711)$$

Appendix G: Calculating Φ, a Matrix of Legendre Polynomials Evaluated at Different Ages or Time Periods

The matrix $\mathbf{\Phi}$ is of order t (the number of days in milk or ages) by k (where k is the order of fit) with element $\phi_{ij} = \phi_j(a_t)$ equals the jth Legendre polynomial evaluated at the tth standardized age or days in milk (DIM). Thus, a_t is the tth DIM or age standardized to the interval for which the polynomials are defined. Kirkpatrick *et al.* (1990, 1994) used Legendre polynomials that span the interval −1 to +1. Defining d_{min} and d_{max} as the first and latest DIM on the trajectory, DIM d_t can be standardized to a_t as:

$$a_t = -1 + 2(d_t - d_{min}) / (d_{max} - d_{min})$$

In matrix notation, $\mathbf{\Phi} = \mathbf{M\Lambda}$, where $\mathbf{M}$ is the matrix containing the polynomials of the standardized DIM values and $\mathbf{\Lambda}$ is a matrix of order k containing the coefficients of Legendre polynomials. The elements of $\mathbf{M}$ can be calculated as $m_{ij} = (a_i^{(j-1)}, i = 1,\ldots t; j = 1,\ldots k)$. For instance, given that k = 5 and that t = 3 (three standardized DIM), $\mathbf{M}$ is:

$$\mathbf{M} = \begin{bmatrix} 1 & a_1 & a_1^2 & a_1^3 & a_1^4 \\ 1 & a_2 & a_2^2 & a_2^3 & a_2^4 \\ 1 & a_3 & a_3^2 & a_3^3 & a_3^4 \end{bmatrix}$$

Using the fat-yield data in Table 9.1 as an illustration, with ten DIM, the vector of standardized DIM is:

$$\mathbf{a}' = [-1.0\ -0.7778\ -0.5556\ -0.3333\ -0.1111\ 0.1111\ 0.3333\ 0.5556\ 0.7778\ 1.0]$$

and $\mathbf{M}$ is:

$$\mathbf{M} = \begin{bmatrix} 1.0000 & -1.0000 & 1.0000 & -1.0000 & 1.0000 \\ 1.0000 & -0.7778 & 0.6049 & -0.4705 & 0.3660 \\ 1.0000 & -0.5556 & 0.3086 & -0.1715 & 0.0953 \\ 1.0000 & -0.3333 & 0.1111 & -0.0370 & 0.0123 \\ 1.0000 & -0.1111 & 0.0123 & -0.0014 & 0.0002 \\ 1.0000 & 0.1111 & 0.0123 & 0.0014 & 0.0002 \\ 1.0000 & 0.3333 & 0.1111 & 0.0370 & 0.0123 \\ 1.0000 & 0.5556 & 0.3086 & 0.1715 & 0.0953 \\ 1.0000 & 0.7778 & 0.6049 & 0.4705 & 0.3660 \\ 1.0000 & 1.0000 & 1.0000 & 1.0000 & 1.0000 \end{bmatrix}$$

DOI: 10.1079/9781800620506.appx

Next, the matrix **Λ** of Legendre polynomials needs to be computed. The *j*th Legendre polynomial evaluated at age $t(P_j(t))$ can in general be evaluated by the formula given by Abramowitz and Stegun (1965). In general, for the *j* integral:

$$P_j(t) = \frac{1}{2^j}\sum_{r=0}^{j/2}\frac{(-1)^r(2j-2r)!}{r!(j-r)!(j-2r)!}t^{j-2r}$$

where $j/2 = (j-1)/2$ if *j* is odd. The first five Legendre polynomials therefore are:

$$P_0(t) = 1; P_1(t) = t; P_2(t) = \frac{1}{2}(3t^2 - 1)$$

$$P_3(t) = \frac{1}{2}(5t^3 - 3t); \text{ and } P_4(t) = \frac{1}{8}(35t^4 - 30t^2 + 3)$$

The normalized value of the *j*th Legendre polynomial evaluated at age t $(\varphi_j(t))$ can be obtained as:

$$\phi_j(t) = \sqrt{\frac{2n+1}{2}}P_j(t)$$

Thus:

$$\phi_0(t) = \sqrt{\frac{1}{2}}\,P_0(t) = 0.7071; \;\phi_1(t) = \sqrt{\frac{3}{2}}\,P_1(t) = 1.2247(t)$$

$$\phi_2(t) = \sqrt{\frac{5}{2}}\,P_2(t) = 2.3717(t^2) - 0.7906; \;\phi_3(t) = \sqrt{\frac{7}{2}}\,P_3(t) = 4.6771(t^3) - 2.8067(t)$$

$$\text{and } \phi_4(t) = \sqrt{\frac{9}{2}}\,P_4(t) = 9.2808(t^4) - 7.9550(t^2) + 0.7955$$

Therefore, for $t = 5$ in Example 9.1, **Λ** is:

$$\mathbf{\Lambda} = \begin{bmatrix} 0.7071 & 0.0000 & -0.7906 & 0.0000 & 0.7955 \\ 0.0000 & 1.2247 & 0.0000 & -2.8067 & 0.0000 \\ 0.0000 & 0.0000 & 2.3717 & 0.0000 & -7.9550 \\ 0.0000 & 0.0000 & 0.0000 & 4.6771 & 0.0000 \\ 0.0000 & 0.0000 & 0.0000 & 0.0000 & 9.2808 \end{bmatrix}$$

and **Φ** = **MΛ** is:

$$\mathbf{\Phi} = \begin{bmatrix} 0.7071 & -1.2247 & 1.5811 & -1.8704 & 2.1213 \\ 0.7071 & -0.9525 & 0.6441 & -0.0176 & -0.6205 \\ 0.7071 & -0.6804 & -0.0586 & 0.7573 & -0.7757 \\ 0.7071 & -0.4082 & -0.5271 & 0.7623 & 0.0262 \\ 0.7071 & -0.1361 & -0.7613 & 0.3054 & 0.6987 \\ 0.7071 & 0.1361 & -0.7613 & -0.3054 & 0.6987 \\ 0.7071 & 0.4082 & -0.5271 & -0.7623 & 0.0262 \\ 0.7071 & 0.6804 & -0.0586 & -0.7573 & -0.7757 \\ 0.7071 & 0.9525 & 0.6441 & 0.0176 & -0.6205 \\ 0.7071 & 1.2247 & 1.5811 & 1.8704 & 2.1213 \end{bmatrix}$$

References

Abramowitz, M. and Stegun, I.A. (1965) *Handbook of Mathematical Functions*. Dover, New York.

Aguilar, I. and Misztal, I. (2008) Technical note: recursive algorithm for inbreeding coefficients assuming nonzero inbreeding of unknown parents. *Journal of Dairy Science* 91, 1669–1672.

Aguilar, I., Misztal, I., Johnson, D.L., Legarra, A., Tsuruta, S. and Lawlor, T.J. (2010) Hot topic: a unified approach to utilize phenotypic, full pedigree, and genomic information for genetic evaluation of Holstein final score. *Journal of Dairy Science* 93, 743–752.

Albuquerque, L.G. and Meyer, K. (2001) Estimates of covariance functions for growth from birth to 630 days in Nelore cattle. *Journal of Animal Science* 79, 2776–2789.

Aliloo, H., Pryce, J.E., González-Recio, O., Cocks, B.G., Goddard, M.E. and Hayes, B.J. (2017) Including nonadditive genetic effects in mating programs to maximize dairy farm profitability. *Journal of Dairy Science* 100, 1203–1222.

Andersen, S. and Pedersen, B. (1996) Growth and food intake curves for grouphoused gilts and castrated male pigs. *Animal Science* 63, 457–464.

Anderson, T.W. (1958) *An Introduction to Multivariate Statistical Analysis*. Wiley, New York.

Bastiaansen, J.W.M., Bovenhuis, H., Lopes, M., Silva, F., Megens, H. and Calus, M.P.L. (2014) SNP effects depend on genetic and environmental context. In: *Proceedings of the 10th World Congress on Genetics Applied to Livestock Production*, 17–22 August, Vancouver.

Berger, P.J., Luecke, G.R. and Hoekstra, J.A. (1989) Iterative algorithms for solving mixed model equations. *Journal of Dairy Science* 72, 514–522.

Bergsma, R., Kanis, E., Knol, E.F. and Bijma, P. (2008) The contribution of social effects to heritable variation in finishing traits of domesticated pigs (*Sus scrofa*). *Genetics* 178, 1559–1570.

Bijma, P., Muir, W.M. and Van Arendonk, J.A.M. (2007a) Multilevel selection 1: quantitative genetics of inheritance and response to selection. *Genetics* 175, 277–288.

Bijma, P., Muir, W.M., Ellen, E.D., Wolf, J.B. and Van Arendonk, J.A.M. (2007b) Multilevel selection 2: estimating the genetic parameters determining inheritance and response to selection. *Genetics* 175, 289–299.

Bohmanova, J., Miglior, F., Jamrozik, J., Misztal, I. and Sullivan, P.G. (2008) Comparison of random regression models with Legendre polynomials and linear splines for production traits and somatic cell score of Canadian Holstein cows. *Journal of Dairy Science* 91, 3627–3638.

Boichard, D., Bonaiti, B., Barbat, A. and Mattalia, S. (1995) Three methods to validate the estimation of genetic trend for dairy cattle. *Journal of Dairy Science* 78, 431–437.

Boichard, D., Fritz, S., Rossignol, M.N., Boscher, M.Y., Malafosse, A. and Colleau, J.J. (2002) Implementation of marker-assisted selection in French dairy cattle. In: *Proceedings of the 7th World Congress on Genetics Applied to Livestock Production*, August, Montpellier, France. Communication no. 22-03.

Bolormaa, S., Pryce, J.E., Zhang, Y., Reverter, A., Barendse, W., Hayes, B.J. and Goddard, M.E. (2015) Non-additive genetic variation in growth, carcass and fertility traits of beef cattle. *Genetics Selection Evolution* 47, 26.

Bradford, H.L., Pocrnic, I., Fragomeni, B.O., Lourenco, D.A.L. and Misztal, I. (2017) Selection of core animals in the Algorithm for Proven and Young using a simulation model. *Journal of Animal Breeding and Genetics* 134, 545–552.

Brotherstone, S. and Hill, W.G. (1991) Dairy herd life in relation to linear type traits and production. 2 Genetic analyses for pedigree and non-pedigree cows. *Animal Production* 53, 289–297.

Brotherstone, S., Veerkamp, R.F. and Hill, W.G. (1997) Genetic parameters for a simple predictor of the lifespan of Holstein-Friesian dairy cattle and its relationship to production. *Animal Science* 65, 31.
Brotherstone, S., White, I.M.S. and Meyer, K. (2000) Genetic modelling of daily milk yield using orthogonal polynomials and parametric curves. *Animal Science* 70, 407–415.
Bulmer, M.G. (1980) *The Mathematical Theory of Quantitative Genetics*. Clarendon Press, Oxford, UK.
Cameron, N.D. (1997) *Selection Indices and Prediction of Genetic Merit in Animal Breeding*. CAB International, Wallingford, UK.
Chen, C.Y., Misztal, I., Aguilar, I., Legarra, A. and Muir, W.M. (2011) Effect of different genomic relationship matrices on accuracy and scale. *Journal of Animal Science* 89, 2673–2679.
Cheverud, J.M. and Moore, A.J. (1994) Quantitative genetics and the role of the environment provided by relatives in the evolution of behaviour. In: Boake, C.R.B. (ed.) *Quantitative Genetics of Behavioral Evolution.* University of Chicago Press, Chicago, Illinois, pp. 60–100.
Christensen, O.F. (2012) Compatibility of pedigree-based and marker-based relationship matrices for single-step genetic evaluation. *Genetics Selection Evolution* 44, 37.
Christensen, O.F. and Lund, M. (2010) Genomic prediction when some animals are not genotyped. *Genetics Selection Evolution* 42, 2.
Christensen, O.F., Madsen, P., Nielsen, B., Ostersen, T. and Su, G. (2012) Single-step methods for genomic evaluation in pigs. *Animal* 6, 1565–1571.
Christensen, O.F., Madsen, P., Nielsen, B. and Su, G. (2014) Genomic evaluation of both purebred and crossbred performances. *Genetics Selection Evolution* 46, 23.
Christensen, O.F., Legarra, A., Lund, M.S. and Su, G. (2015) Genetic evaluation for three way crossbreeding. *Genetics Selection Evolution* 47, 98.
Christensen, O.F., Nielsen, B., Su, G., Xiang, T., Madsen, P., Ostersen, T., Velander, I. and Strathe, A.B. (2019) A bivariate genomic model with additive, dominance and inbreeding depression effects for sire line and three-way crossbred pigs. *Genetics Selection Evolution* 51, 45.
Cockerham, C.C. (1954) An extension of the concept of partitioning hereditary variance for analysis of covariances among relatives when epistasis is present. *Genetics* 39, 859–882.
Cox, D.R. (1972) Regression models and life tables. *Journal of the Royal Statistical Society B* 34, 187–220.
De Boer, I. and Hoeschele, I. (1993) Genetic evaluation methods for populations with dominance and inbreeding. *Theoretical and Applied Genetics* 86, 245–258.
Dekkers, J.C.M. (2007) Marker-assisted selection for commercial crossbred performance. *Journal of Animal Science* 85, 2104–2114.
Dempster, A.P., Laird, N.M. and Rubin, D.B. (1977) Maximum likelihood from incomplete data via EM algorithm. *Journal of the Royal Statistical Society B* 39, 1–38.
Ducrocq, V. (1992) Solving animal model equations through an approximate incomplete Cholesky decomposition. *Genetics Selection Evolution* 24, 193–209.
Ducrocq, V. (1997) Survival analysis, a statistical tool for longevity data. *48th Annual Meeting of the European Association for Animal Production*, 25–28 August, Vienna.
Ducrocq, V. (2000) *Survival Analysis Applied to Animal Breeding*. Station de Génétique Quantitative et Appliqueé. Institut National de la Recherche Agronomique. F-&8352 Jouy-en-Josas Cedex, France.
Ducrocq, V. and Besbes, B. (1993) Solutions of multiple trait animal models with missing data on some traits. *Journal of Animal Breeding and Genetics* 110, 81–89.
Ducrocq, V. and Solkner, J. (1998) Implementation of a routine breeding value evaluation for longevity of dairy cows using survival analysis techniques. In: *Proceedings of the 6th World Congress on Genetics Applied to Livestock Production*, January, Armidale, Australia, pp. 359–362.
Ducrocq, V., Quaas, R.L., Pollak, E.J. and Casella, G. (1988a) Length of productive life of dairy cows. 1. Justification of a Weibull model. *Journal of Dairy Science* 71, 3061–3070.

Ducrocq, V., Quaas, R.L., Pollak, E.J. and Casella, G. (1988b) Length of productive life of dairy cows. 2. Variance component estimation and sire evaluation. *Journal of Dairy Science* 71, 3071–3079.
Emik, L.O. and Terrill, C.E. (1949) Systematic procedures for calculating inbreeding coefficients. *Journal of Heredity* 40, 51–55.
Everett, R., Keown, J.R. and Clapp, E.E. (1976) Relationships among type, production and stayability in Holstein sire evaluation. *Journal of Dairy Science* 59, 1277–1285.
Falconer, D.S. and Mackay, T.F.C. (1996) *Introduction to Quantitative Genetics*, 4th edn. Longman, Harlow, UK.
Fernando, R.L. (2010) Genomic selection in livestock. Animal breeding and genetics short courses. Summer 2010. Iowa State University, Ames, Iowa.
Fernando, R.L. and Grossman, M. (1989) Marker-assisted selection using best linear unbiased prediction. *Genetics Selection Evolution* 21, 467–477.
Fernando, R.L., Billingsley, R.D. and Gianola, D. (1983) Effects of method of scaling on heritability estimates and sire evaluations for frame size at weaning in Angus cattle. *Journal of Animal Science* 56, 1047–1056.
Fernando, R.L., Dekkers, J.C. and Garrick, D.J. (2014) A class of Bayesian methods to combine large numbers of genotyped and non-genotyped animals for whole-genome analyses. *Genetics Selection Evolution* 46, 50.
Fernando, R.L., Cheng, H., Golden, B.L. and Garrick, D.J. (2016) Computational strategies for alternative single-step Bayesian regression models with large numbers of genotyped and non-genotyped animals. *Genetics Selection Evolution* 48, 96.
Fisher, R.A. (1919) The correlation between relatives on the supposition of Mendelian inheritance. *Transactions of the Royal Society, Edinburgh* 52, 399–433. Available at: https://doi.org/10.1017/S0080456800012163 (accessed 11 March 2023).
Fiske, W.F. and Banos, G. (2001) Weighting factors of sire daughter information in international genetic evaluations. *Journal of Dairy Science* 84, 1759–1767.
Foulley, J.L., Gianola, D. and Thompson, R. (1983) Prediction of genetic merit from data on binary and quantitative variates with an application to calving difficulty, birth weight and pelvic opening. *Genetics Selection Evolution* 15, 401–424.
Fragomeni, B.O., Lourenco, D.A., Tsuruta, S., Masuda, Y., Aguilar, I., Legarra, A., Lawlor, T.J. and Misztal, I. (2015) Hot topic: use of genomic recursions in single-step genomic best linear unbiased predictor (BLUP) with a large number of genotypes. *Journal of Dairy Science* 98, 4090–4094.
Garcia-Baccino, C.A., Legarra, A., Christensen, O.F., Misztal, I., Pocrnic, I., Vitezica, Z.G. and Cantet, R.J.C (2017) Metafounders are related to Fst fixation indices and reduce bias in single step genomic evaluations. *Genetics Selection Evolution* 49, 34.
García-Cortés, L.A. and Toro, M.A. (2006) Multibreed analysis by splitting the breeding values. *Genetics Selection Evolution* 38, 601–615.
Gaynor, R.C., Gorjanc, G. and Hickey, J.M. (2021) AlphaSimR: an R package for breeding program simulations. *G3 (Bethesda, Md.)* 11(2), jkaa017.
Geman, S. and Geman, D. (1984) Stochastic relaxation, Gibbs distributions, and the Bayesian restoration of images. *IEEE Transactions on Pattern Analysis and Machine Intelligence* 6, 721–741.
Gengler, N., Mayeres, P. and Szydlowski, M. (2007) A simple method to approximate gene content in large pedigree populations: application to the myostatin gene in dual-purpose Belgian Blue cattle. *Animal* 1, 21–28.
Gianola, D. (1982) Theory and analysis of threshold characters. *Journal of Animal Science* 54, 1079–1096.
Gianola, D. and Foulley, J.L. (1983) Sire evaluation for ordered categorical data with a threshold model. *Genetics Selection Evolution* 15, 201–224.
Gilmour, A.R., Thompson, R. and Cullis, B. (1995) Average information REML, an efficient algorithm for variance parameter estimation in linear mixed models. *Biometrics* 51, 1440–1450.

Gilmour, A.R., Gogel, B.J., Cullis, B.R., Welham, S.J. and Thompson, R. (2003) *ASREML UserGuide Release 1.0*. VSN International, Hemel Hempstead, UK.
Goddard, M.E. (1985) A method of comparing sires evaluated in different countries. *Livestock Production Science* 13, 321–331.
Goddard, M.E. (1992) A mixed model for the analyses of data on multiple genetic markers. *Theoretical and Applied Genetics* 83, 878–886.
Goddard, M.E. (1999) New technology to enhance genetic improvement of pigs. *Manipulating Pig Production* 7, 44–52.
Goddard M.E. (2009) Genomic selection: prediction of accuracy and maximisation of long term response. *Genetica* 136, 245–252. DOI: 10.1007/s10709-008-9308-0
Goddard, M.E. and Hayes, B.J. (2002) Optimisation of response using molecular data. In: *Proceedings of the 7th World Congress on Genetics Applied to Livestock Production*, August, Montpellier, France. Communication no. 22-01.
González-Diéguez, D., Legarra, A., Charcosset, A., Moreau, L., Lehermeier, C., Teyssedre, S. and Vitezica, Z.G. (2021) Genomic prediction of hybrid crops allows disentangling dominance and epistasis. *Genetics* 218, iyab026.
Goriely, A. (2016) Decoding germline de novo point mutations. *Nature Genetics* 48, 823–824.
Griffing, B. (1967) Selection in reference to biological groups. I. Individual and group selection applied to populations of unordered groups. *Australian Journal of Biological Science* 20, 127–142.
Groeneveld, E. (1990) *PEST User's Manual*. Institute of Animal Husbandry and Animal Behaviour, Federal Agricultural Research Centre, Neustadt, Germany.
Groeneveld, E. and Kovac, M. (1990) A generalised computing procedure for setting up and solving mixed models. *Journal of Dairy Science* 73, 513–531.
Groeneveld, E., Kovac, M. and Wang, T. (1990) Pest, a general purpose BLUP package for multivariate prediction and estimation. In: *Proceedings of the 4th World Congress on Genetics Applied to Livestock Production* 13, 488–491.
Grundy, B., Villanueva, B. and Woolliams, J.A. (1998) Dynamic selection procedures for constrained inbreeding and their consequences for pedigree development. *Genetical Research Cambridge* 72, 159–168.
Guo, Z. and Swalve, H.H. (1997) Comparison of different lactation curve sub-models in test day models. *Proceedings of the Annual Interbull Meeting, Vienna. Interbull Bulletin* 16, 75–79.
Habier, D., Fernando, R.L. and Dekkers, J.C.M. (2007) The impact of genetic relationship information on genome-assisted breeding values. *Genetics* 177, 2389–2397.
Habier, D., Fernando, R.L, Kizikaya, K. and Garrick, D.J. (2011) Extension of the Bayesian alphabet for genomic selection. *BMC Bioinformatics* 12, 186–197.
Haldane, J.B.S. (1919) The combination of linkage values and the calculation of distances between the loci of linked factors. *Journal of Genetics* 8, 299.
Halldorsson, B.V., Eggertsson, H.P., Moore, K.H.S., Hauswedell, H., Eiriksson, O. *et al.* (2022) The sequences of 150,119 genomes in the UK Biobank. *Nature* 607, 732–740. DOI: 10.1038/s41586-022-04965-x
Hayes, B.J. and Daetwyler, H.D. (2013) Genomic selection in the era of genome sequencing. Course notes, March, Piacenza, Italy.
Hayes, B.J. and Daetwyler, H.D. (2019) 1000 Bull Genomes project to map simple and complex genetic traits in cattle: applications and outcomes. *Annual Review of Animal Biosciences* 7(1), 89–102. DOI: 10.1146/annurev-animal-020518-115024
Hayes, B.J., Visscher, P.M. and Goddard, M.E. (2009) Increased accuracy of artificial selection by using the realized relationship matrix. *Genetics Research* 91, 47–60.
Henderson, C.R. (1949) Estimation of changes in herd environment. *Journal of Dairy Science* 32, 709 (abstract).
Henderson, C.R. (1950) Estimation of genetic parameters. *Annals of Mathematical Statistics* 21, 309.

Henderson, C.R. (1953) Estimation of variance and covariance components. *Biometrics* 9, 226–252.
Henderson, C.R. (1963) Selection index and expected genetic advance. In: Hanson, W.D. and Robinson, H.F. (eds) *Statistical Genetics and Plant Breeding*. Publication 982, National Academy of Sciences, National Research Council, Washington DC, pp. 141–163.
Henderson, C.R. (1973) Sire evaluation and genetic trends. In: *Proceedings of the Animal Breeding and Genetics Symposium in Honour of J.L. Lush*. American Society for Animal Science, Blackburgh, Champaign, Illinois, pp. 10–41.
Henderson, C.R. (1975) Best linear unbiased estimation and prediction under a selection model. *Biometrics* 31, 423–447.
Henderson, C.R. (1976) A simple method for computing the inverse of a numerator relationship matrix used in prediction of breeding values. *Biometrics* 32, 69–83.
Henderson, C.R. (1984) *Applications of Linear Models in Animal Breeding.* University of Guelph Press, Guelph, Canada.
Henderson, C.R. (1985) Best linear unbiased prediction of non-additive genetic merits in non-inbred populations. *Journal of Animal Science* 60, 111–117.
Henderson, C.R. and Quaas, R.L. (1976) Multiple trait evaluation using relatives' records. *Journal of Animal Science* 43, 1188–1197.
Henderson, C.R., Kempthorne, O., Searle, S.R. and von Krosigk, C.M. (1959) The estimation of environmental and genetic trends from records subject to culling. *Biometrics* 15, 192–218.
Heringstad, B., Chang, Y.M., Gianola, D. and Klemetsdal, G. (2002) Analysis of clinical mastitis in Norwegian cattle with a longitudinal threshold model. In: *Proceedings of the 7th World Congress on Genetics Applied to Livestock Production*, August, Montpellier, France. Communication no. 20-03.
Hill, W.G., Goddard, M.E. and Visscher, P.M. (2008) Data and theory point to mainly additive genetic variance for complex traits. *PLoS Genetics* 4(2), e1000008.
Hoeschele, I. and VanRaden, P.M. (1991) Rapid inverse of dominance relationship matrices for non-inbred populations by including sire by dam subclass effects. *Journal of Dairy Science* 74, 557–569.
Ibánẽz-Escriche N., Fernando, R.L., Toosi, A. and Dekkers, J.C. (2009) Genomic selection of purebreds for crossbred performance. *Genetics Selection Evolution* 41, 12.
IMSL (1980) *Library Reference Manual.* IMSL, Houston, Texas.
Interbull (2000) *National Genetic Evaluation Programmes for Dairy Production Traits Practised in Interbull Member Countries 1999–2000*. Bulletin 24. Department of Animal Breeding and Genetics, Uppsala, Sweden.
Jairath, L., Dekkers, J.C.M., Schaeffer, L.R., Liu, Z., Burnside, E.B. and Kolstad, B. (1998) Genetic evaluation for herd life in Canada. *Journal of Dairy Science* 81, 550–652.
Jamrozik, J. and Schaeffer, L.R. (1997) Estimates of genetic parameters for a test day model with random regressions for production of first lactation Holsteins. *Journal of Dairy Science* 80, 762–770.
Jamrozik, J., Schaeffer, L.R. and Dekkers, J.C.M. (1997) Genetic evaluation of dairy cattle using test day yields and random regression model. *Journal of Dairy Science* 80, 1217–1226.
Jensen, J. and Madsen, P. (1997) *A User's Guide to DMU. A Package for Analysing Multivariate Mixed Models*. National Institute of Animal Science, Research Center, Foulum, Denmark.
Jensen, J., Wang, C.S., Sorensen, D.A. and Gianola, D. (1994) Bayesian inference on variance and covariance components for traits influenced by maternal and direct genetic effects, using the Gibbs sampler. *Acta Agriculturæ Scandinavica* 44, 193–201.
Kachman, S.D. (1999) Applications in survival analysis. *Journal of Animal Science* 77, 147–153.
Kaplan, E.L. and Meier, P. (1958) Nonparametric estimation from incomplete observations. *Journal of the American Statistical Association* 53, 457–481.
Karaman, E., Su, G., Croue, L. and Lund, M.S. (2021) Genomic prediction using a reference population of multiple pure breeds and admixed individuals. *Genetics Selection Evolution* 53, 46.

Kemp, R.A. (1985) The effects of positive assortative mating and preferential treatment of progeny on the estimation of breeding values. PhD thesis (unpublished), University of Guelph, Guelph, Canada.

Kempthorne, O. and Nordskog, A.W. (1959) Restricted selection indexes. *Biometrics* 15, 10–19.

Kennedy, B.W. (1989) *Animal Model BLUP*. Erasmus Intensive Graduate Course, Trinity College, Dublin.

Kennedy, B.W., Schaeffer, L.R. and Sorensen, D.A. (1988) Genetic properties of animal models. *Journal of Dairy Science* 71, 17–26.

Kirkpatrick, M. and Meyer, K. (2004) Direct estimation of genetic principal components: simplified analysis of complex phenotypes. *Genetics* 168, 2295–2306.

Kirkpatrick, M., Lofsvold, D. and Bulmer, M. (1990) Analysis of the inheritance, selection and evolution of growth trajectories. *Genetics* 124, 979–993.

Kirkpatrick, M., Hill, W.G. and Thompson, R. (1994) Estimating the covariance of traits during growth and ageing, illustrated with lactation in dairy cattle. *Genetics Research Cambridge* 64, 57–69.

Legarra, A. and Misztal, I. (2008) Technical note: computing strategies in genomewide selection. *Journal of Dairy Science* 91, 360–366.

Legarra, A., Aguilar, I. and Misztal, I. (2009) A relationship matrix including full pedigree and genomic information. *Journal of Dairy Science* 92, 4656–4663.

Legarra, A., Christensen, O.F., Aguilar, I. and Misztal, I. (2014) Single Step, a general approach for genomic selection. *Livestock Science* 166, 54–65.

Legarra, A., Christensen, O.F., Vitezica, Z.G., Aguilar, I. and Misztal, I. (2015) Ancestral relationships using metafounders: finite ancestral populations and across population relationships. *Genetics* 200, 455–468.

Lidauer, M., Stranden, I., Mäntysaari, E.A, Pösö, J. and Kettunen, A. (1999) Solving large test-day models by iteration on the data and preconditioned conjugate gradient. *Journal of Dairy Science* 82, 2788–2796.

Lidauer, M., Matilainen, K., Mäntysaari, E.A. and Stradén, I. (2011) Mixed model equations Solver, MiX99. Biometrical Genetics, MTT Agrifood Research, Finland.

Liu, Z., Reinhardt, F. and Reents, R. (2002) The multiple trait effective daughter contribution method applied to approximate reliabilities of estimated breeding values from a random regression test day model for genetic evaluation in dairy cattle. In: *Proceedings of the 7th World Congress Applied to Livestock Production*, August, Montpellier, France. Communication no. 20-15.

Liu, Z., Seefried, F.R., Reinhardt, F., Rensing, S., Thaller, G. and Reents, R. (2011) Impacts of both reference population size and inclusion of a residual polygenic effect on the accuracy of genomic prediction. *Genetics Selection Evolution* 43, 19.

Liu, Z., Goddard, M.E., Reinhardt, F. and Reents, R. (2014) A single-step genomic model with direct estimation of marker effects. *Journal of Dairy Science* 97, 5833–5850.

Liu, Z., Goddard, M.E., Hayes, B.J., Reinhardt, F. and Reents, R. (2016) Technical note: equivalent genomic models with a residual polygenic effect. *Journal of Dairy Science* 99, 2016–2025.

Lo, L.L., Fernando, R.L., and Grossman, M. (1993) Genetic covariance between relatives in multibreed populations: additive model. *Theoretical and Applied Genetics* 87, 423–430.

Lynch, M. (2016) Mutation and human exceptionalism: our future genetic load. *Genetics* 202(3), 869–875. DOI: 10.1534/genetics.115.180471

Lynch, M. and Walsh, B. (1998) *Genetics and Analysis of Quantitative Traits*. Sinauer Associates, Sunderland, Massachusetts. ASIN: B00QAVJKKO.

Madgwick, P.A. and Goddard, M.E. (1989) Genetic and phenotypic parameters of longevity in Australian dairy cattle. *Journal of Dairy Science* 72, 2624–2632.

Makgahlela, M.L., Mäntysaari, E.A., Strandén, I., Koivula, M., Nielsen, U., Sillanpää, M. and Juga, J. (2013) Across breed multi-trait random regression genomic predictions in the Nordic Red dairy cattle. *Journal of Animal Breeding and Genetics* 130, 10–19.

Mantysaari, E.A. (1999) Derivation of multiple trait reduced random regression model for the first lactation test day records of milk, protein and fat. In: *Proceedings 50th Annual Meeting of EAAP*, August, Zurich, Switzerland.
Maron, M.J. (1987) *Numerical Analysis*. Macmillan Publishing Company, New York.
Masuda, Y., Tsuruta, S., Bermann, M., Bradford, H.L. and Misztal, I. (2021) Comparison of models for missing pedigree in single-step genomic prediction. *Journal of Animal Science* 99, skab019.
Masuda, Y., VanRaden, P.M., Tsuruta, S., Lourenco, D.A.L. and Misztal, I. (2022) Invited review: unknown-parent groups and metafounders in single-step genomic BLUP. *Journal of Dairy Science* 105, 923–939.
Meijering, A. and Gianola, D. (1985) Linear versus nonlinear methods of sire evaluations for categorical traits: a simulation study. *Genetics Selection Evolution* 17, 115–132.
Mészáros, G., Pálos, J., Ducrocq, V. and Sölkner, J. (2010) Heritability of longevity in Large White and Landrace sows using continuous time and grouped data models. *Genetics Selection Evolution* 42, 13.
Mészáros, G., Sölkner, J. and Ducrocq, V. (2013) The Survival Kit: software to analyse possible survival data including correlated random effects. *Computer Methods and Programs in Biomedicine* 110, 503–510.
Meuwissen, T.H.E. (1997) Maximising the response of selection with a predefined rate of inbreeding. *Journal of Animal Science* 75, 934–940.
Meuwissen, T.H.E. and Luo, Z. (1992) Computing inbreeding coefficients in large populations. *Genetics Selection Evolution* 24, 305–313.
Meuwissen, T.H., Hayes, B. and Goddard, M.E. (2001) Prediction of total genetic value using genome-wide dense marker maps. *Genetics* 157, 1819–1829.
Meyer, K. (1989) Approximate accuracy of genetic evaluation under an animal model. *Livestock Production Science* 21, 87–100.
Meyer, K. (1999) Estimates of genetic and phenotypic covariance functions for post-weaning growth and mature weight of beef cows. *Journal of Animal Breeding and Genetics* 116, 181–205.
Meyer, K. (2005) Genetic principal components for live ultrasound scan traits of Angus cattle. *Animal Science* 81, 337–345.
Meyer, K. (2007) Multivariate analyses of carcass traits for Angus cattle fitting reduced rank and factor analytic models. *Journal of Animal Breeding and Genetics* 124, 50–64.
Meyer, K. (2009) Factor-analytic models for genotype × environment type problems and structured covariance matrices. *Genetics Selection Evolution* 41, 21–31.
Meyer, K. and Hill, W.G. (1997) Estimation of genetic and phenotypic covariance functions for longitudinal or 'repeated' records by restricted maximum likelihood. *Livestock Production Science* 47, 185–200.
Meyer, K. and Tier, B. (2003) Approximate prediction covariances among multiple estimated breeding values for individuals. In: *Proceedings of the Annual Interbull Meeting, Rome. Interbull Bulletin* 31, 133–136.
Misztal, I. (2006) Properties of random regression models using linear splines. *Journal of Animal Breeding and Genetics* 123, 74–80.
Misztal, I. (2016) Inexpensive computation of the inverse of the genomic relationship matrix in populations with small effective population size. *Genetics* 202, 401–409.
Misztal, I. and Gianola, D. (1988) Indirect solution of mixed model equations. *Journal of Dairy Science* 71(Suppl 2), 99–106.
Misztal, I., Legarra, A. and Aguilar, I. (2009) Computing procedures for genetic evaluation including phenotypic, full pedigree, and genomic information. *Journal of Dairy Science* 92, 4648–4655.
Misztal, I., Aguilar, I., Legarra, A., Tsuruta, S., Johnson, D.L. and Lawlor, T.J. (2010) A unified approach to utilize phenotypic, full pedigree and genomic information for genetic evaluation. In: *Proceedings of the 9th World Congress Applied to Livestock Production*, Leipzig, Germany. Communication no. 0050.

Misztal, I., Tsuruta, S., Aguilar, I., Legarra, A., VanRaden, P.M. and Lawlor, T.J. (2013a) Methods to approximate reliabilities in single-step genomic evaluation. *Journal of Dairy Science* 96, 647–654.

Misztal, I., Vitezica, Z.G., Legarra, A., Aguilar, I. and Swan, A.A. (2013b) Unknown-parent groups in single-step genomic evaluation. *Journal of Animal Breeding and Genetics* 130, 252–258.

Misztal, I., Legarra, A. and Aguilar, I. (2014) Using recursion to compute the inverse of the genomic relationship matrix. *Journal of Dairy Science* 97, 3943–3952.

Moore, A.J., Brodie Jr, E.D. and Wolf, J.B. (1997) Interacting phenotypes and the evolutionary process. I. Direct and indirect genetic effects of social interactions. *Evolution* 51, 1352–1362.

Moser, G., Khatkar, M.S., Hayes, B.J. and Raadsma, H.W. (2010) Accuracy of direct genomic values in Holstein bulls and cows using subsets of SNP markers. *Genetics Selection Evolution* 42, 37.

Mrode, R. (2014) *Linear Models for the Prediction of Animal Breeding Values*, 3rd edn. CAB International, Wallingford, UK.

Mrode, R.A. and Swanson, G.J.T. (1995) Comparison of the efficiency of the repeatability model on observed or transformed yields to a multi-variate analysis. *Summaries of Paper, British Society of Animal Science Winter Meeting*, March, paper no. 129.

Mrode, R.A. and Swanson, G.J.T. (1999) Simplified equations for evaluations of bulls in the Interbull international evaluation system. *Livestock Production Science* 62, 43–52.

Mrode, R.A. and Swanson, G.J.T. (2004) Calculating cow and daughter yield deviations and partitioning of genetic evaluations under a random regression model. *Livestock Production Science* 86, 253–260.

Muir, W.M. (1996) Group selection for adaptation to multiple-hen cages: selection program and direct responses. *Poultry Science* 75, 447–458.

Muir, W.M. (2005) Incorporation of competitive effects in forest tree or animal breeding programs. *Genetics* 170, 1247–1259.

Pander, B.L., Hill, W.G. and Thompson, R. (1992) Genetic parameters of test day records of British Holstein heifers. *Animal Production* 55, 11–21.

Patry, C. and Ducrocq, V. (2011) Evidence of biases in genetic evaluations due to genomic preselection in dairy cattle. *Journal of Dairy Science* 94, 1011–1020.

Patterson, H.D. and Thompson, R. (1971) Recovery of inter-block information when block sizes are unequal. *Biometrika* 58, 545–554.

Phocas, F. and Ducrocq, V. (2006) Discrete vs continuous time survival analysis of productive life of Charolais cows. In: *Proceedings of the 8th World Congress Applied to Livestock Production*, Belo Horizonte, Brazil. Communication no. 03-13.

Piles, M., Garreau, H., Rafel, O., Larzul, C., Ramon, J. and Ducrocq, V. (2006) Survival analysis in two lines of rabbits selected for reproductive traits. *Journal of Animal Science* 84, 1658–1665.

Pocrnic, I., Lourenco, D.A., Masuda, Y., Legarra, A. and Misztal, I. (2016) The dimensionality of genomic information and its effect on genomic prediction. *Genetics* 203, 573–581.

Pocrnic, I., Lindgren, F., Tolhurst, D., Herring, W.O. and Gorjanc, G. (2022) Optimisation of the core subset for the APY approximation of genomic relationships. *Genetics Selection Evolution* 54, 76.

Prentice, R.L. and Gloeckler, L.A. (1978) Regression analysis of grouped survival data with application to breast cancer data. *Biometrics* 34, 57–67.

Ptak, E. and Schaeffer, L.R. (1993) Use of test day yields for genetic evaluation of dairy sires and cows. *Livestock Production Science* 34, 23–34.

Quaas, R.L. (1976) Computing the diagonal elements of a large numerator relationship matrix. *Biometrics* 32, 949–953.

Quaas, R.L. (1984) *Linear Prediction in BLUP School Handbook: Use of Mixed Models for Prediction and Estimation of (Co)variance Components.* Animal Genetics and Breeding Unit, University of New England, NSW, Australia.

Quaas, R.L. (1988) Additive genetic model with groups and relationships. *Journal of Dairy Science* 71, 1338–1345.

Quaas, R.L. (1995) Fx Algorithms. An unpublished note.
Quaas, R.L. and Pollak, E.J. (1980) Mixed model methodology for farm and ranch beef cattle testing programs. *Journal of Animal Science* 51, 1277–1287.
Quaas, R.L. and Pollak, E.J. (1981) Modified equations for sire models with groups. *Journal of Dairy Science* 64, 1868–1872.
Robertson, A. and Rendel, J.M. (1954) The performance of heifers got by artificial insemination. *Journal of Agricultural Science, Cambridge* 44, 184–192.
Ronningen, K. and Van Vleck, L.D. (1985) Selection index theory with practical applications. In: Chapman, A.B. (ed.) *General and Quantitative Genetics*. World Animal Science, A4, Elsevier Science Publishers, Oxford, UK.
Ros-Freixedes, R., Valente, B.D., Chen, C.-Y., Herring, W.O., Gorjanc, G., Hickey, J.H. and Johnsson, M. (2022) Rare and population-specific functional variation across pig lines. *Genetics Selection Evolution* 54, 39. Available at: https://doi.org/10.1186/s12711-022-00732-8 (accessed 11 March 2023).
Schaeffer, L.R. (1984) Sire and cow evaluation under multiple trait models. *Journal of Dairy Science* 67, 1567–1580.
Schaeffer, L.R. (1994) Multiple-country comparison of dairy sires. *Journal of Dairy Science* 77, 2671–2678.
Schaeffer, L.R. (2004) Application of random regression models in animal breeding. *Livestock Production Science* 86, 35–45.
Schaeffer, L.R. and Dekkers, J.C.M. (1994) Random regression in animal models for test-day production in dairy cattle. In: *Proceedings of the 5th World Congress Applied to Livestock Production*, Guelph, Canada, pp. 443–446.
Schaeffer, L.R. and Kennedy, B.W. (1986) Computing solutions to mixed model equations. In: *Proceedings of the 3rd World Congress on Genetics Applied to Livestock Production*, Lincoln, Nebraska.
Schaeffer, L.R. and Wilton, J.W. (1987) RAM computing strategies and multiple traits. Prediction of genetic value for beef cattle. In: *Proceedings Workshop II*, Winrock, Kansas City, Missouri.
Schaeffer, L.R., Wilton, J.W. and Thompson, R. (1978) Simultaneous estimation of variance and covariance components from multitrait mixed model equations. *Biometrics* 34, 199–208.
Schaeffer, L.R., Jamrozik, J., Kistemaker, G.J. and van Doormaal, B.J. (2000) Experience with a test day model. *Journal of Dairy Science* 83, 1135–1144.
Searle, S.R. (1982) *Matrix Algebra Useful for Statistics*. John Wiley & Sons, New York.
Sevillano, C.A., Vandenplas, J., Bastiaansen, J.W.M., Bergsma, R. and Calus, M.P.L. (2017) Genomic evaluation for a three-way crossbreeding system considering breed-of-origin of alleles. *Genetics Selection Evolution* 49, 75.
Sigurdsson, A. and Banos, G. (1995) Dependent variables in international sire evaluations. *Acta Agriculturæ Scandinavica* 45, 209–217.
Silió, L., Rodríguez, M., Fernández, A., Barragán, C., Benítez, R., Óvilo, C. and Fernandez, A.I. (2013) Measuring inbreeding and inbreeding depression on pig growth from pedigree or SNP-derived metrics. *Journal of Animal Breeding and Genetics* 130, 349–60.
Simm, G., Pollott, G., Mrode, R., Houston, R. and Marshall, K. (2021) *Genetic Improvement of Farmed Animals*. CAB International, Wallingford, UK.
Solberg, T.R., Sonesson, A.K., Woolliams, J.A., Ødegard, J. and Meuwissen, T.H.E. (2009) Persistence of accuracy of genomic-wide breeding values over generations when including a polygenic effect. *Genetics Selection Evolution* 41, 53.
Sorensen, D.A. and Gianola, D. (2002) *Likelihood, Bayesian, and MCMC Methods in Quantitative Genetics*. Springer-Verlag, New York.
Sorensen, D.A. and Kennedy, B.W. (1983) The use of the relationship matrix to account for genetic drift variance in the analysis of genetic experiments. *Theoretical and Applied Genetics* 66, 217–220.
Sorensen, D.A., Andersen, S. and Gianola, D. (1995) Bayesian inference in threshold models using Gibbs sampling. *Genetics Selection Evolution* 27, 229–249.

Strandén, I. and Garrick, D.J. (2009) Derivation of equivalent computing algorithms for genomic predictions and reliabilities of animal merit. *Journal of Dairy Science* 92, 2971–2975.

Stranden, I. and Lidauer, M. (1999) Solving large mixed linear models using preconditioned conjugate gradient iteration. *Journal of Dairy Science* 82, 2779–2787.

Strandén, I. and Mäntysaari, E.A. (2013) Use of random regression model as an alternative for multibreed relationship matrix. *Journal of Animal Breeding and Genetics* 130, 4–9.

Su, G., Christensen, O.F., Ostersen, T., Henryon, M. and Lund, M.S. (2012) Estimating additive and non-additive genetic variances and predicting genetic merits using genome-wide dense single nucleotide polymorphism markers. *PLoS ONE* 7(9), e45293.

Szyda, J., Liu, Z., Reinhardt, F. and Reents, R. (2003) Incorporation of QTL information into routine estimation of breeding values for German Holstein dairy cattle. In: *Proceedings of the 54th Annual Meeting of the European Association for Animal Production*, Rome.

Tarrés, J., Bidanel, J.P., Hofer, A. and Ducrocq, V. (2006) Analysis of longevity and exterior traits on Large White sows in Switzerland. *Journal of Animal Science* 84, 2914–2924.

Taskinen, M., Mäntysaari, E.A. and Strandén, I. (2017) Single-step SNP-BLUP with on-the-fly imputed genotypes and residual polygenic effects. *Genetics Selection Evolution* 49, 36.

Ter Braak, C.J.F., Boer, M.P. and Bink, M.C.A.M. (2005) Extending Xu's Bayesian model for estimating polygenic effects using markers of the entire genome. *Genetics* 170, 1435–1443.

The R Development Core Team (2010) *R: A Language and Environment for Statistical Computing*. R Foundation for Statistical Computing, Vienna.

Thompson, R. (1977a) The estimation of heritability with unbalanced data. II. Data available on more than two generations. *Biometrics* 33, 497–504.

Thompson, R. (1977b) Estimation of quantitative genetic parameters. In: Pollak, E., Kempthorne, O. and Bailey, T.B. (eds) *Proceedings of the International Conference on Quantitative Genetics*, Iowa State University Press, Ames, Iowa, pp. 639–657.

Thompson, R. (1979) Sire evaluation. *Biometrics* 35, 339–353.

Thompson, R. and Meyer, K. (1986) A review of theoretical aspects in the estimation of breeding values for multi-trait selection. *Livestock Production Science* 15, 299–313.

Thompson, R., Cullis, B.R., Smith, A.B. and Gilmour, A.R. (2003) A sparse implementation of the average information algorithm for factor analytic and reduced rank variance models. *Australian and New Zealand Journal of Statistics* 45, 445–459.

Teissier, M., Larroque, H., Brito, L.F., Rupp, R., Schenkel, F.S. and Robert-Granié, C. (2020) Genomic predictions based on haplotypes fitted as pseudo-SNP for milk production and udder type traits and SCS in French dairy goats. *Journal of Dairy Science* 103, 11559–11573.

Tier, B. and Solkner, J. (1993) Analysing gametic variation with an animal model. *Theoretical and Applied Genetics* 85, 868–872.

Toro, M.A. and Varona, L. (2010) A note on mate allocation for dominance handling in genomic selection. *Genetics Selection Evolution* 42, 33.

Tyriseva, A.M., Meyer, K., Freddy, F., Ducrocq, V., Jakobsen, J., Lidauer, M.H. and Mantysaari, E. (2011a) Principal component approach in variance component estimation in international sire evaluation. *Genetics Selection Evolution* 43, 21.

Tyriseva, A.M., Meyer, K., Freddy, F., Ducrocq, V., Jakobsen, J., Lidauer, M.H. and Mantysaari, E. (2011b) Principal component and factor analytic models in international sire evaluation. *Genetics Selection Evolution* 43, 33.

Van Arendonk, J.A.M., Tier, B. and Kinghorn, B.P. (1994) Use of multiple genetic markers in prediction of breeding values. *Genetics* 137, 319–329.

Vandenplas, J., Calus, M.P.L., Sevillano, C.A., Windig, J.J. and Bastiaansen, J.W.M. (2016) Assigning breed origin to alleles in crossbred animals. *Genetics Selection Evolution* 48, 61.

VanRaden, P.M. (1992) Accounting for inbreeding and crossbreeding in genetic evaluation of large populations. *Journal of Dairy Science* 75, 3136–3144.

VanRaden, P.M. (2007) Genomic measures of relationship and inbreeding. *Interbull Bulletin* 37, 33–36.

VanRaden, P.M. (2008) Efficient methods to compute genomic predictions. *Journal of Dairy Science* 91, 4414–4423.
VanRaden, P.M. and Klaaskate, E.J.H. (1993) Genetic evaluation of length of productive life including predicted longevity of live cows. *Journal of Dairy Science* 76, 2758–2764.
VanRaden, P.M., Olson, K.M., Wiggans, G.R., Cole, J.B. and Tooker, M.E. (2011) Genomic inbreeding and relationships among Holsteins, Jerseys, and Brown Swiss. *Journal of Dairy Science* 94, 5673–5682.
VanRaden, P.M. and Wiggans, G.R. (1991) Derivation, calculation and use of national animal model information. *Journal of Dairy Science* 74, 2737–2746.
VanRaden, P.M., Van Tassell, C.P., Wiggans, G.R., Sonstegart, T.S., Schnabel, R.D., Taylor, J.F. and Schenkel, F.S. (2009) Invited review. Reliability of genomic predictions for North American Holsteins bulls. *Journal of Dairy Science* 92, 16–24.
VanRaden, P.M., Tooker, M.E., Chud, T.C.S., Norman, H.D., Megonigal Jr, J.H., Haagen, I.W. and Wiggans, G.R. (2020) Genomic predictions for crossbred dairy cattle. *Journal of Dairy Science* 103, 1620–1631.
Varona, L., Legarra, A., Toro, M.A. and Vitezica, Z.G. (2018) Non-additive effects in genomic selection. *Frontiers in Genetics* 9, 78.
Varona, L., Legarra, A., Toro, M.A. and Vitezica, Z.G. (2022) Genomic prediction methods accounting for nonadditive genetic effects. In: Ahmadi, N. and Bartholomé, J. (eds) *Genomic Prediction of Complex Traits: Methods in Molecular Biology* Vol. 2467, 219–243. Humana, New York.
Veerkamp, R.F., Brotherstone, S. and Meuwissen, T.H.E. (1999) Survival analysis using random regression models. *Interbull Bulletin* 21, 36–40.
Veerkamp, R.F., Hill, W.G., Stott, A.W., Brotherstone, S. and Simm, G. (1995) Selection for longevity and yield in dairy cows using transmitting abilities for type and yield. *Animal Science* 61, 189–197.
Visscher, P.M. (1991) Estimation of genetic parameters in dairy cattle using an animal model and implications for genetic evaluation. PhD thesis (unpublished), University of Edinburgh, Edinburgh.
Visscher, P.M., Thompson, R., Yazdi, H., Hill, W.G. and Brotherstone, S. (1999) Genetic analysis of longevity data in the UK: present practice and considerations for the future. *Interbull Bulletin* 21, 16–23.
Vitezica, Z.G., Aguilar, I., Misztal, I. and Legarra, A. (2011) Bias in genomic predictions for populations under selection. *Genetics Research (Cambridge)* 93, 357–366.
Vitezica, Z.G., Varona, L. and Legarra, A. (2013) On the additive and dominant variance and covariance of individuals within the genomic selection scope. *Genetics* 195, 1223–1230.
Vitezica, Z.G., Reverter, A., Herring, W. and Legarra, A. (2018) Dominance and epistatic genetic variances for litter size in pigs using genomic models. *Genetics Selection Evolution* 50, 71.
Wang, C.S., Rutledge, J.J. and Gianola, D. (1993) Marginal inferences about variance components in a mixed linear model using Gibbs sampling. *Genetics Selection Evolution* 25, 41–62.
Wang, C.S., Rutledge, J.J. and Gianola, D. (1994) Bayesian analysis of mixed linear models via Gibbs sampling with an application to litter size in Iberian pigs. *Genetics Selection Evolution* 26, 91–115.
Wang, Y., Miller, S.P., Schenkel, F.S., Wilton, J.W. and Boettcher, P.J. (2002) Performance of a linear-threshold model to evaluate calving interval and birth weight in a multibreed beef population. In: *Proceedings of the 7th World Congress on Genetics Applied to Livestock Production,* August, Montpellier, France. Communication no. 17-10.
Wang, H., Misztal, I., Aguilar, I., Legarra, A. and Muir, W.M. (2012) Genome-wide association mapping including phenotypes from relatives without genotypes. *Genetics Research* 94, 73–83.
Wei, M., van der Werf, J.H.J. and Brascamp, E.W. (1991) Relationship between purebred and crossbred parameters: II. Genetic correlation between purebred and crossbred performance under the model with two loci. *Journal of Animal Breeding and Genetics* 108, 262–269.
Weller, J.I. (2001) *Quantitative Trait Loci Analysis in Animals*. CAB International, Wallingford, UK.

Wellmann, R. and Bennewitz, J. (2012) Bayesian models with dominance effects for genomic evaluation of quantitative traits. *Genetics Research (Cambridge)* 94, 21–37.
Westell, R.A. and VanVleck, L.D. (1987) Simultaneous genetic evaluation of sires and cows for a large population of dairy cattle. *Journal of Dairy Science* 70, 1006–1017.
Westell, R.A., Quaas, R.L. and VanVleck, L.D. (1988) Genetic groups in an animal model. *Journal of Dairy Science* 71, 1310–1318.
White, I.M.S., Thompson, R. and Brotherstone, S. (1999) Genetic and environmental smoothing of lactation curves with cubic splines. *Journal of Dairy Science* 82, 632–638.
Wiggans, G.R., VanRaden, P.M. and Powell, R.L. (1992) A method for combining United States and Canadian bull proofs. *Journal of Dairy Science* 75, 2834–2839.
Wiggans, G.R., Misztal, I. and Van Tassell, C.P. (2003) Calving ease (co)variance components for a sire-maternal grandsire threshold model. *Journal of Dairy Science* 86, 1845–1848.
Willham, R.L. (1963) The covariance between relatives for characters composed of components contributed by related individuals. *Biometrics* 19, 18–27.
Wolf, J.B., Brodie, E.D., Cheverud, J.M., Moore, A.J. and Wade, M.J. (1998) Evolutionary consequences of indirect genetic effects. *Trends in Ecological Evolution* 13, 64–69.
Wright, S. (1922) Coefficients of inbreeding and relationship. *American Naturalist* 56, 330–338.
Xiang, T., Christensen, O.F., Vitezica, Z.G. and Legarra, A. (2016) Genomic evaluation by including dominance effects and inbreeding depression for purebred and crossbred performance with an application in pigs. *Genetics Selection Evolution* 48, 92.
Xu, S.Z. (2003) Estimating polygenic effects using markers of the entire genome. *Genetics* 163, 789–801.
Zaabza, H.B., Mäntysaari, E.A. and Strandén, I. (2020) Snp_blup_rel: software for calculating individual animal SNP-BLUP model reliabilities. *Agricultural and Food Science* 29, 297–306.
Zaabza, H.B., Mäntysaari, E.A. and Strandén, I. (2021) Estimation of individual animal SNP-BLUP reliability using full Monte Carlo sampling. *Journal of Dairy Science Communications* 2, 137–141.

Index